FISH PHYSIOLOGY

VOLUME XI
The Physiology of Developing Fish

Part A
Eggs and Larvae

CONTRIBUTORS

D. F. ALDERDICE

J. H. S. BLAXTER

RANDAL K. BUDDINGTON

THOMAS A. HEMING

THOMAS P. MOMMSEN

PETER J. ROMBOUGH

H. von WESTERNHAGEN

PATRICK J. WALSH

KENJIRO YAMAGAMI

FISH PHYSIOLOGY

Edited by

W. S. HOAR
DEPARTMENT OF ZOOLOGY
UNIVERSITY OF BRITISH COLUMBIA
VANCOUVER, BRITISH COLUMBIA, CANADA

D. J. RANDALL
DEPARTMENT OF ZOOLOGY
UNIVERSITY OF BRITISH COLUMBIA
VANCOUVER, BRITISH COLUMBIA, CANADA

VOLUME XI
The Physiology of Developing Fish

Part A
Eggs and Larvae

ACADEMIC PRESS, INC.
Harcourt Brace Jovanovich, Publishers
San Diego   New York   Berkeley   Boston
London   Sydney   Tokyo   Toronto

ACADEMIC PRESS, INC.
1250 Sixth Avenue, San Diego, California 92101

United Kingdom Edition published by
ACADEMIC PRESS INC. (LONDON) LTD.
24–28 Oval Road, London NW1 7DX

Library of Congress Cataloging in Publication Data
(Revised for vol. 11)

Hoar, William Stewart, Date
   Fish physiology.

   Vols. 8-   edited by W. S. Hoar [et al.]
   Includes bibliographies and indexes.
   Contents: v. 1. Excretion, ionic regulation, and
metabolism — v. 2. The endocrine system —   — v. 11.
The physiology of developing fish. pt. A. Eggs and
larvae. pt. B. Viviparity and posthatching juveniles
(2 v.)
      1. Fishes—Physiology—Collected works.
I. Randall, D. J.   II. Conte, Frank P., Date
III. Title.
QL639.1.H6   597'.01   76-84233
ISBN 0—12—350433—3 (v. 11, pt. A) (alk. paper)

PRINTED IN THE UNITED STATES OF AMERICA

88 89 90 91   9 8 7 6 5 4 3 2 1

CONTENTS

7.   Mechanisms of Hatching in Fish
      Kenjiro Yamagami

CONTRIBUTORS

Numbers in parentheses indicate the pages on which the authors' contributions begin.

D. F. ALDERDICE *(163), Department of Fisheries and Oceans, Fisheries Research Branch, Pacific Biological Station, Nanaimo, British Columbia, Canada V9R 5K6*

J. H. S. BLAXTER *(1), Dunstaffnage Marine Research Laboratory, Oban, Argyll PA34 4AD, Scotland*

RANDAL K. BUDDINGTON *(407), Department of Physiology, University of California, Los Angeles, California 90024*

THOMAS A. HEMING *(407), Pulmonary Division, Department of Internal Medicine, University of Texas Medical Branch, Galveston, Texas 77550-2780*

THOMAS P. MOMMSEN *(347), Department of Zoology, University of British Columbia, Vancouver, British Columbia, Canada V6T 2A9*

PETER J. ROMBOUGH *(59), Zoology Department, Brandon University, Brandon, Manitoba, Canada R7A 6A9*

H. VON WESTERNHAGEN *(253), Biologische Anstalt Helgoland (Zentrale), D-2000 Hamburg 52, Federal Republic of Germany*

PATRICK J. WALSH *(347), Rosenstiel School of Marine and Atmospheric Science, University of Miami, Miami, Florida 33149*

KENJIRO YAMAGAMI *(447), Life Science Institute, Sophia University, Chiyoda-ku, Tokyo 102, Japan*

PREFACE

Dramatic changes occur in the physiology of most animals during their development. Among the vertebrates, birds are entirely oviparous, live for variable periods in a cleidoic egg, and show fundamental alterations in excretion, nutrition, and respiration at the time of hatching. In contrast, the eutherian mammals are all viviparous, depending on the maternal circulation and a specialized placenta to provide food, exchange gases, and discharge wastes. The physiology of both mother and fetus is highly specialized during gestation and changes fundamentally at the time of birth. Fishes exemplify both the oviparous and the viviparous modes of development, with some examples that are intermediate between the two. In these two volumes, we present reviews of many, but not all, aspects of development. The chapters in Part A relate to the physiology of eggs and larvae: different patterns of larval development osmotic and ionic regulation, gas exchange, effects of pollutants, vitellogenesis, the absorption of yolk, and the mechanisms of hatching. Chapters in Part B deal with maternal–fetal relations, meristic variation, smolting salmonids, the ontogeny of behavior, and the development of sensory systems.

The editors wish to thank the authors for their cooperation and dedication to this project and also to express their deep appreciation to the many reviewers whose careful readings and constructive criticisms have greatly improved the final presentations.

W. S. HOAR
D. J. RANDALL

CONTENTS OF OTHER VOLUMES

Volume XA

1

PATTERN AND VARIETY IN DEVELOPMENT

J. H. S. BLAXTER

Dunstaffnage Marine Research Laboratory
Oban, Argyll PA34 4AD, Scotland

I. INTRODUCTION

Present interest in the development of fish has a twofold basis. First, the factors that determine brood strength and so recruitment to commercial fisheries may well operate during the early life history stages. Second, the practice of aquaculture and the range of species used are expanding rapidly. Improvements in hatchery techniques during the last two decades have made it possible to rear almost any species, even the halibut *Hippoglossus hippoglossus* (V. Øiestad, personal communication) if fertile eggs are available. Manipulation of

1

spawning time has also meant that a steady supply of larvae—for example, of turbot *Scophthalmus maximus* and northern anchovy *Engraulis mordax*—can be maintained throughout the year, even with temperate species. Experimental material is thus readily available for the study of optimum rearing conditions, nutrition and growth, critical periods in development, toxicity testing, and the like, all germane to the assessments of the fishery biologist and the practices of the fish farmer. This material is also available to experimental embryologists and to physiologists interested in the ontogeny of organ systems, or to behaviorists interested in the ontogeny of behavior. Great insight into function is possible in larval stages that lack certain organs or have them only partially developed. In many species, it is the transparency of the larvae, their lack of hemoglobin, their simple intestinal tracts, undifferentiated skeleton, and incompletely developed nervous system and sense organs that can make them especially useful as experimental animals.

The progress of work on the developmental biology of fish since the early 1970s can be traced in a series of symposia and other meetings. International Early Life History Symposia were held in 1973, 1979, and 1984 in Scotland, the United States, and Canada (Blaxter, 1974; Lasker and Sherman, 1981; Marliave, 1985). A symposium on the "The Ontogeny and Systematics of Fishes," dedicated to the memory of E. H. Ahlstrom, was held in California in 1983 (Moser *et al.*, 1984), and larval fish conferences are held annually by the American Fisheries Society (e.g., Hubbs, 1986).

By way of introduction to this volume of "Fish Physiology," a brief account will first be given of recent advances in techniques of rearing, general life history stages, and terminology. It is intended that variation of experimental material and the sources of variation will be an underlying but continuing theme. Thus diversity of development, parental effects on the young, and the effects of captivity, starvation, and fixation will be discussed. Finally, the development of structure will be related to the development of function with particular reference to possible critical periods in ontogeny.

II. REARING TECHNIQUES

Over the past few years improvements in techniques for rearing marine fish have increased the number of species available for experiment and for aquaculture. These techniques are summarized by

Kinne (1977), Blaxter (1981), and Hunter (1984). The greatest advance has been the use of very small food items, especially of the rotifer *Brachionus plicatilis*, but also the naked dinoflagellate *Gymnodinium splendens* and other organisms such as *Mytilus* trochophores and sieved natural zooplankton, as a food source in the very young stages when the size of the mouth is limiting (see page 34). Interest is now increasing in the use of "green-water" culture, where the larvae are maintained in fairly high densities of algae such as *Chlorella*, which may damp out metabolite fluctuations, perhaps improve oxygenation, and provide a secondary food source for the larvae (Houde, 1977; Morita, 1985). In the future the use of compounded diets, especially in the form of microcapsules small enough to be eaten whole, may provide a further breakthrough. Appelbaum (1985) reared Dover sole *Solea solea* entirely on compounded diets and cited other similar successful work on plaice *Pleuronectes platessa*, vendace *Coregonus albula*, sea bass *Dicentrarchus labrax*, turbot, catfish *Clarias gariepinus*, and the Atlantic silverside *Menidia menidia*. The best survival rate of sole on an artificial diet was obtained when live brine shrimp *Artemia nauplii* were provided for the first 10 days of feeding. *Artemia* is certainly still the staple live food, both in experimental and applied fish culture (Sorgeloos, 1980). It has become increasingly obvious that *Artemia* from different sources can vary in quality—for example, in fatty acid "profile"—and success and failure in the past may have hinged on this hitherto unappreciated factor (van Ballaer *et al.*, 1985; Kanazawa, 1985). Dabrowski (1984) reviewed work on some of the nutritional aspects of rearing fish larvae and the relevance of digestive processes.

Other factors in rearing, such as optimum food density, stocking density, type of tank, light, and other environmental conditions, have now been established. Production of eggs out of season by the use of artificial photoperiods, temperature, and hormone injections have also greatly improved the availability of larvae year-round (see Lam, 1982).

One of the most striking advances has been the improvement in survival and growth when larvae are reared in the absence of predators in large-scale facilities, or "mesocosms," in the form of large on-shore tanks, large plastic-walled cylinders sited in sheltered coastal waters, or impounded coastal bays or lagoons (Kvenseth and Øiestad, 1984; Øiestad *et al.*, 1985; Gamble *et al.*, 1985; Morita, 1985; Paulsen *et al.*, 1985; Sturmer *et al.*, 1985). Atlantic cod *Gadus morhua*, turbot, and red drum *Sciaenops ocellatus* have been reared with unprecedented success.

III. PROGRESS AND DIVERSITY OF DEVELOPMENT

Some comprehensive keys have recently appeared in the literature that give a good insight into the variety of eggs and larvae and their development. Russell (1976) describes the eggs and planktonic stages of British marine fishes, Fahay (1983) the ichthyoplankton of the western North Atlantic, and Auer (1982) that of the Lake Michigan region of the Great Lakes Basin. The early life history stages of fish and their characteristics are discussed more generally by Blaxter (1969), Hempel (1979), and Kendall *et al.* (1984). Great variety exists from species to species and, in particular, the size and extent of differentiation when the young fish first becomes free-living is of considerable significance for its chance of survival.

During the final ovarian maturation of the eggs of marine fish such as the Atlantic cod, whiting *Merlangius merlangus*, haddock *Melanogrammus aeglefinus*, and plaice, there is a massive uptake of water and concomitant reduction in protein phosphate (Craik and Harvey, 1984a). This influx of water, such that the water content may reach as high as 92% of the egg weight, is an adaptation to pelagic life because the egg fluids are hypotonic and make the eggs buoyant. Freshwater fish with demersal eggs, like the rainbow trout *Salmo gairdneri*, powan *Coregonus lavarcticus*, and pike *Esox lucius*, do not show these changes.

The initial buoyancy of pelagic fish eggs, like those of the flounder *P. flesus*, depends on the salinity in which the female is kept before spawning (Solemdal, 1973). Females from low salinities tend to produce eggs that are neutrally buoyant at lower salinities, buoyancy, of course, not only being affected by water content but also by the osmolarity of the egg fluid.

There is great variety in the reproductive styles of fish (Table I). In most species the eggs develop independently, but there are many instances of parental care (Breder and Rosen, 1966). In littoral species such as cottids, blennies, and gobies, this takes the form of guarding the eggs, but nests may be built with one or other parent guarding and often ventilating the nest. Mouth brooding of eggs and larvae is found in cichlids such as tilapia and in ariid catfish. Other species have evolved ovoviviparity or viviparity, the former where the eggs develop within the female, the latter where nourishment is provided via "placental" structures (trophotaenia) within the female (see Wourms and Grove, volume XIB). Recently Ridley (1978) and Blumer (1979) summarized parental care and its evolutionary significance. While care is more common by the male parent (in 61 families), care by the female occurs in 41 families. Care by the male is clearly linked to the prevalence of external fertilization in fish and generally to polygamy and male territoriality.

The morphological characteristics of fish eggs are described by Russell (1976), Ahlstrom and Moser (1980), and Matarese and Sandknop (1984). Typically marine eggs are single, buoyant, and with a modal diameter of about 1 mm (although the range is from 0.6 to 4.0 mm). Most freshwater fish lay demersal eggs with a modal diameter somewhat greater than 1 mm. The eggs may merely rest on the sub-

Table I

Classification of Reproductive Styles[a]

A. Nonguarders
   1. Open and substratum spawners
      a. Pelagic spawners
      b. Rock and gravel spawners with pelagic larvae
      c. Rock and gravel spawners with benthic larvae
      d. Nonobligatory plant spawners
      e. Obligatory plant spawners
      f. Sand spawners
      g. Terrestrial spawners, in damp conditions

   2. Brood hiders
      a. Beach spawners; above waterline at high tides
      b. Annual spawners; eggs estivate
      c. Rock and gravel spawners
      d. Cave spawners
      e. Spawners in live invertebrates

B. Guarders
   1. Substratum spawners
      a. Pelagic spawners; at surface of hypoxic waters
      b. Above-water spawners; male splashes clutch
      c. Rock spawners
      d. Plant spawners

   2. Nest spawners
      a. Froth nesters
      b. Miscellaneous substratum and materials nesters
      c. Rock and gravel nesters
      d. Glue-making nesters
      e. Plant material nesters
      f. Sand nesters
      g. Hole nesters
      h. Anemone nesters; at base of host

(continued)

Table 1 (*Continued*)

C. Bearers
    1. External bearers
        a. Transfer brooders; eggs carried before deposition
        b. Auxiliary brooders; adhesive eggs carried on skin under fins etc.
        c. Mouth brooders
        d. Gill-chamber brooders
        e. Pouch brooders

    2. Internal bearers
        a. Facultative internal bearers; occasional internal fertilization of normally oviparous fish, eggs rarely retained long
        b. Obligate lecithotrophic live bearers; no maternal–embryonic nutrient transfer
        c. Matrotrophous oophages and adelphophages; one or a few eggs developing at expense of other eggs or embryos
        d. Viviparous trophoderms; nutrition partially or entirely from female via "placental" structures

[a] Adapted from Balon (1981a).

stratum, or have some means of attachment such as adhesive threads or a supporting pedestal. In species such as salmonids the eggs are buried in the gravel, and the grunion *Leuresthes tenuis* lays its eggs intertidally in the sand. Other types of demersal eggs are found in some littoral marine species and, in the more offshore Atlantic herring, capelin *Mallotus villosus* and Pacific cod *Gadus macrocephalus*.

While teleosts usually have round eggs, most engraulids have eggs that are ellipsoidal, thought to be an adaptation to reduce cannibalism by the filter-feeding parents after spawning. Other families like the gobies have slightly flattened eggs, and demersal eggs are sometimes irregular in shape. Oviparous elasmobranchs have eggs of unusual shapes (the "mermaid's purse") with tendrils for attachment. Tendrils are also found in the silverside *Atherinopsis* and the gar *Belone,* while the flying fish *Oxyporhamphus* has spines (Boehlert, 1984). It is easy to understand the adaptive value of tendrils in distantly related species, but it is much more difficult to explain the ornamentations of the chorion. Most teleosts have a smooth surface to the chorion, but the unrelated inshore dragonet *Callionymus* and bathypelagic gonostomatid *Maurolicus muelleri* have chorions with hexagonal facets, and the flatfish *Pleuronichthys coenosus* has a chorion with very many small facets.

The yolk is usually translucent, unpigmented, and homogeneous in texture, but may be segmented in primitive species like the pilchard *Sardina pilchardus* and sprat *Sprattus sprattus*. In some soleids the segmentation is confined to the periphery of the yolk, and in other species like the jack mackerel *Trachurus symmetricus* segmentation appears progressively during early development. Most commonly, pelagic fish eggs have a single oil globule in the yolk. Of a total of 515 species checked by Ahlstrom and Moser (1980), 60% had one oil globule, 25% had no oil globule, and 15% had multiple oil globules. The oil globule, when single, usually lies at the vegetal pole in Marine fish larvae. It is generally thought that the oil globules are a specialized form of nourishment and have a minimal effect on buoyancy.

Following activation or fertilization the egg absorbs water, the perivitelline space forms, and the chorion hardens. The perivitelline space is usually narrow but is wide in some "primitive" species such as the pilchard, in some eels, and in unrelated species like the striped bass *Morone labrax* and long rough dab *Hippoglossoides platessoides*. Cleavage is meroblastic in hagfish, elasmobranchs, and teleosts, although in the lampreys it is holoblastic but with the formation of micro- and macromeres. In primitive groups like the bowfin *Amia*, gar *Lepisosteus*, and sturgeon *Acipenser*, cleavage is intermediate or semiholoblastic. The embryo develops as a blastodisc at the animal pole. The periphery of the blastodisc overgrows the yolk (epiboly), eventually enclosing it to form a gastrula but leaving an opening, the blastopore. The embryonic axis forms by a process of convergence and concentration in relation to the dorsal lip of the blastopore at the neurula stage but the quantity of yolk influences the timing of such events. The head and eye cups are soon identifiable and the trunk lengthens and separates from the yolk sac. The heart functions well before hatching, and in some demersal eggs a vitelline circulation can be seen within the yolk sac. Examples of the development of a marine egg (dab) and a freshwater egg (rainbow trout) are given in Figs. 1 and 2.

Before hatching, the embryo becomes very active and the chorion is softened as a result of enzymes secreted by hatching glands. The degree of differentiation of the newly hatched larva depends very much on the species and egg size, and the incubation period depends on these factors and on temperature (see Fig. 18). In many marine pelagic species the mouth and jaws are not formed, the eye is not pigmented, the yolk sac is huge, and a primordial finfold runs around the trunk in the median position. Apart from a few melanophores, the larva is very transparent. All newly hatched larvae have free neuromasts on the head and trunk, and otoliths are present in the otic cap-

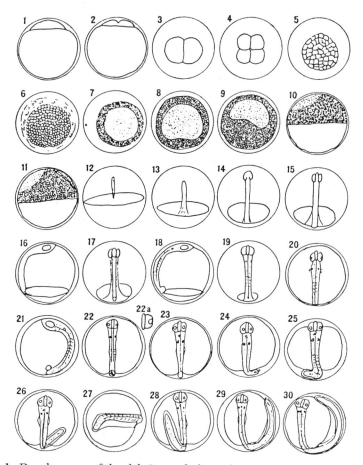

Fig. 1. Development of the dab *Limanda limanda*, using Apstein's stages. [From Russell (1976), with permission of Academic Press.]

sule. Other marine species hatch with the alimentary system nearly functional and with pigmented eyes. Some larvae are very advanced, and in loricariids the dorsal and caudal fin are partly developed at hatching (Fuiman, 1984); in flying fish, flexion of the notochord (which precedes caudal fin formation) actually occurs before hatching. In the salmonids—for example, rainbow trout (Fig. 2)—although the yolk sac is still large, the larva (alevin) is better developed and especially the vascular system and vitelline circulation are conspicuous with the blood containing hemoglobin. The young of cichlid and ariid mouth brooders are also further developed and adapted to early life within the parental mouth. In ovoviviparous and viviparous species

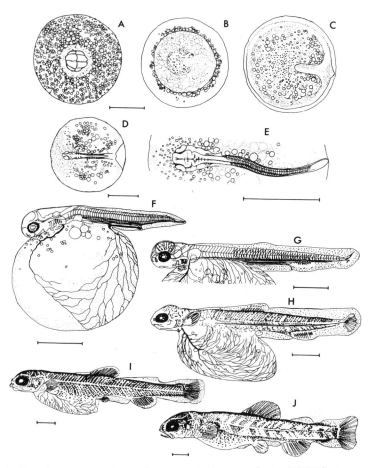

Fig. 2. Development of the rainbow trout *Salmo gairdneri*. (A) 8-Blastomeres. (B) Early embryo apparent, one-third epiboly. (C) 0–5 Somites, one-half epiboly. (D) Otic placodes, three-fourths epiboly. (E) Caudal bud with 10–20 somites, total somites 51–58, heart beating. (F) Posterior cardinal veins formed, choroid of eye pigmented. (G) Near hatching, pelvic fins develop. (H) Hatched alevin, first anal and dorsal fin rays. (I,J) Later alevin stages as yolk is resorbed. Scale bars 2 mm long. [Redrawn from Vernier (1969).]

(Amoroso, 1960) the young may hatch effectively as postmetamorphic juveniles.

Examples of early life history stages, those of the jack mackerel, northern anchovy, and Pacific hake, are illustrated in Figs. 3 and 4. The changing shape of the larvae is clearly shown, with the impor-

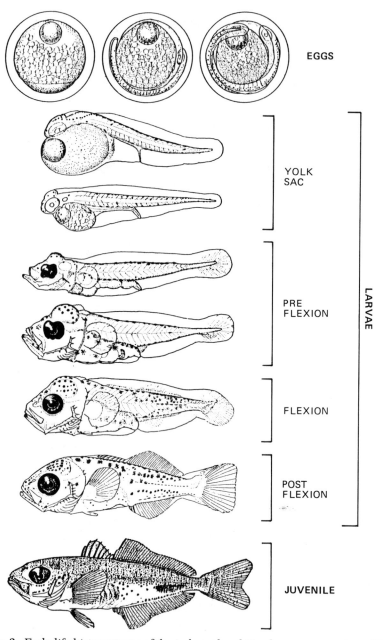

Fig. 3. Early life history stages of the jack mackerel *Trachurus symmetricus*. [From the original drawings of Ahlstrom and Ball in Kendall *et al.* (1984), with permission of the American Society of Ichthyologists and Herpetologists.]

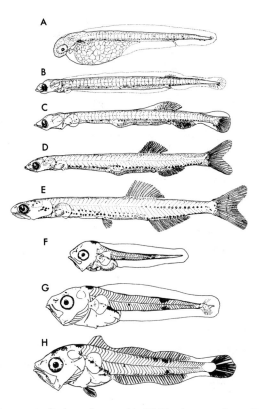

Fig. 4. Development of teleost larvae. (A–E) Northern anchovy *Engraulis mordax*, 2.5, 7.5, 11.5, 18.4, 31.0 mm. [Redrawn from Kramer and Ahlstrom (1968).] (F–H) Pacific hake *Merluccius productus*, 4.3, 7.7, 11.0 mm. [Redrawn from Ahlstrom and Counts (1955).]

tance of flexion of the notochord and development of the caudal fin being emphasized, with implications for improved swimming.

The duration of the yolk-sac period depends on both species and temperature but also on egg size (see also p. 17). The argentine *Argentina silus* and the halibut have egg diameters of 3.0–3.5 mm. Unexpectedly, the newly hatched larvae are very undeveloped but the halibut takes 50 days to resorb its yolk (at 5.3°C) and reaches a length of 11.5 mm (Blaxter *et al.*, 1983a) and the argentine reaches a prodigious length of 17 mm on its yolk supply (Russell, 1976).

During the yolk-sac period the mouth and gut and the eyes become functional to allow the larva to switch from endogenous to exogenous nutrition. The subsequent larval period ranges from a few days to some months (and even 2–3 years in eels), depending on tempera-

ture and species. During this time the larva is likely at least to double
its length and to increase its weight by 10 to 100 times. Transient
characters, such as spines, may appear (Fig. 5), which are presumably
antipredator adaptations. Other bizarre structures, such as eyestalks,
elongated fin rays, or tentacles, may also appear, often as larval characters (Fig. 5), to be lost later in development. The cobitid *Misgurnus
fossilis* even has external gill filaments for a time (Fuiman, 1984).

The importance of allometric growth during larval development
has been emphasized by Fuiman (1983) and Strauss and Fuiman
(1985). In some species relative growth intensity follows a U-shaped
gradient along the body with fastest growth in the caudal region,

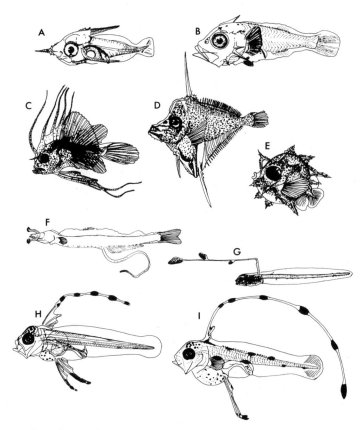

Fig. 5. Teleost larvae showing spines and other processes. (A) *Holocentrus vexilla-
rius* 5.0 mm. (B) *Sebastes macdonaldi* 9.0 mm. (C) *Lophius piscatorius* 26 mm. (D)
Acanthurid 7 mm. (E) *Ranzania laevis* 2.8 mm. (F) *Myctophum aurolaternatum* 26 mm.
(G) *Carapus acus* 3.8 mm. (H) *Trachipterus* sp. 7.6 mm. (I) *Zu cristatus* 6.5 mm. [Re-
drawn from Moser (1981).]

linked to an increase in the propulsive area of the body. Growth is also fast in the head region, where elaboration of feeding and respiratory functions may be taking place. In sculpins, however, the caudal region grows more slowly than the rest of the body and growth is fastest in the head region.

Progressive differentiation of adult characters (such as fin rays and skeleton) occurs. The larvae eventually pass through a process of metamorphosis to the juvenile stage. This process may be rather abrupt or it may be prolonged. Typically the blood becomes pigmented, scales and pigment appear on the body surface, the meristic characters such as fin rays are complete, and the body shape becomes like the adult. The juvenile appears as a small adult. In flatfish, metamorphosis is a remarkable process as the fish starts to change from the bilaterally symmetrical larva to an asymmetrical juvenile lying on one (abocular or blind) side (Fig. 6). Changes take place to the skull and sense organs and, in particular, the eye of the abocular side migrates across the top of the skull.

By way of summary, Table II lists some of the characteristics of a few species to emphasize the diversity that is to be found. The young of many elasmobranchs, with a long incubation period, effectively hatch as juveniles, albeit with a yolk sac. In species with parental care the early larvae may also be advanced or "precocial." This variety makes it difficult to categorize early life histories in a neat and convincing way.

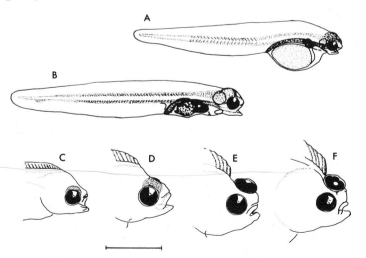

Fig. 6. Development and metamorphosis in the plaice *Pleuronectes platessa*. (A) Yolk-sac larva, 6.6 mm. (B) Larva at first feeding, 7.3 mm. (C–F) Stages of metamorphosis with eye migration; scale bar 2 mm. [Redrawn from Ryland (1966).]

Table II

Early Life History Characteristics[a]

Species	Common name	Egg diameter (mm)	Hatching length (mm)	Weeks from fertilization to			Temperature range (°C)
				Hatch	First feed	Metamorphosis	
Gadus morhua	Cod	1.1–1.9	3–4	1.4–2.9	1.9–3.6	4.2–5.0	4–12
Pleuronectes platessa	Plaice	1.7–2.2	6–7	2.5	4.0	10–12	7–11
Scomber scombrus	Mackerel	1.0–1.4	3–4	0.8–1.5	1.3–2.0	11–13	9–15
Scophthalmus maximus	Turbot	0.9–1.2	2–3	0.6–0.9	1.1–1.7	5–6	13–18
Engraulis mordax	Northern anchovy	0.6–0.7 ×1.3–1.4[b]	2–3	0.4–0.5	0.9–1.1	11–15	13–15
Clupea harengus	Herring	0.9–1.7	5–8	1.0–3.0	2.0–4.5	12–24	6–14
Hippoglossus hippoglossus	Halibut	3.0–3.2	6–7	2.0–3.0	9.0–10.0	?15–16	4–7
Acanthurus triostegus	Convict surgeonfish	0.7	1.7	0.15	0.7	?	26
Oreochromis (=Tilapia) mossambicus	Tilapia	1.7–2.2	4	0.6–0.7	1.7–2.6	?	28
Oryzias latipes	Medaka	1.0–1.3	4–5	1.5–2.0	2.0–2.4	Not clearcut	20–25
Salmo salar	Salmon	5–6	15–25	20–22	26–28	Not clearcut	1–7
Scyliorhinus caniculus	Spotted dogfish	65 (long)	100	24–32	28–36	28–36	4–12
Squalus acanthias	Spur dogfish	24–32	240–310	~104	~104	<104	4–12

[a] Data from Blaxter (1969), Blaxter *et al.* (1983a), Howell (1979), J. R. Hunter and C. Kimbrell (personal communication), Iversen and Danielssen (1984), Jones (1972), Kuhlmann *et al.* (1981), Rana (1985), and Russell (1976).

[b] Eggs are ellipsoidal, minor and major axes given.

IV. TERMINOLOGY OF EARLY LIFE HISTORY STAGES

A terminology is important both for understanding the literature and for brevity in describing development. A good terminology should be as simple as possible (and so easily remembered) and linked to both form and function. The production of a generally accepted terminology is a current issue in ichthyology, and some of the varied attempts to produce standardization are shown in Fig. 7.

The problems are discussed by Snyder and Holt (1983). The difficulties lie in producing a terminology that embraces all species and all patterns of development in fish, and it almost becomes an intellectual challenge to achieve this. Some workers favor a large number of stages, others very few; one point of view suggests terminology based on size alone, another that ecological considerations should be paramount. Some workers (e.g., Balon, 1984; see Fig. 22) use the term "embryo" to cover the period from fertilization to first feeding and consider hatching to be a relatively insignificant process. While it is certainly true that the change from endogenous to exogenous food supply is a major hurdle for the organism to overcome, it should not be forgotten that eggs cannot avoid predators although hatched larvae can. Many species of fish hatch in a very well developed state, especially where ovoviviparity, viviparity, or other parental care is involved, or where the incubation period within the egg is long; other species hatch in a much earlier state of development. It is difficult to resolve a nomenclature to cover such a wide variation in ontogeny.

The present author prefers to use the term "embryo" only to the point of hatching, does not accept the terms "prelarva" and "postlarva," which suggest stages before and after a larval stage, and uses the term "larva" to cover development from hatching to metamorphosis and the term "juvenile" from metamorphosis to first spawning. Terms such as "fingerling" or "young-of-the-year" are unsatisfactory: the former can hardly be applied to very short fish or the latter to species with a short generation time. A simplistic approach to terminology may well require additional qualification to be given, such as "yolk-sac" larva, or it may have to be made clear that some species hatch in an advanced state of development. This is in broad agreement with Kendall *et al.* (1984), who also favor dividing the larval stage into "preflexion," "flexion," and "postflexion" substages, referring to the turning up of the notochord tip during the first stages of development of the caudal fin (Fig. 3). Since flexion is accompanied by rather rapid development of other characters such as the fin rays and change of body shape, as well as a dramatic improvement in

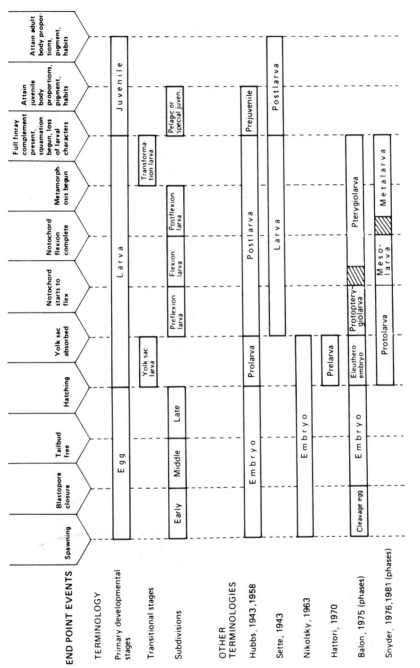

Fig. 7. Terminology of life history stages. [From Kendall *et al.* (1984), in which see original references; with permission of the American Society of Ichthyologists and Herpetologists.]

locomotor ability, this seems a valuable elaboration of the terminology.

V. EGG SIZE AND EGG QUALITY

A. Egg Size

The influence of initial egg size on subsequent survival and development has important ecological implications. Although a large yolk sac may reduce locomotor performance, plentiful yolk is likely to extend the period for switching from endogenous to exogenous feeding. Any factor that might provide larger larvae is likely to reduce the dangers from predation, since large larvae are likely to be able to escape faster and to be less susceptible generally to small predators. Furthermore, any factors that might cause early improvements in growth and size are likely to reduce the range of predators more rapidly.

Interspecifically, egg size and fecundity tend to be inversely related (see Table 1 in Blaxter, 1969). Egg size also influences the rate of development, those species with large eggs having a longer incubation time. While the small eggs of many fecund marine species hatch within a few hours or days (see Fig. 18), salmonids may take several weeks to hatch, and the very large eggs of elasmobranchs, several months. The ovoviviparous dogfish *Squalus acanthias* has a gestation period of about 2 years.

The effect of intraspecific egg size on larval size and survival has been investigated. Egg size does not seem to affect incubation time (at any one temperature) either in Atlantic herring (Blaxter and Hempel, 1963), Atlantic salmon (Kazakov, 1981), Arctic charr *Salvelinus alpinus* (Wallace and Aasjord, 1984), tilapia *Oreochromis* (= *Tilapia*) *mossambicus* (Rana, 1985), or the orangethroat darter *Etheostoma spectabile* (Marsh, 1986). However, egg size does influence larval size, with larger larvae, at first feeding, being produced from larger eggs. This was shown for Atlantic herring by Blaxter and Hempel (1963), for Atlantic salmon by Kazakov (1981) and Thorpe *et al.* (1984), for Chinook salmon *Oncorhynchus tshawytscha* by Fowler (1972), for Arctic charr by Wallace and Aasjord (1984), for rainbow trout by Gall (1974) and Springate and Bromage (1985), for tilapia by Rana (1985), for dace *Leuciscus leuciscus* by Mann and Mills (1985), and for the orangethroat darter by Marsh (1986). Furthermore, the larvae from larger eggs live longer on their yolk reserves (Blaxter and Hempel, 1963; Theilacker, 1981; Mann and Mills, 1985; Marsh, 1986; Fig. 8B), although the effect on ultimate survival may not always be established (Fig. 8A).

Wallace and Aasjord (1984) found the effect of egg size on the length of Arctic charr alevins was still clearly evident 140 days post-hatching; Springate *et al.* (1985) also found that the greater size of rainbow trout fry from females fed on a high ration could be followed for 2–3 months (Fig. 9). Glebe *et al.* (1979) showed that the length of fry of Atlantic salmon from four New Brunswick rivers was still related to the original egg diameter some 8 months later (Fig. 10). K. J. Rana (personal communication) reared tilapia and found that the weight of 20-, 40-, and 60-day-old fish was significantly related to the original egg size, although the degree of association decreased over the 20- to 60-day period. Thorpe *et al.* (1984) and Springate and Bromage (1985), however, working on Atlantic salmon and rainbow trout, respectively, concluded that the benefits of large egg size were soon lost during subsequent growth.

Intraspecifically, there are also many factors that control egg size. These are:

1. Parental size. Larger females produce larger eggs in Atlantic salmon *Salmo salar* (see review by Thorpe *et al.*, 1984). In addition, females that spend longer at sea (and so are larger) also produce larger eggs (Kazakov, 1981). In Atlantic herring, Hempel and Blaxter (1967) found that the eggs of first-time spawners were the smallest but that there was no influence of parental size on egg weight in repeat spawners.

2. Spawning group. There is an extensive literature showing how salmonid eggs vary in size from different rivers (Bagenal, 1971; Thorpe *et al.*, 1984). For example, Glebe *et al.* (1979) measured the calorific value of Atlantic salmon eggs from four New Brunswick rivers and found a range from 1067–1576 J/egg.

3. Season. The influence of spawning group is closely linked to different spawning seasons that may occur within a species. This is most dramatically shown in the herring (Fig. 11), where different seasons may have egg dry weights varying by a factor of four, with the largest eggs being produced in the winter and spring and smallest in the summer and fall. This general intra-specific trend toward smaller eggs as the spring season progresses was shown in a wide range of species by Bagenal (1971) and for clupeoids by Blaxter and Hunter (1982) (Figs. 11 and 12).

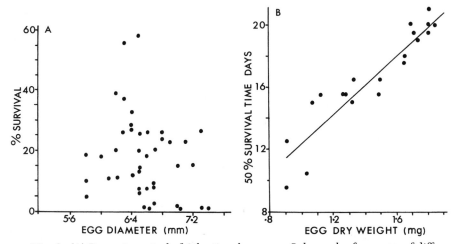

Fig. 8. (A) Percent survival of Atlantic salmon parr *Salmo salar* from eggs of different diameter. Each point refers to a different female. There is no significant correlation between egg size and survival. [Redrawn from Glebe *et al.* (1979).] (B) The time to 50% survival from hatching of unfed tilapia *Oreochromis* (*Tilapia*) *mossambicus* fry related to the mean dry weight of the eggs. Each point refers to a different clutch of eggs. The regression is significant ($r = 0.923$, d.f. $= 23$, $p < 0.01$). [Redrawn from Rana (1985).]

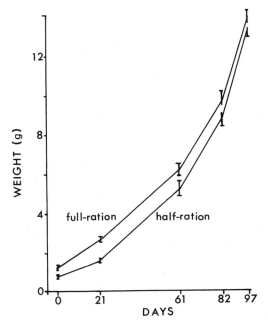

Fig. 9. The weight (mean ± S.E.) of rainbow trout *Salmo gairdneri* fry derived from eggs of females held on half and full rations. Day 0 refers to first feeding, so that the difference in weight is still detectable 97 days later. [Redrawn from Springate *et al.* (1985).]

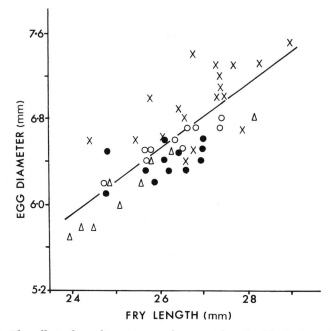

Fig. 10. The effect of egg diameter on subsequent length of fry in four New Brunswick (Eastern Canada) Atlantic salmon *Salmo salar* stocks. The four stocks are shown by different symbols. Each point refers to a different female. [Redrawn from Glebe *et al.* (1979).]

4. Diet. The way in which the diet of the female affects fecundity and egg size needs clarification. Much may depend on the phase of the egg maturation cycle during which an experimental diet is given. Diets may delay or accelerate spawning, so allowing more or less time for material to be laid down in the egg, or diets may affect the processes of atresia. The most common effect of starvation or overcrowding is to reduce fecundity, and for good feeding or low stocking density, to increase fecundity (Wooton, 1979). Concomitant changes of egg size often, but not always, occur; with poor feeding, egg size is sometimes increased (Wooton, 1979) and sometimes decreased, for example, in haddock (see Hislop *et al.*, 1978) and in rainbow trout (Springate *et al.*, 1985).

Despite the extensive experimental work on egg size, it is not clear the extent to which changes of fecundity or egg size have adaptive value in the wild and whether egg survival can be enhanced or not.

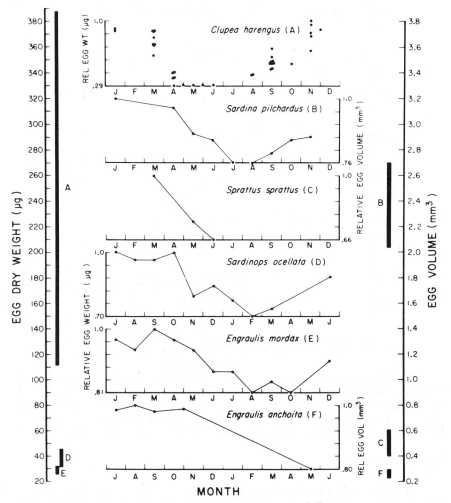

Fig. 11. Relative changes of egg weight or volume in different spawning seasons (center graphs) where 1 = maximum egg weight or volume for the season. Species from the southern hemisphere (*S. ocellata* and *E. anchoita*) are offset by 6 months. The separate ordinates show the range in absolute weight or volume of eggs. Note the enormous range of egg weight in herring spawning in different seasons compared with other species. [From Blaxter and Hunter (1982), in which see original references; with permission of Academic Press.]

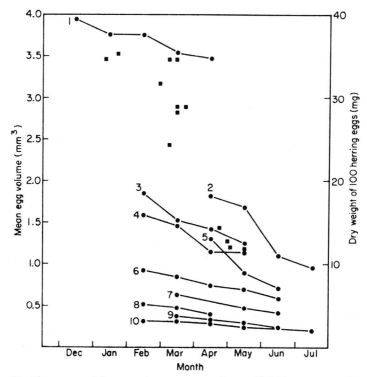

Fig. 12. The seasonal decrease in mean egg volume of (1) *Pleuronectes platessa*, (2) *Trigla gurnardus*, (3) *Melanogrammus aeglefinus*, (4) *Gadus morhua*, (5) *Solea solea*, (6) *Merlangius merlangus*, (7) *Sprattus sprattus*, (8) *Platichthys flesus*, (9) *Rhinonemus cimbrius*, and (10) *Limanda limanda*. Square symbols show dry weight of herring eggs. [From Bagenal (1971), with permission of The Fisheries Society of the British Isles.]

B. Egg Quality

A key to early success in rearing is the production of eggs of good quality. If broodstock are maintained in captivity, as is so often the case in aquaculture, it is desirable to strip the fish soon after ovulation, unless the holding conditions will allow natural spawning. It is certainly possible for ovulated eggs to remain in the body cavity of the female for many days, often culminating in their resorption into the maternal tissues. Craik and Harvey (1984b,c) found that the eggs of rainbow trout *S. gairdneri* held for 30 or more days postovulation underwent changes of water content and biochemical components that must have reflected a deterioration and decomposition of the yolk. They also found that the percentage of eggs that hatched fell sharply when the eggs were held for 18 days postovulation. Springate *et al.* (1984) found maximum egg and fry survival if rainbow trout were stripped 4–6 days after ovulation. In contrast, Mollah and Tan (1983) found that the eggs of catfish *Clarias macrocephalus* only remained viable for up to 10 h postovulation at 26–31°C. The need to fertilize eggs at an optimal time is particularly important in cases of hand stripping, or if spawning is induced by hormones. Flounder (*Limanda yokahamae*) eggs remain in good condition for 2–3 days after ovulation, depending on the hormone used for inducing spawning (Hirose *et al.*, 1979).

The postovulatory decline in egg viability seems much more significant than small differences in dry weight, protein, and lipid content of eggs from different rainbow trout females, which may also be related to hatching success. Possibly a more important biochemical component of salmonid eggs is the carotenoid. Craik (1985) reviewed the extensive Japanese and Russian literature on the role of carotenoids in development. They may act as precursors of vitamin A and play a part in respiration; they are certainly a source of pigment for the chromatophores. It is likely that they have to be above a critical concentration to give high hatch rates.

Variation in egg quality may arise in other ways. Billard *et al.* (1981) review the effect of stress and other environmental factors on teleost reproduction. Adverse feeding, temperature, light, and water quality, as well as crowding, can cause low fertilization and hatching rates. In the red sea bream *Chrysophrys major* a low-protein, phosphorus-deficient diet produced eggs with poor hatchability and deformed hatched larvae (Watanabe *et al.*, 1984). In other species, such as the rainbow trout, ayu *Plecoglossus altivelis*, and carp *Cyprinus carpio*, deficiencies of fatty acids and vitamin E in the diet of the female may effect egg viability (T. Watanabe, 1985). Whipple *et al.* (1981) found that striped bass migrating through polluted water showed poor condition of the parents, low fecundity, and low egg viability, as well as a high level of lesions and parasites.

Poor quality can also be manifested in other forms. In Atlantic cod, poor-quality eggs may show irregular shape, abnormal fertilization, soft chorions, or negative buoyancy (Kjørsvik and Lønning, 1983). In a unique study of a wild stock, Kjørsvik *et al.* (1984) found that 6–60% of cod eggs had abnormal mitoses or chromosome aberrations.

VI. THE EFFECT OF STARVATION

During starvation a number of morphological and chemical changes occur. There is a progressive collapse of the larval head and trunk in Atlantic herring and plaice (Ehrlich *et al.*, 1976) and in jack mackerel (Theilacker, 1978), so that body weight or head height relative to the length can be used as an index of starvation. The condition factor (dry weight/length^3) has been used on sea-caught Atlantic herring larvae (Blaxter, 1971) to assess their nutritional condition. A U-shaped relationship is found between condition factor and length (see Fig. 17) because the larval body is denser when yolk is present and later as ossification occurs. Chenoweth (1970) and Ehrlich *et al.* (1976) therefore used relative condition factor (dry weight/length^b) where b is the slope of the regression line relating weight to length (i.e., weight $\propto$ length ^{b}), which prevents the right-hand arm of the U appearing.

Ehrlich *et al.* (1976) and Y. Watanabe (1985) also found changes in the height of the gut epithelia and shrinkage of the liver. O'Connell (1976) used 11 histological criteria to assess starvation in northern anchovy larvae and found the appearance of the pancreas, trunk muscle and liver cytoplasm gave the most reliable results. Theilacker (1978) later used histological characteristics of the pancreas and gut to estimate the extent of starvation of the larvae of jack mackerel.

There were also changes in biochemical components in plaice and herring larvae during starvation (Ehrlich, 1974a,b). Percentage of water increased and percentage of triglyceride, carbohydrate, and carbon decreased. Percentage nitrogen decreased in plaice but not in herring. Recently Buckley (1981) and Clemmesen (1987) have shown that the RNA/DNA ratio drops dramatically in starving winter flounder, herring, and turbot larvae.

The time over which starvation takes place depends on a number of factors. Theilacker and Dorsey (1980) and McGurk (1984) summarized the time to reach the point-of-no-return (PNR; Blaxter and Hempel, 1963), or irreversible starvation, in some 25 marine species if they failed to feed once the endogenous yolk supply had been exhausted. There were very large differences, ranging from 3.5 days postfertilization in *Anchoa mitchilli* at 28°C to 36.5 days postfertilization for Pacific herring at 6°C. Generally the time to the PNR for first-feeding larvae increases for species with long incubation periods and at low temperatures (Fig. 13) and when the eggs are large (page 18). Once feeding is established the time to the PNR increases with age, as shown in Fig. 14, presumably because of greater body reserves.

Although the morphological and chemical effects of starvation during development are now fairly well established, it is not certain how much these changes are reflected in locomotor performance and behavior. Blaxter and Ehrlich (1974) found a decrease in sinking rate (or an increase in buoyancy) in plaice and Atlantic herring larvae as a result of increasing hypotonicity of body fluids and a decrease in body

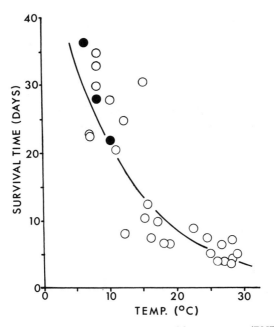

Fig. 13. The time from fertilization to irreversible starvation (PNR) of unfed fish larvae of 25 species at different temperatures (open circles). The black circles are larvae of Pacific herring *Clupea pallasi*. [Redrawn from McGurk (1984).]

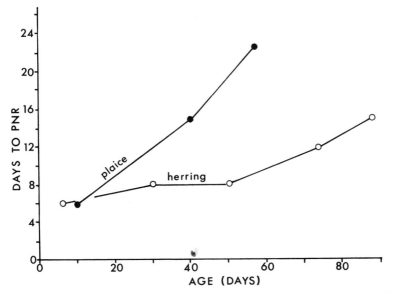

Fig. 14. The number of days to irreversible starvation (PNR) of different aged plaice *Pleuronectes platessa* and Atlantic herring *Clupea harengus* larvae. Temperatures 7.5–11.5°C (Blaxter and Ehrlich, 1974).

protein during starvation (Fig. 15). Activity was maintained for some days and tended to decrease drastically only late in the period of starvation, presumably as a way to maintain food-searching ability as long as possible, even if it was energetically expensive. Huse and Skiftesvik (1985) found that starving turbot larvae swam faster for shorter periods and searched for food less efficiently than feeding larvae.

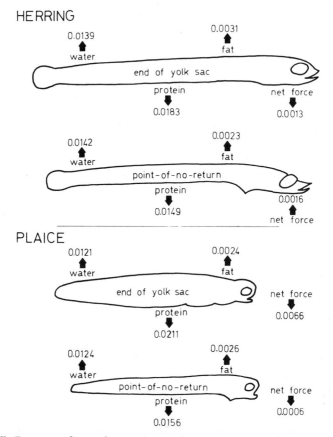

Fig. 15. Buoyancy forces due to chemical components in Atlantic herring *Clupea harengus* and plaice *Pleuronectes platessa* larvae at the end of the yolk-sac stage and at the point-of-no-return (PNR). All forces are dynes/wet weight (mg); arrows show direction of forces. Note that the larvae become increasingly buoyant as they starve. [From Blaxter and Ehrlich (1974), with permission of Springer Verlag.]

VII. THE EFFECT OF CAPTIVITY

The effect of captivity on growth and morphology was summarized by Blaxter (1976). A common phenomenon is the "size hierarchy" (sometimes called depensation of growth)—that is, an increasing *range* of size as the larvae grow. Some examples are given in Fig. 16. Whether size hierarchies occur in natural conditions is uncertain, since variations in egg size and spawning time may obscure the phenomenon, and selective predation on small individuals may limit the lower end of the size range. The increase in length range with age may be due to a natural variation early in development. For example, a 10% variability in length of 10–11 mm early in development would result in a range from 50 to 55 mm later. The increase of range in length is often well beyond such considerations, and it seems likely that size hierarchies are a tank phenomenon caused by competition for food, social dominance of some individuals in crowded conditions, and lack of predation on smaller individuals. Thus Eaton and Farley

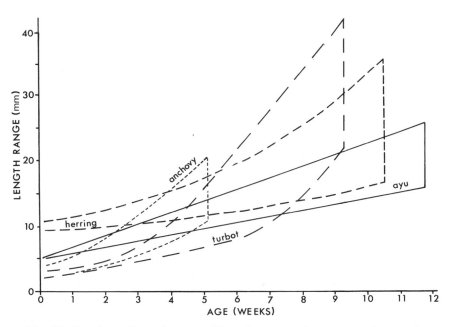

Fig. 16. Size hierarchies. Increase of length range with age in northern anchovy *Engraulis mordax*, Atlantic herring *Clupea harengus*, turbot *Scophthalmus maximus*, and ayu *Plecoglossus altivelis* reared in the laboratory. [Redrawn from Blaxter (1981) and Tanaka (1973).]

(1974) found a reduction in the size hierarchy of the zebrafish *Danio* (= *Brachydanio*) *rerio* fed an adequate diet. After 32 days, larvae fed on a low ration ranged from 5 to 12.5 mm (mean 9.5 mm), and after 39 days another group fed on a high ration ranged from 16.0 to 17.5 mm (mean 16.9 mm).

Rearing in captivity also tends to produce shorter, fatter fish, with high condition factors (Fig. 17) and growth abnormalities such as foreshortened snouts, neoplasms of the head, and failure of eye migration and fin development in flatfish. Arthur (1980) found that the hearts of northern anchovy larvae, as determined by length of the ventricle,

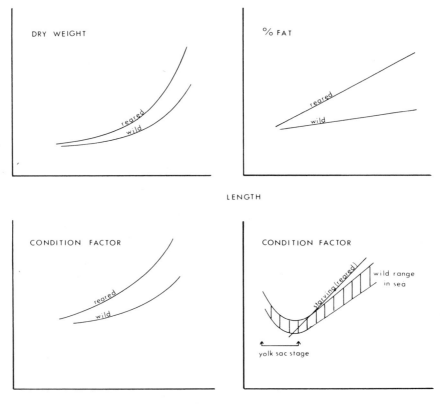

Fig. 17. A general comparison of changes in dry weight, fat, and condition factor of reared and "wild" fish as they grow, based mainly on Atlantic herring; see text for details. The change of condition factor (shown bottom right) in the sea results from initial loss of yolk, causing a fall, followed by progressive ossification, causing a rise. Condition factors of reared starved larvae tend to be higher than "wild" larvae, showing that starvation criteria cannot be obtained satisfactorily from captive larvae.

were as much as 40% longer in laboratory-reared compared with sea-caught individuals. Excessive contact with the walls of the tank, other forms of stress, and social factors can be implicated in causing these abnormalities.

The effect of the volume of the rearing container was investigated by Theilacker (1980b) using jack mackerel larvae reared at the same density in 10-l and 100-l circular tanks. The beneficial effect of the larger volume was apparent after only 4–5 days of feeding, when the larvae were larger and in better nutritional condition. The dramatic effect of using rearing facilities on the mesocosm scale was shown by Morita (1985) and Paulsen et al. (1985), who reported excellent growth and survival of Pacific herring larvae and turbot larvae respectively in 20-m^3 on-shore tanks. Sturmer et al. (1985) found similar beneficial effects using red drum larvae in large on-shore tanks and Kvenseth and Øiestad (1984) and Øiestad et al. (1985) using cod larvae in a coastal impoundment of 60,000 m^3.

High density of larvae may also create problems other than those associated with physical interactions. Crowded conditions may cause the production of inhibitory substances. Although originally demonstrated in amphibians, such substances may be implicated in population control of the guppy Lebistes (Poecilia) reticulatus, whitecloud mountain minnow Tanichthys albonubes, tiger barb Barbus tetrazona, rainbow trout, and ciscoes Leucichthys sp. (Rose, 1960). Laale and McCallion (1968) found that the development of zebrafish could be arrested before gastrulation by the use of supernatant homogenates produced from other zebrafish embryos. Such effects are likely to be relevant in the wild only to species that live in crowded conditions or in stagnant water.

In biochemical terms, hatchery fish often have a higher percentage fat content than their wild counterparts but lower percentage protein and ash (Table 1 in Blaxter, 1976) (Fig. 17). The biochemical and morphological changes occurring in hatchery fish can be reflected in their behavior and especially in their locomotor performance. Crowding induces stress and aggression that can be partly alleviated by adequate food, for example, in the medaka Oryzias latipes (Magnuson, 1962) and Atlantic salmon (Symons, 1968). Bams (1967) found that the fry of wild migrant sockeye salmon Oncorhynchus nerka could stem a water current and avoid predation better than hatchery fry, although mainly this was due to the fact that they were bigger.

As the inverse of stress and crowding, sensory deprivation may be implicated as a feature of hatcheries or rearing tanks (Blaxter, 1970). Developing fish may not be subjected to the normal interplay of light and shade, nor to the very high light intensities appertaining in the wild. They may not be able to practice avoidance or other responses in the absence of a natural substratum, typical water currents, or predators. Such deprivation will be particularly harmful where learning is involved in the development of behavior patterns. The problems are well demonstrated by attempts to establish hatchery-reared plaice in the sea (Blaxter, 1976). Survival was negligible, and it is likely that inadequately developed predator-avoidance, burying behavior, and feeding mechanisms were to blame.

A number of workers have assessed the effect of blinding, or rearing in darkness, on the eye and optic tectum of fish larvae. Pflugfelder (1952) found that unilateral blinding of newly hatched swordtails *Xiphophorus* and guppy *Lebistes* caused a reduction in the development of the contralateral optic tectum, mainly by a decrease in volume of the ganglion cells. Experiments in darkness confirmed that the effect was caused by lack of visual input rather than by degeneration products released from dying axons of the optic tract. Blaxter (1970) could find no retinal degeneration in dark-reared Atlantic herring larvae. Zeutzius *et al.* (1984) found that dark-rearing of tilapia *Sarotherodon* (= *Tilapia*) *mossambicus* did not affect the normal outgrowth of the nerve fibers of the optic tectum into the retina. It did, however, reduce the optic layer and the differentiation of the synapses, where the number of synaptic vesicles increased.

Dark-rearing can affect behavior. Blaxter (1970) found that newly hatched Atlantic herring larvae were very inactive when returned to the light after rearing in the dark and subsequently showed a high mortality. Zeutzius and Rahmann (1984) found that dark-reared larvae of tilapia failed to swim up after yolk resorption; visual acuity was impaired after 20–30 days in the dark, and after 50 days no optokinetic nystagmus was present. Effects on body weight and length were minimal, although the increase in body depth was substantial compared with control fish.

Sensory deprivation may operate at a more subtle level. For example, both Breder and Halpern (1946) and Shaw (1960, 1961) had difficulty in rearing isolated individuals of zebrafish and *Menidia*. Survival was poor, and in *Menidia* schooling was retarded when isolates were brought together.

VIII. THE EFFECT OF FIXATION ON SHRINKAGE

The interpretation of developmental events related to size, and estimates of growth rate and condition factor, will depend on whether live or fixed material is used. Larvae lacking ossification, or those with a long thin body form, will be especially prone to shrinkage, not only from fixatives, but also as a result of capture by plankton net if sampled from the wild. This problem has been addressed in the larvae of Californian sardine *Sardinops caerulea* by Farris (1963), in Atlantic herring by Blaxter (1971), in Pacific herring by Schnack and Rosenthal (1977–1978) and Hay (1981) and northern anchovy, Pacific mackerel *Scomber japonicus,* jack mackerel *Trachurus symmetricus,* and Pacific barracuda *Sphyraena argentea* by Theilacker (1980a) and Theilacker and Dorsey (1980), and in southern flounder *Paralichthys lethostigma* by Tucker and Chester (1984). The degree of shrinkage depends on the concentration and osmolality of the fixative, the period of fixation, and the age of the larvae. A shrinkage of 5–10% is normally experienced in long thin larvae, especially when young, but may be as little as 2% in older stages. Capture by net, either simulated (Blaxter, 1971; Theilacker, 1980a) or after release of larvae in the path of a net (Hay, 1981), followed by fixation, may cause shrinkage as great as 20%, or even more if fixation is delayed. The effect is much less serious in older larvae with ossified skeletons.

Theilacker and Dorsey (1980) also mention the loss of dry weight following fixation. Formalin fixation caused a 30% loss in weight of larval Pacific sardines and fixation in ethyl alcohol a 30–80% loss in Pacific mackerel larvae.

Methods of preservation and curation are discussed by Lavenberg *et al.* (1984), who recommend final preservation in 70% ethanol to obviate problems of buffering acid fixatives such as formalin. Problems of buffering are, however, trivial compared with the amount of shrinkage in formalin or alcohol.

IX. RATE OF DEVELOPMENT

The rate of development is clearly under genetical control. Within a species it is most strongly influenced by temperature, and this is exemplified by data on days to hatch at different temperatures in 13 species (Fig. 18). Herzig and Winkler (1986) give further data on three cyprinids. Blaxter (1969) and Herzig and Winkler (1986) discuss the mathematical relationship between temperature and incubation and the use of Q_{10} and other temperature coefficients. In particular Q_{10} is found to vary with range of temperature, and it may be that optimum temperatures for development, where hatching rates are highest, take place where the Q_{10} is between 2 and 3.

Temperature can also influence size at hatching, efficiency of yolk utilization, growth, feeding rate, time to metamorphose, behavior and

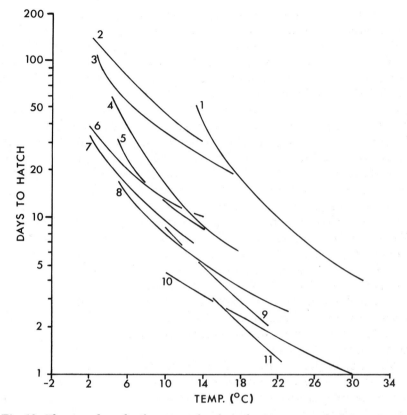

Fig. 18. The time from fertilization to hatching for 11 species of teleost related to temperature. Data originally cited in Blaxter (1969) except where stated. (1) Desert pupfish *Cyprinodon macularius*, (2) brook trout *Salvelinus fontinalis*, (3) rainbow trout *Salmo gairdneri*, (4) smelt *Osmerus eperlanus*, (5) Atlantic herring *Clupea harengus*, (6) plaice *Pleuronectes platessa*, (7) Pacific cod *Gadus macrocephalus*, (8) rockling *Enchelyopus cimbrius*, (9) mackerel *Scomber scombrus*, (10) grey mullet *Mugil cephalus* (Nash and Kuo, 1975), and (11) striped bass *Morone saxatilis*.

swimming speed, digestion and gut evacuation rates, and metabolic demand. Temperature also has indirect effects on larvae through the oxygen capacity of water, its viscosity, and phytoplankton blooms (Theilacker and Dorsey, 1980). Much of the data on marine fish are summarized by these authors and in more general terms by Kinne (1963) and Brett (1970). In particular, Brett (pp. 524–527) gives a comprehensive table of upper and lower lethal temperatures of embryonic and postembryonic stages of marine and estuarine fish.

The structural response to temperature is discussed by Garside

(1970) with particular reference to the effect of temperature on the interaction of differentiation and growth. Salmonids and clupeids incubated at high temperatures tend to weigh less at hatching. The best known effect is on meristic characters—counts of serial structures such as vertebrae, fin rays, scales, and gill rakers, which are labile (within limits) and susceptible to various environmental factors (Tåning, 1952; Barlow, 1961; Blaxter, 1969; also see Lindsey, volume XIB). Until recently it has not been clear what adaptive advantage might exist for varying numbers of meristic characters. It has been shown that individuals of a particular species with more vertebrae are often longer, and it seems plausible that they might also be more flexible and able to swim faster. Swain and Lindsey (1984) have recently shown that there was selective predation for vertebral number in young sticklebacks *Gasterosteus aculeatus* preyed on by sunfish *Lepomis gibbosus*. The survival of 8.2-mm-long sticklebacks was 1.3–1.7 times greater for fish with 31 vertebrae than with 32. This effect was not found with 8.9-mm-long sticklebacks, nor was there any influence of temperature.

Early development can be enhanced hormonally. Dales and Hoar (1954) treated the eggs of chum salmon *O. keta* with thyroxine and thiourea, an antithyroid compound. Thyroxine accelerated growth of the body wall and pectoral fins, increased guanine deposition, decreased pigmentation, caused exophthalmia, but reduced the rate of increase of body length. Thiourea decreased guanine deposition and also decreased the rate of growth in length. More recently, it has been shown that immersion in thyroxine accelerates growth in larval tilapia *Sarotherodon* (= *Tilapia*) *mossambicus*, carp *Cyprinus carpio*, and milkfish *Chanos chanos* and enhances survival in tilapia and carp (Lam, 1980; Lam and Sharma, 1985; Lam *et al.*, 1985). Further information—for example, on the influence of hormones on smoltification of salmonids—will be found in Hoar (volume XIB).

X. ORGAN SYSTEMS

A. Alimentary System

Many species hatch without a mouth, but this develops rapidly to allow for the transfer from endogenous yolk to exogenous food. Feeding in many species is a predatory act requiring vision, and feeding does not occur in the dark, especially in the very young stages (Blaxter, 1981, 1986; Hunter, 1980, 1981). The size or gape of the mouth at first feeding, and therefore the size of food that can be taken, is crucial for survival at the end of the yolk-sac stage (Fig. 19A,B). Experiments show that the size of food taken is related to the gape of the jaw and that both increase with age of the larvae.

Length and complexity of the alimentary tract increase as the larvae grow. Dabrowski (1984) describes three groups of fish larvae based on the morphology of the alimentary tract and gut enzymes. Most species have early larvae without a functional stomach or gastric glands; salmonids, on the other hand, have a functional stomach before changing to external feeding. Tanaka (1973) gives a full account of the development of the alimentary tract in 21 Japanese marine and freshwater species. At first feeding there are no pharyngeal teeth and few, if any, taste buds. The esophagus has longitudinal folds and mucous cells; the intestine and rectum are lined with columnar epithelium, and it is likely that most digestion occurs here. Cilia are present in the gut of early clupeoid and salangid larvae and may help to pass food along the gut (Iwai and Rosenthal, 1981). The liver, gallbladder, and pancreas are also formed early. In many species, but not in salmonids like the rainbow trout, the stomach and pyloric caeca develop late in larval life as the pattern and quantity of feeding changes. These subsequent processes are well described by O'Connell (1981) in the northern anchovy and Govoni (1980) in the spot *Leiostomus xanthurus*.

The length of the gut influences the passage time for food. For example, in roach *Rutilus rutilus* larvae the food is retained for only 2.5 h at 20°C, whereas in the adult it is 6 h (Hofer, 1985). The time for digestion and resorption and for the recovery of digestive enzymes is therefore reduced in the larval stages. Dabrowski (1984) reviewed work on the appearance of digestive enzymes during development. Clark *et al.* (1985) found an increase in protease activity of Dover sole from the age of 24 days up to the adult stage. Similarly, Lauff and Hofer (1984) found a progressive increase in activity and number of proteolytic enzymes with age in whitefish *Coregonus* hybrids, in rainbow trout, and in roach. Higher proteolytic activity in the roach could be correlated with the lack of a stomach. In the rainbow trout the well-developed digestive tract, which is differentiated into a stomach, pyloric caeca, and a short intestine at first feeding, may compensate for a lower level of proteolytic activity.

The relatively underdeveloped state of the alimentary system may explain why most species have carnivorous larvae, although later, when the gut lengthens, they may become herbivorous. It may also

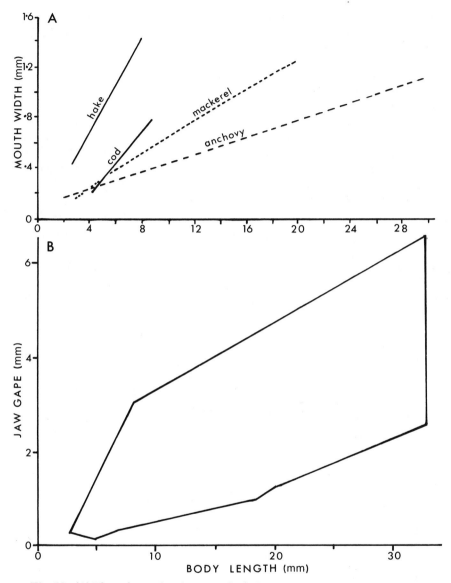

Fig. 19. (A) The relationship between body length and width of the mouth in the larvae of hake *Merluccius merluccius,* cod *Gadus morhua,* Pacific mackerel *Scomber japonicus,* and anchovy *Engraulis ringens.* [Redrawn from Hunter (1980).] (B) "Profile" of the body length: vertical jaw gape relationship in the larvae of 19 teleost species. [Redrawn from Shirota (1970).]

explain why artificial food is less satisfactory for young larvae, since live food contains exogenous enzymes that may aid digestion (Dabrowski and Glogowski, 1977).

B. Respiratory System

By first feeding most species have larvae with gill slits and gill arches (Tanaka, 1973), but gill filaments develop later. The cobitid *Misgurnus fossilis* has external gill filaments for a time (Fuiman, 1984). Harder (1954) gave a particularly thorough account for the herring, where the filaments first appear at a body length of 20 mm several weeks after hatching. De Silva (1974) measured the gill area of both herring and plaice larvae during early development and related them to the surface area of the body (Fig. 20). In the early stages it is clear that respiration is cutaneous. The larval heart is present even before hatching and pumps a colorless body fluid around an as yet unknown

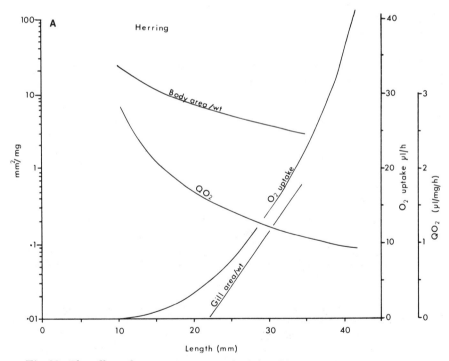

Fig. 20. The effect of size on oxygen uptake (μl O$_2$/h), QO$_2$ (μl O$_2$/mg dry weight/h), gill area (mm^2/mg wet weight) and body area (mm^2/mg wet weight) in (A) Atlantic herring *Clupea harengus* and (B) plaice *Pleuronectes platessa* larvae. [Data from De Silva and Tytler (1973) and De Silva (1974).]

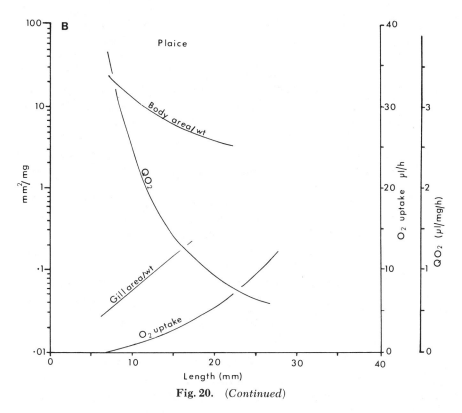

Fig. 20. (*Continued*)

vascular system. Weihs (1980b) found that yolk-sac larvae of northern anchovy were more active when the oxygen saturation of the water dropped below 60%, suggesting the requirement to renew the ambient water close to the body when oxygen is limiting.

In relating size to respiration the following theoretical conclusions may be drawn:

Total body area for cutaneous respiration $\propto$ body length^2
Body weight $\propto$ body length^3
Respiratory area per unit weight $\propto$ body length^{-1}
O_2 requirement $\propto$ body weight$^{0.8}$ or body length$^{2.4}$ (see Winberg, 1960)
Specific oxygen requirement (QO_2) $\propto$ body weight$^{-0.2}$ or body length$^{-0.6}$

Thus the body surface area per unit weight declines as the body length^{-1}, whereas the oxygen requirement per unit weight (QO_2) declines less rapidly, as body length$^{-0.6}$. This would lead to a critical situation without the development of gills to increase the respiratory surface area (Fig. 20).

Although the blood of herring and plaice and many other species does not appear pink until metamorphosis, its precursors, or related substances such as myoglobin, can be identified histochemically soon after hatching. The circulating body fluids were reported to be acellular until about 16 mm in the herring and 10 mm length in the plaice (De Silva, 1974), but Hickey (1982) found unidentified cells 8–14 μm in diameter in newly hatched larvae of both species and an efficient wound-healing mechanism in herring, plaice, and Atlantic salmon larvae.

In the walleye *Stizostedion vitreum* the circulatory system has been described by McElman and Balon (1979). The blood becomes red before hatching, although the development is not precocial. In the precocial larvae of salmonids and elasmobranchs, hemoglobin may be present, and the blood pink, at hatching. In these species the body size is large enough for cutaneous respiration to be inadequate and so increase in oxygen-carrying capacity of the blood is essential, even if the gills are in a rudimentary state.

C. Locomotor System

Most species hatch with V-shaped myotomes acting against the notochord as a hydrostatic skeleton. Additional myotomes may be added posteriorly, but the final number is attained during the early larval period. Generally the myotomes, which comprise white muscle, become progressively more complex in shape and interdigitate with adjacent myotomes. The red muscle develops initially as a myotube, a superficial cylindrical sheath around the body, e.g., in northern anchovy (O'Connell, 1981), zebrafish (van Raamsdonk *et al.*, 1982), herring (Batty, 1984), and red sea bream *Pagrus major* (Matsuoka and Iwai, 1984). In *Coregonus* sp. (Forstner *et al.*, 1983) the red muscle extends as a thin layer dorsally and ventrally from the lateral line. In all these species it later concentrates in a strip in the midlateral position on the flank (Fig. 21).

Larvae usually hatch with a primordial median finfold; the median fins often first appear as a discontinuity in the margin of the finfold; a few fin rays then appear, which gradually increase in number and size. In species with a homocercal tail the caudal fin develops after the tip of the notochord turns up (flexion, see p. 15). Lateral fins, used for stability, maneuvering, and sometimes for propulsion, develop differently. Pectoral buds, fin-like structures that lack rays, are often present at hatching. Pelvic buds and fin rays develop later.

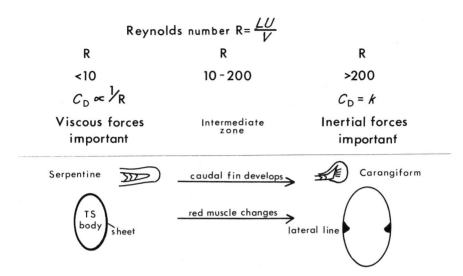

Fig. 21. Changes in hydrodynamic considerations as fish larvae grow related to the Reynolds number (R), where L is body length, U is swimming speed, and V is the kinematic viscosity (see Webb, 1975). An inertial regime exists where R > 200 and a viscous regime where R < 10, with an intermediate zone between. The drag coefficient (C_D) also changes. At the same time, the larvae change from a serpentine to a carangiform swimming mode as the tail flexion occurs and the red muscle develops from a myotube surrounding the whole body in a thin sheet (TS) to a strip situated along the midlateral position on the flank.

A number of workers (Webb, 1975; Weihs, 1980a; Batty, 1984; Webb and Weihs, 1986; Blaxter, 1986) have considered the "locomotor regime" of larvae of different size (Fig. 21). Where the Reynolds number (R) is below 10 (in very small larvae), viscous forces are paramount and continuous high-speed swimming is energetically efficient. Where R is greater than 200, inertial forces are more important and beat-and-glide swimming is more efficient. Since R depends on body length and velocity, the hydrodynamic regime changes as the larvae grow or alter their swimming speed. Often they may occupy an intermediate zone between the viscous and inertial regime. During later development there are, in many species, increases in the surface area for propulsion as a result of the appearance of the caudal fin and allometric growth (Fuiman, 1983).

Linked to swimming is the buoyancy of larvae. Although unimportant for demersal larvae, pelagic larvae can potentially waste much energy in maintaining their position in the water column. As larvae grow and the skeleton ossifies they become heavier (Fig. 17) and tend more and more to negative buoyancy. Conversely, if they starve, they will tend to neutral buoyancy (Fig. 15). The larvae of many species fill the swimbladder soon after hatching, or at the end of the yolk resorption, probably by swallowing air at the surface (Doroshev *et al.*, 1981). This stage, sometimes called "swim-up," can be critical for successful later development, e.g., in the turbot. The time of appearance of the swimbladder varies widely between species. For example, the herring fills its swimbladder at a length of about 30 mm, while the northern anchovy inflates its swimbladder at about 10 mm (Hunter and Sanchez, 1976) and the menhaden *Brevoortia tyrannus* at 13 mm (Hoss and Blaxter, 1982). In the physostomatous northern anchovy and menhaden there is a diel rhythm, the larvae filling their swimbladders by swallowing air at the surface at night (Hunter and Sanchez, 1976; Hoss and Phonlor, 1984).

D. Sense Organs

1. THE EYE

In some species the eyes are free of pigment at hatching and histological examination confirms that they must be nonfunctional. By first feeding the eyes are pigmented, and vision plays a major part in feeding. Of 10 teleost families examined by Blaxter and Staines (1970), eight had larvae at first feeding with a pure-cone retina, and only an anguillid and a macrourid, caught in deep water, had a pure-rod retina. A pure-cone retina at first feeding was also found in the northern anchovy by O'Connell (1981) and in the goldfish by Johns (1982). In zebrafish, rods can be identified 9 days after hatching (Branchek and Bremiller, 1984), and at hatching in Pacific salmon, *Oncorhynchus* spp. (Ali, 1959). In the advanced young of the viviparous guppy *Poecilia reticulata*, Kunz *et al.* (1983) found a well-differentiated duplex retina even before birth. Retinomotor movements of the masking pigment and the photoreceptors usually develop concomitantly with the rods (Ali, 1959; Blaxter and Staines, 1970; Kunz and Ennis, 1983; Neave, 1984) so that the process of light-and-dark adaptation is linked to the establishment of a duplex retina. The development of visual performance is described by Blaxter (1986), improvements in acuity being the most noticeable feature.

2. MECHANORECEPTORS

Free neuromasts are present at hatching in all species examined (Iwai, 1967, 1980), usually on the head and sometimes on the trunk. Disler (1971) gives a very detailed account of the development of the free neuromasts and lateral line of many freshwater species, including the sturgeon *Acipenser stellatus*, Pacific salmon *O. keta*, and some percids and cyprinids. Generally the initial number of free neuromasts is low but they proliferate and may become regularly arranged along the flank. In marine species similar systems are found in gadids (Fridgeirsson, 1978), northern anchovy (O'Connell, 1981), Atlantic herring (Blaxter *et al.*, 1983b), halibut (Blaxter *et al.*, 1983a), plaice and turbot (Neave, 1986), and spotted bass *Micropterus punctatus* (Kokkala and Hoyt, 1985).

The lateral line canals almost invariably develop some time after hatching: at 17 mm in menhaden, 18–20 mm in northern anchovy, 24–26 mm in Atlantic herring, 8 mm in the turbot, 10 mm in the plaice, and 12 mm in spotted bass. Thus young larvae have very incomplete mechanoreceptors. The development of the inner ear is not well known, except that larvae have one or more pairs of otoliths at hatching, which must give them a basic perception of posture.

3. CHEMORECEPTORS

Olfactory pits are described in the early larval stages of Atlantic herring by Dempsey (1978), tilapia by Iwai (1980), northern anchovy by O'Connell (1981) walleye *Stizostedion vitreum* by Elston *et al.* (1981), carp by Appelbaum (1981), and striped bass *Morone saxatilis* by Bodammer (1985). Kokkala and Hoyt (1985) described taste buds in larval spotted bass and Iwai (1980) found them between 1 and 14 days posthatching in tilapia, pond smelt *Hypomesus transpacificus*, goldfish, sea bass *Lateolabrax japonicus*, puffer *Fugu niphobles*, flatfish *Kareius bicoloratus*, and red sea bream *Pagrus major*.

XI. STRUCTURE AND FUNCTION

Clearly most larvae, apart from some highly developed ovoviviparous or viviparous species, or species with very large eggs, go through a massive increase in complexity while free-swimming. Since structures are often absent or incompletely developed, the associated behaviors are also absent or poorly developed. When relating structure to function, mention should be made of the theory propounded by Balon (1981b, 1984) that ontogeny is saltatory, meaning that development proceeds by a series of rather rapid changes in both structure and function with relatively prolonged intervals in between, during which a more-or-less steady state exists as the organism prepares itself for the next rapid change. Thus development does not proceed by a continuous accumulation of small changes. As an example, Balon cites a cyprinid, the bream *Abramis ballerus* (Fig. 22).

A number of studies link the development of structure with function. One of the best examples is found in the work of Hunter and Coyne (1982) on northern anchovy (Fig. 23), where age and length are related to the development of the sensory, respiratory, digestive, and locomotor systems and associated behavior and ability to withstand starvation. Fukuhara (1985) gives a similarly comprehensive account of the functional morphology of the red sea bream *Pagrus major*.

Allen *et al.* (1976) related the development of swallowing behavior, avoidance responses, and shoaling specifically to the development of the pro-otic bullae and swimbladder of Atlantic herring larvae (Fig. 24). In a somewhat similar fashion, Kawamura and Ishida (1985) compared the development of the whole sensory system of the flounder *Paralichthys olivaceus* to primary orientation, feeding, migration, and activity (Fig. 25). Considerable differences in morphological and

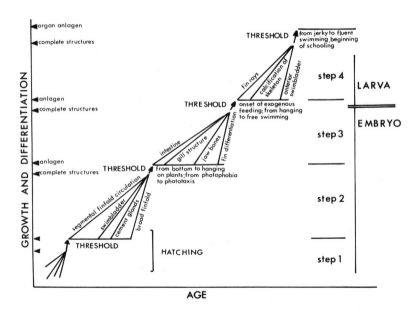

Fig. 22. A scheme of consecutive steps in the ontogeny of the Danubian bream *Abramis ballerus,* demonstrating saltatory development. During each step various structures grow and differentiate at different rates but are completed and become functional at the same time, at the end of the step, thus enabling the larvae to make substantial and rapid changes in behavior. Note according to the author's terminology that the embryo changes to the larva at first feeding, after step 3. [Modified and redrawn from Balon (1984).]

behavioral events can be seen in these two unrelated species; the herring, with an extended larval period metamorphosing into a pelagic schooling species, the flounder, a flatfish with a shorter larval period ending in settlement.

Balon (1980) gives a detailed but similar style of summary of developmental events in five species of charr *Salvelinus*. Despite their close relationship, they show substantial differences in the timing of the appearance of both morphological and behavioral features such as fins, melanophores, branchial respiration and swimbladder filling. The ontogeny of behavior, especially in salmonids and cichlids, is discussed by Noakes (1978) and Noakes and Godin (volume XIB). These groups are of particular interest because much of the early life history may be passed in gravel beds or under the care of a parent. Such a lifestyle may enhance the protection of the young but it imposes other problems, such as the avoidance of abrasion or cannibal-

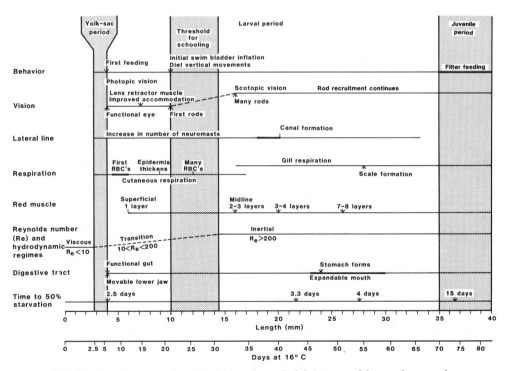

Fig. 23. Developmental events during the early life history of the northern anchovy *Engraulis mordax*. RBC, Red blood cells; time to 50% starvation is the number of days after which 50% of unfed larvae died (equivalent to point-of-no-return). For further discussion of Reynolds numbers see Fig. 21. [From Hunter and Coyne (1982).]

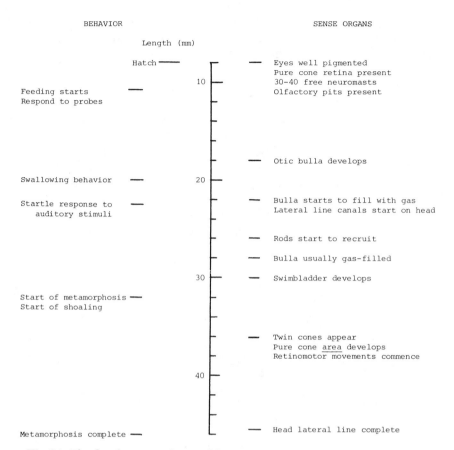

Fig. 24. The development of parts of the acoustic system and swimbladder of Atlantic herring *Clupea harengus* in relation to behavior. [Adapted from Allen *et al.* (1976).]

ism, which are quite different from free-living fish larvae. Both the behaviorist and the physiologist need to be aware of the ecology and reproductive habits of their experimental material before making evaluations of their data.

XII. CRITICAL PERIODS

The high fecundity of most fish and the low survival rate of their offspring imply a high mortality from a number of possible causes: inherited defects, egg quality, starvation, disease, predation. Whether these sources of mortality occur continuously or sporadically or

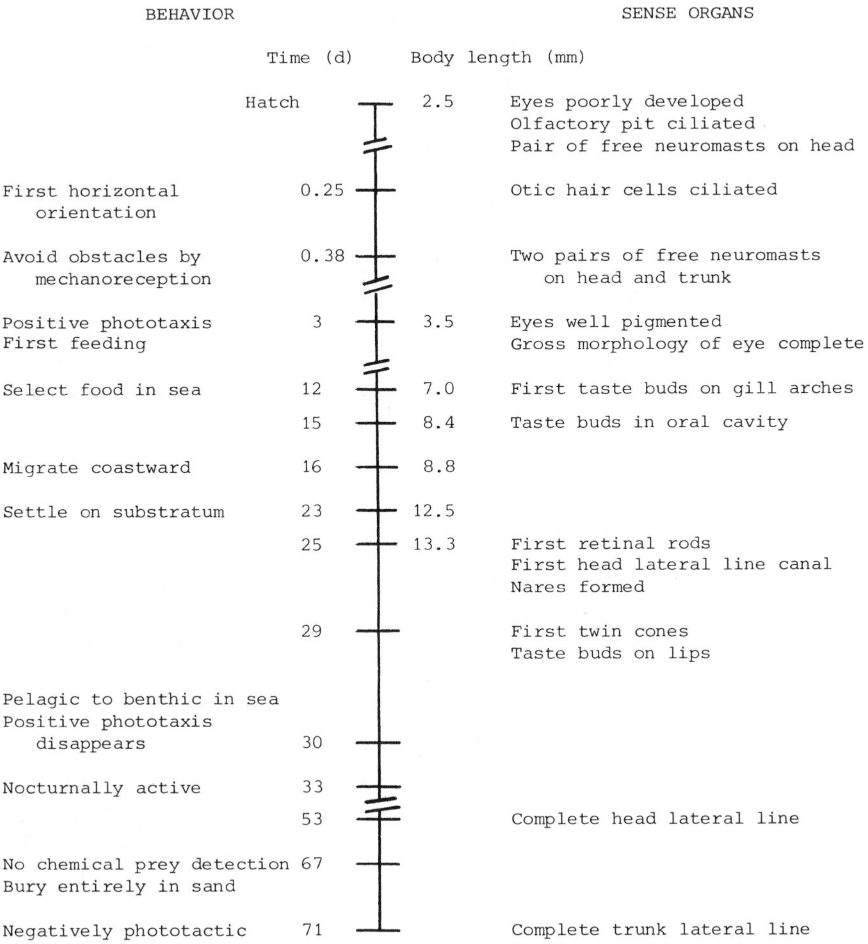

BEHAVIOR   SENSE ORGANS

Time (d)   Body length (mm)

Hatch   2.5   Eyes poorly developed
Olfactory pit ciliated
Pair of free neuromasts on head

First horizontal   0.25   Otic hair cells ciliated
orientation

Avoid obstacles by   0.38   Two pairs of free neuromasts
mechanoreception   on head and trunk

Positive phototaxis   3   3.5   Eyes well pigmented
First feeding   Gross morphology of eye complete

Select food in sea   12   7.0   First taste buds on gill arches

                  15   8.4   Taste buds in oral cavity

Migrate coastward   16   8.8

Settle on substratum   23   12.5

                  25   13.3   First retinal rods
First head lateral line canal
Nares formed

                  29   First twin cones
Taste buds on lips

Pelagic to benthic in sea
Positive phototaxis
    disappears   30

Nocturnally active   33

                  53   Complete head lateral line

No chemical prey detection   67
Bury entirely in sand

Negatively phototactic   71   Complete trunk lateral line

Fig. 25. The development of the sense organs and behavior in the flounder *Paralichthys olivaceus*. [Adapted from Kawamura and Ishida (1985).]

whether there are particularly "critical periods" of high mortality is often uncertain. The previous sections and Figs. 22–25 illustrate well the accretion of new structures and functions as larvae grow until the full adult repertoire (apart from reproductive behavior) is achieved at the onset of the juvenile stage. At the present time it is not possible to conclude the extent to which development is saltatory (Balon, 1981b, 1984) or gradual. Such a conclusion requires a thorough study of physiology and behavior as well as anatomy.

Because the larvae lack certain behavioral responses, and because they are going through a massive morphogenesis, it seems almost inevitable that critical periods arise through which larvae have to pass to allow development to proceed. These critical periods are especially related to feeding and predation but also to respiration. Some potentially critical periods may be listed as follows:

1. Hatching. This depends on the production of hatching enzymes to break down the tough chorion that protects the embryo from the rigors of wave action or the pressures and abrasion within or on a spawning substratum (see Blaxter, 1969, for a discussion).

2. First-feeding. Both high mortality at first-feeding in rearing experiments and considerations of brood survival under natural conditions have suggested in the past that one of the main phases of high mortality occurs when the larvae change from endogenous to exogenous sources of food (sometimes called the mixed feeding stage). Obviously relevant to this thesis is the gape of the jaw (see Fig. 19) in relation to the size of prey available and the quantity and quality of the prey. May (1974), Vladimirov (1975), and Blaxter (1984) discuss this problem and conclude that it is likely that mortality often takes place rather steadily over the early life history. Mortality rates are sometimes as high in the egg stage of pelagic marine fish as in the larval stage (5–20% per day), suggesting that predation can predominate over starvation on some occasions. Furthermore, rearing in the absence of predators can lead to excellent rates of survival (see page 3). First-feeding as the major critical period in development, while sometimes being applicable, should thus be treated with caution as a general concept.

3. Respiration. The disadvantageous decline in body surface area with size in relation to oxygen requirements has already been discussed (Fig. 20). There is some evidence of a high phase of mortality associated with early development of the gill filaments, for example, in herring (De Silva, 1974).

4. Swim-up. The first-filling of the swimbladder (and of the bulla system in clupeoids) is essential for maintaining buoyancy and other functions such as hearing associated with a gas-filled swimbladder. In many species, such as sea bass and sea bream (Chatain, 1986), the swimbladder appears soon after hatching. In the physostomatous salmonids, such as brown trout, rainbow trout, and whitefish *C. clupeaformis*, the swimbladder is filled by swallowing air at the surface at the end of yolk resorp-

tion (Tait, 1960). If access to the surface is prevented, the pneumatic duct remains open and the swimbladder can be filled later. The motivation to fill the swimbladder is very strong; in experiments *Salvelinus (Cristivomer) namaycush* will swim up for at least 270 m without fatigue to fill the swimbladder if it is empty. In physoclists failure to fill the swimbladder can lead to abnormal behavior and sometimes to death (e.g., in mullet, sea bass, and turbot) (Doroshev *et al.*, 1981). In clupeoids (Blaxter and Batty, 1985) avoidance responses to sound stimuli fail to develop unless the bulla contains gas.

5. Metamorphosis. In those species where a marked metamorphosis occurs at the end of the larval stage, a number of major changes occur. Increasing conspicuousness, as the transparent larval form is lost, makes it essential for fish to develop other protective mechanisms, such as schooling or burying behavior, in order to avoid predation. Protective coloration mechanisms also develop as the scales, pigment, and new chromatophores appear. Single fish must be vulnerable at this time, especially in the pelagic habitat, and survival in a schooling species may well depend on early successful aggregation with conspecifics. At or near metamorphosis many marine species move inshore to seek nursery grounds and flatfish need to settle on an appropriate substratum such as sand.

XIII. CONCLUSIONS

The aim of this chapter is to give the less initiated reader, especially potential experimentalists, an account of the variety of material now available within the early life history stages of fish. This variety has its attractions and its disadvantages. Fish eggs and larvae present a wealth of new and interesting problems related to development. In the case of marine fish larvae, new sources of material, hitherto neglected because of difficulties in rearing, are now available. The research worker needs, however, to be aware of the pace at which events are occurring. Body shape is changing rapidly, new external structures are being added or modified, and new internal organs are appearing or being elaborated. For example, a sensory physiologist needs to have some insight into the state of development of the sense organ and its connections to the central nervous system, a student of osmoregulation needs to know the state of development of the kidney, and a nutritionist requires some knowledge of the repertoire and activity of the digestive enzymes. Major changes in organs can occur within the duration of an experiment. Staging of experimental material is therefore paramount.

Within a species, the viability of the material can depend on parental effects such as egg size and egg quality. Successful experimentation often requires some expertise in husbandry and handling to ensure that the material is in good condition and performing optimally. Small size may demand sensitive and delicate techniques.

One of the main rewards for perseverance is that behavioral and physiological experiments can be done in the absence of certain organs or structures and again as such organs and structures develop. Invasive or ablation techniques can thus be avoided and experiments can be done on intact animals.

ACKNOWLEDGMENTS

I am extremely grateful to Dr. L. A. Fuiman, at present on a National Science Foundation Post Doctoral Fellowship at the Scottish Marine Biological Association, Oban, for reading and commenting on this chapter in draft form and for making many constructive and useful comments to improve it.

The following have been most helpful in correspondence, allowing me to use drawings, or in other ways providing information: Prof. E. K. Balon, Dr. B. D. Glebe, Prof. R. D. Hoyt, Dr. J. R. Hunter, Dr. A. W. Kendall, Jr., Dr. H. G. Moser, Dr. K. Rana, Prof. J. S. Ryland, Dr. D. E. Snyder, Dr. J. E. Thorpe, and Dr. J.-M. Vernier. I am also grateful to Catriona Stewart, who drew several of the figures.

REFERENCES

Ahlstrom, E. H., and Counts, R. C. (1955). Eggs and larvae of the Pacific hake *Merluccius productus*. *Fish. Bull.* **56**, 295–329.

Ahlstrom, E. H., and Moser, H. G. (1980). Characters useful in identification of pelagic marine fish eggs. *Rep. Calif. Coop. Oceanogr. Fish. Invest.* **21**, 121–131.

Ali, M. A. (1959). The ocular structure, retinomotor and photobehavioral responses of juvenile Pacific salmon. *Can. J. Zool.* **37**, 965–996.

Allen, J. M., Blaxter, J. H. S., and Denton, E. J. (1976). The functional anatomy and development of the swimbladder–inner ear–lateral line system in herring and sprat. *J. Mar. Biol. Assoc. U.K.* **56**, 471–486.

Amoroso, E. C. (1960). Viviparity in fishes. *Symp. Zool. Soc. London* **1**, 153–181.

Appelbaum, S. (1981). Zum Geruchsorgan der Karpfenlarven (*Cyprinus carpio*). *Zool. Anz.* **206**, 227–233.

Appelbaum, S. (1985). Rearing of the Dover sole *Solea solea* (L) through its larval stages using artificial diets. *Aquaculture* **49**, 209–221.

Arthur, D. K. (1980). Differences in heart size between ocean-caught and laboratory-grown larvae of the northern anchovy *Engraulis mordax* Girard. *J. Exp. Mar. Biol. Ecol.* **43**, 99–106.

Auer, N. A., ed. (1982). "Identification of Larval Fishes of the Great Lakes Basin with Emphasis on the Lake Michigan Drainage," Great Lakes Fish. Comm. Spec. Publ. No. 82-3. University of Michigan: Ann Arbor.

Bagenal, T. B. (1971). The interrelationship of the size of fish eggs, the date of spawning and the production cycle. *J. Fish Biol.* **3**, 207–219.

Balon, E. K. (1980). Comparative ontogeny of charrs. *In* "Charrs: Salmonid Fishes of the Genus *Salvelinus*" (E. K. Balon, ed.), pp. 703–720. Junk, The Hague.

Balon, E. K. (1981a). Additions and amendments to the classification of reproductive styles in fishes. *Environ. Biol. Fishes* **6**, 377–389.

Balon, E. K. (1981b). Saltatory processes and altricial to precocial forms in the ontogeny of fishes. *Am. Zool.* **21**, 573–596.

Balon, E. K. (1984). Reflections on some decisive events in the early life of fishes. *Trans. Am. Fish. Soc.* **113**, 178–185.

Bams, R. A. (1967). Differences in performance of naturally and artificially propagated sockeye salmon migrant fry as measured with swimming and predation tests. *J. Fish. Res. Board Can.* **24**, 1117–1153.

Barlow, G. (1961). Causes and significance of morphological variation in fishes. *Syst. Zool.* **10**, 105–117.

Batty, R. S. (1984). Development of swimming movements and musculature of larval herring (*Clupea harengus*). *J. Exp. Biol.* **110**, 217–229.

Billard, R., Bry, C., and Gillet, C. (1981). Stress, environment and reproduction in teleost fish. *In* "Stress and Fish" (A. D. Pickering, ed.), pp. 185–208. Academic Press, London.

Blaxter, J. H. S. (1969). Development: Eggs and larvae. *In* "Fish Physiology" (W. S. Hoar and D. J. Randall, eds), Vol. 3, pp. 177–252. Academic Press, New York.

Blaxter, J. H. S. (1970). Sensory deprivation and sensory input in rearing experiments. *Helgol. Wiss. Meeresunters.* **20**, 642–654.

Blaxter, J. H. S. (1971). Feeding and condition of Clyde herring larvae. *Rapp. P.-V. Reunions Cons. Int. Explor. Mer* **160**, 129–136.

Blaxter, J. H. S., ed. (1974). "The Early Life History of Fish." Springer-Verlag, Berlin and New York.

Blaxter, J. H. S. (1976). Reared and wild fish—How do they compare? *In* "Proceedings of the Tenth European Symposium on Marine Biology" (G. Persoone and E. Jaspers, eds.), Vol. 1, pp. 11–26. Inst. Mar. Sci. Res., Bredene, Belgium.

Blaxter, J. H. S. (1981). The rearing of larval fish. *In* "Aquarium Systems" (A. D. Hawkins, ed.), pp. 303–323. Academic Press, London.

Blaxter, J. H. S. (1984). Ontogeny, systematics and fisheries. *In* "Ontogeny and Systematics of Fishes" (H. G. Moser *et al.*, eds.), Am. Soc. Ichthyol. Herpetol., Spec. Publ. No. 1, pp. 1–6. Allen Press, Lawrence, Kansas.

Blaxter, J. H. S. (1986). The development of sense organs and behavior in teleost larvae with special reference to feeding and predator avoidance. *Trans. Am. Fish. Soc.* **115**, 98–114.

Blaxter, J. H. S., and Batty, R. S. (1985). The development of startle responses in herring larvae. *J. Mar. Biol. Assoc. U.K.* **65**, 737–750.

Blaxter, J. H. S., and Ehrlich, K. F. (1974). Change in behavior during starvation of herring and plaice larvae. *In* "The Early Life History of Fish" (J. H. S. Blaxter, ed.), pp. 575–588. Springer-Verlag, Berlin and New York.

Blaxter, J. H. S., and Hempel, G. (1963). The influence of egg size on herring larvae (*Clupea harengus* L). *J. Cons., Cons. Int. Explor. Mer* **28**, 211–240.

Blaxter, J. H. S., and Hunter, J. R. (1982). The biology of the clupeoid fishes. *Adv. Mar. Biol.* **20**, 1–223.

Blaxter, J. H. S., and Staines, M. (1970). Pure-cone retinae and retinomotor responses in larval teleosts. *J. Mar. Biol. Assoc. U.K.* **50**, 449–460.

Blaxter, J. H. S., Danielssen, D., Moksness, E., and Øiestad, V. (1983a). Description of the early development of the halibut *Hippoglossus hippoglossus* and attempts to rear the larvae past first feeding. *Mar. Biol. (Berlin)* **73**, 99–107.

Blaxter, J. H. S., Gray, J. A. B., and Best, A. C. G. (1983b). Structure and development of the free neuromasts and lateral line system of the herring. *J. Mar. Biol. Assoc. U.K.* **63**, 247–260.

Blumer, L. S. (1979). Male parental care in bony fishes. *Q. Rev. Biol.* **54**, 149–161.

Bodammer, J. E. (1985). A morphological study on the olfactory organ of striped bass (*Morone saxatilis*) larvae at the time of yolk sac absorption. *Abstr. 9th Ann. Larval Fish Conf. Am. Fish. Soc. 1985.*

Boehlert, G. W. (1984). Scanning electron microscopy. *In* "Ontogeny and Systematics of Fishes" (H. G. Moser *et al.*, eds.), Am. Soc. Ichthyol. Herpetol., Spec. Publ. No. 1, pp. 43–48. Allen Press, Lawrence, Kansas.

Branchek, T., and Bremiller, R. (1984). The development of photoreceptors in the zebrafish *Brachydanio rerio*. I. Structure. *J. Comp. Neurol.* **224**, 107–115.

Breder, C. M., Jr., and Halpern, F. (1946). Innate and acquired behavior affecting the aggregation of fishes. *Physiol. Zool.* **19**, 154–190.

Breder, C. M., Jr., and Rosen, D. R. (1966). "Modes of Reproduction in Fishes." Natural History Press, Garden City, New York.

Brett, J. R. (1970). 3. Temperature 3.3. Animals 3.32. Fishes. Functional responses. *In* "Marine Ecology" (O. Kinne, ed.), Vol. 1, Part 1, pp. 515–560. Wiley (Interscience), London.

Buckley, L. J. (1981). Biochemical changes during ontogenesis of cod (*Gadus morhua* L.) and winter flounder (*Pseudopleuronectes americanus*) larvae. *Rapp. P.-V. Reun, Cons. Int. Explor. Mer* **178**, 547–552.

Chatain, B. (1986). La vessie natatoire chez *Dicentrarchus labrax* et *Sparus auratus*. Aspects morphologiques du développement. *Aquaculture* **53**, 303–311.

Chenoweth, S. B. (1970). Seasonal variations in condition of larval herring in Boothbay Area of the Maine Coast. *J. Fish. Res. Board Can.* **27**, 1875–1879.

Clark, J., MacDonald, N. L., and Stark, J. R. (1985). Development of proteases and an examination of procedures for analysis of elastase activity in Dover sole (*Solea solea*). *In* "Nutrition and Feeding of Fish" (C. B. Cowey, A. M. Mackie, and J. G. Bell, eds.), pp. 217–221. Academic Press, London.

Clemmesen, C. M. (1987). Laboratory studies on RNA/DNA ratios of starved and fed herring (*Clupea harengus*) and turbot (*Scophthalmus maximus*) larvae. *J. Cons., Cons. Int. Explor. Mer* **43**, 122–128.

Craik, J. C. A. (1985). Egg quality and egg pigment content in salmonid fish. *Aquaculture* **47**, 61–88.

Craik, J. C. A., and Harvey, S. M. (1984a). Biochemical chang͘ ͘ccurring during final maturation of eggs of some marine and freshwater teleoṣ̣s. *J. Fish Biol.* **24**, 599–610.

Craik, J. C. A., and Harvey, S. M. (1984b). Biochemical changes associated with over-ripening of the eggs of rainbow trout *Salmo gairdneri* Richardson. *Aquaculture* **37**, 347–357.

Craik, J. C. A., and Harvey, S. M. (1984c). Egg quality in rainbow trout: The relation between egg viability, selected aspects of egg composition, and time of stripping. *Aquaculture* **40**, 115–134.

Dabrowski, K. (1984). The feeding of fish larvae: Present "state of the art" and perspectives. *Reprod. Nutr. Dev.* **24**, 807–833.

Dabrowski, K., and Bardega, R. (1984). Mouth size and recommendation of feed size preferences in three cyprinid fish. *Aquaculture* **40**, 41–46.

Dabrowski, K., and Glogowski, J. (1977). Studies on the role of exogenous proteolytic enzymes in digestion processes in fish. *Hydrobiologia* **54**, 129–134.

Dales, S., and Hoar, W. S. (1954). Effects of thyroxine and thiourea on the early development of chum salmon (*Oncorhynchus keta*). *Can. J. Zool.* **32**, 244–251.

Dempsey, C. H. (1978). Chemical stimuli as a factor in feeding and intraspecific behaviour of herring larvae. *J. Mar. Biol. Assoc. U.K.* **58**, 739–747.

De Silva, C. (1974). Development of the respiratory system in herring and plaice larvae. *In* "The Early Life History of Fish" (J. H. S. Blaxter ed.), pp. 465–485. Springer Verlag, Berlin and New York.

De Silva, C., and Tytler, P. (1973). The influence of reduced environmental oxygen on the metabolism and survival of herring and plaice larvae. *Neth. J. Sea Res.* **7**, 345–362.

Disler, N. N. (1971). "Lateral Line Sense Organs and Their Importance in Fish Behavior." Israel Program for Scientific Translations, Jerusalem.

Doroshev, S. I., Cornacchia, J. W., and Hogan, K. (1981). Initial swim bladder inflation in the larvae of physoclistous fishes and its importance for larval culture. *Rapp. P.-V. Reun., Cons. Int. Explor. Mer* **178**, 495–500.

Eaton, R. C., and Farley, R. D. (1974). Growth and reduction of depensation of zebra fish *Brachydanio rerio* reared in the laboratory. *Copeia* pp. 204–209.

Ehrlich, K. F. (1974a). Chemical changes during growth and starvation of larval *Pleuronectes platessa*. *Mar. Biol. (Berlin)* **24**, 39–48.

Ehrlich, K. F. (1974b). Chemical changes during growth and starvation of herring larvae. *In* "The Early Life History of Fish" (J. H. S. Blaxter, ed.), pp. 301–323. Springer-Verlag, Berlin and New York.

Ehrlich, K. F., Blaxter, J. H. S., and Pemberton, R. (1976). Morphological and histological changes during the growth and starvation of herring and plaice larvae. *Mar. Biol. (Berlin)* **35**, 105–118.

Elston, R., Corazza, L., and Nickum, J. G. (1981). Morphology and development of the olfactory organ in larval walleye *Stizostedion vitreum*. *Copeia* pp. 890–893.

Fahay, M. P. (1983). Guide to the early stages of marine fishes occurring in the western North Atlantic, Cape Hatteras to the southern Scotian shelf. *J. Northwest. Atl. Fish. Sci.* **4**, 1–423.

Farris, D. A. (1963). Shrinkage of sardine (*Sardinops caerulea*) larvae upon preservation in buffered formalin. *Copeia* pp. 185–186.

Forstner, H., Hinterleitner, S., Mahr, K., and Wieser, W. (1983). Towards a better definition of "metamorphosis" in *Coregonus* sp.: Biochemical, histological and physiological data. *Can. J. Fish. Aquat. Sci.* **40**, 1224–1232.

Fowler, L. G. (1972). Growth and mortality of fingerling chinook salmon as affected by egg size. *Prog. Fish-Cult.* **34**, 66–69.

Fridgeirsson, E. (1978). Embryonic development of five species of gadoid fishes in Icelandic waters. *Rit Fiskideildar* **5**, 1–68.

Fuiman, L. A. (1983). Growth gradients in fish larvae. *J. Fish Biol.* **23**, 117–123.

Fuiman, L. A. (1984). Ostariophysi: Development and relationships. *In* "Ontogeny and Systematics of Fishes" (H. G. Moser *et al.*, eds.), Am. Soc. Ichthyol. Herpetol., Spec. Publ. No. 1, pp. 126–137. Allen Press, Lawrence, Kansas.

Fukuhara, O. (1985). Functional morphology and behavior of early life stages of the red sea bream. *Bull. Jpn. Soc. Sci. Fish.* **51**, 731–743.

Gall, G. A. E. (1974). Influence of size of eggs and age of female on hatchability and growth of rainbow trout. *Calif. Fish Game* **60**, 26–35.

Gamble, J. C., MacLachlan, P., and Seaton, D. D. (1985). Comparative growth and development of autumn and spring spawned Atlantic herring larvae reared in large experimental ecosystems. *Mar. Ecol.: Prog. Ser.* **26**, 19–33.

Garside, E. T. (1970). 3. Temperature. 3.3. Animals. 3.32. Fishes. Structural responses. *In* "Marine Ecology" (O. Kinne, ed.), Vol. 1, Part 1, pp. 561–573. Wiley (Interscience), London.

Glebe, B. D., Appy, T. D., and Saunders, R. L. (1979). "Variation in Atlantic Salmon (*Salmo salar*) Reproductive Traits and their Implications in Breeding Programs," Pap. M:23 (mimeo). Anadromous and Catadromous Fish Comm., Int. Counc. Explor. Sea, Copenhagen.

Govoni, J. J. (1980). Morphological, histological and functional aspects of alimentary canal and associated organ development in larval *Leiostomus xanthurus*. *Rev. Can. Biol.* **39**, 69–80.

Harder, W. (1954). Die Entwicklung der Respirationsorgane beim Hering. *Z. Anat. Entwicklungsgesch.* **118**, 102–123.

Hay, D. E. (1981). The effects of capture and fixation on gut contents and body size of Pacific herring larvae. *Rapp. P.-V. Reun., Cons. Int. Explor. Mer* **178**, 395–400.

Hempel, G. (1979). "Early Life History of Marine Fish: The Egg Stage." Univ. of Washington Press, Seattle.

Hempel, G., and Blaxter, J. H. S. (1967). Egg weight in Atlantic herring (*Clupea harengus* L.). *J. Cons., Cons. Int. Explor. Mer* **31**, 170–195.

Herzig, A., and Winkler, H. (1986). The influence of temperature on the embryonic development of three cyprinid fishes, *Abramis brama*, *Chalcalburnus chalcoides mento* and *Vimba vimba*. *J. Fish Biol.* **28**, 171–181.

Hickey, G. M. (1982). Wound healing in fish larvae. *J. Exp. Mar. Biol. Ecol.* **57**, 149–168.

Hirose, K., Machida, Y., and Donaldson, E. M. (1979). Induced ovulation of Japanese flounder (*Limanda yokohamae*) with human chorionic gonadotrophin and salmon gonadotropin, with special reference to changes in quality of eggs retained in the ovarian cavity after ovulation. *Bull. Jpn. Soc. Sci. Fish.* **45**, 31–36.

Hislop, J. R. G., Robb, A. P., and Gauld, J. A. (1978). Observations on effects of feeding level on growth and reproduction in haddock *Melanogrammus aeglefinus* (L.) in captivity. *J. Fish. Biol.* **13**, 85–98.

Hofer, R. (1985). Effects of artificial diets on the digestive processes of fish larvae. *In* "Nutrition and Feeding in Fish" (C. B. Cowey, A. M. Mackie, and J. G. Bell, eds.), pp. 213–216. Academic Press, London.

Hoss, D. E., and Blaxter, J. H. S. (1982). Development and function of the swimbladder–inner ear–lateral line system in the Atlantic menhaden *Brevoortia tyrannus* (Latrobe). *J. Fish Biol.* **20**, 131–142.

Hoss, D. E., and Phonlor, G. (1984). Field and laboratory observations on diurnal swimbladder inflation-deflation in larvae of gulf menhaden *Brevoortia patronus*. *Fish. Bull.* **82**, 513–517.

Houde, E. (1977). Food concentrations and stocking density effects on survival and growth of laboratory-reared larvae of bay anchovy *Anchoa mitchilli* and lined sole *Achirus lineatus*. *Mar. Biol. (Berlin)* **43**, 333–341.

Howell, B. R. (1979). Experiments on the rearing of larval turbot, *Scophthalmus maximus* L. *Aquaculture* **18**, 215–225.

Hubbs, C. (1986). American Fisheries Society. Ninth larval fish conference. *Trans. Am. Fish. Soc.* **115**, 98–171.

Hunter, J. R. (1980). The feeding behavior and ecology of marine fish larvae. *In* "Fish Behavior and its Use in the Capture and Culture of Fishes" (J. E. Bardach, J. J. Magnuson, R. C. May, and J. M. Reinhart, eds.), pp. 287–330. ICLARM, Manila, Philippines.

Hunter, J. R. (1981). Feeding ecology and predation of marine fish larvae. *In* "Marine Fish Larvae, Morphology, Ecology and Relation of Fisheries" (R. Lasker, ed.), Washington Sea Grant Program, pp. 34–87. Univ. of Washington Press, Seattle.

Hunter, J. R. (1984). Synopsis of culture methods for marine fish larvae. *In* "Ontogeny and Systematics of Fishes" (H. G. Moser *et al.*, eds.), Am. Soc. Ichthyol. Herpetol., Spec. Publ. No. 1, pp. 24–27. Allen Press, Lawrence, Kansas.

Hunter, J. R., and Coyne, K. M. (1982). The onset of schooling in northern anchovy larvae *Engraulis mordax*. *Rep. Calif. Coop. Oceanogr. Fish. Invest.* **23**, 246–251.

Hunter, J. R., and Sanchez, C. (1976). Diel changes in swim bladder inflation of the larvae of the northern anchovy *Engraulis mordax*. *Fish. Bull.* **74**, 847–855.

Huse, I., and Skiftesvik, A. B. (1985). "Qualitative and Quantitative Behaviour Studies in Starving and Feeding Turbot (*Scophthalmus maximus* L.) Larvae," Pap. F:38 (mimeo). Maricult. Comm., Int. Counc. Explor. Sea, Copenhagen.

Iversen, S. A., and Danielssen, D. S. (1984). Development and mortality of cod (*Gadus morhua* L.) eggs and larvae in different temperatures. *In* "The Propagation of Cod *Gadus morhua* L." (E. Dahl, D. S. Danielssen, E. Moksness, and P. Solemdal, eds.), Part 1, pp. 49–65. Inst. Mar. Res., Flødevigen Biol. Stn., Arendal, Norway.

Iwai, T. (1967). Structure and development of lateral line cupulae in teleost larvae. *In* "Lateral Line Detectors" (P. H. Cahn, ed.), pp. 27–44. Indiana Univ. Press, Bloomington.

Iwai, T. (1980). Sensory anatomy and feeding of fish larvae. *In* "Fish Behavior and its Use in the Capture and Culture of Fishes" (J. E. Bardach, J. J. Magnuson, R. C. May, and J. M. Reinhart, eds.), pp. 124–145. ICLARM, Manila, Philippines.

Iwai, T., and Rosenthal, H. (1981). Ciliary movements in guts of early clupeoid and salangid larvae. *Mar. Ecol.: Prog. Ser.* **4**, 365–367.

Johns, P. R. (1982). Formation of photoreceptors in larval and adult goldfish. *J. Neurosci.* **2**, 178–198.

Jones, A. (1972). Studies on egg development and larval rearing of turbot *Scophthalmus maximus* L., and brill *Scophthalmus rhombus* L., in the laboratory. *J. Mar. Biol. Assoc. U.K.* **52**, 965–986.

Kanazawa, A. (1985). Essential fatty acid and lipid requirement of fish. *In* "Nutrition and Feeding in Fish" (C. B. Cowey, A. M. MacKie, and J. G. Bell, eds.), pp. 281–298. Academic Press, London.

Kawamura, G., and Ishida, K. (1985). Changes in sense organ morphology and behaviour with growth in the flounder *Paralichthys olivacus*. *Bull. Jpn. Soc. Sci. Fish.* **51**, 155–165.

Kazakov, R. V. (1981). The effect of the size of Atlantic salmon *Salmo salar* L. eggs on embryos and alevins. *J. Fish Biol.* **19**, 353–360.

Kendall, A. W., Jr., Ahlstrom, E. H., and Moser, H. G. (1984). Early life history stages of fishes and their characters. *In* "Ontogeny and Systematics of Fishes" (H. G. Moser *et al.*, eds.), Am. Soc. Ichthyol. Herpetol., Spec. Publ. No. 1, pp. 11–22. Allen Press, Lawrence, Kansas.

Kinne, O. (1963). The effects of temperature and salinity on marine and brackish water animals. I. Temperature. *Oceanogr. Mar. Biol.* **1**, 301–340.

Kinne, O. (1977). Cultivation of animals. Pisces. *In* "Marine Ecology" (O. Kinne, ed.), Vol. 3, Part 2, pp. 968–1004. Wiley (Interscience), London.

Kjørsvik, E., and Lønning, S. (1983). Effects of egg quality on normal fertilization and early development of the cod *Gadus morhua* L. *J. Fish Biol.* **23**, 1–12.

Kjøsvik, E., Stene, A., and Lønning, S. (1984). Morphological, physiological and genetical studies of egg quality in cod (*Gadus morhua* L.). *In* "The Propagation of Cod *Gadus morhua* L." (E. Dahl, D. S. Danielssen, E. Moksness, and P. Solemdal, eds.), Part 1, pp. 67–86. Inst. Mar. Res., Flødevigen Biol. Stn., Arendal, Norway.

Kokkala, I., and Hoyt, R. D. (1985). The development of anatomical features and the early ecology of the spotted bass. *Abstr. 9th Ann. Larval Fish Conf. Am. Fish. Soc.*, *1985*.

Kramer, D., and Ahlstrom, E. H. (1968). Distributional atlas of fish larvae in the California Current region: Northern anchovy *Engaulis mordax* Girard, 1951 through 1965. *Calif. Coop. Oceanogr. Fish. Invest. Atlas* **9**, 1–269.

Kuhlmann, D., Quant, G., and Witt, U. (1981). Rearing of turbot larvae (*Scophthalmus maximus* L.) on cultured food organisms and post-metamorphosis growth on natural and artificial food. *Aquaculture* **23**, 183–196.

Kunz, Y. W., and Ennis, S. (1983). Ultrastructural diurnal changes of the retinal photoreceptors in the embryo of a viviparous teleost (*Poecilia reticulata* P.). *Cell Differ.* **13**, 115–123.

Kunz, Y. W., Ennis, S., and Wise, C. (1983). Ontogeny of the photoreceptors in the embryonic retina of the viviparous guppy, *Poecilia reticulata* (Teleostei). *Cell Tissue Res.* **230**, 469–486.

Kvenseth, P. G., and Øiestad, V. (1984). Large scale rearing of cod fry on the natural food production in an enclosed pond. *In* "The Propagation of Cod *Gadus morhua* L." (E. Dahl, D. S. Danielssen, E. Moksness, and P. Solemdal eds.), Part 2, pp. 645–655. Inst. Mar. Res., Flødevigen Biol. Stn., Arendal, Norway.

Laale, H. W., and McCallion, D. J. (1968). Reversible developmental arrest in the embryo of the zebra fish *Brachydanio rerio. J. Exp. Zool.* **167**, 117–123.

Lam, T. J. (1980). Thyroxine enhances larval development and survival in *Sarotherodon* (*Tilapia*) *mossambicus* Ruppell. *Aquaculture* **21**, 287–291.

Lam, T. J. (1982). Applications of endocrinology to fish culture. *Can. J. Fish. Aquat. Sci.* **39**, 111–137.

Lam, T. J., and Sharma, R. (1985). Effects of salinity and thyroxine on larval survival, growth and development in the carp *Cyprinus carpio. Aquaculture* **44**, 201–212.

Lam, T. J., Juario, J. V., and Banno, J. (1985). Effect of thyroxine on growth and development in post-yolk-sac larvae in milkfish *Chanos chanos. Aquaculture* **46**, 179–184.

Lasker, R., and Sherman, K., eds. (1981). "The Early Life History of Fish: Recent Studies," Rapp. P.-V. Reun., Cons. Int. Explor. Mer, Vol. 178. Int. Counc. Explor. Sea, Copenhagen.

Lauff, M., and Hofer, R. (1984). Proteolytic enzymes in fish development and the importance of dietary enzymes. *Aquaculture* **37**, 335–346.

Lavenberg, R. J., McGowen, G. E., and Woodsum, R. E. (1984). Preservation and curation. *In* "Ontogeny and Systematics of Fishes" (H. G. Moser *et al.*, eds.), Am. Soc. Ichthyol. Herpetol., Spec. Publ. No. 1, pp. 57–59. Allen Press, Lawrence, Kansas.

McElman, J. F., and Balon, E. K. (1979). Early ontogeny of walleye, *Stizostedion vitreum*, with steps of saltatory development. *Environ. Biol. Fishes* **4**, 290–348.

McGurk, M. D. (1984). Effects of delayed feeding and temperature on the age of irreversible starvation and on the rates of growth and mortality of Pacific herring larvae. *Mar. Biol. (Berlin)* **84**, 13–26.

Magnuson, J. J. (1962). An analysis of aggressive behavior, growth and competition for food and space in the medaka *Oryzias latipes. Can. J. Zool.* **40**, 313–363.

Mann, R. H. K., and Mills, C. A. (1985). Variations in the sizes of gonads, eggs and larvae of the dace *Leuciscus leuciscus*. *Environ. Biol. Fishes* **13**, 277–287.

Marliave, J. B., ed. (1985). International Symposium on the Early Life History of Fish and 8th Larval Fish Conference, Vancouver 1984. *Trans. Am. Fish. Soc.* **114**, 445–621.

Marsh, E. (1986). Effects of egg size on offspring fitness and maternal fecundity in the orangethroat darter *Etheostoma spectabile* (Pisces: Percidae). *Copeia* pp. 18–30.

Matarese, C., and Sandknop, E. M. (1984). Identification of fish eggs. *In* "Ontogeny and Systematics of Fishes" (H. G. Moser *et al.*, eds.), Am. Soc. Ichthyol. Herpetol., Spec. Publ. No. 1, pp. 27–31. Allen Press, Lawrence, Kansas.

Matsuoka, M., and Iwai, T. (1984). Development of the myotomal musculature in the Red Sea bream. *Bull Jpn. Soc. Sci. Fish.* **50**, 29–35.

May, R. C. (1974). Larval mortality in marine fishes and the critical period concept. *In* "The Early Life History of Fish" (J. H. S. Blaxter, ed.), pp. 3–19. Springer-Verlag, Berlin and New York.

Mollah, M. F. A., and Tan, E. S. P. (1983). Viability of catfish (*Clarias macrocephalus* Gunther) eggs fertilized at varying post-ovulation times. *J. Fish. Biol.* **22**, 563–566.

Morita, S. (1985). History of the herring fishery and review of artificial propagation techniques for herring in Japan. *Can. J. Fish. Aquat. Sci.* **42**, Suppl., 222–229.

Moser, H. G. (1981). Morphological and functional aspects of marine fish larvae. *In* "Marine Fish Larvae, Morphology, Ecology and Relation to Fisheries" (R. Lasker ed.), Washington Sea Grant Program, pp. 90–131. Univ. of Washington Press, Seattle.

Moser, H. G., Richards, W. J., Cohen, D. M., Fahay, M. P., Kendall, A. W., Jr., and Richardson, S., eds. (1984). "Ontogeny and Systematics of Fishes," Am. Soc. Ichthyol. Herpetol., Spec. Publ. No. 1. Allen Press, Lawrence, Kansas.

Nash, C. E., and Kuo, C.-M. (1975). Hypotheses for problems impeding the mass propagation of grey mullet and other finfish. *Aquaculture* **5**, 119–133.

Neave, D. A. (1984). The development of the retinomotor reactions in larval plaice (*Pleuronectes platessa* L.) and turbot (*Scophthalmus maximus* L.). *J. Exp. Mar. Biol. Ecol.* **76**, 167–175.

Neave, D. A. (1986). The development of the lateral line system in plaice (*Pleuronectes platessa* L.) and turbot (*Scophthalmus maximus* L.). *J. Mar. Biol. Assoc. U.K.* **66**, 683–693.

Noakes, D. L. G. (1978). Ontogeny of behavior in fishes: A survey and suggestions. *In* "The Development of Behavior, Comparative and Evolutionary Aspects" (G. M. Burghardt and M. Bekoff, eds.), pp. 103–125. Garland STPM Press, New York.

O'Connell, C. P. (1976). Histological criteria for diagnosing the starving condition in early post yolk sac larvae of the northern anchovy *Engraulis mordax* Girard. *J. Exp. Mar. Biol. Ecol.* **25**, 285–312.

O'Connell, C. P. (1981). Development of organ systems in the northern anchovy *Engraulis mordax* and other teleosts. *Am. Zool.* **21**, 429–446.

Øiestad, V., Kvenseth, P. G., and Folkvord, A. (1985). Mass production of Atlantic cod juveniles *Gadus morhua* in a Norwegian saltwater pond. *Trans. Am. Fish. Soc.* **114**, 590–595.

Paulsen, H., Munk, P., and Kiørboe, T. (1985). "Extensive Rearing of Turbot Larvae (*Scophthalmus maximus* L.) on Low Concentrations of Natural Plankton," Pap. F:33 (mimeo). Maricult. Comm., Int. Counc. Explor. Sea, Copenhagen.

Pflugfelder, O. (1952). Weitere volumetrische Untersuchungen über die Wirkung der Augenexstirpation und der Dunkelhaltung auf das Mesencephalon und die

Pseudobranchien von Fischen. *Wilhelm Roux' Arch. Entwicklungsmech. Org.* **145**, 549–560.

Rana, K. J. (1985). Influence of egg size on the growth, onset of feeding, point-of-no-return, and survival of unfed *Oreochromis mossambicus* fry. *Aquaculture* **46**, 119–131.

Ridley, M. (1978). Paternal care. *Anim. Behav.* **26**, 904–932.

Rose, S. M. (1960). A feedback mechanism of growth control in tadpoles. *Ecology* **41**, 188–199.

Russell, F. S. (1976). "Eggs and Planktonic Stages of Marine Fishes." Academic Press, London.

Ryland, J. S. (1966). Observations on the development of larvae of the plaice *Pleuronectes platessa* L. in aquaria. *J. Cons., Cons. Int. Explor. Mer* **30**, 177–195.

Schnack, D., and Rosenthal, H. (1977/1978). Shrinkage of Pacific herring larvae due to formalin fixation and preservation. *Ber. Dtsch. Wiss. Komm. Meeresforsch.* **26**, 222–226.

Shaw, E. (1960). The development of schooling behavior in fishes. *Physiol. Zool.* **33**, 79–86.

Shaw, E. (1961). The development of schooling in fishes. II. *Physiol. Zool.* **34**, 263–272.

Shirota, A. (1970). Studies on the mouth size of fish larvae (in Japanese). *Bull. Jpn. Soc. Sci. Fish.* **36**, 353–368.

Snyder, D. E., and Holt, J. G. (1983). Terminology Workshop. (Mimeo) Univ. Texas Mar. Sci. Inst. Port Aransas, Texas.

Solemdal, P. (1973). Transfer of Baltic flatfish to a marine environment and the long term effects on reproduction. *Oikos* **15**, 268–276.

Sorgeloos, P. (1980). The use of brine shrimp *Artemia* in aquaculture. *In* "The Brine Shrimp *Artemia*" (G. Persoone, P. Sorgeloos, O. Roels, and E. Jaspers, eds.), Vol. 3, pp. 25–46. Universa Press, Wetteren, Belgium.

Springate, J. R. C., and Bromage, N. R. (1985). Effects of egg size on early growth and survival in rainbow trout (*Salmo gairdneri* Richardson). *Aquaculture* **47**, 163–172.

Springate, J. R. C., Bromage, N. R., Elliott, J. A. K., and Hudson, D. L. (1984). The timing of ovulation and stripping and their effects on the rates of fertilization and survival to eyeing, hatch and swim-up in the rainbow trout (*Salmo gairdneri* R.). *Aquaculture* **43**, 313–322.

Springate, J. R. C., Bromage, N. R., and Cumaranatunga, P. R. T. (1985). The effects of different ration on fecundity and egg quality in the rainbow trout (*Salmo gairdneri*). *In* "Nutrition and Feeding in Fish" (C. B. Cowey, A. M. Mackie, and J. G. Bell, eds.), pp. 371–393. Academic Press, London.

Strauss, R. E., and Fuiman, L. A. (1985). Quantitative comparisons of body form and allometry in larval and adult Pacific sculpins (Teleostei: Cottidae). *Can. J. Zool.* **63**, 1582–1589.

Sturmer, L. N., McCarthy, C. E., and Rutledge, W. P. (1985). Hatchery production of red drum fingerlings in Texas. Abstr. *9th Annu. Larval Fish Conf. Am. Fish. Soc., 1985.*

Swain, D. P., and Lindsey, C. C. (1984). Selective predation for vertebral number of young sticklebacks *Gasterosteus aculeatus*. *Can. J. Fish. Aquat. Sci.* **41**, 1231–1233.

Symons, P. E. K. (1968). Increase in aggression and in strength of the social hierarchy among juvenile Atlantic salmon deprived of food. *J. Fish. Res. Board Can.* **25**, 2387–2401.

Tait, J. S. (1960). The first filling of the swim bladder in salmonids. *Can. J. Zool.* **38**, 179–187.

Tanaka, M. (1973). Studies in the structure and function of the digestive system of teleost larvae. D. Agric. Thesis, Kyoto University, Japan.

Tåning, A. V. (1952). Experimental study of meristic characters in fishes. *Biol. Rev. Cambridge Philos. Soc.* **27**, 169–193.

Theilacker, G. H. (1978). Effect of starvation on the histological and morphological characteristics of jack mackerel *Trachurus symmetricus* larvae. *Fish. Bull.* **76**, 403–414.

Theilacker, G. H. (1980a). Changes in body measurements of larval northern anchovy *Engraulis mordax* and other fishes due to handling and preservation. *Fish. Bull.* **78**, 685–692.

Theilacker, G. H. (1980b). Rearing container size affects morphology and nutritional condition of larval jack mackerel *Trachurus symmetricus*. *Fish. Bull.* **78**, 789–791.

Theilacker, G. H. (1981). Effect of feeding history and egg size on the morphology of jack mackerel, *Trachurus symmetricus*. larvae. *Rapp. P.-V. Reun. Cons. Int. Explor. Mer* **178**, 432–440.

Theilacker, G. H., and Dorsey, K. (1980). Larval fish diversity, a summary of laboratory and field research. *Intergov. Oceanogr. Comm. (IOC) Workshop Rep.* **28**, 105–142.

Thorpe, J. E., Miles, M. S., and Keay, D. S. (1984). Development rate, fecundity and egg size in Atlantic salmon *Salmo salar* L. *Aquaculture* **43**, 289–305.

Tucker, J. W., Jr., and Chester, A. J. (1984). Effects of salinity, formalin concentration and buffer on quality of preservation of southern flounder (*Paralichthys lethostigma*) larvae. *Copeia*, pp. 981–988.

van Ballaer, E., Amat, F., Hontoria, F., Léger, P., and Sorgeloos, P. (1985). Preliminary results on the nutritional evaluation of ω3-HUFA-enriched *Artemia* nauplii for larvae of the sea bass *Dicentrarchus labrax*. *Aquaculture* **49**, 223–229.

van Raamsdonk, W., van't Veer, L., Veeken, K., Heytin, C., and Pool, C. W. (1982). Differentiation of muscle fiber types in the teleost *Brachydanio rerio*, the zebrafish. *Anat. Embryol.* **164**, 51–62.

Vernier, J.-M. (1969). Table chronologique du développement embryonnaire de la truite arc-en-ciel, *Salmo gairdneri* Rich. 1836. *Ann. Embryol. Morphog.* **2**, 495–520.

Vladimirov, V. I. (1975). Critical periods in the development of fishes (translated from Russian). *J. Ichthyol.* **15**(6), 851–868.

Wallace, J. C., and Aasjord, D. (1984). An investigation of the consequences of egg size for the culture of Arctic charr, *Salvelinus alpinus* (L.). *J. Fish. Biol.* **24**, 427–435.

Watanabe, T. (1985). Importance of the study of broodstock nutrition for further development of aquaculture. *In* "Nutrition and Feeding in Fish" (C. B. Cowey, A. M. MacKie, and J. G. Bell, eds.), pp. 394–414. Academic Press, London.

Watanabe, Y. (1985). Histological changes in the liver and intestine of freshwater goby larvae during short-term starvation. *Bull. Jpn. Soc. Sci. Fish.* **51**, 707–709.

Watanabe, T., Arakawa, T., Kitajima, C., and Fujita, S. (1984). Effect of nutritional quality of broodstock diets on reproduction of red sea bream. *Bull. Jpn. Soc. Sci. Fish.* **50**, 495–501.

Webb, P. W. (1975). Hydrodynamics and energetics of fish propulsion. *Bull. Fish. Res. Board Can.* **190**, 1–158.

Webb, P. W., and Weihs, D. (1986). Functional locomotor morphology of early life history stages. *Trans. Am. Fish. Soc.* **115**, 115–127.

Weihs, D. (1980a). Energetic significance of changes in swimming modes during growth of larval anchovy *Engraulis mordax*. *Fish. Bull.* **77**, 597–604.

Weihs, D. (1980b). Respiration and depth control as possible reasons for swimming of northern anchovy *Engraulis mordax* yolk-sac larvae. *Fish. Bull.* **78**, 109–117.

Whipple, J., Eldridge, M., Benville, P., Bowers, M., Jarvis, B., and Stapp, N. (1981). The effect of inherent parental factors on gamete condition and viability in striped bass *Morone. Rapp. P.-V. Reun., Cons. Int. Explor. Mer* **178**, 93–94.

Winberg, G. G. (1960). Rate of metabolism and food requirements of fishes. *Fish. Res. Board Can. Transl.* **194**.

Wooton, R. J. (1979). Energy costs of egg production and environmental determinants of fecundity in teleost fish. *Symp. Zool. Soc. London* **44**, 133–159.

Zeutzius, I., and Rahmann, H. (1984). Influence of dark-rearing on the ontogenetic development of *Sarotherodon mossambicus* (Cichlidae Teleostei). I. Effects of body weight, body growth pattern, swimming activity and visual acuity. *Exp. Biol.* **43**, 77–85.

Zeutzius, I., Probst, W., and Rahmann, H. (1984). Influence of dark-rearing on the ontogenetic development of *Sarotherodon mossambicus* (Cichlidae, Teleostei). II. Effects on allometric and growth relations and differentiation of the optic tectum. *Exp. Biol.* **43**, 87–96.

RESPIRATORY GAS EXCHANGE, AEROBIC METABOLISM, AND EFFECTS OF HYPOXIA DURING EARLY LIFE

PETER J. ROMBOUGH

Zoology Department
Brandon University
Brandon, Manitoba, Canada R7A 6A9

I. INTRODUCTION

The basic mechanisms involved in respiratory gas exchange in juvenile and adult fish are fairly well established (see reviews by Jones and Randall, 1978; Randall, 1982; Randall *et al.*, 1982; Randall and Daxboeck, 1984). The study of oxygen metabolism in older fish, similarly, is well advanced (reviewed by Fry, 1957, 1971; Beamish, 1978; Brett and Groves, 1979; Tytler and Calow, 1985). In contrast, relatively little is known of respiratory gas exchange and energy usage during early life. This arises not so much from a lack of effort on the part of researchers—in excess of 500 papers dealing with various aspects of oxygen supply and demand during early life have been published in the last 20 years—but rather from the lack of a systematic approach to the problem. The aim of this review is to collate the large amounts of data that are currently available and to fit it into a conceptual framework that can be used as the basis for future investigations.

II. RESPIRATORY GAS EXCHANGE

Analytical models provide a useful framework for the study of respiratory gas exchange. The cascade model, in particular, has been used widely to describe various aspects of vertebrate respiratory function (Dejours, 1981; Piiper, 1982; Weibel, 1984; DiPrampero, 1985). In this model, respiratory gases are viewed as passing through a series of resistances, each of which is correlated with a specific process or structure. The overall resistance of the system is the sum of the individual resistances and, under steady-state conditions, overall flow through the system is equal to the flow through each of the elements. The model is especially useful in helping to define the nature and magnitude of the various resistances in the respiratory pathway and the partial pressure gradients necessary to overcome them. Such analysis is complicated for early life stages because of changes in respiratory rate and the nature and relative importance of resistances during development. These problems, while formidable, are not insurmountable, as evidenced by the successful application of the cascade model to the study of gas transport in mammalian (Dejours, 1981) and avian embryos (Dejours, 1981; Piiper and Scheid, 1984). Unfortunately, there is currently insufficient information to apply the model rigorously to the study of gas exchange during teleost ontogeny. However, enough is known of gas exchange in developing fish to use the cascade model in a more general way, that is, as a guide to help identify and describe the major resistances during development.

In many ways gas exchange is very similar in embryos and larvae, particularly once organogenesis is complete. In both stages gas exchange is primarily cutaneous. Branchial exchange typically becomes dominant only near the end of the larval period. However, in spite of this similarity, it is often convenient to treat embryos and larvae separately because of the major impact the egg capsule (zona radiata) has on gas exchange during embryonic life. The egg capsule, in addition to acting as a significant barrier in its own right, creates two other barriers, the external boundary layer and the perivitelline fluid, that together have an equal or greater impact (Fig. 1).

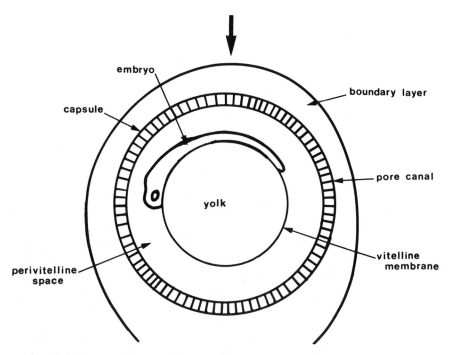

Fig. 1. Schematic diagram showing the major resistances to respiratory gas exchange for fish embryos (not in proportion). The heavy arrow indicates the direction of water flow. The actual shape of the trailing edge of the boundary layer will depend on water velocity and egg size.

A. The Boundary Layer

The laminar boundary layer is a semistagnant region of water adjacent to the egg surface where oxygen is depleted and metabolic wastes accumulate. The boundary layer actually has no outer limit but for practical purposes is usually defined as the distance from the egg surface where local conditions are equivalent to 99% of free-stream conditions (Vogel, 1981). For the laminar flow regimes that eggs are typically exposed to (Reynolds numbers < 75; Johnson, 1980), the thickness of the boundary layer is proportional to egg size and inversely proportional to water velocity. Daykin (1965) indicated that the thickness of the solute boundary layer, d_s, is less than the thickness of the velocity boundary layer and can be estimated from the Sherwood number (Sh) and egg diameter (d) using the equation

$$d_s = d(\mathrm{Sh}^{-1}) \tag{1}$$

The Sherwood number is a dimensionless number dependent on two other dimensionless numbers, the Reynolds number (Re) and the Schmidt number (Sc). For spherical eggs, Sh can be estimated as

$$Sh = 2.0 + 0.8(B{\cdot}Re)^{1/2} Sc^{1/3} \qquad (2)$$

where $Re = \mu d v^{-1}$, $Sc = v D^{-1}$, and B is the ratio of the interstitial and bulk velocities (Daykin, 1965; Johnson, 1980). Here μ is bulk water velocity, d is egg diameter, v is kinematic viscosity, and D is the diffusion coefficient. Interstitial velocity is equal to bulk velocity for isolated eggs. According to Eq. (2), the thickness of the oxygen boundary layer would be about 0.02 and 0.05 cm, respectively, for single eggs with diameters of 0.1 and 0.5 cm in a 100 cm h^{-1} current. Increasing current velocity to 1000 cm h^{-1} would reduce boundary layer thickness to approximately 0.0008 and 0.02 cm, respectively. The higher metabolic (Winnicki, 1968; DiMichele and Powers, 1984a) and growth (Silver *et al.*, 1963; Shumway *et al.*, 1964) rates reported at higher water velocities are probably due to reductions in boundary-layer thickness (Daykin, 1965; Wickett, 1975). Estimating interstitial velocities is a problem for eggs laid in masses or in substrate. If a hexagonal array is assumed, the interstitial velocity in an egg mass averages about 9.1 times the bulk velocity through the mass (Daykin, 1965; Wickett, 1975). Interstitial velocities in substrate can be estimated from porous bed theory as $\mu_i = \mu \varepsilon^{-1}$, where μ_i is interstitial velocity, μ is bulk velocity, and ε is the empirically determined porosity (Johnson, 1980).

The driving force required to overcome the resistance imposed by the boundary layer can be predicted by rearranging the Fick's equation for diffusion through a plane to yield

$$C_1 - C_0 = \dot{V}O_2 d_s (4\pi r^2 D)^{-1} \qquad (3)$$

where C_1 is the free-stream oxygen concentration, C_0 is the oxygen concentration at the egg surface, $\dot{V}O_2$ is the rate of oxygen consumption, and r is the egg radius. This equation is equivalent to the more widely used mass transport equation (Daykin, 1965; Wickett, 1975; Johnson, 1980) in which the reciprocal of the mass transport constant, k^{-1}, replaces $d_s D^{-1}$. The value of k is estimated as $k = (Sh)Dd^{-1}$. Equation (3) indicates that the driving force required to meet metabolic demands is directly proportional to the rate of oxygen consumption and thus increases more or less steadily throughout embryonic development. Daykin (1965) estimated that a partial pressure gradient across the boundary layer of about 52 mm Hg was required to meet the oxygen requirements of chum (*Oncorhynchus keta*) eggs ($r = 0.37$ cm) near hatch at 10°C and a flow rate of 85 cm h^{-1}. Smaller pressure differences are required for smaller eggs because of generally lower metabolic rates and thinner boundary layers. For example, Wickett (1975) estimated that cod (*Gadus macrocephalus*) eggs ($r = 0.05$ cm) incubated at 5°C in a current of 170 cm h^{-1} would require a pressure difference across the boundary layer of only about 17 mm Hg to fully satisfy their oxygen requirements.

Until recently it was assumed that forced convection (i.e., bulk water flow) was the major means of supplying oxygen to eggs. It now appears that under certain circumstances natural (free) convection may be important as well. Embryonic metabolism gives rise to solute concentration gradients across the boundary layer. O'Brien *et al.* (1978) have shown that the oxygen-depleted, carbon dioxide-rich water immediately adjacent to the egg is denser than the well oxygenated, low CO_2 water in the free stream. In still water, this sets up a toroidal flow as the denser solution adjacent to the egg sinks (Fig. 2). Water velocities in the toroid can be relatively high. For example, O'Brien *et al.* (1978) observed an average velocity of 72 cm h^{-1} in the toroid set up by eyed eggs (400 degree-days, 10°C) of coho salmon (*O. kisutch*). At this velocity, natural convection would be about 150 times as effective as simple diffusion in supplying oxygen under "static" conditions. The effectiveness of natural convection can be expected to increase as metabolic rate increases, due to greater depletion of oxygen and accumulation of carbon dioxide in the boundary layer. Thus, natural convection may act as a homeostatic mechanism helping to balance oxygen supply and demand.

Whether natural convection plays a significant role in nature will depend on bulk water velocity and the orientation of the egg mass. Analysis of mixed regimes is complex but, in general, natural convec-

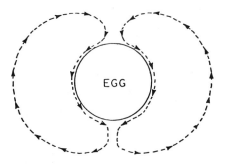

Fig. 2. Stylized depiction of the toroidal flow of water set up around a respiring egg as a result of natural convection in the absence of bulk water movements. [From O'Brien *et al.* (1978).]

tion will play a greater role than forced convection at low bulk velocities [see Vogel (1981, pp. 178–195) on how the ratio of the Grashof
number and the square of the Reynolds number can be used to estimate the relative importance of the two forces]. In nature, bulk velocities are often low. For example, the bulk flow in many spawning beds
is less than the toroidal velocity set up by coho eggs in still water
(O'Brien et al., 1978). Similarly, interstitial velocities in egg masses of
species such as lumpfish (Cyclopterus lumpus) and lingcod
(Ophiodon elongatus) can be expected to be rather low. It has been
suggested, but not demonstrated, that voids in such masses may provide avenues for natural convection to supply oxygenated water to the
interior (O'Brien et al., 1978; Giorgi and Congleton, 1984). Natural
convection will be most effective if eggs are oriented so that the
heavier oxygen-depleted water can sink and its movement is not op-
posed by forced convection (O'Brien et al., 1978; Johnson, 1980).

B. The Egg Capsule

The egg capsule traditionally has been viewed as the major barrier
to diffusive gas exchange during embryonic life. This may seem obvious, but the empirical evidence supporting this viewpoint is actually
rather scanty. Strongest support comes from observations that incipient limiting oxygen tensions (P_c) drop significantly on removal of the
capsule (Hayes et al., 1951; Rombough, 1986, 1987). If metabolic rate
and the drop in P_c are known, the diffusion coefficient of the capsule
(D_c) can be calculated using Fick's equation for diffusion through a
plane [Eq. (3)]. Several investigators have done this (Hayes et al.,
1951; Alderdice et al., 1958; Daykin, 1965; Wickett, 1975), but the
value most often cited, 0.18×10^{-5} cm^2 s^{-1}, is that worked out by
Wickett (first presented in Daykin, 1965) using data provided by
Hayes et al. (1951) for Atlantic salmon (Salmo salar). Wickett (1975)
recognized that his estimate was based on rather sketchy data and
indicated that standard diffusion tests should be conducted to check
its validity. Unfortunately, this has not been done for teleosts, although Diez and Davenport (1987) recently used a Krogh-type diffusion chamber to estimate the oxygen diffusion coefficient of the egg
case of the dogfish, Scyliorhinus canicula. Interestingly, the value
they arrived at, 0.285×10^{-5} cm^2 s^{-1}, is similar to Wickett's value for
Atlantic salmon. However, given the structural differences between
the dogfish egg case and the salmon capsule, this cannot be taken as
confirmation of Wickett's value.

Wickett's value for D_c is about one-tenth that of water. Wickett (1975) pointed out that the surface area of the radial pores that penetrate the capsule is similarly about one-tenth the total surface area of the capsule and speculated that this may indicate that diffusion takes place primarily through the pore canals rather than through the capsule matrix. If this is true, doubt is cast on the validity of the practice of some investigators (e.g., Daykin, 1965; Wickett, 1975; Kamler, 1976; DiMichele and Powers, 1984) applying diffusion coefficients, calculated for one species, to another unrelated species. There is considerable variation among teleosts in capsule structure (Lønning, 1972; Stehr and Hawkes, 1979; Groot and Alderdice, 1985). Even in closely related salmonids, pore area can vary anywhere between 7% and 30% of the total surface area (Groot and Alderdice, 1985).

Recent evidence suggests that the capsule may not be as great a barrier to diffusive gas exchange as was supposed previously. Berezovsky et al. (1979) used microelectrodes to measure the dissolved oxygen profile across the capsule, perivitelline fluid, and vitelline membrane of the recently fertilized loach eggs (Misgurnis fossilis). Surprisingly, they recorded very little drop in oxygen concentration across the capsule. In contrast, there was a gradual decline in oxygen tension across the perivitelline fluid and a sharp drop across the vitelline membrane (Fig. 3). Sushko (1982) reported a similar oxygen profile for loach eggs incubated in helium–oxygen and nitrogen–oxygen gas mixtures. Observations made by Alderdice et al. (1984) may provide an explanation for why the capsule appears to offer relatively little resistance to gas exchange. Alderdice et al. (1984) observed that the hydrostatic pressure exerted on the capsule of steelhead (S. gairdneri) was considerably less than the osmotic pressure of the perivitelline fluid and calculated an effective filtration pressure of about −62 mm Hg driving water into the perivitelline space. They reasoned that according to Starling's hypothesis this should lead to the movement of water into the perivitelline space. Since the egg is in volume equilibrium, this must be balanced by an equal outflow by filtration. Exactly how this would be accomplished is not clear, since the capsule is not a linear structure like a capillary. Alderdice et al. (1984) suggested that the capsule may act like a balloon with microsieves in its wall. Increasing internal pressure would be accompanied by volume expansion. As the capsule expanded the pores would enlarge and more water would flow out. Volume and tension increases thus would be self-limiting. Alderdice et al. (1984) proposed that this process would tend to facilitate respiratory gas exchange and point to radiotracer studies (Potts and Rudy, 1969; Loeffler and Lovtrup, 1970;

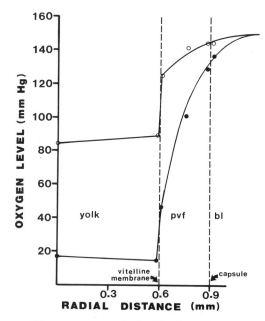

Fig. 3. Oxygen profiles across recently fertilized roach eggs measured using platinum microelectrodes. Open symbols, normal eggs; closed symbols, eggs in which rate of oxygen uptake was stimulated 3.3-fold using 10^{-4} M 2,4-dinitrophenol; pvf, perivitelline fluid; bl, boundary layer. [After Berezovsky *et al.* (1979).]

Loeffler, 1971) indicating an exchange of water across the capsule equivalent to the volume of the perivitelline fluid every 1–4 min. A connective flux of this magnitude would add to the diffusive flux of oxygen across the capsule but, perhaps more importantly, the currents generated would tend to prevent the establishment of a large concentration gradient across the capsule.

C. The Perivitelline Fluid

As noted earlier, microelectrode studies (Berezowsky *et al.*, 1979; Sushko, 1982) indicated that, at least for early embryos, much of the resistance to gas exchange previously attributed to the capsule actually resides in the perivitelline fluid. The perivitelline fluid can be expected to have an oxygen diffusion coefficient similar to that for water ($\simeq 2.5 \times 10^{-5}$ cm^2 s^{-1}) (Dejours, 1981), but because diffusion distances are much larger the net impact of the perivitelline fluid may be greater than that the capsule. While capsule thicknesses in

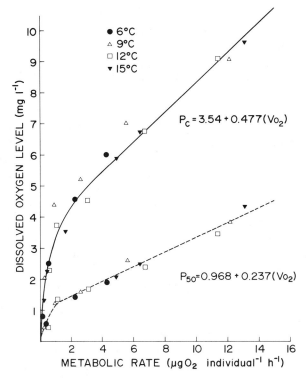

Fig. 4. Relationships between critical dissolved oxygen levels (P_c) and routine metabolic rate and P_{50} values (oxygen concentration at which $\dot{V}O_2 = 0.5\ r\dot{V}O_2$) and routine metabolic rate for steelhead embryos incubated at constant temperatures. Equations are for the linear portions of the curves. [From Rombough (1987).]

salmonids range between 15 and 70 μm (Groot and Alderdice, 1985), the distance across the perivitelline space can be in excess of 500 μm. Absolute distances tend to be smaller in pelagic eggs, but because of a thinner capsule and comparatively larger amount of perivitelline fluid, the relative impact of the perivitelline fluid on gas exchange would be even larger than in salmonids.

The impact of the perivitelline fluid on oxygen uptake can be seen in the rapid increase in P_c values during early development (Rombough, 1987). Critical oxygen tensions for steelhead increase very rapidly in relation to metabolic rate until about the time embryos began to move and stir the perivitelline fluid. Thereafter, P_c values increase more slowly and in direct proportion to metabolic rate (Fig. 4). Thus in a teleological sense, stirring of the perivitelline appears to be necessary if metabolic oxygen demands are to be met. Reznichenko *et al.* (1977) used a modified polarographic electrode to model oxygen exchange across the egg capsule and perivitelline fluid. He found that when the analog of the perivitelline fluid was stirred, the PO_2 at the body (electrode) surface increased while that under the capsule (membrane) decreased. This had the effect of increasing the steepness of the concentration gradient across the capsule and enhancing net oxygen transport. Peterson and Martin-Robichaud (1983) observed that Atlantic salmon embryos began to stir the perivitelline fluid fairly early in development. Trunk movements began abruptly at about 200 degree-days of development with an initial frequency of about 60–120 flexures h^{-1}. Dye studies indicated that trunk flexures resulted in rapid water movement along the trunk and from one side of the perivitelline fluid to the other. These movements were apparently of a respiratory nature, since an unexpected water failure leading to hypoxia resulted in an increase in frequency of trunk flexures. Trunk movements normally decline rather abruptly to a frequency of only 1 every 2–4 h by 350–400 degree-days in Atlantic salmon (Peterson and Martin-Robichaud, 1983). However, by this time the embryo had begun to move its pectoral fins rapidly at a rate of 40–150 min^{-1}. This rate was maintained until hatch. Dye studies indicated that these movements generated a rapid water flow ($\approx$300 cm h^{-1}) in the immediate area of the pectoral fins but were not as effective as trunk flexures in completely mixing the perivitelline fluid. Complete mixing was accomplished by periodic trunk flexures.

Recent studies of amphibians suggest that stirring of the perivitelline fluid may facilitate oxygen transport within an egg mass as well as within individual eggs. Burggren (1985) noted that oxygen partial pressures were higher and carbon dioxide partial pressures were lower in the interior of the egg mass of the frog *Rana palustris* than would be expected if simple diffusion was the only process involved. Frog embryos use cilia to stir the perivitelline fluid. Burggren (1985) suggested that the currents generated by movement of the cilia could lead to oxygen being transported to the interior of the egg mass by convection as well as diffusion. Oxygen would diffuse across the capsule at the surface of the egg closest to the outside of the egg mass. This mass of oxygen-rich water would then be moved by ciliary action to the opposite side of the egg where oxygen then would diffuse outward across the capsule toward the center of the egg mass. An oxygen molecule thus could be passed from egg to egg in a manner somewhat analogous to water being passed along a bucket brigade. Carbon dioxide would pass in the opposite direction.

This appears to be a plausible mechanism for supplying oxygen to the interior of teleost as well as amphibian egg masses. Many teleosts lay large, compact masses of eggs. For example, the egg mass of lingcod may be up to 5 liters in volume (Giorgi and Congleton, 1984). At present, there is not enough information on oxygen and current profiles within such egg masses to adequately test the "bucket-brigade" hypothesis. However, there is some circumstantial evidence that processes in addition to bulk water flows may be involved. Giorgi and Congleton (1984) noted that while oxygen concentrations in the center of a lingcod egg mass declined rather sharply following cessation of current flow, levels did not decline to zero as expected but stabilized at about 10% air saturation. Davenport (1983) similarly indicated that oxygen levels in the egg mass of lumpfish declined more slowly than expected when aeration ceased. These observations are intriguing but say little about the mechanisms involved. As discussed previously, these observations can be explained equally as well by natural convection as by the "bucket-brigade" hypothesis. They do, however, suggest that egg masses do not depend solely on forced convection to meet metabolic oxygen demands.

Braum (1973) noted that the deeper eggs in the egg mass of species such as herring (*Clupea harengus*) are threatened with asphyxia as a result of poor water circulation. He suggested that the perivitelline fluid could function as an oxygen reservoir to tide embryos over short periods of anoxia. This is unlikely to be of much significance. The amount of oxygen in the perivitelline fluid, assuming it is 100% saturated, is only sufficient to meet the oxygen requirements of advanced embryos for 1–2 min. This calculation assumes that oxygen would not diffuse back out of the capsule—which of course it would under hypoxic conditions—and that there is no convective exchange between the perivitelline fluid and the surrounding water—which is likely.

The perivitelline fluid provides the immediate environment for the developing embryo, and it is the gas concentration in the perivitelline fluid—not the surrounding water—that is of physiological significance. As predicted by the mass transport equation, Eq. (3), oxygen concentrations in the perivitelline fluid decline progressively as development proceeds. Assuming relatively constant capsule conductance and ambient PO_2, the only way the rising metabolic demands associated with tissue growth can be met is by an increase in the driving force across the capsule. This necessitates a reduction in the PO_2 of the perivitelline fluid. Berezovsky et al. (1979) demonstrated such a drop in perivitelline fluid PO_2 when the metabolic rate of loach embryos was stimulated by low concentrations of dinitrophenol (Fig. 3). A decline in perivitelline fluid PO_2 is also implied by the gradual increase in P_c that was seen during the course of steelhead development (Fig. 4; Rombough, 1987). Recently, Diez and Davenport (1987) showed that the PO_2 of the fluid in the egg case of the dogfish declined as development proceeded. Finally, similar declines in PO_2 have been well documented for reptilian and avian eggs (Dejours, 1981). Bird eggs in particular have been studied extensively, and since many of the structures in bird and fish eggs can be considered analogous, the type of relationships seen in bird eggs probably apply to fish eggs as well. For example in the hen egg, oxygen levels in the air space, which is analogous to the perivitelline fluid, decrease as metabolic rate increases (Wangensteen, 1972). This increases the diffusion gradient across the shell, which like the teleost capsule is pierced by tiny pores, and automatically ensures a greater rate of diffusive flux. It does so, however, at the expense of arterial PO_2 levels which gradually decline as development proceeds. Blood gas relationships have not been examined in fish embryos, but if the analogy with bird eggs holds, they probably follow a similar pattern.

A decrease in perivitelline fluid PO_2 late in embryonic development appears to be the trigger that initiates hatching in at least some teleosts. If advanced embryos are placed in hypoxic water, premature hatching occurs (Yamagami, 1981; DiMichele and Powers, 1984a; Ishida, 1985). Conversely, hatching can be delayed more or less indefinitely under hyperoxic conditions (Taylor et al., 1977; DiMichele and Taylor, 1980; Ishida, 1985). Low oxygen levels do not appear to act directly on the hatching glands. Studies involving various anesthetics suggest the response is mediated by the central nervous system (Ishida, 1985). The location of the oxygen sensor is not known.

Hatching can be regarded as an adaptive response to physiological hypoxia. Escape from the confines of the egg capsule reduces the ambient oxygen level required to meet metabolic requirements by 30–50 mm Hg (Rombough, 1987). However, removal of the capsule does not alter the basic mechanisms involved in respiratory gas exchange.

D. Cutaneous Gas Exchange

Respiratory gas exchange in fish, and indeed in all vertebrates, is initially cutaneous. As development proceeds there is a gradual increase in the relative importance of gills, although in many species the skin remains the major site of gas exchange throughout the embryonic and larval periods. Recent evidence indicates that even in adults the skin may persist as an important site for respiratory gas exchange

(Kirsch and Nonnotte, 1977; Lomholt and Johansen, 1979; Steffenson and Lomholt, 1985; Feder and Burggren, 1985).

Studies of gas exchange during the early life stages of teleosts have tended to be descriptive. As a result, most of what we know of respiratory mechanisms has been inferred from studies of the morphology of what are assumed to be respiratory structures. Morphological adaptations to facilitate gas exchange appear early in development. Boulekbache and Devillers (1977) suggested that the function of the microvilli present on the outer surfaces of blastomeres of rainbow trout (*S. gairdneri*) was to increase the surface area for respiratory gas exchange. In many species well-developed vascular networks form just under the skin during early organogenesis (Fig. 5). These capillary beds are often associated with specialized cutaneous structures, such as an enlarged yolk sac, expansive medial finfolds, or enlarged pectoral fins, that greatly increase the surface area available for gas exchange. Detailed descriptions of such specialized structures are provided by Taylor (1913), Sawaya (1942), Wu and Liu (1942), Kryzanowsky (1934), Smirnov (1953), Soin (1966), Balon (1975), Lanzing (1976), McElman and Balon (1979), Liem (1981) and Hughes *et al.* (1986), among others.

The degree to which embryonic and larval respiratory structure are elaborated varies widely among species. Several authors have sug-

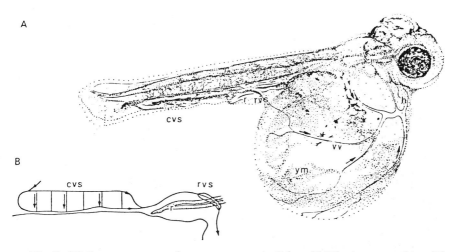

Fig. 5. (A) Cutaneous gas exchange structures in 5-day-old *Tilapia mossambica*. (B) Schematic diagram showing blood flow through caudal and rectal vascular systems: cvs, caudal vascular system; h, heart; r, rectum; rvs, rectal vascular system; vv, vitelline vein; ym, yolk mass. [From Lanzing (1976).]

gested that this reflects variations in oxygen levels in spawning habitat and can be used as the basis for a functional classification system (Kryzanowsky, 1934; Soin, 1966; Balon, 1975). It is beyond the scope of the current discussion to examine the merits of such systems, but attention will be drawn to one characteristic that these authors have considered particularly important, that is, whether the species lays pelagic or demersal eggs. In the marine and temperate freshwater environments, pelagic embryos tend to have relatively poorly developed capillary plexes near the body surface. Respiratory pigments (hemoglobin, myoglobin, perhaps carotenoids) usually do not appear until late in development, and gills may not become functional until near the end of the larval period (Balon, 1975). Pelagic eggs tend to be small and are normally found in well-oxygenated waters. As a result, it has been suggested that oxygen is not normally a limiting factor and thus specialized respiratory structures are not necessary (Hempel, 1979). In contrast, demersal eggs are usually larger and often are exposed to relatively low oxygen concentrations for extended periods. According to Balon (1975), this has resulted in selection for extensive vascularization of the body surface and the elaboration of specialized cutaneous gas exchange structures. Such structures tend to develop early and often persist throughout the larval period. Respiratory pigments appear early, and gills become functional soon after hatch. It should be recognized, though, that as with most generalizations in zoology, there are exceptions. For example, the Indian air-breather *Anabas testudineus* lays small pelagic eggs, but the body surface is well vascularized, and respiratory pigments appear early in development (Hughes *et al.*, 1986). These adaptations are not particularly surprising when one realizes that the eggs are laid in the very oxygen-poor waters of tropical swamps, and that it is only in the surface layer that oxygen levels are high enough to sustain embryonic development.

The effective surface area, the length of the diffusive pathway, the magnitude of the partial pressure gradient between the water and blood, the amount of blood perfusing the structure, and the convective movement of water past the structure are among the most important factors influencing the performance of respiratory gas exchangers. On the basis of these characteristics, cutaneous gas exchange in fish embryos and larvae would appear to be highly efficient. As mentioned previously, specialized exchange structures comprise a relatively large fraction of total body surface area in many species. For example, the well-vascularized medial and paired fins of larval herring (*C. harengus*) and plaice (*Pleuronectes platessa*) account for about 40% of total surface area at hatch (DeSilva and Tytler, 1973). In addition, the surface/volume ratio of most larvae is large because of their small absolute size. Total surface area for a 1.6 mg carp (*Cyprinus carpio*) larvae is in the order of 12,000 mm^2 g^{-1}. In contrast, total surface area of a 350 mg juvenile is only about 1400 mm^2 g^{-1} (Oikawa and Itazawa, 1985).

Cutaneous diffusion distances have been estimated for only a few species, but the available evidence indicates that distances are only slightly greater than lamellar diffusion distances in juveniles and adults. The skin is only two cells thick over most of the body surface in young larvae (Lasker, 1962; Jones *et al.*, 1966; Roberts *et al.*, 1973). Lasker (1962) reported that the skin thickness of larval sardine (*Sardinops caerislea*) ranged from 1.7 μm on the finfold to 3.0 μm on the lateral portion of the trunk. Jones *et al.* (1966) estimated a minimum skin thickness in larval herring (*C. harengus*) of about 2.3 μm. The actual length of the diffusive pathway is somewhat greater. Many, particularly pelagic, species have a relatively thick fluid layer between the dermis and epidermis: 5.0 μm in the case of larval plaice (*P. platessa;* Roberts *et al.*, 1973). In addition, distances associated with diffusion across capillary walls and through the plasma should be taken into account. Even when this is done, distances remain relatively small. Webb and Brett (1972a) measured a mean distance of 4.7 μm from the surface of the skin to the center of blood capillaries in embryos of two species of viviparous seaperch (*Rhacohilus vacca* and *Embioteca lateralis*). Liem (1981) estimated 8–15 μm for the total thickness of the water–blood barrier in larval *Monopterus*. These distances are considerably less than the cutaneous diffusion distances of larval amphibians. Burggren and Mwalukoma (1983) estimated a blood–water barrier of 20–50 μm for larval bullfrog (*R. catesbeiana*). In this species up to 60% of total gas exchange takes place across the skin (Burggren and West, 1982). The skin of larval teleosts is probably at least as effective as an organ of gas exchange, given the shorter diffusion distances and, in many cases, more elaborate vascularization.

Cutaneous gas exchange in other vertebrates frequently suffers from a relatively small partial pressure gradient across the skin as a result of central mixing of oxygenated and deoxygenated blood prior to transit to the skin (Burggren, 1984). This problem appears to be minimized in many teleosts. Cutaneous gas exchange structures—for example, the caudal and rectal vascular systems of larval tilapia (Fig. 5) or the vitelline circulation of salmonids—typically receive blood that has already passed through at least a portion of the systemic circulation. Although no in situ measurements have been made, the blood entering the exchange structures should be comparatively poor in oxygen thus maximizing the partial pressure gradient between the blood and the water.

Extremely little is known of the cardiovascular physiology of teleost embryos. Cutaneous exchange obviously would be more effective if blood flow could be regulated. McElman and Balon (1979) noted that the amount of blood passing through cutaneous capillary beds varied during the development of walleye (*Stizostedion vitreum*) and suggested that this implied a mechanism for shunting blood so as to optimize gas exchange. Vascular recruitment as a result of higher systemic blood pressure caused by increased heart rate during periods of physiological hypoxia was proposed. However, there may be differences between ontogenetic changes in blood flow patterns and compensatory changes due to transient alterations in oxygen supply or demand. Reflex control of the amount of blood perfusing cutaneous gas exchange structures may well occur, capillary recruitment in response to hypoxia has been demonstrated in amphibian larvae (Burggren, 1984), but it has yet to be demonstrated in teleost embryos or larvae.

E. Respiratory Pigments

As mentioned previously, the stage at which hemoglobin first appears is highly variable. In many demersal species, such as salmonids, large numbers of pigmented erythrocytes are evident well before hatch. This is thought to be an adaptation to hypoxic conditions (Balon, 1975). In contrast, hemoglobin may not appear in the circulation of pelagic species, such as herring, until after metamorphosis (De Silva, 1974). Lack of hemoglobin has been proposed as a mechanism to limit predation by making pelagic larvae less conspicuous. However, it simply may be that hemoglobin is not required in well-oxygenated waters. Holeton (1971) reported that rainbow trout larvae showed little distress when their hemoglobin was poisoned by carbon monoxide. Similarly, Iuchi (1985) reported that rainbow trout larvae survived to the fry stage after having their erythrocytes destroyed by treatment with phenylhydrazine. Indeed, it may not even be necessary for small larvae to have a functioning circulatory system. Burggren (1984) points out that the so-called "cardiac lethal" larval mutant of the amphibian *Ambystoma,* in which the heart forms but fails to beat, is able to survive after hatching for many days in well-oxygenated water.

The evidence to date indicates that embryonic and larval hemoglobins are structurally and functionally distinct from those of juveniles and adults. Iuchi and Yamagami (1969) reported a gradual change in the electrophoretic banding pattern for the hemoglobins of rainbow trout during the period between hatch and gravel emergence. Similar shifts in electrophoretic banding patterns have been observed in Homasu salmon (*O. rhodurus*) and brook trout (*Salvelinus fontenalis;* Iuchi *et al.*, 1975) and in coho salmon (*O. kisutch;* Giles and Vanstone, 1976). Distinct embryonic and adult hemoglobins also have been reported for several viviparous species (Ingermann and Terwilliger, 1981a,b, 1982, 1984; Ingermann *et al.*, 1984; Weber and Hartvig, 1984; Hartvig and Weber, 1984).

Iuchi (1973b) compared the chemical and physiological properties of larval and adult hemoglobins of rainbow trout. Both were tetrameric, but larval hemoglobin displayed a higher oxygen affinity, less of a Bohr effect, virtually no Root effect, and greater cooperativity at physiological pH than adult hemoglobin. Larval and adult hemoglobins had P_{50} values of 31 mm Hg and 57.5 mm Hg, respectively, at pH 7.2 and 25°C. The Bohr effect ($\Delta \log P_{50}$/pH) was 0.023 for larval hemoglobin but 0.64 for adult hemoglobin. The oxygen capacity of larval hemoglobin was virtually unaffected by pH, while a drop to pH 6.5 reduced the oxygen capacity of adult hemoglobins to 50% of that at pH 8.0. Slopes of Hill plots were $n = 2.33$ and $n = 1.62$, respectively, at pH 7.2 for larval and adult hemoglobins. The high oxygen affinity and pH independence of larval hemoglobins are clearly advantageous to embryos and larvae exposed to the oxygen-poor, low-pH and high-CO_2 environments of the perivitelline fluid and interstices of the redd.

The shift in electrophoretic banding patterns suggests that the greater oxygen affinity of embryonic and larval blood is due primarily to intrinsic differences in the structure of the hemoglobins rather than to changes in concentrations of modulators of hemoglobin (Hb) affinity. This was shown to be the case for the viviparous eelpout *Zoarces viviparous* (Weber and Hartvig, 1984; Hartvig and Weber, 1984). Weber and Hartvig (1984) reported that fetal hemoglobin had a higher O_2 affinity (P_{50} values of 9 mm Hg and 23 mm Hg, respectively, at pH 7.5 and 10°C), reduced Bohr effect, and greater cooperativity than adult hemoglobin in nucleoside triphosphate-free preparations. Measurement of intraerythrocyte nucleoside triphosphate (NTP) concentrations revealed no significant difference in the NTP/Hb ratios of fetal and adult blood. Differences in modulator concentrations, though, do appear to be important in some species. Ingermann and Terwilliger

(1981b, 1982, 1984) reported that part of the reason for the higher O_2 affinity of the hemoglobin of fetal seaperch *E. lateralis* was a lower NTP/Hb ratio. ATP was the most abundant modulator (82% total NTP), but there was also a significant amount of GTP (18% of total NTP) present within fetal erythrocytes. Indirect evidence suggests cofactors may modulate the O_2 affinity of larval hemoglobins in some oviparous species as well. DiMichele and Powers (1982) attributed differences in hatching times of different lactate dehydrogenase (LDH) genotypes of *Fundulus heteroclitus* to the ability to deliver oxygen to tissues. The LDH genotype (LDH B^aB^a) that hatched earliest was also the genotype that had the highest concentration of ATP in their erythrocytes as adults, and presumably as embryos. Increased ATP concentrations would lower hemoglobin O_2 affinity and thus reduce O_2 delivery to tissues near hatch when PO_2 levels in the perivitelline fluid are low. This would trigger release of the hatching enzyme and lead to early hatch.

Iuchi and Yamamoto (1983) demonstrated that the shift from embryonic–larval hemoglobins to juvenile hemoglobins in rainbow trout was the result of erythrocyte replacement. Erythrocytes containing embryonic hemoglobins were formed in the extraembryonic blood islands and intermediate cell mass beginning about one-third through embryonic development. These hemopoietic centers ceased production shortly after hatch, and centers in the kidney and spleen began to produce morphologically and antigenically distinct erythrocytes containing juvenile-type hemoglobins. Similar shifts in the site of hemopoiesis have been reported for Atlantic salmon (Vernidub, 1966) and anglefish *Pterophyllum scalare* (Al-Adhami and Kunz, 1976). Hemoglobin switching based on erythrocyte replacement may turn out to be widespread in lower vertebrates. Kobel and Wolff (1983) recently reported that the shift from embryonic to larval type hemoglobins in the amphibian *Xenopus borealis* also was associated with a shift in the site of erythropoiesis.

It has been suggested that pigments other than hemoglobin may be involved in respiratory gas exchange. De Silva (1974) noted that myoglobin was formed prior to hemoglobin in herring larvae and indicated that this may reflect a respiratory role during early development. This hypothesis has not been tested. Other authors have proposed a respiratory role for the carotenoid pigments (Smirnov, 1953; Volodin, 1956; Balon, 1975, 1984; Mikulin and Soin, 1975; Soin, 1977; Czeczuga, 1979). This hypothesis is based largely on circumstantial evidence, namely, that eggs of species containing high concentrations of carotenoids are often relatively large and, thus, are faced with relatively large diffusion distances. The problem of gas exchange is compounded for some of these species by the fact that they must develop in comparatively low oxygen environments. Unfortunately, there is little experimental evidence to indicate that carotenoids actually aid in oxygen transport under such conditions. In fact, there is now considerable evidence indicating that low carotenoid levels in eggs, normally rich in carotenoids, do not significantly reduce survival (Steven, 1949; Craik, 1985; Craik and Harvey, 1986; Tveranger, 1986). Craik (1985) speculates that carotenoids may play some minor, as yet undefined, role in oxygen transport but that present evidence does not warrant acceptance of carotenoid-based respiration as an established fact, as Balon (1975, 1984) would suggest.

We have seen how the movement of water past the egg and stirring of the perivitelline fluid enhance embryonic gas exchange. Adequate ventilation of body surfaces is similarly necessary for efficient gas exchange after hatch. Fish have adopted a number of tactics to ensure that this occurs. Some, such as nest-fanning, mouth-brooding, and wriggler-hanging, involve the parents. Others involve behavioral and physiological adaptations on the part of the larvae.

Hunter (1972) noted that newly hatched anchovy (*Engralis mordax*) exhibited regular bouts of swimming that did not appear to be associated with feeding or predator avoidance. He suggested that swimming might have a respiratory significance, presumably by enhancing ventilation of the body surface. This hypothesis was examined in some detail by Weihs (1980, 1981). Weihs (1980) pointed out that because of their small size, anchovy larvae existed in a viscous environment (i.e., low Reynolds number, Re). Consequently, both the larva and its immediate surroundings tend to be transported together by oceanic currents. A nonswimming larva thus would remain in the same mass of water and gradually deplete the available oxygen. Weihs (1980) developed a mathematical model for diffusive oxygen uptake by the larvae and estimated that oxygen would become limiting for a stationary day-old larva at concentrations below 63% air saturation (ASV). He then tested this prediction by observing larval swimming behavior at various oxygen concentrations. As predicted, both the frequency and duration of swimming significantly increased at oxygen levels below 60% ASV (Fig. 6).

Weihs' (1980) study indicates that the limiting step in cutaneous gas exchange, at least in anchovy larvae, is the convective flow of water past the exchange surface. It should be remembered that anchovy larvae live in a comparatively oxygen-rich environment. It would thus appear to be even more important for larvae inhabiting

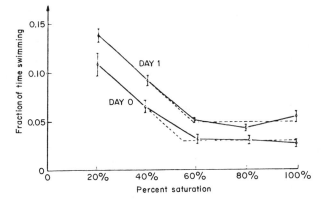

Fig. 6. Fraction of time spent actively swimming as a function of ambient oxygen concentration for newly hatched (day 0) and 24-h-old (day 1) northern anchovy larvae. [From Weihs (1980).]

hypoxic environments to be able to generate large convective flows. This is accomplished in larvae of the Austrailian lungfish (*Neocera-todes forsteri*) by means of cilia that direct a water current posteriorly along the body surface (Whiting and Bone, 1980). Larvae of many warm-water teleosts use their pectoral fins to create a similar water flow. Liem (1981) reported that larvae of the air-breathing fish *Monop-terus albus* use large, well-vascularized pectoral fins to direct a flow of relatively oxygen-rich surface water backward along their body surface. The yolk sac and caudal region of *Monopterus* larvae are also extensively vascularized. Microscopic observations indicated that surficial blood flow in these regions ran in the opposite direction to the flow of water generated by the pectoral fins. When larvae were placed in a tube with water flowing in the normally anterior to posterior direction the oxygen extraction efficiency was about 41%. When the direction of water flow was reversed, extraction efficiency dropped to about 20% (Fig. 7). Liem (1981) pointed out that on this basis the whole larva could be considered a functional analog of the gill lamellae of adult fish. He noted that similar elaborate vascular networks and mobile pectoral fins were common in larvae of other species inhabiting hypoxic waters and speculated that such countercurrent flow mechanisms might be widespread. It should be noted, though, that pectoral fin movements are not always associated with ventilating cutaneous gas exchange surfaces. Peterson (1975) found that, contrary to his expectations, the rapid pectoral fin movements of Atlantic salmon alevins did not direct water over the well-vascularized yolk sac. Instead, they appeared to be involved in drawing water over the gills.

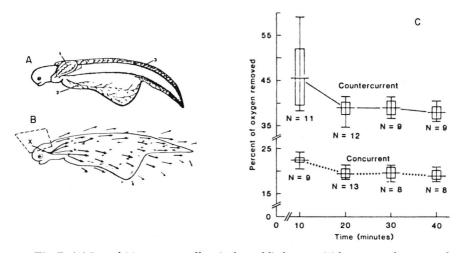

Fig. 7. (A) Larval *Monopterus albus* (4 days old) showing (1) large vascular pectoral fins, (2) well-developed capillary network in the yolk sac, and (3) well-vascularized unpaired medial fin. (B) Schematic representation of water currents generated by movement of pectoral fins. The stippled area shows the region from which water is drawn by the pectoral fins. (C) The effectiveness of oxygen extraction from water flowing in an anterior to posterior direction (countercurrent to blood flow) or in a posterior to anterior direction (concurrent to blood flow). The flow rate was 130 ml l^{-1}. (Reprinted with permission from Liem, 1981; copyright © AAAS.)

F. Branchial Gas Exchange

The larval period is characterized by a shift in the site of respiratory gas exchange from the skin to the gills. Unfortunately, relatively little is known of the physiology of this transition. It is known that species vary in the stage at which gills first appear and the speed with which they are elaborated. For example, gill arches begin to form in rainbow trout shortly after gastrulation and by hatch are complete with functional filaments and secondary lamellae (Morgan, 1974a,b). Arctic char (*Salvelinus alpinus*) also possess filaments at hatch but secondary lamellae do not begin to form until about 8 days posthatch (at 6.5°C; McDonald and McMahon, 1977). In smallmouth bass (*Micopterus dolomieu*), filaments do not appear until about 7 days posthatch (at 16°C; Coughlan and Gloss, 1984). Lamellae begin to form about 4 days later, about the time the larvae becomes free-swimming. Filaments and secondary lamellae first appear on gill arches of herring and plaice larvae about midway through the larval period, at body lengths of 10 and 8 mm, respectively (De Silva, 1974).

Such description provides relatively little information about the relative importance of the gills in larval gas exchange. This question can be answered best by directly measuring gas fluxes across the gills and skin. This has been done for amphibian larvae (Burggren and West, 1982; Burggren, 1984) but not for teleost larvae, even though the techniques developed for amphibians would be relatively easy to apply to some of the larger fish larvae. In the absence of such partitioning studies, the relative importance of gills and skin in fish larvae must be inferred from the rather limited amount of information available on the morphometrics of the two structures. As mentioned previously, the factors of particular importance in determining the efficiency of gas exchange structures are total surface area, diffusion distance, partial pressure gradients, and the convective supplies of blood and water.

De Silva (1974) reported that, at hatch, the gills of both herring and plaice accounted for an insignificant portion (<0.5%) of total body surface area. The surface area of the gills expanded rapidly during larval development; gill area weight exponents were 3.36 and 1.59, respectively. However, even at the end of the larval period, gill surface area was only a relatively small fraction ($\simeq$10% for herring and <5% for plaice) of the total surface area over which diffusive gas exchange could take place. Similarly, the surface area of the gills of newly hatched carp larvae accounted for < 0.2% of the total area available for gas exchange (gills, fins, body surface) (Oikawa and Itazawa, 1985). Gills grew rapidly, with allometric mass exponents of 7.066 between 1.6 and 3.0 mg and 1.222 between 3.0 and 200 mg. However, even for a 140 mg postlarva, the surface area of the gills remained less than one-third of the total potential exchange area. Thus, morphometric analysis would suggest that in these species the skin remains the most important organ for gas exchange throughout larval development. Gills appear to be relatively more important in salmonid larvae, but even here, cutaneous exchange remains significant until long after swim-up. McDonald and McMahon (1977) reported that early development of Arctic char gills was characterized by a rapid increase in filament number and size followed by an increase in the number and size of the secondary lamellae. At 15 days posthatch lamellae only accounted for about 10% of gill surface area, while by 47 days posthatch (at 6.5°C) they comprised 23% of total gill area. However, this was only about half the lamellar contribution (45% total gill area) in yearling char. McDonald and McMahon (1977) calculated that on the basis of body weight Arctic char larvae would require a lamellar surface area about four times that actually measured 47 days posthatch to meet metabolic oxygen demands if the gills were the only site of gas exchange.

Only a few measurements have been made of the thickness of the blood–water diffusion barrier in the gill of larval fish. Morgan (1974b) reported that the thickness of the blood–water pathway in rainbow trout decreased gradually from 11.1 μm at hatch to 6.8 μm 102 days posthatch. This decrease was due primarily to a shift from a multilayer respiratory epithelium at hatch to one that was only two cells thick. There was also a modest decrease in the thickness of the basement membrane. A reduction in the thickness of the diffusion barrier is probably typical of obligate water-breathing larvae. In contrast, lamellar diffusion distances increase as development proceeds in bimodal breathers. In *Anabas testudineus*, diffusion distances increased from 1–2 μm prior to air breathing to 4–8 μm after the start of air breathing (Mishra and Singh, 1979). In *Channa striatus*, diffusion distances increased similarly from 4 to 9 μm during the transition to air breathing (Prasad and Prasad, 1984). This increase in diffusion distances is probably an adaptation to prevent lamellar collapse while the gills are out of water.

Development of branchial blood vessels has been described for many species (Solewski, 1949; Markiewicz, 1960; Morgan, 1974a,b; McElman and Balon, 1979; Paine and Balon, 1984, among others). Much less is known about the functioning of the branchial circulation. Morgan (1974b) suggested that rainbow trout larvae might not be able to regulate blood flow through specific secondary lamellae. This was based on histological observations that pillar cells did not appear to be contractile. Overall flow to the gills presumably can be regulated by alterations in heart rate. Holeton (1971) and McDonald and McMahon (1977) noted that larvae responded to moderate hypoxia by increasing heart rate.

Holeton (1971) noted that ventilatory movements of newly hatched rainbow trout were weak and poorly coordinated and suggested that the respiratory pump was not fully functional at hatch. Peterson (1975), similarly, observed that opercular movements were infrequent and irregular in newly hatched Atlantic salmon. In contrast, Morgan (1974b) reported that the buccal and opercular pumps of rainbow trout were fully functional at hatch. The cause of these apparently contradictory results may lie in the oxygen concentrations at which the observations were made. McDonald and McMahon (1977) noted that the buccal and opercular movements of newly hatched Arctic char were infrequent and poorly coordinated in normoxic water, but if the larvae were placed in hypoxic water, ventilatory movements become stronger and fully coordinated. McDonald and McMahon (1977) proposed that this indicated that the larvae possessed the necessary neural and musculoskeletal mechanisms to irrigate their gills effectively but did not do so unless oxygen was limiting. In well-oxygenated water, oxygen requirements could be satisfied by cutaneous exchange at less cost. This raises the question of whether branchial gas exchange is cost-effective for very small larvae. Small larvae reside in a viscous environment (low Re). As a result, fluid accelerated by the buccal and opercular pumps would rapidly come to halt between cycles. Irrigating the gills using the buccal and opercular pumps thus can be expected to be relatively expensive and will be avoided by small larvae in favor of less expensive cutaneous exchange under normoxic conditions. A low Reynolds number environment may be the reason Atlantic salmon larvae use pectoral fin movements to assist in moving water over the gills. Pectoral fin movements are rapid (≈ 100 cycles min^{-1}; Peterson, 1975) and can be expected to produce a more or less constant flow, which at low Re may be more efficient than a pulsatile flow.

It has been suggested that larvae experience respiratory distress during the transition from cutaneous to branchial gas exchange. Iwai and Hughes (1977) reported that a preliminary study of the morphometrics of gill development in the black sea bream (*Acanthopagrus schlegi*) indicated a decline in the surface/volume ratio of the gills 5–8 days after hatch. They speculated that this would produce a situation unfavorable to gas exchange and might be the reason for high larval mortality during the so-called "critical period." McElman and Balon (1979) suggested that walleye larvae similarly may experience respiratory distress leading to mortality during the period when blood flow through the vitelline circulation is in the process of being reduced but before the gills are fully functional. Additional support for respiratory involvement in critical-period mortality in some species is provided by observations that critical oxygen levels rise (DeSilva and Tytler, 1973; Davenport, 1983) and resistance to hypoxia decreases (DeSilva and Tytler, 1973; Spoor, 1977, 1984) midway through the larval period. These observations will be discussed later in more detail.

III. AEROBIC METABOLISM

A. Measurement Techniques

Many factors influence the metabolic rate of embryos and larvae. It is virtually impossible to control all the factors when measuring metabolic rates and, as a result, the measurement technique itself often significantly affects the rate that is recorded. Three general techniques have been used to measure metabolic rates: indirect calorimetry, bioenergetic analysis, and direct calorimetry. Each of these techniques will be examined briefly with regard to those aspects that are most pertinent to their use with fish embryos and larvae.

1. INDIRECT CALORIMETRY

Indirect calorimetry is by far the most widely used technique. In virtually all cases the rate of energy production is estimated by measuring the rate of oxygen consumption. It is possible theoretically to estimate energy production by measuring the rate of carbon dioxide production, but because of practical difficulties this approach is rarely used. The few instances in which CO_2 production has been measured have been in conjunction with measurements of oxygen consumption. This allows calculation of respiratory quotients and provides some insight into energy sources during development (see Devillers, 1965; Kamler, 1976; Gnaiger, 1983a).

A variety of indirect methods have been used with fish embryos and larvae. The most widely used method seems to be the manometric technique using Warburg- or Gilson-type respirometers. The method is relatively easy to master and because of the standard design of the respirometers, results of different studies can be compared directly. However, the technique has a number of disadvantages, the most serious being the fact that respirometers are agitated. This disturbs the animals and may seriously affect the rate of oxygen consumption (Gnaiger, 1983b). It is usually assumed that shaking elevates metabolic rate (e.g., Eldridge et al., 1977) but several investigators have speculated that in some species shaking may depress rather than enhance larval metabolism (Hunter and Kimbrell, 1980; Theilacker and Dorsey, 1980; Solberg and Tilseth, 1984). The impact of shaking can be reduced by only agitating flasks just prior to readings (e.g., DeSilva et al., 1986), but if this is done, continuous readings of oxygen uptake cannot be made and much information is lost. Another problem with the method is the fairly long time required for equilibrium to occur, and as a result early responses may be unreliable. It is often suggested that because the system is closed, metabolic waste products accumulate and may influence oxygen uptake. Given the relative insensitivity of fish embryos and larvae to metabolic byproducts, this is probably not a major problem.

In a few instances, Cartesian divers have been used to measure oxygen consumption. The technique is very sensitive, and accuracies in the order of $\pm 10^{-5}$ μl h^{-1} have been reported (Davenport, 1976; Gnaiger, 1983b). The problem is that divers are not readily available

and require a good deal of skill to operate. As a result, their application has been limited. In addition, measurements involving active
larvae are difficult because of the relatively small size of the divers.

The Winkler method has been used frequently in the past to measure oxygen utilization. It has the advantages of being inexpensive,
simple, and reasonably accurate. It allows for a variety of respirometer
designs and can be applied to either flow-through or closed systems.
The major disadvantage of the technique is that it does not permit
continuous monitoring of oxygen levels and in recent years has been
largely superceeded by the use of polarographic oxygen sensors.

The general use of polarographic oxygen sensors to measure rates
of oxygen consumption in the aquatic environment has been reviewed
thoroughly (Hitchman, 1978; Forstner, 1983; Gnaiger, 1983b). Davenport (1976) reported that with care, accuracies in the order of ± 0.02 μl
h^{-1} are possible. Polarographic sensors can be used in conjunction
with virtually any type of respirometer. This flexability allows respirometers to be designed to suit the particular circumstances. For example, Dabrowski (1986) recently designed a circular respirometer to
measure active metabolism in larvae. A banded paper drum moving
around the respirometer induces activity, which is monitored visually.
Polarographic electrodes allow oxygen uptake to be monitored continuously and correlated with activity. Both closed and flow-through systems have been used to measure oxygen uptake by fish embryos and
larvae. As Forstner (1983) points out, each has advantages. Closed
systems are easier to construct and accuracies are usually greater because only oxygen concentration has to be monitored (it is assumed
that time is recorded in all cases). In flow-through tests both oxygen
concentrations and flow must be monitored and accuracy may suffer.
On the other hand, flow-through systems provide a more constant
environment and eliminate the possibility of metabolic waste products influencing metabolic rate. In addition, long-term trends in oxygen uptake can be investigated. Gnaiger (1983b) points out that the
choice of respirometer design requires a compromise between factors
such as sensitivity, time resolution, a more or a less disturbed environment, and ease of construction. Forstner (1983) described an intermittent-flow system suitable for fish eggs or larvae that combines attrib-
utes of both flow-through and closed systems.

Plastics are frequently used in the construction of respirometers.
Researchers should be aware that plastics vary considerably in their
permeability to gases, and some, such as silastic [$P = 60 \times 10^{-9}$ cc cm
(s cm^2 cm Hg)$^{-1}$] and acrylic [$P = 27 \times 10^{-9}$ cc cm (s cm^2 cm Hg)$^{-1}$],
are relatively permeable to oxygen (see the Cole-Parmer 1987–1988
catologue for a listing of N_2, O_2, and CO_2 permeability coefficients for

the more common plastics). As a result, significant amounts of oxygen may diffuse into or out of the system if the walls of the respirometer are thin. This is particularly a problem with plastic tubing because of the high surface-to-volume ratio. Plastics can be a problem even when chamber walls are relatively thick, since they have a tendency to act as "sponges" and thus dampen response times. Eriksen and Feldmeth (1967) indicated that the permeability of plastics could be reduced by coating them with water-soluble silicone (e.g., Siliclad, Clay-Adams Co.). The efficacy of this procedure has not been documented, but even if it does not significantly reduce gas permeability, it may still be worthwhile because of its effectiveness in inhibiting bubble formation. When all is said, probably the best procedure is to follow Forstner's (1983) recommendation and use glass or stainless steel instead of plastic wherever possible.

Control of flow is often overlooked when respirometers are designed. As we have seen, low flows reduce oxygen uptake in eggs by effectively increasing the thickness of the boundary layer (Winnicki, 1968; Wickett, 1975; DiMichele and Powers, 1984a,b). In addition, adequate flow is required to ensure rapid mixing between the organisms and the oxygen sensor to avoid the problem of excessive lag. Flow also influences oxygen uptake by larvae, but the situation is more complex than is the case for eggs. Low flows can reduce uptake by effectively increasing boundary layer thickness, while high flows may elevate rates of oxygen consumption by increasing activity. In flow-through respirometers, flows can be regulated by simply adjusting the head of the water entering the system. In closed systems, some type of pump is required. The choice of a suitable pump is not a trivial task. Centrifugal pumps often experience cavitation, which can lead to bubble formation. Reciprocating or peristaltic pumps cause pressure changes in the system that can influence the operation of the oxygen sensor. Low-speed gear pumps avoid these problems and would appear to be one of the most effective ways of regulating flows in closed-system respirometers (Rombough, 1987).

Dalla Via (1983) addresses the problem of bacterial contamination of respirometers as a source of error when measuring low rates of oxygen uptake such as those typical of fish eggs. He points out that the use of antibiotics is not particularly effective because of their potential toxicity to the test organism and the fact that many take an appreciable length of time to reduce bacterial populations. In some instances if bacteria are eliminated, fungal growth merely increases. Dalla Via (1983) suggests that perhaps the most effective way to deal with the problem is to run blank controls and subtract bacterial uptake from that of the test animal. This procedure compensates for microbes growing on the respirometer and in the water, but it does not take into account epibiotic contamination of the test organisms. Eggs in particular can harbor large numbers of bacteria (Yoshimizu *et al.*, 1980). Giorgi and Congleton (1984) tried to account for oxygen consumed by microbes growing on lingcod eggs by excising the capsule and measuring its rate of oxygen uptake separately. This technique indicated that between 10 and 24% of total oxygen consumption was due to epibiotic contamination.

2. BIOENERGETIC ANALYSIS

Rates of oxygen consumption can be estimated from energetic analysis of embryonic and larval growth on the basis of the relationship

$$A = G + R + U \tag{4}$$

where A is the amount of energy assimilated, G is the energy equivalent of tissue elaborated, R is the energy equivalent of the amount of oxygen consumed, and U is the amount of energy lost in excretory products. The rate of oxygen consumption is estimated by rearranging the equation to yield $R = A - (G + U)$. Most investigators have used the dry-weight method of Toetz (1966) and Laurence (1969). Caloric equivalents are given by Elliot and Davison (1975), Brett and Groves (1979), and Gnaiger (1983a). Balanced energy budgets are rarely used as the sole means of estimating rates of oxygen consumption. Their main use has been to check the accuracy of metabolic rates measured by other methods (Gray, 1926; Smith, 1947; Laurence, 1969, 1973, 1978; Johns and Howell, 1980; Gruber and Wieser, 1983; Houde and Schekter, 1983; Rombough, 1987). Investigators cannot be sure that measured rates of oxygen consumption are representative of those in nature. Fry (1957) showed that simply placing fish in a respirometer significantly elevated their metabolic rate for a considerable length of time. However, if energy budgets can be shown to balance when measured rates of oxygen are substituted for R in Eq. (4), the investigator can be reasonably assured that measured rates are in fact representative of those in the natural environment (or at least in the laboratory rearing system).

Direct calorimetry has been used to estimate metabolic rates of fish embryos and larvae on only a few occasions (Smith, 1947; Gnaiger, 1979; Gnaiger *et al.*, 1981). The technique measures total heat production, but this can be converted to oxygen consumption using oxycaloric equivalents and assuming aerobic metabolism. The technique has the advantage of allowing the anaerobic contribution to total metabolism to be estimated if oxygen uptake is measured simultaneously (Gnaiger, 1983a). Descriptions of modern continuous-flow microcalorimeters are provided by Gnaiger (1983a), Knudsen *et al.* (1983), and Pamatmat (1983). According to Gnaiger (1983a), the sensitivity of such calorimeters is in the order of 2 μW, equivalent to about 0.35 μl O_2 h^{-1} assuming aerobic metabolism. This would appear to be adequate for measuring heat production of individual late embryos and larvae. Simultaneous direct and indirect calorimetry holds much promise for future investigations, since measurement of oxygen uptake alone fails to reflect total metabolic activity, particularly during periods of activity or hypoxia. Both these areas are of current interest.

3. OXYGEN UPTAKE DURING EARLY DEVELOPMENT

The early literature concerning the respiratory metabolism of fish embryos and larvae has been reviewed by Needham (1931, 1942), Smith (1957, 1958), Devillers (1965), and Blaxter (1969). More recent but less comprehensive reviews are provided by Kamler (1976) and Hempel (1979). According to Needham (1931), the measurement of respiratory rates during early development began in 1896 with the publication of a monograph by Bataillon. Since then, hundreds of studies have been conducted, yet our understanding of many of the more basic aspects of aerobic metabolism in fish embryos and larvae is still rudimentary. For instance, controversy persists as to the manner in which metabolic rates vary during development and the relationship between metabolic rate and body size. Activity is known to have a profound effect on the rate of oxygen consumption, particularly in larvae, yet only recently have there been attempts to quantify the relationship. Fish vary considerably in their spawning habitats. Temperature and dissolved oxygen are probably the two most important environmental variables, yet their effects on metabolism during early life have been documented in only a few species. A host of other factors, such as carbon dioxide concentration, pH, salinity, light level, group size, and food intake, may also significantly affect metabolic rates, but our knowledge in these areas is extremely limited. The aim of this section of the review is to summarize what is known about the effects of the more important biotic and abiotic factors on the rate of aerobic metabolism of fish embryos and larvae.

Before beginning this examination, it would be useful to make a few general comments. Literature reports of oxygen uptake during early life are most often expressed on an individual basis. This is perfectly adequate for some purposes, such as illustrating general trends or calculating the total amount of oxygen consumed during specific phases of development. It is not particularly appropriate, though, if the aim is to compare species, different studies with the same species, or even different treatments within the same study, because of the overwhelming effect body mass has on oxygen consumption. The effect of body size can be largely eliminated by expressing oxygen uptake on a mass-specific basis ($\dot{V}O_2/M$; also termed metabolic intensity). Unfortunately, relatively few investigators have done so. Even in those cases in which metabolic rate is purported to be expressed on a mass-specific basis, care must be taken in interpreting results. Many investigators have used total mass (i.e., mass of the embryo plus yolk or in some cases mass of the embryo, yolk, egg capsule, and perivitelline fluid) instead of just the mass of formed tissue to calculate metabolic intensity. This results in considerable underestimation of metabolic intensities early in development when the ratio of nonrespiring yolk to metabolically active tissue is large. Care also should be taken to distinguish between values calculated on the basis of wet and dry masses. Another problem in interpreting the literature is uncertainty as to the level of metabolism measured. Rates of oxygen uptake can vary 3- to 10-fold during early life, depending on the level of activity. Activity levels are rarely given in the literature.

B. Biotic Factors

1. STAGE OF DEVELOPMENT

a. Metabolic Rate. The most important factor affecting metabolic rate is the mass of actively metabolizing tissue, which under given conditions is a direct reflection of the stage of development. Rates of oxygen consumption ($\dot{V}O_2$) shortly after fertilization are low, in the range of 3.7 ng h^{-1} (*Gadus morhua*; Davenport *et al.*, 1979) to 70 ng h^{-1} (*S. gairdneri*; Czihak *et al.*, 1979), depending on species and water temperature. These values increase on average about 20-fold as a result of tissue growth during the period from fertilization to hatch, although the literature suggests a great deal of variation among species. Kaushik *et al.* (1982) indicated that there was only about a 3-fold increase in $\dot{V}O_2$ during embryonic development of carp (*Cyprinus carpio*), while Lukina (1973) reported a 50-fold increase for chum salmon (*O. keta*). Relative increases in $\dot{V}O_2$ tend to be greater for larger eggs and at higher temperatures. Dabrowski *et al.* (1984) indicated that oxygen consumption at hatch can be predicted if initial egg mass is known using the equation

$$\dot{V}O_2 = 0.1334M^{0.6634}T^{0.7143} \tag{5}$$

in which $\dot{V}O_2$ is μl O_2 h^{-1}, M is mg dry mass egg^{-1}, and T is °C. This equation was based on data for 11 species. It is open to question whether this relationship can be applied generally, but it is interesting that $\dot{V}O_2$ at hatch was not directly proportional to egg mass (i.e., metabolic mass exponent, $b < 1.0$). Dabrowski *et al.* (1984) attributed this to a higher proportion of metabolically inactive yolk in larger eggs.

The rate of oxygen consumption continues to increase after hatch—for fed larvae in a roughly exponential fashion (Laurence, 1973; Kamler, 1976). There are few reports of relative increases during the larval period, but they would appear to be of roughly the same order of magnitude as during the embryonic period. DeSilva and Tytler (1973) indicated that the rate of oxygen uptake of plaice and herring at metamorphosis was about 12.5 and 35 times, respectively, the rate of hatch. In many studies the larvae are not fed. If this is the case, larval $\dot{V}O_2$ typically increases about 3-fold, although again there is considerable species variation, before endogenous food supplies become limiting and metabolic rates decline. For unfed larvae, a parabolic model rather than an exponential model usually provides the best fit (up to maximum $\dot{V}O_2$) for data relating $\dot{V}O_2$ and developmental period (Hayes *et al.*, 1951; Rombough, 1987).

The general trend of increasing metabolic rates is well established, but some controversy remains as to whether there are significant deviations from this trend associated with specific developmental events. The events that have been considered most likely to sharply alter the rate at which oxygen is taken up are fertilization, gastrulation, the formation of the embryonic circulation, hatching, and the transition from cutaneous to branchial gas exchange. Early investigators suggested that there was a sharp increase in $\dot{V}O_2$ associated with fertilization (Boyd, 1928; Needham, 1931). More recently studies indicate that this is not the case and that $\dot{V}O_2$ actually increases rather gradually after fertilization (Nakano, 1953; Hishida and Nakano, 1954; Kamler, 1972, 1976; Czihak *et al.*, 1979). Trifonova (1937) reported that there were large fluctuations in $\dot{V}O_2$ during gastrulation and neurulation in the perch (*Perca fluviatus*). The reliability of these observations was later disputed (Hayes *et al.*, 1951; Winnicki, 1968). However, several more recent studies have also shown fluctuations in $\dot{V}O_2$ during this period, although not of the magnitude reported by Trifonova (Stelzer *et al.*, 1971; Kamler, 1972, 1976; Hamor and Garside, 1976). These studies have also tended to support early observations (Hyman, 1921; Amberson and Armstrong, 1933) that $\dot{V}O_2$ increases sharply as soon as the embryonic circulation is established. Many studies have indicated a leveling off or even a decrease in $\dot{V}O_2$ shortly before hatch (Amberson and Armstrong, 1933; Hayes *et al.*, 1951; Hamdorf, 1961; Stelzer *et al.*, 1971; Braum, 1973; DiMichele and Powers, 1984a,b). This is usually attributed to the capsule limiting the rate at which oxygen can be delivered to the embryo. Conversely, sharp increases in metabolic rate immediately after hatch (Eldridge *et al.*, 1977; Davenport and Lønning, 1980; Cetta and Capuzzo, 1982; Davenport, 1983; Solberg and Tilseth, 1984) have been attributed to the fact that the capsule no longer restricts oxygen supply or activity. The magnitude of the increases in $\dot{V}O_2$ associated with removal of the capsule is highly variable. Eldridge *et al.* (1977) reported a roughly 10-fold increase in Pacific herring (*Clupea harengus pallasi*). Davenport and Lønning (1980) observed only about a 40–60% increase in cod. They suggested that part of the reason for the relatively small increase was that the egg capsule of cod is comparatively thin ($\simeq 7\ \mu$m) and thus is less of a barrier to diffusion than the capsules of other species. The capsule normally does not appear to restrict oxygen consumption to all in some species. Robertson's (1974) data indicate little change in $\dot{V}O_2$ of southern pigfish (*Congiopodus leucopaecilus*) on hatch. Kamler (1976) similarly failed to observe an appreciable change in $\dot{V}O_2$ on hatch in carp (*Cyprinus carpio*). Kaushik *et al.* (1982) and Dabrowski *et al.* (1984) actually reported a small decline in $\dot{V}O_2$ immediately after hatch in carp and whitefish (*Coregonus lavaretus*), respectively. During larval development, the transition from cutaneous to branchial gas exchange has been linked to both an increase (Lukina, 1973) and a decrease (Kamler, 1972, 1976) in $\dot{V}O_2$.

b. Metabolic Intensity. Attempts to link changes in $\dot{V}O_2$ to particular developmental events have, on the whole, proved inconclusive because of the overwhelming effect tissue mass has on oxygen consumption. As pointed out previously, changing tissue mass can be largely compensated for by expressing oxygen consumption on a mass-specific basis. The few studies in which this has been done indicate a more or less consistent pattern during embryonic development. It appears that values increase following fertilization to reach a maxima sometime prior to blastopore closure (Trifonova, 1937; Smith,

1957, 1958) (see Fig. 8). Trifonova (1937) indicated that there were two peaks during early development: one during cleavage and one during epiboly. Most other researchers have reported only a single peak during epiboly, but this may be simply a reflection of insufficient measurements during early development. Metabolic intensity declines following blastopore closure, to reach a minimum about midway through organogenesis (Hayes *et al.*, 1951; Smith, 1957, 1958; Alderdice *et al.*, 1958; Gruber and Wieser, 1983). Thereafter there is a gradual increase to hatch. It is interesting that metabolic intensity continues to increase near hatch while $\dot{V}O_2$ values are either stable or declining. This would suggest that oxygen supplies are maintained to

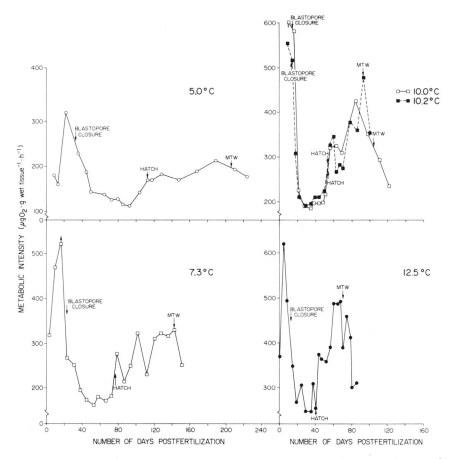

Fig. 8. Metabolic intensities of chinook (*Oncorhynchus tshawytscha*) embryos and larvae incubated at constant temperatures.

existing tissues and that embryos compensate for the fact that the capsule limits oxygen uptake by reducing growth. Smith (1957, 1958) demonstrated that in both brown trout and Atlantic salmon metabolic intensity was directly correlated with specific growth rate.

There is typically a sharp increase in metabolic intensity on hatching. As was the case for $\dot{V}O_2$, this increase can be attributed to removal of the capsule and perivitelline fluid as barriers to diffusion and to enhanced activity (Holliday *et al.*, 1964; Davenport and Lønning, 1980; Gruber and Wieser, 1983; Davenport, 1983). Patterns after hatch are variable. Wieser and Forstner (1986) reported that $\dot{V}O_2/M$ (routine) began to decline exponentially almost immediately in three species of cyprinids (*Rutilis rutilis, Scardinius erythrophthalmus, Leuciscus cephalus*). DeSilva *et al.* (1986) noted a similar decline in the nile tilapia (*Oreochromis niloticus*). A more common pattern is for specific oxygen consumption to continue to increase until about midway through the period of endogenous feeding. This pattern has been reported for a number of salmonids (Smith, 1957, 1958; Gruber and Wieser, 1983; Wieser and Forstner, 1986; Rombough, 1987), sardine (Lasker and Theilacker, 1962), herring (Holliday *et al.*, 1964), cod (Davenport and Lønning, 1980), and lumpfish (Davenport, 1983). In most of these studies the larvae were not fed, so that $\dot{V}O_2/M$ declined rapidly once endogenous food supplies became limiting. For fed larvae, specific oxygen consumption may continue to rise until metamorphosis (Forstner *et al.*, 1983), remain relatively stable (Holeton, 1973), or decline gradually (DeSilva and Tytler, 1973), depending on the species.

2. Mass Relationships

Some investigators, rather than reporting mass-specific rates of oxygen consumption, have chosen to express the relationship between $\dot{V}O_2$ and body mass (M) in terms of the allometric power function

$$\dot{V}O_2 = aM^b \qquad (6)$$

where a and b are constants. Such allometric relationships have long interested physiologists, although the underlying biological principles remain rather obscure. Expressing oxygen consumption in this fashion has the disadvantage that potentially important fluctuations in $\dot{V}O_2$ tend to be obscured by the overall trend. Nonetheless, the relationship is widely used, particularly for larvae.

The metabolic mass exponents [b, Eq. (6)] calculated for larvae are often compared to Winberg's (1956) generalized value of $b = 0.80$.

However, Winberg's value was based primarily on studies involving juvenile and adult fish, and there is no reason to assume that the same exponent applies to earlier life stages. Indeed, ontogenic variations in the value of the metabolic mass exponent are widespread throughout the animal kingdom (Zeuthen, 1950, 1970; Adolph, 1983; Wieser, 1984). In virtually all cases, values decrease as development proceeds. This generalization appears to hold for many fish species. Kamler (1976) recognized four periods during the ontogeny of carp on the basis of different metabolic mass exponents: embryos, prefeeding larvae, feeding larvae, and postlarvae. Values were not calculated for embryos and prefeeding larvae, but graphic presentation of the data indicates that the mass exponents were significantly greater than 1.0. The mass exponent of feeding larvae was approximately unity ($b = 0.97$) while that of postlarvae was typical of juvenile and adult fish ($b = 0.80$).

High mass exponents during early life appear to be widespread, although not universal, among teleosts. Table I lists 16 species for which the larval mass exponent approached or exceeded unity. In addition, mass exponents close to unity can be inferred for larvae of brown trout (Gray, 1926; Wood, 1932), Pacific sardine (*Sardinops caerule;* Lasker and Theilacker, 1962), cod (Davenport and Lønning, 1980), Pacific mackerel (*Scomber japonicus;* Hunter and Kimbrell, 1980), and largemouth bass (*Micropterus salmoides;* Laurence, 1969) from the way in which metabolic intensity varies during development or from graphs showing the relationship between $\dot{V}O_2$ and M. Many investigators, however, have reported metabolic mass exponents for larvae that are not significantly different from those expected for juvenile and adult fish (Table I). Not all such values, though, can be taken as accurately representing the relationship between $\dot{V}O_2$ and M during much of early life. In most cases no attempt was made to differentiate between data collected for prefeeding and feeding larvae. Cetta and Capuzzo (1982) point out that this practice tends to mask any stage-specific differences in the value of the mass exponent. It will probably turn out that in many cases larval mass exponents are actually considerably higher prior to exogenous feeding than literature reports would suggest. For example, careful examination of data presented by Laurence (1978) indicates that $\dot{V}O_2$ tended to increase more rapidly with tissue mass for very young cod and haddock larvae than it did for older larvae. In the one case in which calculations were made using only data for small larvae (haddock, 4°C), it was found that the mass exponent was significantly greater than unity ($b = 1.267$). In contrast, when the data for all larval stages were combined, mass

Table I
Metabolic Rate–Mass Relationships of Fish Larvae[a]

Species	Temp. (°C)	Stage	Size range	Allometric relationship ($\dot{V}O_2 = aM^b$)				Comments	Reference[c]
				a	b	$\dot{V}O_2$ units	M units		
Cyprinus carpio	20	L	< 674 mg	0.64	$b \geq 1.0$ 0.97	μl h^{-1}	mg wet	Slope significantly different for small and large larvae	1
		F	1.2–4.6 g	1.27	0.80				1
Cyprinus carpio	20	L	<1 g	0.60	0.95	μl h^{-1}	mg wet		2
Cyprinus carpio	20	L	<1 g	0.60	0.98	μl h^{-1}	mg wet		3
			>1 g	1.27	0.80				3
Cyprinus carpio[b]	23	L	2–70 mg	1.56	−0.054	mg g^{-1} h^{-1}	mg wet	$\dot{V}O_2/M$ recalculated	4
Abramis brama	20	L	1.6–40 mg	0.45	0.93	ml h^{-1}	g wet		5
Anabas testudineus	28	L	5–50 mg	1.00	0.95	mg h^{-1}	g wet	Recalculated	6
Channa punctatus	28	L	1–10 mg	1.00	0.91	mg h^{-1}	g wet	Recalculated	7
Coregonus sp.[b]	10	L	6–65 mg	14.8	0.12	μmol h^{-1} g^{-1}	mg wet	$\dot{V}O_2/M$, to metamorphosis	8
Acipenser baeri	20	L	11–28 mg	2.4	1.31	mg h^{-1}	g wet	Prolarvae	9
		F	11–390 mg	0.35	0.85			Feeding larvae	
Salvelinus alpinus	2	L	50–124 mg	2.45	1.09	mg h^{-1}	g wet	Yolk-sac larvae	10
Salmo gairdneri	4	L–F	80 mg–7 g	3.1	0.96	μmol h^{-1}	g wet	r$\dot{V}O_2$ a$\dot{V}O_2$	11
				11.0	1.11				11
Salmo gairdneri	12	L–F	80 mg–7 g	6.8	0.93	μmol h^{-1}	g wet	r$\dot{V}O_2$ a$\dot{V}O_2$	11
				19.9	1.14				
Clupea harengus	8	L			1.1	μl h^{-1}	mg dry	Hatch to MLDW	12
Clupea harengus	8	L	0.1–0.7 mg dry	1.19	0.74	μl h^{-1}	mg dry		13

n	Stage	Mass range	a	b			Notes	Ref
13	L	0.1–0.7 mg dry	3.63	1.33				
18	L	0.1–0.7 mg dry	3.33	0.87				13
8	L	0.7–1.2 mg dry	1.21	0.78	μl h^{-1}	mg dry		
13	L	0.07–1.2 mg dry	2.20	0.96				
18	L	0.07–1.2 mg dry	3.29	0.74				
7	L	7–10 μg protein (dry)	0.006	1.03	μl h^{-1}	μg protein (dry)		14
4	L	70–200 mg dry	0.0053	1.27	μl h^{-1}	μg dry	Small larvae only	15
7	L	50–1000 mg dry	0.071	0.68				
9	L	50–1000 mg dry	0.179	0.55				
26	L	8.9–424 μg dry	0.0077	0.98	μl h^{-1}	μg dry	Feeding larvae	16
26	L	14.3–248 μg dry	0.014	0.94	μl h^{-1}	μg dry	Feeding larvae	16
20	L	1–50 mg	0.33	1.05	μl h^{-1}	mg wet	Recalculated	17
19	L	28–1000 μg dry	7.09	1.00	μg h^{-1}	mg wet		18
24	L	28–1000μg dry	6.47	1.03		mg dry		
20	L	0.8–79 mg dry	0.31	$b \leq 0.8$ 0.82	ml h^{-1}	g wet		5

Species (left column, top to bottom): *Pleuronectes platessa*, *Pseudopleuronectes americanus*, *Melanogrammus aeglefinus*, *Anchoa mitchilli*, *Achirus lineatus*, *Hypophthalmichthys molitrix*, *Sparus aurata*, *Stizostedion lucioperca*

(continued)

Table I (*Continued*)

Species	Temp. (°C)	Stage	Size range	a	b	$\dot{V}O_2$ units	M units	Comments	Reference[c]
				under: Allometric relationship ($\dot{V}O_2 = aM^b$)					
Perca fluviatilis	20	L	1.5–32 mg	0.29	0.78	ml h⁻¹	g wet		5
Rutilis rutilis	20	L	1.6–62 mg	0.29	0.82	ml h⁻¹	g wet		5
Heteropneustes fossilis	28	L	2–20 mg	1.00	0.88	mg h⁻¹	g wet		19
Morone saxatalis	18	E–L	50–1500 µg	0.028	0.72	µl h⁻¹	µg dry	Recalculated	20
Lebeo calbasu	28	L	100–300 mg	0.50	0.84	mg h⁻¹	g wet		21
Oreochromis niloticus	30	F	1–10 mg?	9.68	0.42	µl h⁻¹	mg dry		22
Clupea harengus	10	L	0.1–10 mg	1.88	0.82	µl h⁻¹	mg dry	Recalculated	23
Pleuronectes platessa	10	L	0.5–10 mg	1.67	0.65	µl h⁻¹	mg dry		23
Pseudopleuronectes americanus	7	F	8–150 µg protein	0.016	0.78	µl h⁻¹	µg protein dry		14
Gadus morhua	4	L	50–1000 µg	0.018	0.71	µl h⁻¹	µg dry		15
	7	L	50–1000 µg	0.017	0.78	µl h⁻¹	µg dry		15
	10	L	50–1000 µg	0.054	0.69				15
Archosargus rhomboidalis	26	L	18–66 µg	0.018	0.84	µl h⁻¹	µg dry		16

[a] Abbreviations: L, larvae; F, fry; MLDW, maximum larval dry weight.

[b] Regression describes relationship between metabolic intensity ($\dot{V}O_2/M$) and tissue mass.

[c] References: (1) Kamler, 1976; (2) Winberg and Khartova, 1953 (cited in Kamler, 1976) (3) Kamler, 1972; (4) Kaushik and Dabrowski, 1983; (5) Kudrinskaya, 1969; (6) Mishra and Singh, 1979; (7) Singh *et al.*, 1982; (8) Forstner *et al.*, 1983; (9) Khakimullin, 1985; (10) Holeton, 1973; (11) Wieser, 1985; (12) Blaxter and Hempel, 1966; (13) Almatar, 1984; (14) Cetta and Capuzzo, 1982; (15) Laurence, 1978; (16) Houde and Schekter, 1983; (17) Mukhamedova, 1977; (18) Quantz and Tandler, 1982; (19) Sheel and Singh, 1981; (20) Eldridge *et al.*, 1982; (21) Durve and Sharma, 1977; (22) DeSilva *et al.*, 1986; (23) De Silva and Tytler, 1973.

exponents were found to be typical of those seen in much older fish ($b = 0.55$–0.78). It should be recognized, however, that mass exponents are not always high, even for very young larvae. For example, a careful study by DeSilva and Tytler (1973) showed the metabolic mass exponents of both herring and plaice larvae to be considerably less than unity (0.82 and 0.65, respectively). The reason why some species have such relatively low mass exponents as larvae has not been established.

Several reasons have been advanced to explain high metabolic mass exponents during early development. Kamler (1976) suggested that the transition from the high larval value to the lower juvenile value in carp was the result of morphological changes affecting gas exchange. This idea was expanded upon by Pauly (1981). Pauly (1981) proposed that the underlying factor limiting metabolic rate and hence growth in fish was the availability of oxygen, rather than food or some intrinsic control mechanism. He reasoned that if oxygen was not limiting, metabolic rate should be directly proportional to body mass. To support his contention that oxygen is not limiting during early life, he noted that gill surface area of larval herring and plaice increased at a much faster rate than body mass (gill area mass exponents of 3.36 and 1.59, respectively; De Silva, 1974). This was rather an unfortunate choice of species, since in both herring and plaice the larval metabolic mass exponent is significantly less than unity (0.82 and 0.65, respectively; De Silva and Tytler, 1973). This does not necessarily disprove Pauly's basic hypothesis (i.e., that the ability to supply oxygen to tissues limits metabolic rate) but indicates that more sophisticated analysis of total exchange capacity, perhaps similar to that done by Ultsch (1973) for lungless salamanders, is required to adequately test the hypothesis. As discussed earlier, cutaneous gas exchange is extremely important in larvae and must be included in any analysis of diffusing capacity. In both herring and plaice cutaneous surface area expands at a slower rate than body mass (mass exponents of 0.58 and 0.50, respectively; De Silva, 1974). Thus, the total diffusing capacity of these larvae, in fact, may be expanding at about the same rate as metabolism. It must be pointed out that not all investigators agree with this hypothesis. In particular, Oikawa and Itazawa (1985) recently presented data for carp that they contend show no direct relationship between respiratory surface area and resting metabolism.

Forstner *et al.* (1983) attributed the high mass exponent of coregonid larvae to a preponderence of red muscle and concommitant high activity of oxidative enzymes. The decline in the value of the mass exponent following metamorphosis was attributed to a major reorganization of metabolism in which glycolytic activity becomes progressively more important (Forstner *et al.*, 1983; Hinterleitner *et al.*, 1987). There are, however, major differences among species in the developmental trajectories of oxidative and glycolytic enzymes (Hinterleitner *et al.*, 1987), and it is difficult to believe that high oxidative enzyme activity per se is sufficient to explain high mass exponents early in life. For example, salmonids generally display high mass exponents up to at least the end of the yolksac stage (Table I, Wieser and Forstner, 1986; Rombough, 1987), yet activity is borne almost entirely by white muscle (Forstner *et al.*, 1983), and levels of oxidative enzymes are very low (Hinterleitner *et al.*, 1987).

Quantz and Tandler (1982) attributed the high metabolic mass exponent of larval gilthead seabream (*Sparus aurata*) to their high feeding rate. They speculated that since the larvae were feeding more or less continuously, they should have a high specific dynamic action (heat increment). Noting that juvenile and adult fish had a mass exponent near unity for active metabolism (Brett and Groves, 1979), they suggested that it was not surprising that mass exponents for larvae were of a similar magnitude.

The allometric equations relating $\dot{V}O_2$ and M frequently have been used to compare metabolic levels of different species or ecological groupings. Konstantinov (1980) points out that such comparisons are only valid if weight exponents are equal. This severely restricts the use of such equations. Metabolic weight exponents are seldom the same for different species. Even in the same species, values for b vary during development and are influenced by factors such as temperature (Laurence, 1978; Konstantinov, 1980; Almatar, 1984) and activity (De Silva and Tytler, 1973; Wieser, 1985).

3. ACTIVITY

As for juveniles and adults (Brett, 1970), activity is the single most important factor influencing the metabolic intensity of larvae (Blaxter, 1969). The relationship between activity and oxygen consumption has been investigated in some detail for juvenile and adult fish (Beamish, 1964, 1978; Brett, 1964, 1970, 1972; Brett and Groves, 1979; Fry, 1971). Yet only recently has there been much interest in the earlier life stages, although there are compelling reasons to suspect that energetic relationships may be significantly different from those of older fish. The body musculature and supporting metabolic machinery develop gradually during the embryonic and larval periods and in many species do not assume typical juvenile patterns until after metamorphosis (Johnston, 1982; Forstner et al., 1983; Batty, 1984; Wieser et al., 1985). In addition, juveniles and larvae are exposed to different hydrodynamic regimes at routine swimming speeds. In larvae, routine activities occur in an intermediate hydrodynamic environment where both resistive and inertial forces are important (Webb and Weihs, 1986). As larvae grow or during burst swimming they move into the adult hydrodynamic regime, where inertial forces dominate. In many species this transition is accompanied by a change in swimming style from continuous movement of body and tail to a beat and glide pattern (Hunter, 1972, 1981; Weihs, 1980; Batty, 1984). To further confound matters, there are species-specific differences in development of muscle types and biochemical pathways (Wieser et al., 1985; Hinterleitner et al., 1987).

The study of the physiological energetics of the early life stages has been hampered by the lack of clear definitions for the various levels of metabolism associated with activity. In juvenile and adult fish, energy expenditure is normally described in terms of standard ($s\dot{V}O_2$), routine ($r\dot{V}O_2$) and active ($a\dot{V}O_2$) metabolism (see Brett and Groves, 1979, for definitions). These terms have been applied to the early stages, but it should be noted that conditions are somewhat different during early life. In older fish standard metabolism refers to the postabsorptive state, a condition that is obviously not met during endogenous feeding. In addition, the standard metabolic rate of embryos and larvae includes a sizeable growth component so that, unlike in older fish, metabolic rates can be depressed considerably below the so-called standard level without affecting survival. What most investigators consider standard metabolism during early life is actually simply the metabolic rate under conditions of minimal neuromuscular activity. The vast majority of investigators have attempted to measure the embryonic and larval equivalent of routine metabolism. This is a rather nebulous term, since, as Wieser (1985) points out, it can vary at least twofold in response to a variety of intrinsic and extrinsic factors. For embryos and larvae, routine metabolism probably can be defined best as the average rate of aerobic metabolism under normal rearing conditions. Active metabolism in older fish usually refers to sustained activity. The young of many species, however, are not capable of sustained activity. As a result, most estimates of active metabolism during early life are based on burst activity. It is a moot point whether such results are comparable with values for juveniles and adults.

Absolute aerobic scope, defined as the difference between $a\dot{V}O_2$ and $s\dot{V}O_2$, is of particular interest to physiologists because it represents the amount of energy available to a fish to cover the cost of various biological activities. Unfortunately, it has proved difficult to estimate absolute scope during early life, although a new type of metabolic chamber recently described by Dabrowski (1986) may make such determinations easier in the future. It has proved somewhat easier to estimate selected portions of absolute scope, in particular the difference between $a\dot{V}O_2$ and $r\dot{V}O_2$ (termed the relative scope; Wieser, 1985) and the difference between $r\dot{V}O_2$ and $s\dot{V}O_2$ (termed the routine scope; Beamish, 1964). It thus is possible to obtain rough estimates of absolute scope indirectly by adding values reported for routine and relative scopes.

Some investigators have used the ratios of metabolic rates at the various activity levels, instead of their differences, as a way of expressing scope. The ratios $a\dot{V}O_2/s\dot{V}O_2$, $a\dot{V}O_2/r\dot{V}O_2$, and $r\dot{V}O_2/s\dot{V}O_2$ are referred to, respectively, as absolute factorial scope, relative factorial scope, and routine factorial scope. This is convenient for comparative purposes but does not tell the whole tale, since it gives no indication of absolute costs, which are of profound ecological importance. As will be discussed later, scopes and factorial scopes can change in response to various intrinsic and extrinsic factors. It should be recognized, however, that because of the manner in which they are calculated, they may not always change in the same direction. For instance, if $a\dot{V}O_2$ and $s\dot{V}O_2$ decline in parallel, absolute scope will remain constant but absolute factorial scope will increase.

Absolute aerobic scope ($a\dot{V}O_2 - s\dot{V}O_2$) is routinely estimated for juvenile and adult fish by forcing them to swim against a current at progressively faster speeds. The oxygen uptake at maximum sustained swimming speed is taken to represent $a\dot{V}O_2$, while extrapolation of the power–performance curve back to zero swimming speed yields $s\dot{V}O_2$. Ivlev (1960a) appears to have been the only investigator to have applied this technique to very young fish. He reported that young Atlantic salmon (400 mg wet mass) were capable of sustaining a rate of oxygen uptake 22.5 times their standard rate. This value is sometimes cited as the degree of metabolic expansibility that can be expected for fish larvae (e.g., Blaxter, 1969). However, more recent evidence suggests that this is a gross overestimation and that absolute factorial scopes actually range from about 2.5 to 10, depending on species, fish size, and temperature (Table II).

Dabrowski (1986) recently used a modification of the standard power–performance procedure to estimate aerobic scope for salmon alevins and coregonid fry. The fish were induced to swim at progressively faster speeds by making use of their optomotor reaction to a moving background. Swimming speeds were monitored and correlated with rates of oxygen uptake. For young Atlantic salmon, the maximum $\dot{V}O_2$ recorded was about 3.3 times the value estimated for zero activity. The absolute factorial scope of *Coregonus schinizi* fry was somewhat lower, about 2.5.

A less common way of estimating aerobic scope in juvenile and adult fish is to take the difference between the maximum and minimum $\dot{V}O_2$ recorded while fish are held in a respirometer for an extended period. The value obtained is correctly termed the aerobic scope for spontaneous activity but, at least in some species, it appears to approximate absolute scope (Ultsch *et al.*, 1980). Several investigators have applied this technique to fish larvae. Holliday *et al.* (1964) indicated that there was approximately a 10-fold difference between the minimum and maximum $\dot{V}O_2$ of herring larvae. For sardine larvae the maximum difference in $\dot{V}O_2$ recorded was about 3.5-fold (Lasker and Theilacker, 1962), while in winter founder (*Pseudopleuronecter americanus*) the maximum difference was only about 3.0-fold (Cetta and Capuzzo, 1982). It is not clear why the apparent scope of herring should be so much greater than that of the other species, but it is interesting that larvae of Pacific herring (*Clupea pallasi*) also appear capable of increasing their metabolic rate about 10-fold (Eldridge *et al.*, 1977). In the case of the Pacific herring, it was stress caused by the shaking of the respirometer, rather than spontaneous activity, that led to elevated metabolic rates.

As mentioned previously, it has proved difficult to induce larvae to swim in a respirometer and, as a consequence, other techniques have had to be developed to estimate $s\dot{V}O_2$ and $a\dot{V}O_2$. The most common method for estimating $s\dot{V}O_2$ has been to anesthetize the larvae (Holliday *et al.*, 1964; DeSilva and Tytler, 1973; Davenport and Lønning, 1980; DeSilva *et al.*, 1986). This technique indicates routine factorial scopes ($r\dot{V}O_2/s\dot{V}O_2$) in the range of 1.4–3.3, depending on species and stage of development (Table II). Another method has been to assume that the capsule severely restricts activity just before hatch (Davenport and Lønning, 1980; Davenport, 1983). Davenport and Lønning (1980) demonstrated that, at least in cod, the metabolic rate of embryos just before hatch was not significantly different from that of anaesthetized larvae shortly after hatch. Assuming that the metabolic rate of unanesthetized larvae shortly after hatch is representative of $r\dot{V}O_2$, this method gives routine factorial scopes for cod (Davenport and Lønning, 1980) and lumpfish (Davenport, 1983) of about 2.0.

The evidence is rather sketchy, but it appears that the difference between $r\dot{V}O_2$ and $s\dot{V}O_2$ (routine scope) tends to decrease as larvae mature (DeSilva and Tytler, 1973; Davenport and Lønning, 1980; De-

Table II

Scope for Activity

Species	Stage	Temp. (°C)	Scope	Technique	Reference
			Routine factorial scope ($r\dot{V}O_2/s\dot{V}O_2$)		
Clupea harengus	E	8	2.1	$r\dot{V}O_2$/anesthetized $\dot{V}O_2$	Holliday et al. (1964)
	L	8	1.4	$r\dot{V}O_2$/anesthetized $\dot{V}O_2$	Holliday et al. (1964)
	L	8	2.5	Posthatch $\dot{V}O_2$/prehatch $\dot{V}O_2$	Holliday et al. (1964)
	L	10	1.9	$r\dot{V}O_2$/anesthetized $\dot{V}O_2$	De Silva and Tytler (1973)
Pleuronectes plattesa	L	10	1.6	$r\dot{V}O_2$/anesthetized $\dot{V}O_2$	De Silva and Tytler (1973)
Gadus morhua	L	5	2.2	$r\dot{V}O_2$/anesthetized $\dot{V}O_2$	Davenport and Lonning (1980)
	L	5	2.0	Posthatch $\dot{V}O_2$/prehatch $\dot{V}O_2$	Davenport and Lonning (1980)
Gadus morhua	L	5	1.6	$\dot{V}O_2$ in light/$\dot{V}O_2$ in dark	Solberg and Tilseth (1984)
Cyclopterus lumpus	L	5	2.0	Posthatch $\dot{V}O_2$/prehatch $\dot{V}O_2$	Davenport (1983)
Oreochromis niloticus	L	30	3.5	$r\dot{V}O_2$/anesthetized $\dot{V}O_2$	De Silva et al. (1986)
Salmo salar	L	20	1.5	Extrapolated from power–performance curve	Ivlev (1960a)
Salvelinus alpinus	L	4	2.6	Posthatch $\dot{V}O_2$/Prehatch $\dot{V}O_2$	Gruber and Wieser (1983)

			Relative factorial scope ($a\dot{V}O_2/r\dot{V}O_2$)		
Alburnus alburnus	L	20	9.7[a]	Power–performance relationship	Ivlev (1960b)
Salmo gairdneri	L	4	2.7	Forced bursts	Wieser *et al.* (1985)
	L	12	1.9	Forced bursts	Wieser *et al.* (1985)
	L	20	~1.8	Forced bursts	Wieser *et al.* (1985)
Cyprinids (3 species)	L	12	2.4	Forced bursts	Wieser and Forstner (1986)
	L	20	2.1	Forced bursts	Wieser and Forstner (1986)
			Spontaneous factorial scope		
Sardinops caerulea	E	14	2.8	Observed max. $\dot{V}O_2$/min. $\dot{V}O_2$	Lasker and Theilacker (1962)
	L	14	3.5	Observed max. $\dot{V}O_2$/min. $\dot{V}O_2$	Lasker and Theilacker (1962)
Pseudopleuronectes americanus	L	7	3.0	Observed max. $\dot{V}O_2$/min. $\dot{V}O_2$	Cetta and Capuzzo (1982)
Clupea harengus	L	8	10.4	Observed max. $\dot{V}O_2$/min. $\dot{V}O_2$	Holliday *et al.* (1964)
Clupea harengus pallasi	L	12.5	10.0	Posthatch max. $\dot{V}O_2$/prehatch $\dot{V}O_2$	Eldridge *et al.* (1977)
Salmo salar	E	10.0	3.0	Observed max. $\dot{V}O_2$/min. $\dot{V}O_2$	Hayes *et al.* (1951)
Salvelinus alpinus	L	8	5.0	Posthatch max. $\dot{V}O_2$/prehatch $\dot{V}O_2$	Gruber and Wieser (1983)
			Absolute factorial scope ($a\dot{V}O_2/s\dot{V}O_2$)		
Salmo salar	L	20	22.5[a]	Power–performance relationship	Ivlev (1960a).
Salmo salar	L	22	3.3	Power–performance relationship	Dabrowski (1986)
Coregonus sp.	F	14	2.5	Power–performance relationship	Dabrowski (1986)

[a] Unrealistically high, see text.

103

Silva *et al.*, 1986). For example, the routine scope of larval herring and plaice at metamorphosis, measured as the difference between un-anaesthetized and anaesthetized $\dot{V}O_2$, was only about 25% of that at hatch (DeSilva and Tytler, 1973). The decrease in routine scope reflected a more rapid decline in $r\dot{V}O_2$ than in $s\dot{V}O_2$.

Wieser and his co-workers (Wieser, 1985; Wieser *et al.*, 1985; Wieser and Forstner, 1986) have attempted to obtain estimates of active metabolism by forcing larvae (using electrical or mechanical stimulation) to swim at burst speeds for short periods (30–60 s). The main driving force behind such activity was shown to be anaerobic (Wieser *et al.*, 1985), but it was felt that the maximum rate of oxygen uptake during activity or the first few minutes of recovery (the response time of the system was not fast enough to say precisely when $\dot{V}O_2$ was maximal) approached $a\dot{V}O_2$. Using the average rate of oxygen uptake prior to activity for $r\dot{V}O_2$, this technique gave estimates of 1.9–2.7 and 2.4–2.9 for the relative factorial scope ($a\dot{V}O_2/r\dot{V}O_2$) of young rainbow trout (Wieser *et al.*, 1985) and larval cyprinids (Wieser and Forstner, 1986), respectively. Assuming that $a\dot{V}O_2$ is about twice $s\dot{V}O_2$, absolute factorial scopes would appear to range from about 4 to 6.

In juvenile and adult fish, aerobic scope tends to increase as the fish grows (Brett and Glass, 1973; Wieser, 1985). The pattern is not as obvious for younger fish. The metabolic expansibility of young rainbow trout increased with body mass, more or less as expected (Wieser, 1985; Wieser *et al.*, 1985). For example at 4°C, relative factorial scope increased from about 2.7 for yolk-sac larvae (80–120 mg) to about 5.2 for fry weighing between 3 and 10 g (Wieser *et al.*, 1985). On the other hand, in cyprinids, relative factorial scope was independent of body mass between 1 and 400 mg (Wieser and Forstner, 1986). Wieser and Forstner (1986) suggested that this may reflect the need of very small larvae to avoid the constraints small size normally places on aerobic scope if they are to escape predation.

The influence of temperature on aerobic scope appears to vary depending on the species. As was the case involving the effect of body mass, young rainbow trout follow a pattern similar to that seen in juveniles and adults (Wieser, 1985; Wieser *et al.*, 1985). Routine metabolic rate increases steadily with temperature between 4°C and 20°C (Fig. 9). Up to about 12°C, active metabolism also increases with temperature, but at a faster rate than routine metabolism, so that relative scope increases. At temperatures above 12°C, however, there is a decrease in the rate at which active metabolism expands so that relative scope ($a\dot{V}O_2/r\dot{V}O_2$) remains constant or even declines. The reasons for this decline have not been demonstrated, but it may be that, as in

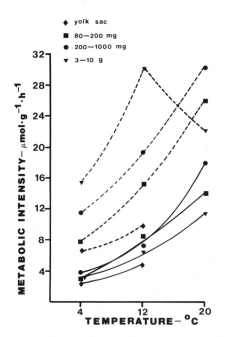

Fig. 9. Temperature dependence of rates of oxygen consumption of four size classes of rainbow trout *Salmo gairdneri*. Solid lines, routine rate; dashed lines, maximum rate during or immediately following forced burst activity. [After Wieser *et al.* (1985).]

adults (Fry, 1971), oxygen limits active metabolism at high ambient temperatures. Relative factorial scope in cyprinids, unlike in rainbow trout, appears to be relatively independent of temperature between 12°C and 24°C (Wieser and Forstner, 1986). This relationship, however, should probably be checked since Wieser and Forstner (1986) used data pooled from three species (*Rutilus rutilus, Leuciscus cephalis, Scardinius erythrophthalmus*), and the apparent temperature independence of aerobic scope in cyprinids may simply reflect compensating differences in temperature optima.

It is usually assumed that $a\dot{V}O_2$ is more or less fixed for a given temperature and size of fish. Thus factors that increase $r\dot{V}O_2$ would be expected to reduce relative scope. Wieser (1985) indicated that this may not be the case. He noted that, within a given size class of rainbow trout, those individuals with a high $r\dot{V}O_2$ also had a high $a\dot{V}O_2$, while individuals with a low $r\dot{V}O_2$ had a low $a\dot{V}O_2$. Wieser (1985) speculated that as yet undefined controlling factors (hormones?) may influence individual metabolism in the same way temperature does.

Thus, an increase in $r\dot{V}O_2$ would result in an increase in $a\dot{V}O_2$ and a concomitant increase in relative scope. This hypothesis appears to be worth testing because if true, fish have a greater flexibility in their metabolic response to stress than previously assumed.

4. ENDOGENOUS RHYTHMS

It is recognized that diurnal variation in rates of oxygen consumption can be a major source of error in energetic analyses of early growth (Houde and Schekter, 1983). Empirical estimates of the magnitude of such variations during early life, though, are scanty. Ryzhkov (1965) reported oxygen uptake by eggs of the Sevan trout (*Salmo ischan*) peaked at about 0700 h and again at about 2100 h. Minimum and maximum values for $\dot{V}O_2$ differed 2- to 3-fold. Holliday *et al.* (1964) present data showing that the $\dot{V}O_2$ of both anesthetized and unanesthetized herring larvae was lower during the "night" than during the "day" even though all measurements were taken under identical indoor lighting conditions. This pattern was confirmed by De Silva and Tytler (1973). Oxygen uptake by both anesthetized and unanesthetized herring larvae declined at "dusk" and then gradually rose to typical daylight levels by "dawn." Again, all tests were conducted under indoor lighting. Minimun night-time values were about 50% of typical daytime values. De Silva *et al.* (1986) recently reported diurnal variation in $\dot{V}O_2$ from larvae and fry of the nile tilapia (*Oreochronis niloticus*). Unlike in herring, $\dot{V}O_2$ peaked twice in the course of a day. For newly hatched larvae, the peaks occurred just after sunset and just before dawn. The $\dot{V}O_2$ varied about 3-fold, with average nighttime rates being somewhat higher than average daytime rates. For older larvae (5 days) and fry (3 weeks), $\dot{V}O_2$ peaked at sunrise and then again at about noon. Rates varied 2- to 3-fold, with average daytime rates being somewhat higher than average nighttime rates. Geffen (1983) examined the effect of various light–dark regimes on oxygen uptake by Atlantic salmon embryos. During a 24-h period, $\dot{V}O_2$ rose to a single peak at about 1200 h when embryos were held in constant darkness. Under a 12:12 light–dark cycle, $\dot{V}O_2$ peaked toward the end of the dark period (1000 h) at about 0600 h. Under a 6:6 light–dark regime (two cycles per day), $\dot{V}O_2$ peaked in the middle of the first dark period, again at about 0600 h. There was no peak during the second dark period. Under all light regimes peaks were rather sharp, with $\dot{V}O_2$ remaining elevated for only 2–6 h. Peak $\dot{V}O_2$ values were from 2 to 4 times the average nonpeak $\dot{V}O_2$.

5. GROUP EFFECT

Metabolic rates of juvenile and adult fish can vary 20–50% depending on the number of individuals in the group (Brett, 1970; Itazawa et al., 1978; Kanda and Itazawa, 1981). The possibility of a similar "group effect" during the early life stages has received considerable attention from Soviet researchers. Most of the earlier studies indicated that metabolic rates were significantly higher for isolated individuals than for groups (Ryzhkov, 1968; Grigor'yeva, 1967; Malyukina and Konchin, 1969; Kudrinskaya, 1969; Korwin Kossakowski et al., 1981). This appeared to hold true for embryos as well as larvae and for schooling and nonschooling species. Reduced oxygen uptake was attributed to reduced motor activity, chemicals released into the water by other fish, or reduction in territoriality. However, Konchin (1971, 1981, 1982) has recently reexamined the "group effect" in several schooling and nonschooling species and concluded that earlier reports of reduced oxygen consumption by individuals in groups were simply the result of oxygen levels in respirometers dropping below the critical level and hence limiting gas exchange. He found that if fish to respirometer volume ratios were kept constant there was no significant difference in oxygen consumption of eggs and larvae of the summer bakhtah (*Salmo ischan*), roach (*Rutilus rutilus*), or pike (*Esox lucius*) as groups or as individuals. The rate of oxygen uptake of isolated individuals was not affected by visual contact with larvae of that particular species. Similarly, "conditioned" water (water in which larvae were previously held) did not affect oxygen uptake if it was aerated.

C. Abiotic Factors

1. TEMPERATURE

Next to stage of development and activity, temperature has the greatest influence on the metabolic rates of embryos and larvae. In spite of this, temperature relationships have been examined in detail in only a few species. In some ways thermal relationships during early life appear to be similar to those in later life. For example, when the logarithms of the routine metabolic rates of embryos and larvae of the various species listed in Table III are plotted against temperature, the resulting curve assumes the convex shape typical of that for juveniles and adults (Krogh's standard curve) (Fig. 10). In other ways, however,

Table III

Routine Metabolic Intensities during Early Life[a]

Species	Stage	Temp. (°C)	Metabolic intensity [µg(g wet wt.)$^{-1}$ h^{-1}]	Technique	Comments	Reference[b]
			Freshwater salmonid			
Salmo salar	L, newly hatched	4.0	164	W		1
	L, newly hatched	14	257	W		1
	E, 19–50 dpf	10.0	205	HG	Mean, n = 11, tissue wt. only	2
	L, 110 mg	9.0	354	W	At yolk absorption	3
Salmo gairdneri	E, latter half	10.0	442	DC	Average	4
	L, 120–200 mg	4.0	96	POS	Tissue wt. only, start free-swimming	5
	L, 120–200 mg	12.0	269	POS		5
	L, 120–200 mg	20.0	448	POS		5
	E, 16.4 mg	10.0	280	W		6
	E–L, 23.2 mg	6.0	200	POS	Mean fertilization to MLWW	7
	E–L, 27.8 mg	9.0	311	POS	Mean	7
	E–L, 24.9 mg	12.0	405	POS	Mean	7
	E–L, 21.6 mg	15.0	548	POS	Mean	7
Salmo trutta	L, 70 mg	10.0	571	W		8
	L, 70 mg	12.0	486	W		8
	L, 70 mg	14.0	514	W		8
	L, 70 mg	16.0	571	W		8
	L, 70 mg	18.0	786	W		8
	E–L	3.0	291	M	Mean, just before hatch to MLWW	9
	E–L	7.0	383	M		9
	E–L	12.0	836	M		9
	L, 45–70 dph	10.0	190	M		10
Salvelinus alpinus	L, newly hatched	8.0	131	POS	DW, 5.1 mg dry wt.	11
	E	4.0	64	POS	DW, tissue wt. only, near hatch	12
	E	8.0	107	POS	DW, tissue wt. only, near hatch	12

Species						Ref.
Oncorhynchus gorbuscha	L, 245 mg	3.6	110	W	At emergence	13
Oncorhynchus keta	L, MLWW	8	929	POS		14
	E, 2.3–29 mg	10	231	W	Mean	15
Oncorhynchus tshawytscha	E–L, 1–500 mg wet	5	182	POS	Mean	16
	E–L, 1–500 mg	7.5	248	POS	Mean	16
	E–L, 1–500 mg	10.0	303	POS	Mean	16
	E–L, 1–500 mg	10.2	339	POS	Mean	16
	E–L, 1–500 mg	12.5	366	POS	Mean	16
Freshwater, nonsalmonid						
Cyprinus carpio	L, 100 mg	20	796	M	Estimated from allometric equation	17
	L, 100 mg	20	782	M	Estimated from allometric equation	18
	L, < 1000 mg	10	350	?	Estimated from allometric equation	19
	L, < 100 mg	20	829	?	Estimated from allometric equation	19
	L, 22–24 dph	20	639	M	Mean, stock differences	20
	L, 5.5–15.8 mg	22	727	M		21
	L, 2.1–11.1	28	1700	M		21
	L, 2.0	23	857	M		22
	L, 2–70 mg	23	1205		Mean	23
Cyprinids (three species combined)	L, 40–400 mg	12	208	POS	Mean	24
Rutilus rutilus	L, 40–400 mg	16	384	POS	Mean	24
	L, 40–400 mg	20	509	POS	Mean	24
	L, 40–400 mg	24	733	POS	Mean	24
	L	21	859	W	Swimbladder filled	25
	L, 2.6–3.6 mg	20	1134	?		26

(continued)

Table III (*Continued*)

Species	Stage	Temp. (°C)	Metabolic intensity [μg(g wet wt.)$^{-1}$ h^{-1}]	Technique	Comments	Reference[b]
Abramis bramis	L, 2.6–2.9	20	876	?		26
Stizostedion lucioperca	L, 2.3–4.6	20	1214	?		26
Perca fluviatis	L, 2.3–3.4	20	1486	?		26
	E, 5–11 dpf	22	1905	W		27
Esox lucius	L	21.5	892	W	Start exogenous feeding	25
Micropterus dolomieui	L, 1–12 dph	20	847	POS	Mean	28
Micropterus salmoides	L, 18–148 g	19	796	W	Mean	29
Morone saxatilis	L, 1.4–2.4 mg	20	1320	DOS	Mean 4–9 dpf	30
Anguilla rostrata	L, 7 dpf	18	1333	M	DW, start feeding	31
	L, 300 mg	23	737	POS	Elver	32
Hyophthalmichthys molitrix	L, 3 mg	20	706	W	Start exogenous feeding	33
Channa punctatus	L, 4.7 mg	28	1846	POS	Bimodal breather	34
Heteopneustes fossilis	L, 5.4 mg	28	1046	POS	Bimodal breather	35
Anabas testudineus	L, < 18.5 mg	28	1255	POS	Bimodal breather	36
Etroplus maculatus	L, ?	27	573	M	DW, prior to free-swimming	37
Lebeo calbasu	L, 100–260 mg	28	680	W		38
Coregonus sp.	L, 10–100 mg	10	640	POS		24
Coregonus lavaretus	L, 8 mg	5	588	W		39

Species	Stage		Value	Type	Notes	Page
Coregonus peled	L, 8 mg	10	650	W		39
	L, 8 mg	15	875	W		39
	L, 4.5 mg	5	667	W		39
	L, 4.5 mg	10	778	W		39
	L, 4.7	15	1894	W		39
Coregonus lavaretus	E, near hatch	12	331	POS	DW	40
Coregonus schinzi	L, 15 dph	12.8	734	POS	DW	40
	E, near hatch	12.5	429	POS	DW	40
Coregonus sp.	L, 15 mg	10	685	POS	At yolk absorption	41
Coregonus schinzi	L, ?	14.5	485	POS		42
Coregonus nasus	L, 6.8 mg	12	1057	W		43
Acipenser baeri	L, 20 mg	20	717	?		44
Oreochromis nilotis	L, feeding	30	1213	M	DW	45
Marine						
Clupea harengus	L, yolk sac	10	499	M	DW	46
	L, first-feeding	8	282	M	DW, 2–10 days post yolk absorption	47
		13	571	M	DW	47
		18	724	M	DW	47
Clupea harengus	L, newly hatched	6	476	M	DW, tissue wt. only	48
	L, newly hatched	8	591	M	DW, tissue wt. only	48
	L, newly hatched	11	800	M	DW, tissue wt. only	48
	L, newly hatched	14	952	M	DW, tissue wt. only	48
	L, sac larvae	10	438	W	DW	49
Clupea harengus pallasi	L, yolk sac	12.5	6343	M	DW, near yolk absorption	50
Gadus morhua	L, yolk absorbed	5	286	POS	DW	51
	L, yolk absorbed	5	278	POS	DW	52
	L, 55 µg dry	5	362	M	DW, at MTW	53
	L, 1 dph	5	381	POS	DW	51
Pleuronectes platessa	L, 5 dph	10	785	M	DW	46
	L, first-feeding	8	265	M	DW	47

(continued)

Table III (*Continued*)

Species	Stage	Temp. (°C)	Metabolic intensity [μg(g wet wt.)$^{-1}$ h^{-1}]	Technique	Comments	Reference[b]
Sardinops caerulea	L, first-feeding	13	436	M	DW	47
	L, first-feeding	18	724	M	DW	47
	L, 70–180 hph	14	476	D	DW, estimated	54
	L	14	253	D	DW, inactive	55
Sardinops sagax	L, 1.8 dph	20	3293	POS	DW(?)	56
Cheilopogon unicolor	L, 4 dph	28	1215	POS	DW(?)	56
Sparus aurata	L, 50–150 μg dry	19	1499	POS	DW	57
	L, 50–150 μg dry	24	1670	POS	DW	57
Engraulis sp.	L	18	857	M(?)	DW	58
Scomber japonicus	L	18	1162	M(?)	DW	58
	L, 3–5 dph	18	1162	W	DW	59
	L, 3–5 dph	22	2171	W	DW, estimated from graph	59
Pseudopleuronectes americanus	L, 1000 μg	2	343	W	DW, estimated from graph	60
	L, 1000 μg	5	495	W	DW	60
	L, 1000 μg	8	743	W	DW	60
Parophrys retulus	E, near hatch	6	560	W		61
Hirundichthys marginatus	L, 1.4 mg	22	1608	POS	DW(?), end endogenous feeding	56
Anchoa mitchilli	L, feeding	26	1352	POS	DW	62
Archosargus rhomboidalis	L, feeding	26	1886	POS	DW	62
Achirus lineatus	L, feeding	28	2248	POS	DW	62
Tautoga onitus	L, 10.9 mg dry	16	5371	M	DW, 5% yolk remaining	63
	L, 11 mg dry	19	4838	M	DW, 2.7% yolk remaining	63

Species						
Lagodon rhomboides	L, 11 mg dry	22	7810	M	63	DW, 10% yolk remaining
	L, 25 mg	15	500	M	64	
Leistomus xanthurus	L, 42 mg	15	500	M	64	
Congiopodus leucopaecilus	L, 18 dph	11.5	541	M	65	
Brevoortia tyrannus	L, 47–55 mg	14	780	POS	66	
	L, 47–55 mg	19	1000	POS	66	
	L, 47–55 mg	24	1555	POS	66	
Viviparous						
Zoarces viviparus	254 mg	5	92	POS	67	10–27 days prepartum
Clinus superciliosus	180–380 mg	11	134	M	68	
	46.3 mg	16	676	M	69	
Rhacochilus racca	389 mg	10	103	W	70	
	3.8 g	18	222	W	70	At parturition

[a] Abbreviations: L, larva; E, embryo; dpf, days postfertilization; dph, days posthatch; hph, hours posthatch; W, Winkler; DC, direct calorimetry; POS, polarographic oxygen sensor; M, manometric; D, diver; Hg, dropping mecury electrode; DW, original data expressed on dry weight basis, converted to wet weight assuming dry weight = 13.3% wet weight; MLWW, maximum larval wet weight; MTW, maximum tissue weight.

[b] References: (1) Lindroth, 1942 (cited in Hayes et al., 1951); (2) Hayes et al., 1951; (3) Tamarin and Komarova, 1970 (4) Smith, 1958; (5) Wieser et al., 1985; (6) Hamor, 1967 (cited in Hamor and Garside, 1975); (7) Rombough, 1987; (8) Penaz and Prokes, 1973; (9) Wood, 1932; (10) Gray, 1926; (11) Gnaiger, 1983b; (12) Gruber and Wieser, 1983; (13) Bailey et al., 1980; (14) Storozhyk and Smirnov, 1982; (15) Alderdice et al., 1958; (16) P. J. Rombough, unpublished data; (17) Kamler, 1976; (18) Winberg and Hartov, 1953 (cited in Kamler, 1972); (19) Kamler, 1972; (20) Jitariu et al., 1971; (21) Korwin-Kossakowski et al., 1981; (22) Kamler et al., 1974; (23) Kaushik and Dabrowski, 1983; (24) Wieser and Forstner, 1986; (25) Konchin, 1981; (26) Kudrinskaya, 1969 (27) Trifonova 1937 (28) Spoor, 1984; (29) Laurence, 1969; (30) Spoor, 1977; (31) Eldridge et al., 1982; (32) Gallagher et al., 1984 (33) Mukhamedova, 1977; (34) Singh et al., 1982; (35) Sheel and Singh, 1981; (36) Mishra and Singh, 1979; (37) Zoran and Ward, 1983; (38) Durve and Sharma, 1977; (39) Prokes, 1973; (40) Dabrowski et al., 1984; (41) Forstner et al., 1983; (42) Dabrowski and Kaushik, 1984; (43) Chernikova, 1964; (44) Khakimullin, 1985; (45) DeSilva et al., 1986; (46) DeSilva and Tytler, 1973; (47) Almata, 1984; (48) Holliday et al., 1964; (49) Marshall et al., 1937; (50) Eldridge et al., 1977; (51) Davenport and Lonning, 1980; (52) Davenport et al., 1979: (53) Solberg and Tilseth, 1984; (54) Lasker, 1962; (55) Lasker and Theilacker, 1962; (56) Klekowski et al., 1980; (57) Quantz and Tandler, 1982; (58) Hunter, 1981; (59) Hunter and Kimbrell, 1980; (60) Laurence, 1975; (61) Alderdice and Forrester, 1971; (62) Houde and Schekter, 1983; (63) Laurence, 1973; (64) Kjelson and Johnson, 1976; (65) Robertson, 1974; (66) Hettler, 1976; (67) Broberg and Kristofferson, 1983; (68) Korsgaard and Andersen, 1985; (69) Veith, 1979; (70) Webb and Brett, 1972b.

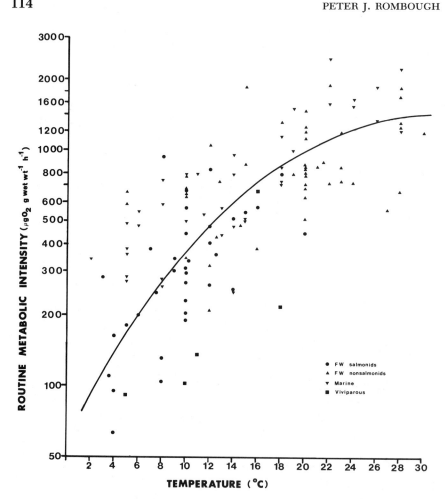

Fig. 10. Relationship between routine metabolic intensities (log scale) and temperature for fish embryos and larvae. Data were taken from Table III.

there are significant differences. In general, temperature changes have a more profound effect during early life. Literature values for the Q_{10} for metabolic rate in embryos and larvae range from about 1.5 (*Coregonus lavaretus;* Prokes, 1973) to 4.9 (*Salvelinus alpinus,* Gruber and Wieser, 1983), with an average value of about 3.0 (Table IV). In juvenile and adult fish, Q_{10} values average about 2.0 (Fry, 1971). The higher values during early life may reflect the fact that embryos and larvae tend to be more stenothermal than juveniles and adults. It is interesting that Q_{10} values for the rate of development are often

Table IV
Q_{10} Values for Embryos and Larvae

Species	Temp. range (°C)	Q_{10}	Reference
Salmo gairdneri	4–12	2.90	Wieser *et al.* (1985)
	12–20	1.80	Wieser *et al.* (1985)
Salmo gairdneri	6–9	4.37	Rombough (1987)
	9–12	2.41	Rombough (1987)
	12–15	2.74	Rombough (1987)
Salmo trutta	3–7	1.98	Wood (1932)
	7–12	4.76	Wood (1932)
Salmo trutta	10–20	1.87	Penaz and Prokes (1973)
Salmo salar	5–10	3.67	Hamor and Garside (1977)
Salvelinus alpinus	4–8	4.9	Gruber and Wieser (1983)
Onchorynchus tshay-	5–7.5	3.45	P. J. Rombough (unpublished)
wtscha	7.5–10	2.23	P. J. Rombough (unpublished)
	10–12.5	2.12	P. J. Rombough (unpublished)
Cyprinus carpio	10–20	2.37	Kamler (1972)
Cyprinids	12–20	3.09	Wieser and Forstner (1986)
Coregonus lavaretus	5–15	1.49	Prokes (1973)
Coregonus peled	5–15	2.84	Prokes (1973)
"Chalcalburnus"	19–23	3.00	Karpenko and Proskurina (1970)
"Taran"	15.5–19	4.68	Karpenko and Proskurina (1970)
Oreochromis niloticus	25–30	2.42	DeSilva *et al.* (1986)
	30–35	2.09	
Clupea harengus	6–8	2.93	Holliday *et al.* (1964)
	8–11	2.75	Holliday *et al.* (1964)
	11–14	1.79	Holliday *et al.* (1964)
Clupea harengus	8–13	1.9	Almatar (1984)
	13–18	3.0	Almatar (1984)
Pleuronectes platessa	8–13	1.8	Almatar (1984)
	13–18	6.4	Almatar (1984)
Pseudopleuronectes americanus	2–5	3.40	Laurence (1975)
	5–8	3.86	Laurence (1975)
Brevoortia tyrannus	14–19	1.64	Hettler (1976)
	19–24	2.40	Hettler (1976)
Tautoga onitus	16–22	1.87	Laurence (1975)
Scomber japonicus	18–22	4.77	Hunter and Kimbrell (1980)

similar to those for metabolism. Johns and Howell (1980) suggested that this similarity may explain why growth efficiency remains relatively constant over a relatively broad temperature range in many species.

The evidence is rather sketchy, but it appears that at least some species of fish may not be capable of thermal acclimation during early life. Clements and Hoss (1977) monitored rates of oxygen consumption of larval flounder (*Paralichthys dentatus* and *P. lethostigwa*) transferred directly from environmental temperatures of 10–12°C to constant temperatures of 10 and 15°C. Rates of oxygen uptake, measured daily for 4 days, did not vary following transfer. This would suggest that thermal acclimation occurs either very rapidly (<1 day) or not at all. The latter is more likely. Hinterleitner *et al.* (1987) found no evidence of metabolic temperature compensation in larval roach or chub (*Levciscus cephalis*). Preliminary analysis of data indicates that steelhead embryos similarly do not show any signs of thermal acclimation upon reciprocal transfer between 5 and 10°C (P. J. Rombough, unpublished data).

The effects of temperature change on VO_2 are sometimes difficult to interpret because of changes in activity levels associated with the temperature change. For example, Gruber and Wieser (1983) attributed the high Q_{10} (4.9) calculated for Arctic char larvae held at 4 and 8°C to what they termed "warm stimulation" of activity at the higher temperature. Unfortunately, activity levels were not measured. Hettler (1976) did measure changes in activity levels of larval menhaden (*Brevoortia tyrannus*) transferred to differe.at temperatures. He found that activity increased significantly as the temperature was raised but, interestingly, this was not reflected in a large increase in $\dot{V}O_2$ ($Q_{10} = 2.1$).

2. DISSOLVED OXYGEN

Dissolved oxygen concentrations obviously greatly influence metabolic rate. The relationship between $\dot{V}O_2$ and oxygen concentration, however, is not simple. Most, if not all, fish can be classified as metabolic regulators on the basis of their standard $\dot{V}O_2$ as juveniles and adults (Beamish, 1964; Fry, 1971; Ultsch *et al.*, 1981). As mentioned previously, it is difficult to measure $s\dot{V}O_2$ during the early life stages, but if $r\dot{V}O_2$ is used as the basis for classification instead of $s\dot{V}O_2$, it appears that embryos and larvae, for the most part, also behave as metabolic regulators. This means that at high oxygen concentrations their metabolic rate is independent of the ambient oxygen concentration, but if oxygen levels are gradually reduced, a point is eventually reached below which metabolic rate becomes dependent on the ambient oxygen concentration. This point, termed the critical oxygen tension (P_c), defines the oxygen concentration required to maintain a particular level of metabolism. It is important to recognize that P_c is not fixed but varies in response to a variety of intrinsic and extrinsic factors. In juvenile and adult fish the two most important factors influencing P_c are activity and temperature (Beamish, 1964; Fry, 1971; Ott et al., 1980). Combined high activity and temperature can result in P_c values near 100% air saturation (Brett, 1970). Activity (Broberg and Kristofferson, 1983) and temperature (Rombough 1986, 1987; Diez and Davenport, 1987) are also important factors influencing P_c during early life. In addition, the stage of development (Lindroth, 1942; Hayes et al., 1951; Rombough, 1986, 1987) and the water flow (Fry, 1971) have profound effects. Temperature and activity influence P_c by altering oxygen demands. Stage of development affects both oxygen demand and supply, while water velocity primarily affects oxygen supply.

Routine P_c (the P_c associated with $r\dot{V}O_2$) is directly dependent on the stage of development and temperature. Values increase more or less steadily throughout embryonic development (Lindroth, 1942; Hayes et al., 1951; Davenport, 1983; Rombough, 1987) and at any given stage of development are greater at higher temperatures (Rombough, 1987) (Fig. 11). The effect of temperature is an indirect one resulting from higher metabolic rates at higher temperatures, as indicated by the fact that when P_c is plotted against $\dot{V}O_2$ all points fall on the same line regardless of incubation temperature (Rombough, 1987) (Fig. 4). At high temperatures, routine P_c for large eggs, such as those of salmonids, may approach 100% air saturation near hatch. Such higher P_c values have led some investigators (e.g., Davenport, 1983; Gruber and Wieser, 1983) to classify teleost embryos as oxyconformers. However, this apparent conformity is simply a consequence of supply problems associated with the presence of the capsule and not intrinsic to the embryo itself. Hatching or artificial removal of the capsule results in an abrupt drop in P_c (Hayes et al., 1951; Gnaiger, 1983b; Gruber and Wieser, 1983; Rombough, 1987).

It must be noted that not all studies have shown fish embryos to be metabolic regulators. Hamor and Garside (1979) noted that $\dot{V}O_2$ was lower at 30% and 50% air saturation than at 100% air saturation at all stages during the embryonic development of Atlantic salmon. This is difficult to explain since P_c values should have been well below 50% ASV during early development (Hayes et al., 1951). It may be that metabolic response to chronic hypoxia is not the same as that to acute

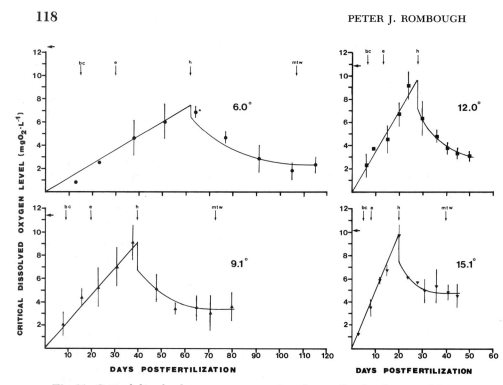

Fig. 11. Critical dissolved oxygen concentrations for steelhead embryos and larvae reared at constant temperatures of 6.0, 9.1, 12.0, and 15.1°C. The horizontal arrow indicates 100% air saturation. The point indicated by an asterisk represents the P_c for unhatched embryos at about 80% hatch; bc, blastopore closure; e, well eyed; h, hatch; mtw, maximum tissue weight. [From Rombough (1987).]

hypoxia. Other studies have indicated that embryos behave as neither true conformers nor true regulators (Davenport, 1983; Gruber and Wieser, 1983). Such inconclusive results may arise, at least in part, from practical difficulties associated with determining P_c (and $\dot{V}O_2$). As we have seen, $\dot{V}O_2$ and hence P_c are dependent on many factors. Excitement and activity in particular are difficult to control. Simply placing fish in a respirometer can significantly elevate their metabolic rate (Fry, 1957). Typically, $\dot{V}O_2$ is high when the fish is first placed in the respirometer and declines as it adapts to the system. Associated with this gradual decline in $\dot{V}O_2$ is a gradual decline in P_c. Metabolic rates should eventually stabilize at $r\dot{V}O_2$, but if the respirometer is operated as a closed system, oxygen levels may drop below routine P_c before this can occur. If this occurs, critical levels will not be apparent and the animal may be classified as an oxyconformer. Even if metabolic rates stabilize before oxygen levels drop below routine P_c, the transition from oxygen independence to dependence may be obscured. Davenport (1983) presents data illustrating this point. Lumpfish eggs or larvae were placed in closed respirometers and $\dot{V}O_2$ was monitored until all oxygen was exhausted. Embryonic $\dot{V}O_2$ declined rapidly at first but, after a while, $\dot{V}O_2$ tended to stabilize even though ambient oxygen levels continued to decline. Oxygen uptake during this period probably approximated $r\dot{V}O_2$. Eventually oxygen levels in the respirometer dropped below routine P_c, and $\dot{V}O_2$ again declined. The problem is that the period of stable $\dot{V}O_2$ is not always easily recognized, particularly later in development when routine P_c is relatively high. Compensatory adjustments, such as changes in activity level or the efficiency of gas exchange, on the part of the embryo or larvae may also make it difficult to discern routine P_c.

Additional problems in estimating P_c using closed respirometers arise from response lags associated with such systems and instability of polarographic sensors caused by pressure changes when the system is closed. Both have the effect of appearing to increase initial $\dot{V}O_2$. Rombough (1987) was able to minimize these problems, and that of initial excitement on being placed in the respirometer, by allowing the fish sufficient time to adapt to the system before closing the respirometer and, in the case of advanced embryos, initiating tests at moderate levels of hyperoxia (110–160% air saturation). This ensured that apparent $\dot{V}O_2$ stablized well before oxygen levels in the respirometer reached routine P_c, allowing P_c to be easily estimated. The technique appears to have been effective since energy budgets calculated on the basis of $\dot{V}O_2$ at the estimated routine P_c balanced [[mean(G + R)]A^{-1} = 102%]. Some of the problems associated with using "closed-system" respirometers to estimate P_c can be avoided by using flow-through systems, but this is very time-consuming, and in rapidly developing species routine P_c may have changed before the tests are completed.

The metabolic response of larvae to declining oxygen levels appears to be more variable than in embryos. As expected, activity is more of a problem, but even taking this into account there appear to be significant differences among species. All species can be expected to show an abrupt decrease in routine P_c at hatch. Rombough (1987) noted that the routine P_c for steelhead larvae continued to decline after hatch up to about midway through yolk absorption. The P_c then remained relatively stable until the end of yolk absorption, when experiments were terminated. Routine P_c was directly dependent on temperature. Levels during the latter half of the period of endogenous feeding ranged from 2.3 mg l^{-1} at 6°C to 4.8 mg l^{-1} at 15°C. Values were not determined for feeding larvae, but they probably continued to decline slowly since routine P_c values during endogenous feeding were somewhat higher than those reported for juvenile salmonids. De Silva and Tytler (1973) reported a different pattern for herring and plaice larvae. In both species P_c levels were lower for younger larvae than for older larvae. In fact, older larvae behaved as oxyconformers up to close to 100% air saturation. After metamorphosis, P_c levels dropped and both species again behaved as metabolic regulators. Davenport (1983) presented data indicating that P_c levels for lumpfish larvae similarly increased during the first 26 days of the larval period (at 5°C). The different patterns reported for steelhead and for herring, plaice, and lumpfish may reflect differences in the rate of transition from cutaneous to branchial gas exchange. The gills of salmonids are fairly well developed at hatch and probably assume an increasingly important role in respiratory gas exchange as development proceeds. In contrast, the gills of larval herring and plaice appear to be of little functional significance until close to metamorphosis. The skin remains the primary site of gas exchange, and because of an increasingly unfavorable surface/volume ratio the ability to extract sufficient oxygen from the environment may become limiting. The reduction in P_c after metamorphosis can be attributed to rapid elaboration of the gills and the appearance of hemoglobin in the blood.

3. SALINITY

The influence of salinity on the rate of oxygen consumption of embryos and larvae has been investigated in only a few species, but results to date suggest that net ionoregulatory costs are negligible once acclimation has occurred. It is not clear whether this is because costs remain fairly constant over a broad range of salinities or because increases in costs are paralleled by proportional decreases in other metabolic processes. Evidence that net ionoregulatory costs are small comes from several sources. Lasker and Theilacker (1962) reported no significant difference in the $r\dot{V}O_2$ of sardine embryos held in half-, full-, and double-strength seawater. Holliday et al. (1964) similarly observed no significant difference in standard metabolic rates of anesthetized herring embryos and newly hatched larvae incubated at constant salinities between 5‰ and 50‰. DeSilva et al. (1986) recently reported no significant difference in either routine or standard (anesthetized) metabolic intensities of larvae of the eurhyaline species *Oreochromis niloticus* reared at salinities between 0 and 18‰.

Abrupt changes in salinity do affect rates of oxygen consumption.

Holliday *et al.* (1964) reported up to a 10-fold increase in oxygen uptake by anesthetized herring embryos and larvae abruptly transferred from 35‰ to 5‰. Rates gradually declined to the pretransfer level over a period of 6–8 h. Holliday *et al.* (1964) indicated that the period of elevated metabolism coincided with the length of time required to restore osmotic imbalance brought about by the abrupt change in salinity. The pattern of oxygen uptake in unanaesthetized larvae was quite different. Transfer from 35‰ to both 15‰ and 5‰ resulted in a decrease in metabolic rate followed by a very gradual increase over the next 24 h to values typical of constant exposure to 35‰. Holliday *et al.* (1964) attributed the initial reduction in $\dot{V}O_2$ to reduced activity associated with buoyancy changes. Almatar (1984) similarly reported a reduction in oxygen uptake rates of herring and plaice yolk-sac larvae transferred from seawater (32‰) to low salinities (5 and 12.7‰). In feeding larvae, however, metabolic rates were elevated on transfer to 12.7‰ compared with control larvae (constant 32‰) and with larvae transferred to 5‰ and 40‰. Almatar (1984) attributed the apparent increase in $\dot{V}O_2$ to the fact that the larvae used in the 12.7‰ test were somewhat smaller than those used in the other tests, although why this should make a difference is not made clear.

4. LIGHT

Within limits, light exerts a direct tonic effect on metabolic rate (MacCrimmon and Kwain, 1969; Hamor and Garside, 1975; Konchin, 1982a; Solberg and Tilseth, 1984). Konchin (1982a) indicated that the $\dot{V}O_2$ of roach (*Rutilus rutilus*) larvae was 18–25% higher in the "light" than in the dark. Solberg and Tilseth (1984) reported that the $\dot{V}O_2$ of larval cod was 25–70% higher in the "light" than in the dark, apparently because of greater activity. Light intensities were not specified in either study. Hamor and Garside (1975) found that $\dot{V}O_2$ of Atlantic salmon embryos only increased in response to increasing light intensities up to about 200 lux; at 250 lux, $\dot{V}O_2$ began to decline.

5. OTHER FACTORS

Very little is known of how potentially important factors such as pH, carbon dioxide concentration, and ration level influence aerobic metabolism during the early life stages. Oxygen uptake by steelhead embryos was not significantly affected by low pH until levels dropped below pH 4.0 (Rombough, 1987). Alderdice and Wickett (1958) reported that the metabolic rate of chum salmon embryos was independent of ambient CO_2 levels below 125 mg l^{-1} (PCO_2 not given). The relative insensitivity of embryos to pH and carbon dioxide is not surprising in light of the weak Bohr and Root effects seen in larval hemoglobins (Iuchi, 1973b). As mentioned previously, Quantz and Tandler (1982) suggested that a high specific dynamic action (SDA) may account for the high routine metabolic rate of larval gilthead seabream. Kaushik and Dabrowski (1983) attempted to measure the SDA of young carp but were not able to detect a significant postprandial increase in $\dot{V}O_2$ until the carp had attained a mass greater than 1 g. Dabrowski (1986) has since developed a more sensitive system for detecting changes in $\dot{V}O_2$ and has used it to estimate the specific dynamic action of Atlantic salmon alevins. In this species, an increase in metabolic rate equivalent to about 30% of standard metabolism and peaking 2.5–3 h postfeeding appears to be attributible to SDA. One of the reasons SDA has been difficult to detect is that it is partially compensated for by a decrease in active metabolism.

The effects of water-borne pollutants on the early stages of fish are dealt with elsewhere in this volume (von Westernhagen; Chapter 4), but it would be apropos to comment briefly on the effect of such substances on embryonic and larval $\dot{V}O_2$. A wide variety of substances, including pesticides (Kamler et al., 1974; Klekowski et al., 1977), metals (Storozhk and Smirnov, 1982; Akberali and Earnshaw, 1984), and hydrocarbons (Eldridge et al., 1977; Davenport et al., 1979; Hose and Puffer, 1984), have been tested for their effect on $\dot{V}O_2$. If a generalization is to be made, it is that results are highly variable and often difficult to interpret. A particular pollutant may have significant effects on one stage but not on another (Davenport et al., 1979). A high concentration of the toxicant may have no effect, while a much lower concentration of the toxicant will produce a significant effect. For instance, oxygen uptake by embryonic grunion (Leuresthes tenuis) was not affected by a high body burden of benzo[a]pyrene but was significantly elevated in response to a low body burden (Hose and Puffer, 1984). Elevated $\dot{V}O_2$ at low exposure levels was attributed to hormesis (overcompensation to inhibitory challenge). Benzo[a]pyrene appears to be somewhat of an exception, and normally a relatively high concentration of a pollutant is required to produce a significant change in $\dot{V}O_2$. Davenport et al. (1979) estimated that because of normal variability in $\dot{V}O_2$ a change of about 40–50% r$\dot{V}O_2$ would be required to be statistically significant. For this reason, $\dot{V}O_2$ would not appear to be a particularly useful indicator of sublethal toxicity for most pollutants, especially since other more sensitive and easier to determine indicators are available (e.g., growth).

IV. EFFECT OF HYPOXIA

A. Environmental Hypoxia

Exposure to low levels of dissolved oxygen during early life can elicit a wide variety of responses, some compensatory, others clearly pathological. The particular response and its magnitude depends on species, stage of development, level of hypoxia, and duration of exposure. Other factors, such as temperature and water flow, also may be important. The most obvious place to begin an examination of the effects of environmental hypoxia is to describe the zone of tolerance—that is, to what levels oxygen can fall before mortality occurs.

1. LETHAL LEVELS

Many studies have shown that fish are extremely sensitive to low levels of dissolved oxygen during early life (see reviews by Doudoroff and Shumway, 1970; European Inland Fisheries Advisory Commission, 1973; Davis, 1975; Alabaster and Lloyd, 1980; Chapman, 1986). Incipient lethal levels, however, remain poorly defined. There are two major reasons for this. The first arises from the failure of most investigators to follow standard bioassay procedures. Experimental conditions are often not adequately controlled (e.g., variable temperature regimes). Sufficient numbers of test levels to obtain anything more than the roughest estimate of lethal levels are rarely used. Control mortality is often not taken into account, or even reported. Levels of significance are seldom given, and comparisons are frequently made on the basis of variable exposure periods. If past performance is any indicator, future investigators would do well to review basic toxicological methods (e.g., Shepard, 1955; Sprague, 1973) before proceeding. The second major reason why incipient lethal levels remain poorly defined is inherent to the early stages themselves. Gottwald (1965) pointed out some of the problems: the slow response of embryos to hypoxia and the difficulty in assessing mortality during the early stages. The most significant difficulty, though, is that the organisms are undergoing profound developmental changes that alter their sensitivity to hypoxia even as it is being assessed. This makes it extremely difficult to determine precise response thresholds.

It is fairly well established that sensitivity to hypoxia tends to increase as development proceeds. Maximum sensitivity occurs during the larval period, with the precise stage depending on the particular species. Increasing sensitivity is implied by mortality patterns during chronic exposures to low levels of dissolved oxygen. For example, larval mortality is often recorded at oxygen levels that permitted survival to hatch (Brungs, 1971; Eddy, 1972; Siefert et al., 1973, 1974; Dudley and Eipper, 1975). Mortality patterns are not always valid indicators of stage sensitivity. Rosenthal and Alderdice (1976) noted that with many toxicants, injury may be sustained at an early stage but not manifested until later in development. In the case of hypoxia, however, studies of acute toxicity also indicate increasing sensitivity as development proceeds. Alderdice et al. (1958) noted that the 7-day LC_{50} for chum embryos (O. keta) increased from 0.4 mg l^{-1} shortly after blastopore closure to 1.0–1.4 mg l^{-1} shortly before hatch (Table V). Likewise, Gottwald (1965) reported that 3-day LC_{50} values for rainbow trout increased from <0.9 mg l^{-1} at blastopore closure to 0.9–2.7 mg l^{-1} near hatch. Studies that have compared sensitivities shortly before and after hatch indicate that the newly hatched larvae are significantly more sensitive (Peterka and Kent, 1976; Spoor, 1977). This is somewhat surprising given the fact that the embryos are effectively exposed to a much lower ambient concentration because of the presence of the capsule. The particular stage at which larvae are most sensitive is highly variable. Bishai (1960) indicated that the sensitivity of young Atlantic salmon and brook trout continued to increase up to at least 80 and 127 days posthatch (at 5°C), respectively. Tamarin and Komarova (1972) reported that the "threshold level" (asphyxiation level in a closed container) of Atlantic salmon increased steadily to a maximum 42–60 days posthatch (at 8°C) and then declined to reach typical juvenile and adult levels 240 days posthatch. De Silva and Tytler (1973) reported that 12-h LC_{50} values for herring larvae increased from 2.8 mg l^{-1} shortly after hatch to reach a peak of 5.1 mg l^{-1} after 5–6 weeks of feeding before declining to 3.1 mg l^{-1} at metamorphosis. Smallmouth bass larvae (Micropterus dolomieui) were least resistant to severe hypoxia (1.0 mg l^{-1}) at about the start of exogenous feeding (9 days posthatch at 20°C; Spoor, 1984). In contrast, largemouth bass larvae (M. salmoides) were least resistant shortly after hatch while still feeding endogenously (3 days posthatch at 20°C; Spoor, 1977). Larval plaice were similarly most sensitive shortly after hatch (De Silva and Tytler, 1973).

Changes in larval sensitivity have been linked to changes in the site and efficiency of respiratory gas exchange (DeSilva and Tytler, 1973; Spoor, 1977). The argument is basically as follows. The skin is the major site of gas exchange during much of early life. Young larvae with relatively low metabolic rates and large surface/volume ratios require a relatively small partial pressure gradient to meet oxygen

Table V
Lower Lethal Levels of Dissolved Oxygen for Teleost Embryos and Larvae[a]

Species	Temperature (°C)	Stage tested	Duration of test	LC50 (mg l⁻¹)	No-effect level (mg l⁻¹)	Comment	Reference
			Chronic exposure, freshwater salmonids				
Salmo gairdneri	10.4	E	f-h (~30 days)	<2.8	3.0–4.5		Shumway et al. (1964)
Salmo gairdneri	10.0	E	f-h (~36 days)	1.6–2.5	1.6–2.5		Silver et al. (1963)
Oncorhynchus tshawytscha	10	E	f-h (~45 days)	1.6–2.5	1.6–2.5		
Oncorhynchus kisutch	9.5	E	f-h (~50 days)	<2.5	2.8–4.1		Shumway et al. (1964)
Salmo salar	5	E	f-h (77 days)	<3.7	6.8–11.9		Hamor and Garside (1976)
	10	E	f-h (43 days)	<3.9	5.8–11.6		
Oncorhynchus kisutch	8.5	E–L	119 days	1.4–2.9	1.4–2.9		Siefert and Spoor (1974)
Salvelinus fontinalis	8	E–L	133 days	2.3–2.9	2.3–2.9		
Salvelinus namaycush	7	E–L	131 days	2.4–4.3	4.3–6.0		Carlson and Siefert (1974)
	10	E–L	108 days	5.6– 10.5	5.6–10.5		
Salvelinus alpinus	4	E–L	93 days	<2.6	<2.6	Near upper lethal temp.	Gruber and Wieser (1983)
	8	E–L	77 days	=3.6	3.6–5.9	Near upper lethal temp.	
			Chronic exposure, freshwater nonsalmonids				
Stizostedion vitreum	12.3	E	f-h (~22 days)	<3.0	3.0–5.0	Walleye	Oseid and Smith (1971a)
Catostomus commersoni	12.3	E	f-h (~15 days)	<3.0	<3.0	White sucker	Oseid and Smith (1971b)

(continued)

Table V (*Continued*)

Species	Temperature (°C)	Stage tested	Duration of test	LC$_{50}$ (mg l^{-1})	No-effect level (mg l^{-1})	Comment	Reference
Morone chy-rops	16	E	f–h (4 days)	<1.8	<1.8		Siefert et al. (1974)
Fundulus heteroclitus	20	E	f–h (20 days)	2.4–4.5	4.5–7.5		Voyer and Hennekey (1972)
Morone saxatilis	20	E	f–h (4 days)		>5.0		Turner and Farley (1971)
Coregonus artedii	2	E	f–h (166 days)	<1.0	1.0	Lake herring	Brooke and Colby (1980)
	4	E	f–h (122 days)	<1.0	1.0–2.0		
	6	E	f–h (84 days)	1.0–2.0	1.0–2.0		
	8	E	f–h (54 days)	2.0–4.0	2.0–4.0		
Micropterus salmoides	15	E	f–h		>2.4	Largemouth bass	Dudley and Eipper (1975)
	20	E	f–h		>2.4		
	25	E	f–h		>2.4		
Cyprinus carpio	25	E	f–h (70 h)	3.0–6.0	6.0–9.0	Carp	Kaur and Toor (1978)
Stizostedion vitreum	17	E–L	20 days	3.4–4.8	3.4–4.8	Walleye	Siefert and Spoor (1974)
Catostomus commersoni	18	E–L	22 days	1.2–2.5	1.2–2.5	White sucker	
Poxomis nigro-maculatus	20	E–L	?	<2.7	<2.7	Black crappie	Siefert and Herman (1977)
Prosopium williamsoni	4	E–L	193 days	3.3–4.6	4.6–6.5	Mountain whitefish	Siefert et al. (1974)
	7	E–L	158 days	3.1–6.0	3.1–6.0		
Micropterus dolomieui	20	E–L	14 days	2.5–4.4	4.4–8.7	Smallmouth bass	
Morone chrys-ops	16	E–L	11 days	<1.8	1.8–3.4	White bass	

Micropterus salmoides	20	E–L	20 days	1.7–3.1	3.1–4.5	Largemouth bass	Carlson and Siefert (1974)
	23	E–L	20 days	1.7–3.0	1.7–4.2		
Ictalurus punctatus	25	E–L	19 days	2.4–4.2	4.2–5.0	Channel catfish	Carlson *et al.* (1974)
	29	E–L	19 days	2.3–3.8	3.8–4.6		
Esox lucius	15	E–L	20 days	2.6–4.9	2.6–4.9	Northern pike	
Pimephales promelas	24	L	30 days	4.02	4.02–5.01		Brungs (1971)
Acute exposure, freshwater species							
Oncorhynchus keta	10	E	7 days	0.4		12 dpf	Alderdice *et al.* (1958)
			7 days	0.6		22 dpf	
			7 days	0.6		32 dpf	
			7 days	1.0–1.4		48 dpf	
Salmo gairdneri	11.5	E	3 days	<0.9	0.9–1.7	Blastopore closure	Gottwald (1965)
			3 days	~0.9	1.7–2.7	Start circulation	
			3 days	0.9–2.7	2.7–4.3	Near hatch	
Salmo salar	5	L	120 h		0.4	Newly hatched	Bishai (1960)
			72 h		0.7	10 dph	
			48 h		0.7	40 dph	
			72 h		2.0	54 dph	
			72 h		3.1	80 dph	
			48 h		2.8	135 dph	
Salmo trutta	5	L	120 h		0.4	Newly hatched	
			72 h		0.7	10 dph	
			72 h		2.0	54 dph	
			72 h		2.4	127 dph	
			48 h		2.3	180 dph	
Coregonus sp.	6.5	L	?	3.3	6.0		Einsele (1965)
	11.5	L	?	3.7	6.9		
	19.0	L	?	5.5	7.5		
Esox lucius	10	E	8 h	~0.3	<0.6	3 days prehatch	Peterka and Kent (1976)
		L	8 h	0.3–1.8	0.3–1.8	1 dph (yolk-sac larvae)	
		L	8 h	1.6–3.5	1.6–3.5	Feeding larvae	

(continued)

Table V (*Continued*)

Species	Temperature (°C)	Stage tested	Duration of test	LC$_{50}$ (mg l^{-1})	No-effect level (mg l^{-1})	Comment	Reference
Micropterus dolomieui	21.5	E	6 h	0.5	0.5–1.8	2 d prehatch	Bishai (1960)
		L	6 h		0.5–2.2	Yolk-sac larvae	
Lepomis macrochirus	26.5	E	4 h		<0.5		
		L	4 h	0.5–3.7	0.5–3.7	1–4 dph (yolk sac)	
Marine species							
Clupea harengus	14	L	24 h		2.2–2.6	Newly hatched	Bishai (1960)
			24 h		4.4	3 dph	
			24 h		5.1	4 dph	
Clupea harengus	8–13	L	12 h	2.8		Yolk-sac larvae	DeSilva and Tytler (1973)
			12 h	4.4		2–3 weeks feeding	
			12 h	5.1		5–6 weeks feeding	
			12 h	4.2		7–8 weeks feeding	
			12 h	3.1		Metamorphosis	
Pleuronectes platessa	8–13	L	12 h	3.9		Yolk-sac larvae	DeSilva and Tytler (1973)
			12 h	3.8		2–3 weeks feeding	
			12 h	3.6		6–7 weeks feeding	
			12 h	2.4		3–4 weeks metamorphosis	
Cyclopterus lumpus	12.7	L	24 h	3.6		24 dph	Bishai (1960)
Ophiodon elongatus	8.6	E	96 h	3.0	4.6–7.7	Lingcod	Giorgi and Congleton (1984)
Chasmodes bosquianus	20.8	L	24 h	2.5		Striped blenny, newly hatched	Saksena and Joseph (1972)
Gobiosoma bosci	20.8	L	24 h	1.3		Naked goby	

Gobiesox strumosus	20.5	L	24 h	0.7–1.2		Skillet fish	Sylvester *et al.* (1975)
Mugil cephalus	20	E	48 h	4.5–5.0	4.5–5.0	Near hatch, striped mullet	
		L	48 h	4.8–5.4	4.8–5.4	Newly hatched	
		L	96 h	6.4–7.9	6.4–7.9	Newly hatched	
Pterosmaris axillaris	13(?)	L	24 h	3.5	4.0 (LC_{10})	First-feeding static test	Brownell (1980)
Pachymetopon blochi		L	24 h	3.2	3.5 (LC_{10})	First-feeding static test	
Sparidae (5 species)		L	24 h	3.6	3.9 (LC_{10})	First-feeding static test	
Trulla capensis		L	24 h	<2.3	—	First-feeding static test	
Congiopodus spinifen		L	24 h	3.6	4.2 (LC_{10})	First-feeding static test	
Gaidropsarus capensis		L	24 h	2.5	2.9 (LC_{10})	First-feeding static test	
				2.1	2.7 (LC_{10})	First-feeding flow-through test	
Heteromycteris capensis		L	24 h	6.9	1.2 (LC_{10})	First-feeding flow-through test	

[a] Abbreviations: E, embryos; L, larvae; dph, days posthatch; f–h, fertilization to hatch.

demands. As the larvae grows its metabolic rate increases, but expansion of the area for cutaneous gas exchange fails to keep pace. If metabolic demands are to be met, the partial pressure gradient across the skin must increase (recall $\dot{V}O_2 = GO_2PO_2$). The result is an increase in P_c (standard) and a reduction in the ability of the larva to tolerate hypoxia. As development proceeds, the gills become progressively more important as a site of respiratory gas exchange. Gills are assumed to be a more efficient organ of gas exchange and, as a result, P_c (standard) values gradually decline. This is reflected in a gradual decrease in sensitivity to hypoxia. Species vary in the timing of the transition from cutaneous to branchial gas exchange, and thus it is not surprising that they vary in the particular stage at which they are most sensitive to hypoxia. De Silva and Tytler (1973) indicated that the stage at which hemoglobin first appears also may be important. Spoor (1984) linked increased resistance of smallmouth bass larvae near the start of exogenous feeding to inflation of the swimbladder. It is unlikely that this represents a true increase in tolerance to hypoxia but more likely reflects the fact that the gas in the swimbladder can act as a temporary reservoir when oxygen is in short supply.

It is generally assumed that the ability of fish to resist hypoxia decreases with increasing temperature (European Inland Fisheries Advisory Commission, 1973; Alabaster and Lloyd, 1980; Chapman, 1986), although Chapman (1986) points out that the data supporting this contention is rather spotty. There is some evidence indicating that, within limits, hypoxic sensitivity may actually vary relatively little with temperature. Davis (1975) found that incipient sublethal response thresholds of adult fish were insensitive to temperature when expressed as a concentration (i.e., mg l^{-1}). This still represents an increase in the driving force required to meet oxygen demands, since, at a given concentration, partial pressures are greater at higher temperatures (because of decreased capacitance). Ultsch et al. (1978), on the other hand, reported that asphyxiation levels were actually lower at 20°C than at 10°C, even when expressed as partial pressures, in adults of six species of darters (Etheostoma).

Few studies have specifically examined the effect of temperature on the ability of embryos and larvae to tolerate a lack of oxygen. Temperature variations within a narrow range (3–5°C) appear to have little effect in many species (Siefert et al., 1974; Carlson and Siefert, 1974; Carlson et al., 1974; Hamor and Garside, 1976). There are some exceptions though. Increasing the temperature from 7 to 10°C resulted in a significant increase in the sensitivity of young lake trout (Salvelinus namaycush) (Carlson and Siefert, 1974). Similarly, Arctic char embryos and larvae were less tolerant of hypoxia at 8 than at 4°C (Gruber and Wieser, 1983). Both species are stenothermal, and in both studies the higher temperature was near the upper limit of the zone of tolerance. Lethal levels for embryonic (Brooke and Colby, 1980) and larval (Einsele, 1965) coregonids increased with temperature, but again increases were greatest at the higher temperatures. These results suggest that lethal oxygen *concentrations* may be relatively independent of temperature within the normal temperature range of a particular species but that at high temperatures, there is a strong likelihood of an additive or synergistic interaction.

High flow rates can be expected to increase the hypoxic tolerance of embryos by reducing the thickness of the boundary layer. The interaction between flow rate and dissolved oxygen level has been well documented using sublethal indicators such as growth (Silver *et al.*, 1963; Shumway *et al.*, 1964). There is less information on its effect on lethality. Shumway *et al.* (1964) reported that mortality among coho and steelhead embryos exposed to low oxygen levels tended to be greater at low flows. Miller (1972) reported that mortality among "unfanned" largemouth bass embryos was greater than among fanned embryos. High embryonic mortality in nature is often associated with low flows (Wickett, 1954; Coble, 1961; Phillips and Campbell, 1961; McNeil, 1966; Taylor, 1971; Hempel and Hempel, 1971; Giorgi, 1981). In most cases, however, flow rate and oxygen concentration varied simultaneously so it is difficult to separate effects due to low oxygen from those due to low flow.

Salinity, like temperature, appears to have little effect on hypoxic tolerance within normal limits. Alderdice and Forrester (1971) indicated that viable hatch of Pacific cod (*Gadus macrocephalus*) was largely independent of oxygen, provided levels were above 2–3 mg l^{-1} within the optimal range of temperatures (3–4.5°C) and salinities (17–23‰). Oxygen requirements tended to increase at higher salinities. Similarly, embryonic survival of pilchard (*Sardinops ocellata*) was largely independent of oxygen levels greater than 2.1 mg l^{-1} within the optimal range of temperature (16–21°C) and salinity (33–36‰) (King, 1977).

There are significant differences in the abilities of different species to tolerate hypoxia. There does not appear to be any phylogenetic pattern, but rather the sensitivity of a particular species seems to reflect oxygen levels in its normal habitat. For example, Spoor (1984) linked differences in habitat selection by larvae of smallmouth bass and largemouth bass to differences in their sensitivity to hypoxia. Chapman (1986) pointed out that while salmonids are more sensitive to hypoxia than most nonsalmonid freshwater fish as juveniles, they are less sensitive than many other groups as embryos and larvae. These observations lend some support to Balon's (1975) concept of reproductive guilds. Data in Table V indicate a mean embryonic–larval LC_{50} of 2.7 mg l^{-1} for salmonids at normal temperatures. The nonsalmonid freshwater species fall into two groups: white sucker, black crappie, white bass, and largemouth bass, with a mean embryonic–larval LC_{50} of 2.3 mg l^{-1}, and walleye, mountain whitefish, smallmouth bass, and pike with a mean embryonic–larval LC_{50} of 3.8 mg l^{-1}. Marine species are generally less tolerant than freshwater species. Among marine species, De Silva and Tytler (1973) attributed the greater resistance of newly hatched herring larvae compared with newly hatched plaice larvae to the fact that the former hatch from demersal eggs and are thus more likely to encounter low oxygen levels. A glaring deficiency in the literature is the absence of any data on the hypoxia tolerance of tropical freshwater species, some of which spawn in virtually anoxic water.

Durborow and Avault (1985) reported significant differences among full-sib families of channel catfish (*Ictalurus punctatus*) in larval resistance to hypoxia. This raises the possibility of selecting strains that are resistant to hypoxia for use in aquaculture.

Shepard (1955) demonstrated that acclimation to low oxygen levels increased the ability of juvenile brook trout to tolerate hypoxia. There are limited data indicating that acclimation increases at least the resistance of larvae. McDonald and McMahon (1977) reported that chronic exposure of Arctic char larvae to low oxygen (2.5 mg l^{-1}) resulted in a 2.5-fold increase in median resistance time in 1.1 mg l^{-1} compared with normoxic-reared larvae.

Moderate levels of hyperoxia have been reported to enhance survival of embryos (Gulidov, 1969, 1974; Gulidov and Popova, 1978) and larvae (Sylvester *et al.*, 1975). At higher concentrations (>300% air saturation), though, oxygen becomes toxic. Species vary in their ability to tolerate hyperoxia. Gulidov (1969) reported significant mortality in pike (*Esox lucius*) eggs incubated in 36.4 mg l^{-1} (336% ASV). No pike embryos hatched in 45.3 mg l^{-1} (418% ASV), apparently because of suppressed neuromuscular activity. Embryos of verkhovka (*Leucospuis delineatus*) (Gulidov, 1974) and roach (*Rutilis rutilis*) (Gulidov and Popova, 1978), on the other hand, both hatched successfully in about 40 mg l^{-1}. In *L. delineatus*, hatching was delayed at high oxygen concentrations. Newly hatched larvae tended to have more body segments than normal, and no red blood cells were present. The initial phase of erythropoeisis was not inhibited, and red blood cells appeared in the circulation at about the normal time. The number of erythrocytes later declined, and red cells were absent at hatch. Absence of red blood cells did not adversely affect survival, which is not surprising given the high ambient oxygen concentration. Gulidov (1974) linked the high tolerance of *L. delineatus* and roach to hyperoxia to the fact that their eggs are frequently laid on vegetation and may thus be exposed to high oxygen concentrations in their natural habitat.

2. SUBLETHAL RESPONSES

Hypoxia elicits a broad spectrum of sublethal responses in embryos and larvae. These include reduced rates of growth and development, morphological changes, behavioral alterations, and a wide variety of metabolic and physiological adjustments. In general, sublethal response thresholds are even more poorly defined than lethal thresholds. The reasons for this are basically the same as those discussed previously: deficient experimental design and changes in the intrinsic sensitivity of the organism. The situation is actually somewhat worse than for studies of lethality. Sublethal response thresholds have tended to be higher than anticipated. Many investigators have chosen inappropriate experimental levels (i.e., too low) and, as a result, have been able to define response thresholds only very broadly (typically as lying somewhere between 30–50 and 100% air saturation), if at all. The following discussion, therefore, will be restricted for the most part to a qualitative description of the sublethal effects of hypoxia.

a. Development and Growth. Developmental velocity and early growth are highly sensitive to reductions in ambient oxygen levels. In many species incipient limiting levels appear to be close to, or even in excess of, 100% air saturation (Silver *et al.*, 1963; Shumway *et al.*, 1964; Eddy, 1972; Gulidov, 1974; Gulidov and Popova, 1978). Reduced growth and delays in development under moderate hypoxia should be regarded as compensatory responses whereby the animal adjusts metabolic demands to match available supply. Growth and development are normally closely linked, although there is some suggestion that developmental rates are less plastic than growth rates (Silver *et al.*, 1963).

Embryonic development is progressively retarded by continuous exposure to low levels of dissolved oxygen (Garside, 1959, 1966; Winnicki, 1968). Delays are typically insignificant during early development at moderate levels (> 20–30% ASV) of hypoxia. During later development, times to defined stages become progressively more delayed compared with normoxia. As expected, delays are more pronounced at lower oxygen levels.

Most investigators have not monitored developmental velocities closely. Instead they have looked at the effect of hypoxia on times to easily identifiable events such as hatch, emergence, or the onset of feeding. The problem with this approach is that times to some of these events are not indicative of the overall effect of hypoxia on developmental rate. For example, time to hatch is widely used as an indicator of sublethal hypoxic stress. Low oxygen levels elicit two responses that have opposing effects on time to hatch. Hypoxia reduces the overall rate of development and thus tends to delay hatching. However, once embryos have reached a certain stage, low oxygen levels initiate the release of hatching enzyme (DiMichele and Taylor, 1980; DiMichele and Powers, 1982; Yamagami *et al.*, 1984) and thus tend to reduce time to hatch. Which effect dominates depends on the particular species. Continuous exposure of coho, brook trout (Siefert and Spoor, 1974), walleye (Oseid and Smith, 1971a), mountain whitefish Siefert *et al.*, 1974), mummichog (*Fundulus heteroclitus*) (Voyer and Hennekey, 1972), lake trout (Garside, 1959; Carlson and Siefert, 1974), and lake herring (Brooke and Colby, 1980) to low oxygen levels resulted in significant delays in time to hatch. Continuous exposure of white sucker (Oseid and Smith, 1971b; Siefert and Spoor, 1974) and sockeye (Brannon, 1965) had no significant effect, while hypoxic incubation of smallmouth bass (Siefert *et al.*, 1974) and largemouth bass (Carlson and Siefert, 1974) caused premature hatch. These results suggest that if embryonic development is relatively rapid, as in smallmouth and largemouth bass, the premature release of hatching enzyme outweighs any general delay in development. The effect of low oxygen on time to emergence is similarly complicated by the fact that moderate levels of hypoxia can act as a directive factor and induce the larvae to leave the substrate prematurely. As a result, hypoxia has been reported variously to delay (Phillips *et al.*, 1966; Mason, 1969) or advance (Witzel and MacCrimmon, 1981; Bailey *et al.*, 1980; Bams, 1983) emergence.

Growth has been the most widely used indicator of hypoxic stress during early life (Silver *et al.*, 1963; Shumway *et al.*, 1964; Oseid and Smith, 1971a,b; Carlson *et al.*, 1974; Siefert *et al.*, 1973, 1974; Siefert and Spoor, 1974; Carlson and Siefert, 1974; Hamor and Garside, 1977; Gruber and Wieser, 1983; Florez, 1972). As mentioned earlier, response thresholds are high. Embryos appear to be more sensitive than larvae, although this has not been well documented. Embryonic growth, at least, is more severely inhibited by low oxygen at higher

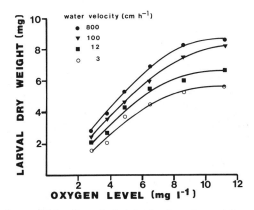

Fig. 12. Mean dry weights of newly hatched coho salmon (*Oncorhynchus kisutch*) larvae reared at various constant oxygen concentrations and bulk water velocities. [After Shumway *et al.* (1964).]

temperatures (Hamor and Garside, 1977; Eddy, 1972; Gruber and Wieser, 1983), as would be predicted from the fact that P_c values tend to increase with temperature. Also, as expected, low flow rates accentuate the limiting effect of low oxygen on embryonic growth (Silver *et al.*, 1963; Shumway *et al.*, 1964; Hamor and Garside, 1977) (Fig. 12). The effect of low flow appears to be more pronounced for larger eggs than for smaller eggs (Silver *et al.*, 1963; Shumway *et al.*, 1964; Brannon, 1965). Apparently the more favorable surface/volume ratios of small eggs make variations in the thickness of the boundary layer less important. High flow rates during the larval period can have the opposite effect and reduce growth efficiency, apparently as a result of enhanced activity levels (Brannon, 1965).

There is some question as to the significance of the reductions in growth noted at moderate levels of hypoxia. Eddy (1972) points out that although low oxygen levels result in smaller salmonid larvae at hatch and greatly extend the period of endogenous feeding, there is relatively little difference in the final size achieved. Growth efficiencies of Atlantic salmon (Hamor and Garside, 1977) and Arctic char (Gruber and Wieser, 1983) embryos were little affected by oxygen levels as low as 20–30% air saturation. Analysis of larval growth is more complicated. The growth efficiency of Atlantic salmon (Hamor and Garside, 1977) and Arctic char (Gruber and Wieser, 1983) larvae was significantly reduced at 20–30% ASV compared with at 100% ASV (Fig. 13). In both species, however, growth efficiency was significantly higher at 50% ASV than at normoxia. This was attributed to

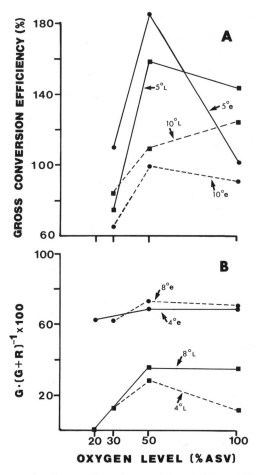

Fig. 13. Efficiency of early growth under conditions of chronic hypoxia. (ASV, air saturation.) (A) Gross conversion efficiencies (weight of tissue produced/weight of yolk consumed during a given period) of embryos (e) and larvae (L) of Atlantic salmon (*Salmo salar*) reared at 5 and 10°C. [After Hamor and Garside (1977).] (B) Growth efficiencies (energy content of tissue elaborated/energy expended on growth and metabolism during a given period) of embryos (e) and larvae (L) of Arctic char (*Salvelinus alpinus*) reared at 4 and 8°C. [after Gruber and Wieser (1983).]

reduced locomotor activity under moderate hypoxia (Hamor and Garside, 1977; Gruber and Wieser, 1983). There is another possibility that serves to illustrate the difficulties in comparing early growth efficiencies. It arises from the fact that the larvae were not all at equivalent stages when the comparisons were made. For example, Hamor and

Garside (1977) terminated their tests when between 80 and 111%(?) of the yolk present at hatch was consumed. Yolk conversion efficiencies decline sharply as yolk reserves near exhaustion. In steelhead, instantaneous yolk conversion efficiencies fall from 45% to 0% between 80 and 100% yolk utilization (Rombough, 1987). It would require only a relatively minor difference in the stages that are compared to produce an apparently significant difference in growth efficiency.

Even if growth efficiency is significantly reduced at low oxygen levels, the environmental significance is not always clear. It is generally assumed that smaller larvae are at a competitive disadvantage. Smaller larvae display slower absolute swimming speeds and hence can search a smaller volume for prey in a given time and may themselves be more susceptible to predation (Giorgi, 1981). Smaller larvae are also less effective at competing for territories (Mason, 1969). However, territories may not always be fully occupied or food always limiting. Mason (1969) observed that coho incubated under hypoxic conditions were smaller at emergence and could not compete successfully for territory with larger normoxic larvae. As a result, they were forced to migrate from the vicinity of the redd but subsequently did well providing the stream was not heavily populated. Bams (1983) reported that chum larvae from Japanese-style incubators were smaller and emerged earier than larvae incubated in upwelling gravel boxes, apparently because of hypoxic conditions in the Japanese-style incubators. These alevins, though smaller at emergence, subsequently grew faster than the alevins from upwelling boxes because of earlier feeding. The hatchery situation may be unusual because of the abundance of food, but, as Bams (1983) points out, it emphasizes that fitness components must be defined carefully and must be viewed in relation to the particular set of conditions.

b. Morphology. Teratogenic effects have been observed at oxygen concentrations close to the lethal level (Silver *et al.*, 1963; Shumway *et al.*, 1964; Garside, 1959; Alderdice *et al.*, 1958; Braum, 1973; Brooke and Colby, 1980). Deformities of the axial skeleton, jaws, and vitelline circulation appear most common. The changes in the vitelline circulation appear to be adaptive. Blood vessels become more finely divided and cover a greater proportion of the yolk. This can be expected to enhance respiratory gas exchange. Burggren and Mwalukoma (1983) noted that chronic exposure to low oxygen levels caused a similar increase in the density of capillaries supplying the skin of larval amphibians. This was associated with a reduction in the thickness of the blood–water diffusion barrier. Diffusion distances have not been reported for fish embryos or larvae chronically exposed to low oxygen, but observations that the yolk sac of hypoxic embryos ruptures readily (Garside, 1959; Brooke and Colby, 1980) suggest that this might be the case. Chronic hypoxia also brings about morphological adjustments in the gills that can be expected to enhance gas exchange. McDonald and McMahon (1977) observed lamellar hypertrophy in Arctic char larvae reared in 2.6 mg l^{-1} (at 6.5°C). Branchial development overall was inhibited as a consequence of a general inhibition of growth, so that by 47 days posthatch hypoxic larvae had 22% fewer filaments and 40% fewer lamellae than normoxic larvae. The surface area of individual lamellae, however, was significantly larger in the hypoxic larvae to the extent that there was no significant difference in total lamellar surface. The fact that the hypoxic larvae were more resistant to lethal oxygen concentrations led McDonald and McMahon (1977) to suggest that other adjustments, such as reduced blood–water distances, increases in the area of lamellar blood spaces, and increased lamellar perfusion, were also involved. Pinder and Burggren (1983) suggested that such morphological adjustments leading to increased conductance of gas-exchange organs was typical of the early stages of lower vertebrates. They contrasted this with the usual adult response to hypoxia of increased blood-carrying capacity and increased hemoglobin oxygen affinity.

c. *Behavior and Physiology.* Larvae respond to acute hypoxia with an increase in random movements (Spoor, 1977, 1984). The level at which they respond appears to be directly related to the lethal level, suggesting they are responding to hypoxemia rather than ambient oxygen levels. Spoor (1984) noted that the oxygen concentration at which smallmouth bass began to become agitated increased steadily during the first 8–10 days after hatch. During this period the length of time the larvae would resist acutely lethal oxygen concentrations also declined progressively. About 12 days posthatch the larvae became less responsive to low oxygen levels. This coincided with increased resistance.

The physiological responses of fish embryos and larvae to acute hypoxia are significantly different from those of juveniles and adults. Most notable is the absence of reflex bradycardia (Fisher, 1942; Holeton, 1971; McDonald and McMahon, 1977). In this regard they are similar to amphibian larvae (West and Burggren, 1982; Feder, 1983a,b; Quinn and Burggren, 1983).

Physiological and behavioral responses designed to facilitate embryonic and larval gas exchange were discussed in a previous section.

Not unexpectedly, such responses tend to intensify during hypoxia. Peterson and Martin-Robichaud (1983) noted that the frequency of trunk movements of Atlantic salmon embryos increased during hypoxia. Holeton (1971) noted that young rainbow trout larvae (1–8 days old) responded to acute exposure to moderate levels of hypoxia with increased pectoral fin movements, increases in rate and amplitude of breathing, and tachycardia. Only when oxygen levels dropped below about 40 mm Hg (at 10°C) did heart rate and breathing rate decline (Fig. 14). Heart rate and breathing rate recovered slowly in restoration of normoxia, unlike the quick recovery seen in adult fish. Peterson (1975) reported that the ventilatory rate also increased in Atlantic salmon larvae in response to acute hypoxia (30–80 mm Hg at 9°C). Increased opercular movements were accompanied by increases in the frequency of pectoral-fin movements. Movements of pectoral fins appear to help draw water across the gills, since ablation of the fins in

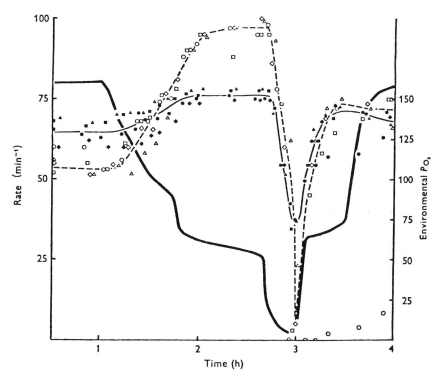

Fig. 14. Effect of acute hypoxia on heart rate (solid line) and breathing rate (dotted line) of rainbow trout larvae 8 days after hatch in 10°C water. Heavy line indicates PO_2 of the water. [From Holeton (1971).]

normoxia results in a significant increase in opercular rate. McDonald and McMahon (1977) indicated that ventilatory and cardiovascular responses during chronic hypoxia were similar to those during acute hypoxia. Heart and circulatory rates of Arctic char larvae remained elevated in 33 mm Hg (at 6.5°C) for at least 47 days after hatch. As in Atlantic salmon (Peterson, 1975), ventilatory movements of newly hatched Arctic char were infrequent and uncoordinated. In contrast, rhythmic and rapid (60 min^{-1}) ventilatory movements were observed in hypoxic larvae within 1 h of hatch. McDonald and McMahon (1977) speculated that cutaneous respiration could provide sufficient oxygen to meet the respiratory demands of newly hatched larvae under normoxic conditions but not during hypoxia. The larvae compensate for a reduced driving force during hypoxia by branchial recruitment, effectively increasing the surface area for respiratory gas exchange (i.e., increasing conductance). Neither Holeton (1971) nor McDonald and McMahon (1977) observed the reflex bradycardia typical of juvenile and adult fish. Holeton (1971) speculated that the basic response to hypoxemia was increased breathing and heart rates and that the bradycardia observed in older fish was a superimposed mechanism designed to balance ventilation–perfusion rates for efficient delivery of oxygen to tissues. McDonald and McMahon (1977) argued that tachycardia was a more appropriate response during early life, since higher cardiac output would lead to opening of vascular channels within the gills and effectively increase the surface area for gas exchange.

d. Metabolism. During severe hypoxia, physiological adjustments may not be sufficient to allow the organism to meet its energy requirements aerobically. Partial compensation can be achieved by channeling energy through glycolytic pathways. Arctic char embryos are able to supply energy at 11–23% the aerobic rate (r$\dot{V}O_2$) using anaerobic pathways (Gnaiger, 1979; Gnaiger *et al.*, 1981). The ability to produce a relatively high proportion of normal energy requirements anaerobically probably accounts for the high anoxic resistance of the embryonic stages, especially when it is remembered that a large proportion of normal metabolism goes toward growth rather than maintenance (Smith, 1957, 1958). Gnaiger *et al.* (1981) indicated that lactic acid was the predominant end product in Arctic char embryos, although small amounts of succinic acid were also produced. There was a transient increase in alanine concentration but no change in volitile fatty acid concentration. It thus appears that the high anoxic tolerance is due to high tolerance of lactic acid, rather than activation of more efficient anaerobic pathways. Recovery from anoxia is slow. Gnaiger (1979) observed that it took up to 48 h after restoration of normoxia for the metabolic rate of Arctic char embryos to return to typical aerobic levels.

The metabolic responses of larvae to acute hypoxia are complex and apparently involve both activation of anaerobic pathways and physiological or biochemical changes leading to more efficient utilization of oxygen. Gnaiger (1983a) monitored changes in total and aerobic energy production of Arctic char larvae following abrupt transfer from 100% to 48% air saturation using simultaneous direct and indirect calorimetry. Heat production dropped rapidly to about half the original rate but over the next 24 h gradually increased again to typical normoxic rates. During this period, total heat production measured directly was greater than that calculated on the basis of oxygen consumption, indicating a significant anaerobic contribution ($\approx$20% midway through recovery). Gnaiger (1983a) termed this period the phase of anoxic compensation, although compensatory changes involving aerobic metabolism were also taking place as evidenced by the gradual increase in the rate of oxygen consumption. After about 24 h, the rate of oxygen consumption stabilized at approximately normoxic levels but total heat production began to decline so that *apparent* aerobic energy production was greater than total energy production. Gnaiger (1983a) termed this the phase of conservative compensation and suggested that it reflected coupled glyconeogenesis utilizing lactic acid accumulated during the period of anoxic compensation. Young salmonids apparently either oxidize or reconvert most of the lactate accumulated during periods of oxygen debt back to glycogen (Wieser *et al.,* 1985). This may not be true in all species. Broberg and Kristoffersson (1983) reported that young of the viviparous species *Zoarces viviparous* excrete a large proportion of accumulated lactate.

B. Physiological Hypoxia

Even under normoxic conditions, young fish may be subjected to hypoxemia as a consequence of developmental events or activity. Several investigators have attempted to link physiological hypoxia, as evidenced by increased lactic acid concentrations, with specific developmental events. It appears that in general there is a rise in lactic acid levels during gastrulation, a drop during early organogenesis, and a gradual rise toward hatch (Hishida and Nakano, 1954; Kamler, 1976; Boulekbache, 1981). Kamler (1976) reported that lactic acid levels continued to rise in carp larvae until near the end of endogenous feeding. The ontogeny of energy metabolism is beyond the scope of this review—recent reviews of this area are provided by Terner (1979) and Boulekbache (1981)—but it is important to recognize that many of the responses seen as a result of environmental hypoxia also may be elicited during normal development.

Wieser *et al.* (1985) recently examined the metabolic responses of young rainbow trout during physiological hypoxia induced by forced swimming. Anaerobic energy production (on a mass basis) was found to be independent of body mass and temperature. It was pointed out previously that maximum aerobic energy production of young rainbow trout increases with body size and temperature, up to a maximum at about 12°C (Wieser *et al.*, 1985; Wieser, 1985). The net result is that anaerobic sources are proportionally more important during larval activity than in older fish. In rainbow trout, anaerobic energy production during a 1-min burst of activity was 6–9 times the aerobic production in yolk sac larvae but only 2–6 times the aerobic production in free-swimming larvae and fry. Anaerobic energy production was derived entirely from depletion of phosphocreatine and ATP reserves during the first 30 s of activity. Glycolysis, as evidenced by increased lactic acid production, began to be important during the second 30 s of activity and during the period of recovery. AMP and ADP concentrations did not increase in response to ATP depletion, indicating the probable presence of a very active AMP deaminase. During recovery, phosphocreatine stores were the first to be replenished (67% control values within 5 min, 80% within 10 min). ATP levels remained low for the first 5 min of recovery but had regained 80% of the control value after 10 min. Wieser *et al.* (1985) compared the energy debts acquired during burst swimming (60 s) with energy liberated during recovery. For yolk-sac larvae the ATP required to replenish body energy stores balanced well with the amount of excess energy ($> r\dot{V}O_2$) supplied by anaerobic and aerobic processes during recovery. Larger fish, however, consumed progressively more energy during recovery than was required to repay the true oxygen debt. This was attributed to poststimulus excitement. Energy production during burst activity has not been reported for other species, but it would not be surprising if significant differences are found given the variations in patterns of muscle development and partitioning of energy between oxidative and glycolytic pathways that are known to occur (Forstner *et al.*, 1983; Wieser *et al.*, 1985; Hintleitner *et al.*, 1987).

V. CONCLUSIONS

In general, the ontogeny of respiration is not as well understood for teleosts as for higher vertebrate classes. However, there are signs that the situation is improving, thanks in large part to technological advances that have made it easier to work with such small and delicate organisms. Culture techniques have improved considerably during the past decade, allowing species that previously were difficult to obtain to be reared conveniently in the laboratory. Polarographic electrodes, including microelectrodes, have become readily available. Advances in electronics have permitted the construction of extremely sensitive microcalorimeters, while methods for assaying metabolite concentrations in small samples have improved greatly.

The last major review of this field was that of Blaxter (1969). Since then, a number of important advances have been made. It is now clearly estabished that the boundary layer represents a major barrier to diffusion (Daykin, 1965; Wickett, 1975; Vogel, 1981). A start has been made in recording PO_2 profiles within the respiring egg (Berezovsky et al., 1979; Sushko, 1982; Diez and Davenport, 1987). Studies by Burgreen and co-workers (e.g., Burggren, 1985) with amphibians and Liem (1981) with fish larvae have brought about a greater appreciation of the sophistication of cutaneous gas-exchange structures. Researchers have begun to examine the transition from cutaneous to branchial gas exchange (e.g., De Silva, 1974; Morgan, 1974a,b; McDonald and McMahon, 1977) and from water breathing to air breathing (Hughes et al., 1986) and the possible ecological implications of these transactions (Iwai and Hughes, 1977). Iuchi (1985) and others have provided details concerning the ontogeny of respiratory pigments in fish. Information on energy partitioning during development is now available for a number of species (e.g., Smith, 1957; Laurence, 1969, 1975, 1977, 1978; Houde and Schekter, 1983; Gruber and Wieser, 1983; Rombough, 1987). Researchers have begun to examine the energetics of larval activity in detail, in terms both of functional morphology (Webb and Weihs, 1986) and of energy production (Wieser, 1985; Wieser et al., 1985; Wieser and Forstner, 1986; Hinterleitner et al., 1987). The physiological and morphological responses of the early stages to hypoxia have been shown to be quite different from those of adult fish (Holeton, 1973; McDonald and McMahon, 1977). The idea of using critical dissolved oxygen levels to predict oxygen requirements and sublethal response thresholds has been advanced (Rombough, 1986, 1987).

Much work, though, remains to be done. Most of what is known about respiration during early life is derived from studies of only a few species. Teleosts are a diverse group, and comparative studies of the physiological and anatomical adaptations that permit survival in hypoxic environments are a necessity. Other potentially productive areas of research include studies of how PO_2 profiles within the egg change during development, the problem of gas transfer in egg masses, and the functional aspects of the transition from cutaneous to branchial gas exchange. So far studies of this transition have been restricted to the structural aspects. The nature of the relationship between $\dot{V}O_2$ and tissue mass during early life remains controversial, and further investigations in this area would seem appropriate. Important aspects of aerobic metabolism in very young fish, such as power–performance relationships and the nature of endogenous rhythms, have received little attention. It would appear to be especially important to be able to relate metabolic rates measured in the laboratory with those that occur in the field. More research needs to be conducted on the physiological, morphological, and metabolic responses of embryos and larvae to chronic hypoxia and the ecological significance of such adjustments. Finally, there is the problem of factor interactions. The recent development of a general multivariate dose-response model (Schnute and Jensen, 1986; Jensen *et al.*, 1986) now would appear to make this problem more tractable.

REFERENCES

Adolph, E. F. (1983). Uptakes and uses of oxygen, from gametes to maturity: An overview. *Respir. Physiol.* **53,** 135–160.

Akberali, H. B., and Earnshaw, M. J. (1984). Copper-stimulated respiration in the unfertilized eggs of the Eurasian perch *Perca fluviatus. Comp. Biochem. Physiol. C* **78C,** 349–352.

Alabaster, J. S., and Lloyd, R. (1980). "Water Quality Criteria for Freshwater Fish. Dissolved Oxygen," pp. 127–142. Food Agric. Org., U. N./Butterworth, London.

Al-Adhami, M. A., and Kunz, Y. W. (1976). Haemopoietic centers in the developing angel fish, *Pterophyllum scalare* (Cuvier and Valenciennes). *Wilhelm Roux's Arch. Dev. Biol.* **179,** 393–401.

Alderdice, D. F., and Forrester, C. R. (1971). Effects of salinity, temperature and dissolved oxygen on early development of the Pacific cod *Gadus macrocephalus. J. Fish. Res. Board Can.* **28,** 883–902.

Alderdice, D. F., and Wickett, W. P. (1958). A note on the response of developing chum salmon eggs to free carbon dioxide in solution. *J. Fish. Res. Board Can.* **15,** 797–799.

Alderdice, D. F., Wickett, W. P., and Brett, J. R. (1958). Some effects of temporary exposure to low dissolved oxygen levels on Pacific salmon eggs. *J. Fish. Res. Board Can.* **15,** 229–249.

Alderdice, D. F., Jensen, J. O. T., and Velsen, F. P. J. (1984). Measurement of hydrostatic pressure in salmonid eggs. *Can. J. Zool.* **62,** 1977–1987.

Almatar, S. M. (1984). Effects of acute changes in temperature and salinity on the oxygen uptake of larvae of herring (*Clupea harengus*) and plaice (*Pleuronectes platessa*). *Mar. Biol. (Berlin)* **80,** 117–124.

Amberson, W. R., and Armstrong, P. B. (1933). The respiratory metabolism of *Fundulus heteroclitus* during embryonic development. *J. Cell. Comp. Physiol.* **2,** 381–397.

Bailey, J. E., Rice, S. D., Pella, J. J., and Taylor, S. G. (1980). Effects of seeding density of pink salmon *Oncorhynchus gorbuscha*, eggs on water chemistry, fry characteristics, and fry survival in gravel incubators. *Fish. Bull.* **78,** 649–658.

Balon, E. K. (1975). Reproductive guilds of fishes: A proposal and definition. *J. Fish. Res. Board Can.* **32,** 821–864.

Balon, E. K. (1984). Patterns in the evolution of reproductive styles in fishes. *In* "Fish Reproduction: Strategies and Tactics" (G. W. Potts and R. J. Wooton, eds.), pp. 35–53. Academic Press, London.

Bams, R. A. (1983). Early growth and quality of chum salmon (*Oncorhynchus keta*) produced in keeper channels and gravel incubators. *Can. J. Fish. Aquat. Sci.* **40,** 499–505.

Batty, R. S. (1984). Development of swimming movements and musculature of larval herring (*Clupea harengus*). *J. Exp. Biol.* **110,** 217–229.

Beamish, F. W. H. (1964). Respiration of fishes with special emphasis on standard oxygen consumption. III. Influence of oxygen. *Can. J. Zool.* **42,** 355–366.

Beamish, F. W. H. (1978). Swimming capacity. *In* "Fish Physiology" (W. S. Hoar and D. J. Randall, eds.), Vol. 7, pp. 101–187. Academic Press, New York.

Berezovsky, V. A., Goida, E. A., Mukalov, I. O., and Sushko, B. S. (1979). Experimental study of oxygen distribution in *Misgurnis fossilis* eggs. *Fiziol. Zh. (Kiev)* **25**(4), 379–389; *Can. Transl. Fish. Aquat. Sci.* No. 5209 (1986).

Bishai, H. M. (1960). The effect of gas content of water on larval and young fish. *Z. Wiss. Zool.* **163,** 37–64.

Blaxter, J. H. S. (1969). Development: Eggs and larvae. *In* "Fish Physiology" (W. S. Hoar and D. J. Randall, eds.), Vol. 3, pp. 178–241. Academic Press, New York.

Blaxter, J. H. S., and Hempel, G. (1966). Utilization of yolk by herring larvae. *J. Mar. Biol. Assoc. U.K.* **46,** 219–234.

Boulekbache, H. (1981). Energy metabolism in fish development. *Am. Zool.* **21,** 377–389.

Boulekbache, H., and Devillers, C. (1977). Etude par microscopie à balayage des modifications de la membrane des blastomères au cours des premiers stades du développement de l'oeuf de la truite (*Salmo irideus* Gibb). *C. R. Hebd. Seances Acad. Sci.* **285,** 917–920.

Boyd, M. (1928). A comparison of the oxygen consumption of unfertilized and fertilized eggs of *Fundulus heteroclitus*. *Biol. Bull. (Woods Hole, Mass.)* **55,** 92–100.

Brannon, E. L. (1965). The influence of physical factors on the development and weight of sockeye salmon embryos and alevins. *Int. Pac. Salmon Fish. Comm. Prog. Rep.* **12,** 1–26.

Braum, E. (1973). Einflüsse Chronischen Exogenen Saurstoffmangels auf die Embryogenese Des Herings (*Clupea harengus*). *Neth. J. Sea Res.* **7,** 363–375 (Engl. Abstr.).

Brett, J. R. (1964). The respiratory metabolism and swimming performance of young sockeye salmon. *J. Fish. Res. Board Can.* **21,** 1183–1226.

Brett, J. R. (1970). Fish—The energy cost of living. *In* "Marine Aquaculture" (W. J. McNeil, ed.), pp. 37–52. Oregon State Univ. Press, Corvallis.

Brett, J. R. (1972). The metabolic demand for oxygen in fish, particularly salmonids, and a comparison with other vertebrates. *Respir. Physiol.* **14,** 151–170.

Brett, J. R., and Glass, N. R. (1973). Metabolic rates and critical swimming speeds of sockeye salmon (*Oncorhynchus nerka*) in relation to size and temperature. *J. Fish. Res. Board Can.* **30,** 379–387.

Brett, J. R., and Groves, T. D. D. (1979). Physiological energetics. *In* "Fish Physiology" (W. S. Hoar, D. J. Randall, and J. R. Brett, eds.), Vol. 8, pp. 279–352. Academic Press, New York.

Broberg, S., and Kristoffersson, R. (1983). Oxygen consumption and lactate accumulation in the intraovarian embryos and young of the viviparous fish *Zoarces viviparus* in relation to decreasing water oxygen concentration. *Ann. Zool. Fenn.* **20,** 301–306.

Brooke, L. T., and Colby, P. J. (1980). Development and survival of embryos of lake herring *Coregonus artedii* at different constant oxygen concentrations and temperatures. *Prog. Fish-Cult.* **42,** 3–9.

Brownell, C. L. (1980). Water quality requirements for 1st feeding in marine fish larvae. 2. pH, oxygen and carbon dioxide. *J. Exp. Mar. Biol. Ecol.* **44,** 285–298.

Brummett, A. R., and Vernberg, W. B. (1972). Oxygen consumption in anterior vs. posterior embryonic shield of *Fundulus heteroclitus*. *Biol. Bull. (Woods Hole, Mass.)* **143,** 296–303.

Brungs, W. A. (1971). Chronic effects of low dissolved oxygen concentrations on the fathead minnow (*Pimephales promelas*). *J. Fish. Res. Board Can.* **28,** 1119–1123.

Burggren, W. (1984). Transition of respiratory processes during amphibian metamorphosis: From egg to adult. *In* "Respiration and Metabolism of Embryonic Vertebrates" (R. S. Seymour, ed.), pp. 31–53. Martinus Nijhoff/Dr. W. Junk Publishers, Dordrecht, The Netherlands.

Burggren, W. (1985). Gas exchange, metabolism, and "ventilation" in gelatinous frog egg masses. *Physiol. Zool.* **58,** 503–514.

Burggren, W., and Mwalukoma, A. (1983). Respiration during chronic hypoxia and hyperoxia in larvae and adult bullfrogs (*Rana catesbeiana*). I. Morphological responses of lungs, skin and gills. *J. Exp. Biol.* **105,** 191–203.

Burggren, W. W., and West, N. H. (1982). Changing respiratory importance of gills, lungs and skin during metamorphosis in the bullfrog *Rana catesbeiana*. *Respir. Physiol.* **47,** 151–164.

Carlson, A. R., and Siefert, R. E. (1974). Effects of reduced oxygen on the embryos and larvae of lake trout (*Salvelinus namaycush*) and largemouth bass (*Micropterus salmoides*). *J. Fish Res. Board Can.* **31,** 1393–1396.

Carlson, A. R., Siefert, R. E., and Herman, L. J. (1974). Effects of lowered dissolved oxygen concentration on channel catfish (*Ictalurus punctatus*) embryos and larvae. *Trans. Am. Fish. Soc.* **103,** 623–626.

Cetta, C. M., and Capuzzo, J. M. (1982). Physiological and biochemical aspects of embryonic and larval development of the winter flounder *Psuedopleuronectes americanus*. *Mar. Biol. (Berlin)* **71,** 327–337.

Chapman, G. (1986). "Ambient Water Quality Criteria for Dissolved Oxygen. Freshwater Aquatic Life," Intern. Rep. U. S. Environmental Protection Agency, Washington, D.C.

Chernikova, V. V., (1964). Variations in the respiration rate of young *Coregonus nasus* (Pallos) with changes in the temperature and the oxygen and carbon dioxide concentration of the water. *In* "Fish Physiology in Acclimation and Breeding" Translated from Russian by the Israel Program for Scientific Translations, 1970 (T. I. Privol'nev, ed.), Keter Press, Jerusalem.

Clements, L. C., and Hoss, D. E. (1977). Effects of acclimation time on larval flounder *Paralichthys* sp. oxygen consumption. *Assoc. Southeast Biol., Bull.* **24**(2), 43.

Coble, D. W. (1961). Influence of water exchange and dissolved oxygen in redds on survival of steelhead trout embryos. *Trans. Am. Fish. Soc.* **90**, 469–474.

Coughlan, D. J., and Gloss, S. P. (1984). Early morphological development of gills in smallmouth bass (*Micropterus dolomieui*). *Can. J. Zool.* **62**, 951–958.

Craik, J. C. A. (1985). Egg quality and egg pigment content in salmonid fishes. *Aquaculture* **47**, 61–88.

Craik, J. C. A., and Harvey, S. M. (1986). The carotenoids of eggs of wild and farmed Atlantic salmon and their changes during development to the start of feeding. *J. Fish Biol.* **29**, 549–565.

Czeczuga, B. (1979). Carotenoids in fish. XIX. Carotenoids in the eggs of *Onchorhynchus keta* (Walbaum). *Hydrobiologia* **63**, 45–47.

Czihak, G., Peter, R., Puschendorf, B., and Grunicke, H. (1979). Some data on the basic metabolism of trout eggs. *J. Fish Biol.* **15**, 185–194.

Dabrowski, K. R. (1986). A new type of metabolism chamber for the determination of active and postprandial metabolism of fish, and consideration of results for coregonid and salmon juveniles. *J. Fish Biol.* **28**, 105–117.

Dabrowski, K., and Kaushik, S. J. (1984). Rearing of coregonid *Coregonus schinzi palea* (Cuv. et Val.) larvae using dry and live food. II. Oxygen consumption and nitrogen excretion. *Aquaculture* **41**, 333–344.

Dabrowski, K., Kaushik, S. J., and Luquet, P. (1984). Metabolic utilization of body stores during the early life of whitefish, *Coregonus lavaretus* L. *J. Fish Biol.* **24**, 721–729.

Dalla Via, G. J. (1983). Bacterial growth and antibiotics in animal respirometry. *In* "Polarographic Oxygen Sensors" (E. Gnaiger and H. Forstner, eds.), pp. 202–218. Springer-Verlag, Berlin and New York.

Davenport, J. (1976). A technique for the measurement of oxygen consumption in small aquatic organisms. *Lab. Pract.* **25**, 693–695.

Davenport, J. (1983). Oxygen and the developing eggs and larvae of the lumpfish, *Cyclopterus lumpus*. *J. Mar. Biol. Assoc. U.K.* **63**, 633–640.

Davenport, J., and Lønning, S. (1980). Oxygen uptake in developing eggs and larvae of the cod, *Gadus morhua*. *J. Fish Biol.* **16**, 249–256.

Davenport, J., Lønning, S., and Saethre, L. J. (1979). The effects of ekofisk oil extract upon oxygen uptake in eggs and larvae of the cod (*Gadus morhua*). *Astarte* **12**, 31–34.

Davis, J. C. (1975). Minimal dissolved oxygen requirements of aquatic life with emphasis on Canadian species. A review. *J. Fish. Res. Board Can.* **32**, 2295–2332.

Daykin, P. N. (1965). Application of mass transfer theory to the problem of respiration of fish eggs. *J. Fish. Res. Board Can.* **22**, 159–171.

Dejours, P. (1981). "Principles of Comparative Respiratory Physiology," 2nd ed. Elsevier/North-Holland Biomedical Press, Amsterdam.

De Silva, C. (1974). Development of the respiratory system in herring and plaice larvae. *In* "The Early Life History of Fish" (J. S. Blaxter, ed.), pp. 465–485. Springer-Verlag, Berlin and New York.

DeSilva, C. D., and Tytler, P. (1973). The influence of reduced environmental oxygen on the metabolism and survival of herring and plaice larvae. *Neth. J. Sea Res.* **7**, 345–362.

DeSilva, C. D., Premawansa, S., and Keemiyahetty, C. N. (1986). Oxygen consumption in *Oreochromis niloticus* (L.) in relation to development, salinity, temperature and time of day. *J. Fish Biol.* **29**, 267–277.

Devillers, C. (1965). Respiration and morphogenesis in the teleostean egg. *Fish Res. Board Can., Transl. Ser.* **3909**, 1–23.

Diez, J. M., and Davenport, J. (1987). Embryonic respiration in the dogfish (*Scyliorhinus canicula* L.) *J. Mar. Biol. Assoc. U.K.* **67**, 249–261.

DiMichele, L., and Powers, D. A. (1982). LDH-B genotype specific hatching times of *Fundulus heteroclitus* embryos. *Nature (London)* **296**, 563–564.

DiMichele, L., and Powers, D. A. (1984a). The relationship between oxygen consumption rate and hatching in *Fundulus heteroclitus*. *Physiol. Zool.* **57**, 46–51.

DiMichele, L., and Powers, D. A. (1984b). Developmental and oxygen consumption rate differences between lactate dehydrogenase-B genotypes of *Fundulus heteroclitus* and their effect on hatching time. *Physiol. Zool.* **57**, 52–56.

DiMichele, L., and Taylor, M. H. (1980). The mechanism of hatching in *Fundulus heteroclitus:* development and physiology. *J. Exp. Zool.* **217**, 73–79.

DiPrampero, P. E. (1985). Metabolic and circulatory limitations to VO₂ max at the whole animal level. *J. Exp. Biol.* **115**, 319–331.

Doudoroff, P., and Shumway, D. L. (1970). Dissolved oxygen requirements of freshwater fishes. *FAO Fish. Tech. Pap.* **86**, 1–291.

Dudley, R. G., and Eipper, A. W. (1975). Survival of largemouth bass embryos at low dissolved oxygen concentrations. *Trans. Am. Fish. Soc.* **104**, 122–128.

Durborow, R. M., and Avault, J. W., Jr. (1985). Differences in mortality among full-sib channel catfish families at low dissolved oxygen. *Prog. Fish-Cult.* **47**, 14–20.

Durve, V. S., and Sharma, M. S. (1977). Oxygen requirements of the early stages of the major carp *Lebeo calbasu*. *J. Anim. Morphol. Physiol.* **24**, 391–393.

Eddy, R. M. (1972). The influence of dissolved oxygen concentration and temperature on the survival and growth of chinook salmon embryos and fry. M.Sc. Thesis, Oregon State University, Corvallis.

Einsele, W. (1965). Problems of fish-larvae survival in nature and the rearing of economically important middle European freshwater fishes. *Calif. Coop. Oceanic. Fish. Invest. Rep.* **10**, 24–30.

Eldridge, M. B., Echeverria, T., and Whipple, J. A. (1977). Energetics of Pacific herring *Clupea harengus pallasi* embryos and larvae exposed to low concentrations of benzene, a mono aromatic component of crude oil. *Trans. Am. Fish. Soc.* **106**, 452–461.

Eldridge, M. B., Whipple, J. A., and Bowers, M. J. (1982). Bioenergetics and growth of striped bass, *Morone saxatilis*, embryos and larvae. *Fish. Bull.* **80**, 461–474.

Elliot, J. M., and Davison, W. (1975). Energy equivalents of oxygen consumption in animal energetics. *Oecologia* **19**, 195–201.

Eriksen, C. H., and Feldmeth, C. R. (1967). A water-current respirometer. *Hydrobiologia* **29**, 495–504.

European Inland Fisheries Advisory Commission (1973). Water quality criteria for European freshwater fish. Report on dissolved oxygen and inland fisheries. Prepared by European Inland Fisheries Advisory Commission Working Party on Water Quality Criteria for European Freshwater Fish. *EIFAC Tech. Pap.* **19**, 1–10.

Feder, M. E. (1983a). Response to acute aquatic hypoxia in larvae of the frog *Rana berlandieri*. *J. Exp. Biol.* **104**, 79–95.

Feder, M. E. (1983b). Effect of hypoxia and body size on the energy metabolism of lungless tadpoles, *Bufo woodhousei*, and air-breathing anuran larvae. *J. Exp. Zool.* **228**, 11–19.

Feder, M. E., and Burggren, W. W. (1985). Cutaneous gas exchange in vertebrates: Design, patterns, control and implications. *Biol. Rev. Cambridge Philos. Soc.* **60**, 1–45.

Fisher, K. C. (1942). The effect of temperature on the critical oxygen pressure for heart beat frequency in embryos of Atlantic salmon and speckled trout. *Can. J. Res., Sect. D* **20**, 1–12.

Florez, F. (1972). Influence of oxygen concentration on growth and survival of larvae and juveniles of the ide *Idus idus. Rep. Inst. Freshwater Res., Drottningholm* **52**, 65–73.

Forstner, H. (1983). An automated multiple-chamber intermittent-flow respirometer. *In* "Polarographic Oxygen Sensors" (E. Gnaiger and H. Forstner, eds.), pp. 111–126. Springer-Verlag, Berlin and New York.

Forstner, H., Hinterleitner, S., Mahr, K., and Wieser, W. (1983). Towards a better definition of "metamorphosis" in *Coregonus* sp: Biochemical, histological and physiological data. *Can. J. Fish. Aquat. Sci.* **40**, 1224–1232.

Fry, F. E. J. (1957). The aquatic respiration of fish. *In* "The Physiology of Fishes" (M. E. Brown, ed.), Vol. 1, pp. 1–64. Academic Press, New York.

Fry, F. E. J. (1971). The effect of environmental factors on the physiology of fish. *In* "Fish Physiology" (W. S. Hoar and D. J. Randall, eds.), Vol. 6, pp. 1–98. Academic Press, New York.

Gallagher, M. L., Kane, E., and Courtney, J. (1984). Differences in oxygen consumption and ammonia production among American elvers (*Anguilla rostrata*). *Aquaculture* **40**, 183–187.

Garside, E. T. (1959). Some effects of oxygen in relation to temperature on the development of lake trout embryos. *Can. J. Zool.* **37**, 689–698.

Garside, E. T. (1966). Effects of oxygen in relation to temperature on the development of embryos of brook trout and rainbow trout. *J. Fish. Res. Board Can.* **23**, 1121–1134.

Geffen, A. J. (1983). The deposition of otolith rings in Atlantic salmon, *Salmo salar* L., embryos. *J. Fish Biol.* **23**, 467–474.

Giles, M. A., and Vanstone, W. E. (1976). Ontogenetic variation in the multiple hemoglobins of coho salmon *Oncorhynchus kisutch* and effect of environmental factors on their expression. *J. Fish. Res. Board Can.* **33**, 1144–1149.

Giorgi, A. E. (1981). "The Environmental Biology of the Embryos, Egg Masses and Nesting Sites of the Lingcod, *Ophiodon elongatus*," NWAFC Processed Rep. 81–06. Northwest and Alaska Fisheries Center, National Marine Fisheries Service, U.S. Department of Commerce, Washington, D.C.

Giorgi, A. E., and Congleton, J. L. (1984). Effect of current velocity on development and survival of lingcod, *Ophiodon elongatus*, embryos. *Environ. Biol. Fishes* **10**, 15–28.

Gnaiger, E. (1979). Direct calorimetry in ecological energetics. Long term monitoring of aquatic animals. *Experientia, Suppl.* **37**, 155–165.

Gnaiger, E. (1983a). Calculation of energetic and biochemical equivalents of respiratory oxygen consumption. *In* "Polarographic Oxygen Sensors" (E. Gnaiger and R. Forstner, eds.), pp. 337–345. Springer-Verlag, Berlin and New York.

Gnaiger, E. (1983b). The twin-flow microrespirometer and simultaneous calorimetry. *In* "Polarographic Oxygen Sensors" (E. Gnaiger and R. Forstner, eds.), pp. 134–166. Springer-Verlag, Berlin and New York.

Gnaiger, E., Lackner, R., Ortner, M., Putzer, V., and Kaufmann, R. (1981). Physiological and biochemical parameters in anoxic and aerobic energy metabolism of embryonic salmonids, *Salvelinus alpinus. Eur. J. Physiol., Suppl.* **391**, R57 (abstr.).

Gottwald, St. (1965). The influence of temporary oxygen deprivation on embryonic development of rainbow trout (*Salmo gairdneri* Rich.). *Z. Fisch. Deren Hilfswiss.* **13**, 63–84.

Gray, J. (1926). The growth of fish. I. The relationship between embryo and yolk sac in *Salmo fario. J. Exp. Biol.* **4**, 215–225.

Grigor'yeva, M. B. (1967). The effect of schooling on the gas exchange of fishes. *In* "Fish Behavior and Reception," pp. 127–132. Nauka Press, Moscow.

Groot, E. P., and Alderdice, D. F. (1985). Fine structure of the external egg membrane of five species of Pacific salmon and steelhead trout. *Can. J. Zool.* **63**, 552–566.

Gruber, K., and Wieser, W. (1983). Energetics of development of the Alpine charr, *Salvelinus alpinus*, in relation to temperature and oxygen. *J. Comp. Physiol.* **149**, 485–493.

Gulidov, M. V. (1969). Embryonic development of the pike, *Esox lucius*, when incubated under different oxygen conditions. *Probl. Ichthyol.* **9**, 841–851.

Gulidov, M. V. (1974). The effect of different oxygen conditions during incubation on the survival and some of the developmental characteristics of the "Verkhova" (*Leucaspius delineatus*) in the embryonic period. *J. Ichthyol. (Engl. Transl.)* **14**(3), 393–397.

Gulidov, M. V., and Popova, K. S. (1978). The influence of increased O_2 concentrations on the survival and hatching of the embryos of the bream *Abramis brama. J. Ichthyol. (Engl. Transl.)* **17**, 174–177.

Hamdorf, K. (1961). The influence of the environmental factors (O_2 partial pressure and temperature) on the embryonic and larval development of the rainbow trout (*Salmo irideus* Gibb). *Z. Vergl. Physiol.* **44**, 523–549.

Hamor, T., and Garside, E. T. (1975). Regulation of oxygen consumption by incident illumination in embryonated ova of Atlantic salmon *Salmo salar. Comp. Biochem. Physiol. A* **52A**, 277–280.

Hamor, T., and Garside, E. T. (1976). Development rates of embryos of Atlantic salmon, *Salmo salar* L., in response to various levels of temperature, dissolved oxygen and water exchange. *Can. J. Zool.* **54**, 1912–1917.

Hamor, T., and Garside, E. T. (1977). Size relations and yolk utilization in embryonated ova and alevins of Atlantic salmon *Salmo salar* in various combinations of temperature and dissolved oxygen. *Can. J. Zool.* **55**, 1892–1898.

Hamor, T., and Garside, E. T. (1979). Hourly and total oxygen consumption by ova of Atlantic salmon, *Salmo salar* L., during embryogenesis, at two temperatures and three levels of dissolved oxygen. *Can. J. Zool.* **57**, 1196–1200.

Hartvig, M., and Weber, R. E. (1984). Blood adaptations for maternal–fetal oxygen transfer in the viviparous teleost, *Zoarces viviparus* L. *In* "Respiration and Metabolism of Embryonic Vertebrates" (R. S. Seymour, ed.), pp. 17–30. Martinus Nijhoff/ Dr. W. Junk Publishers, Dordrect, The Netherlands.

Hayes, F. R., Wilmot, I. R., and Livingstone, D. A. (1951). The oxygen consumption of the salmon egg in relation to development and activity. *J. Exp. Zool.* **116**, 377–395.

Hempel, G. (1979). "Early Life History of Marine Fish. The Egg Stage," Washington Sea Grant Publ. Univ. of Washington Press, Seattle.

Hempel, I., and Hempel, G. (1971). An estimate of mortality in eggs of North Sea herring (*Clupea harengus* L.). *Rapp. P.-V. Reun., Cons. Int. Explor. Mer* **160**, 24–26.

Hettler, W. F. (1976). Influence of temperature and salinity on routine metabolic rate and growth of young Atlantic menhaden. *J. Fish Biol.* **8**, 55–65.

Hinterleitner, S., Platzer, U., and Wieser, W. (1987). Development of the activities of oxidative, glycolytic, and muscle enzymes during early larval life in three families of freshwater fish. *J. Fish Biol.* **30**, 315–316.

Hishida, T. O., and Nakano, E. (1954). Respiratory metabolism during fish development. *Embryologia* **2**, 67–79.

Hitchman, M. L. (1978). "Measurement of Dissolved Oxygen." Wiley, New York.

Holeton, G. F. (1971). Respiratory and circulatory responses of rainbow trout larvae to carbon monoxide and to hypoxia. *J. Exp. Biol.* **55**, 683–694.

Holeton, G. F. (1973). Respiration of Arctic char *Salvelinus alpinus* from a high Arctic lake. *J. Fish. Res. Board Can.* **30**, 717–723.

Holliday, F. G. T., Blaxter, J. H. S., and Lasker, R. (1964). Oxygen uptake of developing eggs and larvae of the herring (*Clupea harengus*). *J. Mar. Biol. Assoc. U.K.* **44**, 711–723.

Hose, J. E., and Puffer, H. W. (1984). Oxygen consumption of grunion *Leuresthes tenuis* embryos exposed to the petroleum hydrocarbon benzo-a-pyrene. *Environ. Res.* **35**, 413–420.

Houde, E. D., and Schekter, R. C. (1983). Oxygen uptake and comparative energetics among eggs and larvae of three subtropical marine fishes. *Mar. Biol. (Berlin)* **72**, 283–293.

Hughes, G. M., Datta Munshi, J. S., and Ojha, J. (1986). Post-embryonic development of water and air-breathing organs of *Anabas testudineus* (Bloch). *J. Fish. Biol.* **29**, 443–450.

Hunter, J. R. (1972). Swimming and feeding behavior of larval anchovy *Engraulis mordax. Fish. Bull.* **70**, 821–838.

Hunter, J. R. (1981). Feeding ecology and predation of marine fish larvae. *In* "Marine Fish Larvae: Morphology, Ecology and Relation to Fisheries" (R. Lasker, ed.), Washington Sea Grant Program, pp. 33–79. Univ. of Washington Press, Seattle.

Hunter, J. R., and Kimbrell, C. A. (1980). Early life history of Pacific mackerel, *Scomber japonicus. Fish. Bull.* **78**, 89–102.

Hyman, L. H. (1921). The metabolic gradients of vertebrate embryos. I. Teleost embryos. *Biol. Bull. (Woods Hole, Mass.)* **40**, 32–72.

Ingermann, R. L., and Terwilliger, R. C. (1981a). Oxygen affinities of maternal and fetal hemoglobins of the viviparous sea perch *Embiotoca lateralis. J. Comp Physiol. B* **142B**, 523–532.

Ingermann, R. L., and Terwilliger, R. C. (1981b). Intraerythrocytic organic phosphates of fetal and adult sea perch *Embiotoca lateralis*. Their role in maternal–fetal oxygen transport. *J. Comp. Physiol. B* **144B**, 253–260.

Ingermann, R. L., and Terwilliger, R. C. (1982). Blood parameters and facilitation of maternal–fetal oxygen transfer in a viviparous fish (*Embiotoca lateralis*). *Comp. Biochem. Physiol.* **73A**, 497–502.

Ingermann, R. L., and Terwilliger, R. C. (1984). Facilitation of maternal–fetal oxygen transfer in fishes: Anatomical and molecular specialization. *In* "Respiration and Metabolism of Embryonic Vertebrates" (R. S. Seymour, ed.), pp. 1–15. Martinus Nijhoff/Dr. W. Junk Publishers, Dordrecht, The Netherlands.

Ingermann, R. L., Terwilliger, R. C., and Roberts, M. S. (1984). Foetal and adult blood oxygen affinities of the viviparous Seaperch, *Embiotoca lateralis. J. Exp. Biol.* **108**, 453–457.

Ishida, J. (1985). Hatching enzyme: Past, present and future. *Zool. Sci.* **2**, 1–10.

Itazawa, Y., Matsumoto, T., and Kanda, T. (1978). Group effects on physiological and ecological phenomena in fish. Part 1. Group effect on the oxygen consumption of the rainbow trout and the medaka. *Bull. Jap. Soc. Sci. Fish.* **44**, 965–970.

Iuchi, I. (1973a). The post-hatching transition of erythrocytes from larval to adult type in the rainbow trout, *Salmo gairdneri irideus. J. Exp. Zool.* **184**, 383–396.

Iuchi, I. (1973b). Chemical and physiological properties of the larval and the adult hemoglobins in rainbow trout, *Salmo gairdneri irideus. Comp. Biochem. Physiol. B* **44B**, 1087–1101.

Iuchi, I. (1985). Cellular and molecular bases of the larval-adult shift of hemoglobins in fish. *Zool. Sci.* **2**, 11–23.

Iuchi, I., and Yamagami, K. (1969). Electrophoretic pattern of larval hemoglobins of the salmonid fish, *Salmo gairdneri irideus*. *Comp. Biochem. Physiol.* **28**, 977–979.

Iuchi, I., and Yamamoto, M. (1983). Erythropoiesis in the developing rainbow trout, *Salmo gardneri irideus:* Histological and immunochemical detection of erythropoietic organs. *J. Exp. Zool.* **226**, 409–417.

Iuchi, I., Suzuki, R., and Yamagami, K. (1975). Ontogenetic expression of larval and adult hemoglobin phenotypes in the intergeneric salmonid hybrids. *J. Exp. Zool.* **192**, 57–64.

Ivlev, V. S. (1960a). "Active Metabolic Intensity in Salmon Fry (*Salmo salar* L.) at Various Rates of Activity," Pap. No. 213 (mimeo). Salmon and Trout Comm., Int. Counc. Explor. Sea, Copenhagen.

Ivlev, V. S. (1960b). On the utilization of food by planktonophage fishes. *Bull. Math. Biophys.* **22**, 371–389.

Iwai, T., and Hughes, G. M. (1977). Preliminary morphometric study on gill development in black sea bream (*Acanthopagrus schlegeli*). *Bull. Jpn. Soc. Sci. Fish.* **43**, 929–934.

Jensen, J. O. T., Schnute, J., and Alderdice, D. F. (1986). Assessing juvenile salmonid response to gas supersaturation using a general multivariate dose-response model. *Can. J. Fish. Aquat. Sci.* **43**, 1694–1709.

Jitariu, P., Badilita, M., and Costea, E. (1971). Oxygen consumption during early stages in some selected breeds of carp in comparison with a local nonselected breed. *Stud. Cercet. Biol., Ser. Zool.* **23**, 213–217.

Johns, D. M., and Howell, W. H. (1980). Yolk utilization in summer flounder (*Paralichthys dentatus*) embryos and larvae reared at two temperatures. *Mar. Ecol.: Prog. Ser.* **2**, 1–8.

Johnson, R. A. (1980). Oxygen transport in salmon spawning gravel. *Can. J. Fish. Aquat. Sci.* **37**, 155–162.

Johnston, I. A. (1982). Physiology of muscle in hatchery raised fish. *Comp. Biochem. Physiol. B* **73B**, 105–124.

Jones, D. R., and Randall, D. J. (1978). The respiratory and circulatory systems during exercise. *In* "Fish Physiology" (W. S. Hoar and D. J. Randall, eds.), Vol. 7, pp. 425–501. Academic Press, New York.

Jones, M. P., Holliday, F. G. T., and Dunn, A. E. G. (1966). The ultrastructure of the epidermis of larvae of herring (*Clupea harengus*) in relation to rearing salinity. *J. Mar. Biol. Assoc. U.K.* **46**, 235–239.

Kamler, E. (1972). Respiration of carp in relation to body size and temperature. *Pol. Arch. Hydrobiol.* **19**, 325–331.

Kamler, E. (1976). Variability of respiration and body composition during early developmental stages of carp. *Pol. Arch. Hydrobiol.* **23**, 431–485.

Kamler, E., Matlak, O., and Srokosz, K. (1974). Further observations on the effect of sodium salt of 2,4-D on early developmental stages of carp. *Cyprinus carpio. Pol. Arch. Hydrobiol.* **21**, 481–502.

Kanda, T., and Itazawa, Y. (1981). Group effect on oxygen consumption and growth of catfish eel *Plotosus anguillaris*. *Bull. Jap Soc. Sci. Fish.* **47**, 341–346.

Karpenko, G. I., and Proskurina, E. S. (1970). Oxygen consumption rates of larvae and fry of *Cyprinidae* in relation to their ecology. *Proc. All-Union Res. Inst. Mar. Fish Oceanog.*, 1970. **4**, 7–13. Translated from Russian by *Fish Res. Board Can., Transl. Ser.* **2109**, 9.

Kaur, K., and Toor, H. S. (1978). Effect of dissolved oxygen on the survival and hatching of eggs of scale carp. *Prog. Fish Cult.* **40**, 35–37.

Kaushik, S. J., and Dabrowski, K. (1983). Postprandial metabolic changes in larval and juvenile carp (*Cyprinus carpio*). *Reprod. Nutr. Dev.* **23**, 223–234.

Kaushik, S. J., Dabrowski, K., and Luquet, P. (1982). Patterns of nitrogen excretion and oxygen consumption during ontogenesis of common carp (*Cyprinus carpio*). *Can. J. Fish. Aquat. Sci.* **39**, 1095–1105.

Khakimullin, A. A. (1985). Levels of standard and basal metabolism in young Siberian sturgeon, *Acipenser baeri* (*Acipenseridae*). *J. Ichthyol.* (*Engl. Transl.*) **24**(5), 29–33.

King, D. P. F. (1977). Influence of temperature, dissolved oxygen and salinity on incubation and early larval development of the south-west African pilchard (*Sardinops ocellata*). *S. Afr. Sea Fish Branch Invest. Rep.* **114**, 1–35.

Kirsch, R., and Nonnotte, G. (1977). Cutaneous respiration in three freshwater teleosts. *Respir. Physiol.* **29**, 339–354.

Kjelson, M. A., and Johnson, G. A. (1976). Further observations of the feeding ecology of post larval pinfish *Cagodon rhomboides* and spot *Leiostomus xanthurus*. *Fish. Bull.* **74**, 423–432.

Klekowski, R. Z., Korde, B., and Kaniuvska-Prus, M. (1977). The effect of sodium salt of 2,4-D on oxygen consumption of *Misgurnus fossilis* during early embryonal development. *Pol. Arch. Hydrobiol.* **24**, 413–422.

Klekowski, R. Z., Opalinski, K. W., and Gorbunova, N. N. (1980). Respiratory metabolism in early ontogeny of 2 Pacific flying fish species *Hirundichthys marginatus* and *Cheilopogon unicolor* and of Pacific sardine *Sardinaps sagax sagax*. *Pol. Arch. Hydrobiol.* **27**, 537–548.

Knudsen, J., Famme, P., and Hansen, E. S. (1983). A microcalorimeter system for continuous determination of the effect of oxygen on aerobic–anaerobic metabolism and metabolite exchange in small aquatic animals and cell preparations. *Comp. Biochem. Physiol. A* **74A**, 63–66.

Kobel, H. R., and Wolff, J. (1983). Two transitions of haemoglobin expression in *Xenopus:* From embryonic to larval and from larval to adult. *Differentiation* **24**, 24–26.

Konchin, V. V. (1971). The rate of oxygen consumption in the early ontogeny of the summer bakhtak [*Salmo ischchan* (Kessler)] when placed in respirometers in groups and singly. *J. Ichthyol.* (*Engl. Transl.*) **11**(6), 916–926.

Konchin, V. V. (1981). Analysis of the rate of gas exchange in young roach, *Rutilus rutilus*. *J. Ichthyol.* (*Engl. Transl.*) **21**(4), 122–132.

Konchin, V. V. (1982). Some characteristics of gaseous interchange in larvae of three species of freshwater fish with differing ecology under conditions of isolation and crowding. *Vestn. Mosk. Univ., Ser. 16: Biol.* **37**(2), 43–49.

Konstantinov, A. S. (1980). Comparative evaluation of the intensity of respiration in fishes. *J. Ichthyol.* (*Engl. Transl.*) **20**(1), 98–104.

Korsgaard, B., and Andersen, F. Q. (1985). Embryonic nutrition, growth and energetics in *Zoarces viviparus* L. as indication of a maternal–fetal trophic relationship. *J. Comp. Physiol. B* **155B**, 437–444.

Korwin-Kossakowski, M., Jowko, G., and Jerierska, B. (1981). The influence of group effect on oxygen consumption of carp (*Cyprinus carpio* L.) larvae. *Rocz. Nauk Roln., Ser. H* **99**, 49–62.

Kryzanowsky, S. G. (1934). Die Atmungsorgane der Fishlarven. *Zool. Jahrb., Abt. Anat. Ontog. Tiere* **58**, 21–61.

Kudrinskaya, O. I. (1969). Metabolic rate in the larvae of pike-perch, perch, carp-bream and roach. *Hydrobiol. J.* **5**(4), 68–72.

Lanzing, W. J. R. (1976). A temporary respiratory organ in the tail of *Tilapia mossambica* fry. *Copeia*, pp. 800–802.

Lasker, R. (1962). Efficiency and rate of yolk utilization by developing embryos and larvae of the Pacific sardine *Sardinops caerula* (Girard). *J. Fish. Res. Board Can.* **19**, 867–875.

Lasker, R., and Theilacker, G. H. (1962). Oxygen consumption and osmoregulation by single Pacific sardine eggs and larvae (*Sardinops caerulea*, Girard). *J. Cons. Int. Explor. Mer* **27**, 25–33.

Laurence, G. C. (1969). The energy expenditure of largemouth bass larvae, *Micropterus salmoides*, during yolk absorption. *Trans. Am. Fish. Soc.* **98**, 398–405.

Laurence, G. C. (1973). Influence of temperature on energy utilization of embryonic and prolarval tautog (*Tautoga onitis*) *J. Fish Res. Board Can.* **30**, 435–442.

Laurence, G. C. (1975). Laboratory growth and metabolism of the winter flounder *Pseudopleuronectes americanus* from hatching through metamorphosis at three temperatures. *Mar. Biol. (Berlin)* **32**, 223–229.

Laurence, G. C. (1977). A bioenergetic model for the analysis of feeding and survival potential of winter flounder, *Pseudopleuronectes americanus*, larvae during the period from hatching to metamorphosis. *Fish. Bull.* **75**, 529–546.

Laurence, G. C. (1978). Comparative growth, respiration and delayed feeding abilities of larval cod (*Gadus morhua*) and haddock (*Melanogrammus aeglefinus*) as influenced by temperature during laboratory studies. *Mar. Biol. (Berlin)* **50**, 1–7.

Liem, K. F. (1981). Larvae of air breathing fishes as countercurrent flow devices in hypoxic environments. *Science* **211**, 1177–1179.

Lindroth, A. (1942). Sauerstoffverbrauch der Fische. II. Verschiedene Entwicklungs— und Alterstadien von Lach und Hecht. *Z. Vergl. Physiol.* **29**, 583–594.

Loeffler, C. A. (1971). Water exchange in the pike egg. *J. Exp. Biol.* **55**, 797–811.

Loeffler, C. A., and Lovtrup, S. (1970). Water balance in the salmon egg. *J. Exp. Biol.* **52**, 291–298.

Lomholt, J. P., and Johansen, K. (1979). Hypoxia acclimation in carp—how it affects O_2 uptake, ventilation and O_2 extraction from water. *Physiol. Zool.* **52**, 38–49.

Lønning, S. (1972). Comparative electron microscopic studies of teleostean eggs with special reference to the chorion. *Sarsia* **49**, 41–48.

Lukina, O. V. (1973). Respiratory rate of the North Okhotsk chum salmon (*Oncorhynchus keta* (Walb)). *J. Ichthyol. (Engl. Transl.)* **13**(3), 425–430.

MacCrimmon, H. R., and Kwain, W. H. (1969). Influence of light on early development and meristic characters in the rainbow trout, *Salmo gairdneri* Richardson. *Can. J. Zool.* **47**, 631–637.

McDonald, D. G., and McMahon, B. R. (1977). Respiratory development in Artic char *Salvelinus alpinus* under conditions of normoxia and chronic hypoxia. *Can. J. Zool.* **55**, 1461–1467.

McElman, J. F., and Balon, E. K. (1979). Early ontogeny of walleye (*Stizostedion vitreum*) with steps of saltatory development. *Environ. Biol. Fishes* **4**, 309–348.

McNeil, W. J. (1966). Effect of the spawning bed environment on reproduction of pink and chum salmon. *Fish. Bull.* **65**, 495–523.

Malyukina, G. A., and Konchin, V. V. (1969). Development of group effects during ontogenetic development of the Baltic salmon and the Sevan trout. *Probl. Ichthyol.* **9**, 292–297.

Markiewicz, F. (1960). Development of blood vessels in branchial arches of the pike (*Esox lucius* L.). *Acta Biol. Cracov. Zool. Ser.* **111**, 163–172.

Marshall, S. M., Nicholls, A. G., and Orr, A. P. (1937). On the growth and feeding of larval and post-larval stages of the Clyde herring. *J. Mar. Biol. Assoc. U. K.* **22**, 245–267.

Mason, J. C. (1969). Hypoxial stress prior to emergence and competition among coho salmon fry. *J. Fish. Res. Board Can.* **26**, 63–91.

Mikulin, A. Ye., and Soin, S. G. (1975). The functional significance of carotenoids in the embryonic development of teleosts. *J. Ichthyol. (Engl. Transl.)* **15**, 749–759.

Miller, R. W. (1972). Three methods for determining dissolved oxygen concentrations near fish embryos. *Prog. Fish-Cult.* **34**, 39–42.

Mishra, A. P., and Singh, B. R. (1979). Oxygen uptake through water during early life of *Anabas testudineus. Hydrobiologia* **66**, 129–134.

Morgan, M. (1974a). The development of gill arches and gill blood vessels of the rainbow trout, *Salmo gairdneri. J. Morphol.* **142**, 351–364.

Morgan, M. (1974b). Development of secondary lamellae of the gills of the trout *Salmo gairdneri* (Richardson). *Cell Tissue Res.* **151**, 509–523.

Mukhamedova, A. F. (1977). The level of standard metabolism of young silver carp, *Hypophthalmichthys molitrix. J. Ichthyol. (Engl. transl.)* **17**, 292–299.

Nakano, E. (1953). Respiration during maturation and at fertilization of fish eggs. *Embryologia* **2**, 21–31.

Needham, J. (1931). "Chemical Embryology." Cambridge Univ. Press, London and New York.

Needham, J. (1942). "Biochemistry and Morphogenesis." Cambridge Univ. Press, London and New York.

O'Brien, R. N., Visaisouk, S., Raine, R., and Alderdice, D. F. (1978). Natural convection: A mechanism for transporting oxygen to incubating salmon eggs. *J. Fish. Res. Board Can.* **35**, 1316–1321.

Oikawa, S., and Itazawa, Y. (1985). Gill and body surface areas of the carp in relation to body mass, with special reference to the metabolism–size relationship. *J. Exp. Biol.* **117**, 1–14.

Oseid, D. M., and Smith, L. L., Jr. (1971a). Survival and hatching of walleye eggs at various dissolved oxygen levels. *Prog. Fish-Cult.* **33**, 81–85.

Oseid, D. M., and Smith, L. L., Jr. (1971b). Survival and hatching of white sucker eggs at various dissolved oxygen levels. *Prog. Fish-Cult.* **33**, 158–159.

Ott, M. E., Heisler, N. H., and Ultsch, G. R. (1980). A re-evaluation of the relationship between temperature and the critical oxygen tension in freshwater fishes. *Comp. Biochem. Physiol. A* **67A**, 337–340.

Paine, M. D., and Balon, E. K. (1984). Early development of the northern logperch, *Percina caprodes semifasciata*, according to the theory of saltatory ontogeny. *Environ. Biol. Fishes* **11**, 173–190.

Pamatmat, M. M. (1983). Simultaneous direct and indirect calorimetry. *In* "Polarographic Oxygen Sensors" (E. Gnaiger and H. Forstner, eds.), pp. 167–175. Springer-Verlag, Berlin and New York.

Pauly, D. (1981). The relationship between gill surface area and growth performance in fish: A generalization of von Bertalanffy's theory of growth. *Meeresforschung* **28**, 251–282.

Penaz, M., and Prokes, M. (1973). Oxygen consumption in the brown trout *Salmo trutta. Zool. Listy* **22**, 181–188.

Peterka, J. J., and Kent, J. S. (1976). "Dissolved Oxygen, Temperature and Survival of Young at Fish Spawning Sites," EPA/600/3-76-113. Environ. Res. Lab., Duluth, Minnesota.

Peterson, R. H. (1975). Pectoral fin and opercular movements of Atlantic salmon (*Salmo salar*) alevins. *J. Fish. Res. Board Can.* **32**, 643–648.

Peterson, R. H., and Martin-Robichaud, D. J. (1983). Embryo movements of Atlantic salmon (*Salmo salar*) as influenced by pH, temperature and state of development. *Can. J. Fish. Aquat. Sci.* **40**, 777–782.

Phillips, R. W., and Campbell, H. J. (1961). The embryonic survival of coho salmon and steelhead trout as influenced by some environmental conditions in gravel beds. *Pac. Mar. Fish. Comm. Rep.* **14**, 60–73.

Phillips, R. W., Campbell, H. J., Hug, W. L., and Claire, E. W. (1966). "A Study of the Effects of Logging on Aquatic Resources, 1960–1966," Prog. Memo. Fish. No. 3. Res. Div., Oregon State Game Comm., Portland.

Piiper, J. (1982). Respiratory gas exchange at lungs, gills and tissues: Mechanisms and adjustments. *J. Exp. Biol.* **100**, 5–22.

Piiper, J., and Scheid, P. (1984). Respiratory gas transport systems: Similarities between avian embryos and lungless salamanders. *In* "Respiration and Metabolism of Embryonic Vertebrates" (R. S. Seymour, ed.), pp. 181–191. Martinus Nijhoff/Dr. W. Junk Publishers, Dordrecht, The Netherlands.

Pinder, A., and Burggren, W. (1983). Respiration during chronic hypoxia and hyperoxia in larval and adult bullfrogs (*Rana catesbeiana*). II. Changes in respiratory properties of whole blood. *J. Exp. Biol.* **105**, 205–213.

Potts, W. T. W., and Rudy, P. P., Jr. (1969). Water balance in the eggs of the Atlantic salmon *Salmo salar. J. Exp. Biol.* **50**, 223–237.

Prasad, M. S., and Prasad, P. (1984). Changes in the water–blood diffusion barrier at the secondary gill lamellae during early life in *Channa striatus* synonym *Ophicephalus striatus. Folia Morphol.* **32**, 200–208.

Prokes, M. (1973). The oxygen consumption and respiratory surface area of the gills of *Coregonus lavaretus* and of *Coregonus peled. Zool. Listy* **22**, 375–384.

Quantz, G., and Tandler, A. (1982). "On the Oxygen Consumption of Hatchery-reared Larvae of the Gilthead Seabream (*Sparus auratus* L.)," Pap. F:7. Maricult. Comm., Int. Counc. Explor. Sea, Copenhagen.

Quinn, D., and Burggren, W. (1983). Lactate production, tissue distribution, and elimination following exhaustive exercise in larval and adult bullfrogs *Rana catesbeiana. Physiol. Zool.* **56**, 597–613.

Randall, D. J. (1982). The control of respiration and circulation in fish during exercise and hypoxia. *J. Exp. Biol.* **100**, 275–288.

Randall, D. J., and Daxboeck, C. (1984). Oxygen and carbon dioxide transfer across fish gills. *In* "Fish Physiology" (W. S. Hoar and D. J. Randall, eds.), Vol. 10, pp. 263–314. Academic Press, New York.

Randall, D. J., Perry, S. F., and Heming, T. A. (1982). Gas transfer and acid/base regulation in salmonids. *Comp. Biochem. Physiol. B* **73B**, 93–103.

Reznichenko, P. N., Solovev, L. G., and Gulidov, M. V. (1977). Simulation of the process of oxygen inflow to the respiratory surfaces of fish embryos by polarographic method. *In* "Metabolism and Biochemistry of Fishes" (G. S. Karzinkin, ed.), pp. 237–248. Indian Natl. Sci. Doc. Cent., New Delhi.

Roberts, R. J., Bell, M., and Young, H. (1973). Studies on the skin of plaice (*Pleuronectes platessa* L.). II. The development of larval plaice skin. *J. Fish Biol.* **5**, 103–108.

Robertson, D. A. (1974). Developmental energetics of the southern pigfish (Teleostei: Congiopodidae). *N. Z. J. Mar. Freshwater Res.* **8**, 611–620.

Rombough, P. J. (1985). Initial egg weight, time to maximum alevin wet weight and optimal ponding times for chinook salmon (*Oncorhynchus tshawytscha*). *Can. J. Fish. Aquat. Sci.* **42**, 287–291.

Rombough, P. J. (1986). Mathematical model predicting the dissolved oxygen require-ments of steelhead (*Salmo gairdneri*) embryos and alevins in hatchery incubators. *Aquaculture* **59**, 119–137.

Rombough, P. J. (1987) Growth, aerobic metabolism and dissolved oxygen require-ments of embryos and alevins of the steelhead trout, *Salmo gairdneri*. *Can. J. Zool.* (in press).

Rosenthal, H., and Alderdice, D. F. (1976). Sublethal effects of environmental stressors, natural and pollutional, on marine fish eggs and larvae. *J. Fish. Res. Board Can.* **33**, 2047–2065.

Ryzhkov, L. P. (1965). The circadian rhythm of oxygen consumption by roe of Sevan trout (*Salmo ishchan*). *Vopr. Ikhtiol.* **5**(2), 378–380.

Ryzhkov, L. P. (1968). The rate of gaseous exchange in the eggs, larvae and fry of the Sevan trout kept in groups or isolated. *Probl. Ichthyol.* **8**, 89–96.

Saksena, V. P., and Joseph, E. B. (1972). Dissolved oxygen requirements of newly hatched larvae of the striped blenny *Chasmodes bosquianus*, the naked goby *Go-biosama bosci* and the skilletfish *Gobiesox strumosus*. *Chesapeake Sci.* **13**, 23–28.

Sawaya, P. (1942). The tail of a fish larva as a respiratory organ. *Nature (London)* **149**, 169–170.

Schnute, J., and Jensen, J. O. T. (1986). A general multivariate dose-response model. *Can. J. Fish. Aquat. Sci.* **43**, 1684–1693.

Sheel, M., and Singh, B. R. (1981). Oxygen uptake through water during early life of *Heteropneustes fossilis*. *Hydrobiologia* **78**, 81–86.

Shepard, M. P. (1955). Resistance and tolerance of young speckled trout (*Salvelinus fontinalis*) to oxygen lack, with special reference to low oxygen acclimation. *J. Fish. Res. Board Can.* **12**, 387–433.

Shumway, D. L., Warren, C. E., and Doudoroff, P. (1964). Influece of oxygen concentra-tion and water movement on the growth of steelhead trout and coho salmon em-bryos. *Trans. Am. Fish. Soc.* **93**, 342–356.

Siefert, R. E., and Herman, L. J. (1977). Spawning success of the black crappie (*Pomoxis nigromaculatus*) at reduced dissolved oxygen concentration. *Trans. Am. Fish. Soc.* **106**, 376–379.

Siefert, R. E., and Spoor, W. A. (1974). Effects of reduced oxygen on embryos and larvae of the white sucker, coho salmon, brook trout and walleye. *In* "The Early Life History of Fish" (J. H. S. Blaxter, ed.), pp. 487–495. Springer-Verlag, Berlin and New York.

Siefert, R. E., Spoor, W. A., and Syrett, R. L. (1973). Effects of reduced oxygen concen-trations on northern pike (*Esox lucius*) embryos and larvae. *J. Fish. Res. Board Can.* **30**, 849–852.

Siefert, R. E., Carlson, A. R., and Herman, L. J. (1974). Effects of reduced oxygen concentrations on the early life stages of mountain whitefish, smallmouth bass, and white bass. *Prog. Fish-Cult.* **36**, 186–190.

Silver, S. J., Warren, C. E., and Doudoroff, P. (1963). Dissolved oxygen requirements of developing steelhead trout and chinook Salmon embryos at different water veloci-ties. *Trans. Am. Fish. Soc.* **92**, 327–343.

Singh, R. P., Prasad, M. S., Mishra, A. P., and Singh, B. R. (1982). Oxygen uptake through water during early life in *Channa punctatus* Pisces *Ophicephaliformes*. *Hydrobiologia* **87**, 211–216.

Smirnov, A. I. (1953). Development of respiratory vessels in the skin of salmonid em-bryo. *Zool. Zh.* **32**, 787–790.

Smith, S. (1947). Studies in the development of the rainbow trout (*Salmo irideus*). I. The heat reproduction and nitrogenous excretion. *J. Exp. Biol.* **23**, 357–378.

Smith, S. (1957). Early development and hatching. *In* "The Physiology of Fishes" (M. E. Brown, ed.), Vol. 1, pp. 323–359. Academic Press, New York.

Smith, S. (1958). Yolk utilization in fishes. *In* "Embryonic Nutrition" (D. Rudnick, ed.), pp. 33–53. Univ. of Chicago Press, Chicago, Illinois.

Soin, S. G. (1966). Development, types of structure and phylogenesis of the vascular system of the vitelline sac in fish embryos, performing respiratory function. *Zool. Zh.* **45**(9), 1382–1397.

Soin, S. G. (1967). Ecological–morphological data on the relationship between carotenoids and the process of embryonal respiration in fish. *In* "Metabolism and Biochemistry of Fishes" (G. S. Karzinkin, ed.), pp. 536–552. Indian Natl. Sci. Docu. Cent., New Delhi, 1977.

Solberg, T., and Tilseth, S. (1984). Growth, energy consumption and prey density requirements in first feeding larvae of cod (*Gadus morhua* L.). *In* "The Propogation of Cod *Gadus morhua* L." (E. Dahl, D. P. Danielssen, E. Moksness, and P. Solemdal, eds.), Part 1, pp. 145–166. Inst. Mar. Res. Flødevigen Biol. Stn., Arendal, Norway.

Solewski, W. (1949). The development of the blood vessels of the gills of the sea trout, *Salmo trutta*, L. *Bull. Acad. Sci. Crac.* **11**, 121–144.

Spoor, W. A. (1977). Oxygen requirements of embryos and larvae of the largemouth bass *Micropterus salmoides*. *J. Fish Biol.* **11**, 77–86.

Spoor, W. A. (1984). Oxygen requirements of larvae of the smallmouth bass *Micropterus dolomieui*. *J. Fish. Biol.* **25**, 587–592.

Sprague, J. B. (1973). The ABC's of pollutant bioassay using fish. *ASTM Spec. Tech. Publ.* **528**, 6–30.

Steffensen, J. F., and Lomholt, J. P. (1985). Cutaneous oxygen uptake and its relation to skin blood perfusion and ambient salinity in the plaice, *Pleuronectes platessa*. *Comp. Biochem. Physiol. A* **81A**, 373–375.

Stehr, C. M., and Hawkes, J. W. (1979). The comparative ultrastructure of the egg membrane and associated pore structures in the starry flounder, *Platichthys stellatus* (Pallas) and pink salmon, *Oncorhynchus gorbuscha* (Walbaum). *Cell Tissue Res.* **202**, 347–356.

Stelzer, R., Rosenthal, H., and Siebers, D. (1971). Influence of 2,4-dinitrophenol on respiration and concentration of some metabolites in embryos of the herring *Clupea harengus*. *Mar. Biol. (Berlin)* **11**, 369–378.

Steven, O. M. (1949). Studies on animal carotenoids. Carotenoids in the reproductive cycle of the brown trout. *J. Exp. Biol.* **26**, 295–302.

Storozhk, N. G., and Smirnov, B. P. (1982). Effect of mercury on the respiration rate of larval pink salmon, *Oncorhynchus gorbuscha*. *J. Ichthyol. (Engl. transl.)* **22**(5), 168–171.

Sushko, B. S. (1982). Microelectrode studies of oxygen tension and transport in loach *Misgurnus fossilis* eggs in helium–oxygen and nitrogen–oxygen media. *Fiziol. Zh. (Kiev)* **28**, 593–597, Can. Transl. Fish. Aquat. Sci. No. 5211 (1986).

Sylvester, J. R., Nash, C. E., and Emberson, C. R. (1975). Salinity and oxygen tolerances of eggs and larvae of Hawaiian striped mullet *Mugil cephalus*. *J. Fish Biol.* **7**, 621–630.

Tamarin, A. E., and Komarova, N. P. (1972). Some data on the respiration of Terek River salmon in relation to the bio-techniques of salmon culture. *Fish Res. Board Can., Transl. Ser.* **2055**, 1–11.

Taylor, F. H. C. (1971). Variation in hatching success in Pacific herring (*Clupea pallasi*) eggs with water depth, temperature, salinity, egg mass thickness. *Rapp. P.-V. Réun., Cons. Int. Explor. Mer* **160**, 34–41.

Taylor, M. (1913). The development of *Synbranchus marmoratus*. *Q. J. Microsc. Sci.* **59**, 1–51.

Taylor, M. H., DiMichele, L., and Leach, G. J. (1977). Egg stranding in the life cycle of the mummichog, *Fundulus heteroclitus*. *Copeia*, pp. 397–399.

Terner, C. (1979). Metabolism and energy conversion during early development. *In* "Fish Physiology" (W. S. Hoar, D. J. Randall, and J. R. Brett, eds.), Vol. 8, pp. 261–278. Academic Press, New York.

Theilacker, G. H., and Dorsey, K. (1980). Larval fish diversity, a summary of laboratory and field research. *In* "Workshop on the Effects of Environmental Variation on the Survival of Larval Pelagic Fishes" (G. D. Sharp, ed.), IOC Workshop Rep. 28, pp. 105–142. UNESCO, New York.

Toetz, D. W. (1966). The change from endogenous to exogenous sources of energy in bluegill sunfish larvae. *Invest. Indiana Lakes Streams* **8**(4), 115–146.

Trifonova, A. N. (1937). La physiologie de la différenciation et de la croissance. I. L'équilibre Pasteur-Meyerhof dans le développement des poissons. *Acta Zool. (Stockholm)* **18**, 375–445.

Turner, J. L., and Farley, T. C. (1971). Effects of temperature, salinity, and dissolved oxygen on the survival of striped bass eggs and larvae. *Calif. Fish Game* **57**, 268–273.

Tveranger, B. (1986). Effect of pigment content in broodstock diet on subsequent fertilization rate, survival and growth rate of rainbow trout (*Salmo gairdneri*) offspring. *Aquaculture* **53**, 85–93.

Tytler, P., and Calow, P. (1985). "Fish Energetics: New Perspectives." p. 349. Croom Helm, London.

Ultsch, G. R. (1973). A theoretical and experimental investigation of the relationships between metabolic rate, body size and oxygen exchange capacity. *Respir. Physiol.* **18**, 143–160.

Ultsch, G. R., Boschung, H., and Ross, M. J. (1978). Metabolism, critical oxygen tension and habitat selection in darters (*Etheostoma*). *Ecology* **59**, 99–107.

Ultsch, G. R., Ott, M. L., and Heisler, N. (1980). Standard metabolic rate, critical oxygen tension, and aerobic scope for spontaneous activity of trout (*Salmo gairdneri*) and carp (*Cyprinus carpio*) in acidified water. *Comp. Biochem. Physiol. A* **67A**, 329–335.

Ultsch, G. R., Jackson, D. C., and Moalli, R. (1981). Metabolic oxygen conformity among lower vertebrates: The toadfish revisited. *J. Comp. Physiol.* **142**, 439–443.

Veith, W. J. (1979). Reproduction in the live bearing teleost *Clinus superciliosus*. *S. Afr. J. Zool.* **14**, 208–211.

Vernidub, M. F. (1966). Composition of the red and white blood cells of embryos of Atlantic salmon and Baltic salmon *Salmo salar* L. and changes in this composition during growth of the organism. *Proc. Murmansk Inst. Mar. Biol.* **12–16**, 139–162, *Fish. Res. Board Can., Transl. Ser.* **1353**.

Vogel, S. (1981). "Life in Moving Fluids. The Physical Biology of Flow." Princeton Univ. Press, Princeton, New Jersey.

Volodin, V. M. (1956). Embryonic development of the autumn Baltic herring and their oxygen requirements during the course of development. *Voprosy Ikhtiol.* **7**, 123–133. *Fish. Res. Board Can. Transl. Ser.* **252**.

Voyer, R. A., and Hennekey, R. J. (1972). Effects of dissolved oxygen on the two life stages of the mummichog. *Prog. Fish-Cult.* **34**, 222–225.

Wangensteen, O. D. (1972). Gas exchange by a bird's embryo. *Respir. Physiol.* **14**, 64–74.

Webb, P. W., and Brett, J. R. (1972a). Respiratory adaptations of prenatal young in the ovary of two species of viviparous seaperch, *Rhacohilus vacca* and *Embioteca lateralis. J. Fish. Res. Board Can.* **29**, 1525–1542.

Webb, P. W., and Brett, J. R. (1972b). Oxygen consumption of embryos and parents and oxygen transfer characteristics within the ovary of 2 species of viviparous seaperch *Rhacohilus vacca* and *Embioteca lateralis. J. Fish. Res. Board Can.* **29**, 1543–1553.

Webb, P. W., and Weihs, D. (1986). Locomotor functional morphology of early life history stages of fishes. *Trans. Am. Fish. Sci.* **115**, 115–127.

Weber, R. E., and Hartvig, M. (1984). Specific fetal hemoglobin underlies the fetal-maternal shift in blood oxygen affinity in a viviparous teleost *Zoarces viviparus. Mol. Physiol.* **6**, 27–32.

Weibel, E. R. (1984). "The Pathway for Oxygen." Harvard Univ. Press, Cambridge, Massachusetts.

Weihs, D. (1980). Respiration and depth control as possible reasons for swimming of northern anchovy *Engraulis mordax* yolk sac larvae. *Fish. Bull.* **78**, 109–117.

Weihs, D. (1981). Swimming of yolk sac larval anchovy *Engraulis mordax* as a respiratory mechanism. *Rapp. P.-V. Reun., Cons. Int. Explor. Mer* **178**, 327.

West, N. H., and Burggren, W. W. (1982). Gill and lung ventilatory responses to steady-state aquatic hypoxia and hyperoxia in the bullfrog tadpole. *Respir. Physiol.* **47**, 165–176.

Whiting, ·H. P., and Bone, Q. (1980). Ciliary cells in the epidermis of the larval Australian dipnoan, *Neoceratodus. J. Linn. Soc. London, Zool.* **68**, 125–137.

Wickett, W. P. (1954). The oxygen supply to salmon eggs in spawning beds. *J. Fish. Res. Board Can.* **11**, 933–953.

Wickett, W. P. (1975). Mass transfer theory and the culture of fish eggs. *In* "Chemistry and Physics of Aqueous Solutions" (W. A. Adams, ed.), pp. 419–434. Electrochem. Soc., Princeton, New Jersey.

Wieser, W. (1984). A distinction must be made between the ontogeny and the phylogeny of metabolism in order to understand the mass exponent of energy metabolism. *Respir. Physiol.* **55**, 1–9.

Wieser, W. (1985). Developmental and metabolic constraints of the scope for activity in young rainbow trout (*Salmo gairdneri*). *J. Exp. Biol.* **118**, 133–142.

Wieser, W., and Forstner, H. (1986). Effects of temperature and size on the routine rate of oxygen consumption and on the relative scope for activity in larval cyprinids. *J. Comp. Physiol.* **156**, 791–796.

Wieser, W., Platzer, U., and Hinterleitner, S. (1985). Anaerobic and aerobic energy production of young rainbow trout (*Salmo gairdneri*) during and after bursts of activity. *J. Comp. Physiol. B* **155B**, 483–492.

Winberg, G. G. (1956). "Rate of Metabolism and Food Requirements of Fishes" Nauch. Tr. Belorussk Gos Univ. Imeni V. I. Lenina, Minsk (Fish. Res. Board Can., Transl. Ser. No. 194, 1960).

Winberg, G. G., and Khartova, L. E. (1953). Rate of metabolism in carp fry. *Dokl. Akad. Nauk SSSR* **89**, 1119–1122.

Winnicki, A. (1968). Respiration of the embryos of *Salmo trutta* L. and *Salmo gairdneri* Rich. in media differing in gaseous diffusion rate. *Pol. Arch. Hydrobiol.* **15**, 23–28.

Witzel, L. D., and MacCrimmon, H. R. (1981). Role of gravel substrate on ova survival and alevin emergence of rainbow trout *Salmo gairdneri. Can. J. Zool.* **59**, 629–636.

Wood, A. H. (1932). The effect of temperature on the growth and respiration of fish embryos (*Salmo fario*). *J. Exp. Biol.* **9**, 271–276.

Wu, H. W., and Liu, C. K. (1942). On the breeding habits and the larval metamorphosis of *Monopterus javanensis. Sinensia* **13**, 1–13.

Yamagami, K. (1981). Mechanisms of hatching in fish: Secretion of hatching enzyme and enzymatic choriolysis. *Amer. Zool.* **21,** 459–471.

Yamagami, K., Yamamoto, M., Iuchi, I., and Taguchi, S. (1984). Retardation of maturation associated and secretion associated ultrastructural changes of hatching gland in the medaka embryos incubated in air. *Annot. Zool. Jap.* **56,** 266–274.

Yoshimizu, M., Kimura, T., and Sakai, M. (1980). Microflora of the embryo and fry of salmonids. *Bull. Jpn. Soc. Sci. Fish.* **46,** 967–975.

Zeuthen, E. (1950). Cartesian diver respirometer. *Biol. Bull. (Woods Hole, Mass.)* **98,** 303–318.

Zeuthen, E. (1970). Rate of living as related to body size in organisms. *Pol. Arch. Hydrobiol.* **17,** 21–30.

Zoran, M. J., and Ward, J. A. (1983). Parental egg care, behavior and farming activity for the orange chromide *Etroplus maculatus. Environ. Biol. Fishes* **8,** 301–310.

3

OSMOTIC AND IONIC REGULATION IN TELEOST EGGS AND LARVAE

D. F. ALDERDICE

Department of Fisheries and Oceans
Fisheries Research Branch
Pacific Biological Station
Nanaimo, British Columbia, Canada V9R 5K6

I. INTRODUCTION

The fishes have evolved to occupy all but a few types of natural waters ranging from low ionic strength ("near-distilled") fresh waters to those with salinities of 80–142.4‰ (Kinne, 1964; Parry, 1966; Griffiths, 1974). Some are restricted to narrow ranges of salinity (stenohaline); others tolerate broad ranges (euryhaline). Some spend a major part of their pre-adult lives in fresh water and migrate to spawn in the

FISH PHYSIOLOGY, VOL. XIA

sea (catadromous); others begin life in fresh water, go to sea as juveniles, grow and mature, and return to spawn in fresh water (anadromous). Some marine forms spawn in estuarine or coastal habitats of reduced salinity; others spend their entire life in oceanic salinities, or in fresh water. There also is much variation within species. The various developmental states—the fertilized egg, larva, juvenile—may have particular salinity optima, including those catadromous and anadromous forms where relatively rapid changes occur in salinity tolerance with juvenile metamorphosis or prespawning migration.

Some fishes are osmoconformers (poikilosmotic); their body fluids tend to follow and conform with changes in osmotic properties of the external medium. Teleost fishes generally are osmoregulators (homoiosmotic); their body fluids remain relatively constant with alteration of the external medium. Many teleosts regulate in the central portion of their ranges of tolerance and conform at the extremities. Hyperosmotic regulators (most freshwater teleosts) maintain body-fluid concentration above that of their external surroundings. Conversely, hypoosmotic regulators (most marine teleosts) maintain body fluid concentration below that of the external medium. Osmoconformers tolerate wider varition in internal concentration; osmoregulators tolerate a wider range of external concentration (Prosser, 1973).

Although the terminology associated with osmosis is thoroughly discussed in most fundamental texts (Dick, 1959a; Potts and Parry, 1964; Dainty, 1965; Florey, 1966), some basic terms are presented here. Osmoticity is a general term referring nonspecifically to the osmotic properties of a solution. Osmotic concentration defines the number of nonpermeating particles per unit volume. The difference between the osmotic concentrations of two solutions separated by a semipermeable membrane is referred to as an osmotic gradient. Water tends to flow by osmosis across a membrane toward the solution of higher osmotic concentration (osmoconcentration). The osmotic activity of a solution is defined as the molar concentration of an ideal nonelectrolyte solution having the same osmotic effects. An osmole is the amount of solute that exerts the same osmotic pressure as 1 mole of ideal nonelectrolyte dissolved in the same volume. An osmotic gradient generates an osmotic pressure (π) that ultimately is just sufficient to prevent further osmotic flow (flux) across a membrane. A 1 molar (osmolar) solution of nonelectrolyte has an osmotic pressure of 22.4 atm (at 0°C) and depresses the freezing point (Δ) of the solution by 1.86°C. A solution with an osmolarity of 1 Osm/l (1000 mOsm/l) contains 1 gram molecular weight (gmw) of solute in 1 l of solution. A solution with an osmolality of 1 Osm/kg (1000 mOsm/kg) contains a gmw of solute per kilogram of solvent (water). Isotonicity is a state of volume equilibrium in which a cell bathed in an aqueous solution neither shrinks nor swells.

The primary basis for regulative capacity can be considered as the semipermeable membrane of the individual cell and its tolerance and response to external and internal osmotic and ionic alteration. A second level of regulative capacity is provided in the development of particular tissues and effector organs, whose cells have specific regulatory functions. A third level, in the intact animal, involves neurosecretory activity, which tends to modulate cell, tissue, or organ functions governing regulation. Osmotic and ionic regulation during early development of teleosts then poses certain questions:

> What are the mechanisms that may provide regulative capacity in the egg, embryo, and larva?
> What patterns of regulation can be identified in particular species?
> When, or at what developmental states, do these regulatory mechanisms become functional?

The first question has received much attention often from studies of oocytes, fertilized eggs, juveniles, or adults of nonteleosts. A wealth of information has been gathered on structure and function of individual mechanisms, yet a full understanding of their integrated action in the intact animal seems near but remains elusive. With respect to the second question, Evans (1979, 1980) suggests that there are three general patterns of regulation among the fishes. One is seen in the myxinid (hagfish) agnathids, who show electrolyte iso-osmolarity with sea water. Renal ion regulation is important, and some ion excretion may occur in the integumental slime. The role of the branchial epithelium is not fully understood, although Evans (1984a) offers intriguing evidence that parallel Na^+/H^+ and Cl^-/HCO_3^- exchange systems, generally associated with NaCl regulation in freshwater vertebrates, are found in the branchial epithelium of these early marine vertebrates. A second pattern involves the elasmobranchs; the main effector organs are the branchial epithelium, the renal complex, and the extrarenal rectal gland. In the elasmobranchs the blood level of NaCl is below that of the external medium, but fluid levels are maintained isosmotic or somewhat hypertonic by the retention of urea and trimethylamine oxide. In a third pattern, found in the teleosts, the principal effector organs are the branchial epithelium, the renal complex, the gut, and (Marshall and Nishioka, 1980) the integument. Similarities with the teleost patterns are found in the chondrosteans (sturgeons), holosteans (gars, bowfin), dipnoans (lungfish), and petromy-zonid (lamprey) agnathids. These patterns refer entirely to the juvenile or adults stages, which brings the third question, basic to this chapter. What processes or tissues have a regulatory function in the egg and embryo, what are the effector mechanisms, and when do they become functional during development?

In considering this question, attention will be directed to the te-leosts. Where relevant data are unavailable, insights will be sought among the nonteleosts. A second constraint arises from the fact that few studies have been conducted on the chronology of development of prejuvenile regulatory processes. Instead, a variety of species and a limited number of development states have been used as a conse-quence either of their local availability, their usefulness in approach-ing a particular question posed, or because of the tradition of particu-lar laboratories. Indeed, initial attention has been centered, with good reason, on a search for understanding of regulatory mechanisms and on the development of improved technique for their examination, not on the order and timing of regulatory events during development.

Studies of osmo- and ionoregulation have arisen from the work in the last century of Van't Hoff, Arrhenius, and De Vries, who described relationships between the gas laws and osmotic pressure. In the inter-val to date, an enormous body of literature on cell, tissue, and whole-animal regulation and regulatory mechanisms has been published, and the basis for enquiry has expanded into a number of diverse but related fields of study. For these reasons much use has been made in this chapter of reviews on regulation that have appeared in the past few decades. That of Potts and Parry (1964) tends to integrate the earlier work. More specific considerations of regulation are found in Parry (1966), Potts (1968, 1976, 1977), Holliday (1969), Hagiwara and Jaffe (1979), Folmar and Dickhoff (1980), Heisler (1980, 1982), Evans (1980, 1984b), Evans *et al.* (1982), Eddy (1982), Hoar and Randall (1984), and Metz and Monroy (1985). An appreciation of the chronol-ogy of development of the theory of particle transport is provided by Florey (1966), Katchalsky and Curran (1965), and Friedman (1986). These were used to establish a perspective and to provide a frame-work from which to ask, more specifically, "at what stage and by what means is regulation achieved in the embryonic teleost?"

An early decision was made to provide an overview of the subject, rather than a review. The decision was based on the number of ex-panding and diverging areas of enquiry relevant to the subject, the enormous size of the literature, the rapidity with which enquiry is moving, and the variability apparent in the different animal groups under investigation that could contribute to understanding in teleosts.

Major advances occurring in recent years in knowledge of regulatory processes are rooted in biochemistry, biophysics, and molecular biology. Of numerous technological advances contributing to this progress, those of electron microscopy, use of isotopes, and most recently electrophysiology and cytochemistry provide outstanding examples. Aspects of the foregoing areas of inquiry will be drawn on, as applicable, in the following sections on regulatory processes in various developmental states of fishes. Attention will focus primarily on biological mechanisms that could influence developmental events elicited through changes in membrane permeability at the cellular level. Where evidence for teleosts is sparse, comparisons will be attempted using available information on invertebrates, amphibians, and mammals.

II. OOGENESIS

In the animal kingdom there are four general classes involving differing temporal relationships between maturation of the female gamete and fertilization (Moreau et al., 1985; Masui, 1985). Hence, there is a logical problem in defining the term oocyte. Of the four classes, class I includes representatives from the nematodes, bivalve molluscs, and echiuroids, and insemination normally occurs at the end of prophase prior to germinal vesicle breakdown (GVBD). In class II, including representative annelids, molluscs, and ascidians, the oocyte normally is inseminated at metaphase I. In class III, which includes *Amphioxus* and most vertebrates, insemination occurs at metaphase II following extrusion of the first polar body. In class IV, including representative coelenterates and sea urchins, insemination occurs at completion of telophase and extrusion of the second polar body. In effect, meiosis in the four classes is arrested, respectively, at the four stages indicated. Notwithstanding the fact that insemination normally occurs at different stages of reductive division, if that stimulus removes the block to completion of meiosis and the oocyte becomes fertilizable, then the fertilizable oocyte will be called an egg. *Activation* is the process by which an arrested oocyte (at prophase, metaphase I or II, or telophase) resumes development following insemination, or some other physical or chemical stimulus. *Fertilization* refers more precisely to the fusion of pronuclei following insemination.

It appears that transmembrane shifts in ion concentration and hormones act as triggers, stimulating a chain of events leading to resumed oocyte development and completion of meiosis. Electrophysiological and cytological studies have been particularly useful in examining these events. In these investigations, substantial changes may be seen in the resting and action potentials of the oocyte plasma membrane. These signal sudden changes in rates of influx or efflux of ions, generally involving Ca^{2+}, K^+, Na^+, and Cl^-, resulting from the electrical gradients generated in association with the established ion concentration gradients across the membrane. The sudden shifts in membrane potential are consistent with rapid changes in the permeability of the oocyte plasma membrane to particular ions. The ion permeation mechanisms or pathways through which these transmembrane fluxes occur in response to particular stimuli are described as *channels*. Morphologically, these channels behave like protein structures with a central pore (Barry and Gage, 1984; Miller *et al.*, 1984). They may be "gated," opening and closing in response to transmembrane ion gradients or by membrane depolarization (Heinz and Grassl, 1984; Loewenstein, 1984). However, while electrically excitable channels, demonstrating action potentials, have been described in oocytes of tunicates, annelids, molluscs, echinoderms, amphibians, and mammals, they have not been unequivocally demonstrated in teleost eggs. Documented evidence on the electrical excitability of teleost oocytes and eggs is sparse; the medaka egg (*Oryzias latipes*) is said to be electrically inexcitable (Hagiwara and Jaffe, 1979).

A. Nonteleosts

In starfish oocytes (Kanatani and Nagahama, 1980; Kanatani, 1985), development of the oocyte is arrested in late prophase; then, just before spawning occurs, meiosis resumes, GVBD occurs, and development proceeds to metaphase II. A first mediator initiating this resumption of meiosis is the hormone *gonad-stimulating substance* (GSS), present in the coelomic fluid. GSS acts on the ovary to produce a second mediator, *maturation-inducing substance* (MIS), identified as 1-methyladenine (1-MeAde). MIS stimulates the production of a *maturation-promoting factor* (MPF) in the oocyte cytoplasm (Kanatani and Nagahama, 1980), and MPF production is amplified by germinal vesicle material. MPF amplification produces GVBD, oocyte maturation, and a fertilizable egg. MPF activity is considered to be Mg^{2+}-dependent and Ca^{2+}-sensitive. Absence of Ca^{2+} delays GVBD in MIS-induced oocyte maturation. A transient stimulation of an Na^+

$-K^+$ electrogenic pump and increased passive Na^+ permeability also are noted (Dorée, 1981) in oocytes treated with MIS. It was concluded that these events are regulated by a transient increase in $[Ca^{2+}]$. MIS appears to act by releasing internally sequestered Ca^{2+}. In considering the resting potential of an oocyte, it must be remembered that ionic composition of the oocyte cytoplasm will be modified by that of the parental body fluid, which itself will be influenced by the composition of the external environment. The starfish oocyte in seawater has a resting potential of -60 to -70 mV (Hagiwara and Jaffe, 1979). The plasma membrane is predominately permeable to K^+; Na^+ or Cl^- permeability is negligible. The resting potential is close to the Nernst estimate on the basis of K^+ activity. In various species in the starfish group there appears to be a decrease in K^+ permeability, or a decrease with respect to other ions such as Na^+, as maturation proceeds. In the sea urchin egg the resting potential ranges between -60 and -80 mV, and it also is K^+-dependent. Significantly, in their review Hagiwara and Jaffe (1979) note that conductance of the oocyte plasma membrane always decreases during oocyte maturation.

In the amphibians, the K^+ ionophore valinomycin was found to trigger GVBD in the *Xenopus* oocyte in the absence of external K^+, which concomitantly resulted in a sharp reduction in intracellular $[K^+]$ (Moreau *et al.*, 1985). In amphibian oocytes, GVBD is accompanied by marked membrane depolarization (Lessman and Marshall, 1984), reduction in number of ouabain-binding sites, and markedly reduced membrane conductance to K^+ and Cl^- ions (Weinstein *et al.*, 1982). In other studies reviewed by Moreau *et al.* (1985) it was concluded that external K^+ deprivation, or reduction of active K^+ influx in ouabain-treated oocytes, facilitated progesterone-induced maturation (Vitto and Wallace, 1976; Wallace and Steinhardt, 1977; Kofoid *et al.*, 1979). In addition, amiloride, which blocks Na^+ channels in the oocyte plasma membrane, prevented reinitiation of meiosis in arrested oocytes. In another study (Robinson, 1979), external current was found to enter the oocyte at the animal pole and exit at the vegetal pole. This current, associated primarily with movement of Cl^- ions, appeared to be regulated by Ca^{2+}. Tracer experiments with *Xenopus* oocytes (O'Connor *et al.*, 1977) have shown a sustained increase in Ca^{2+} uptake, which decreased before GVBD was blocked by maturation-inhibiting drugs. There was a rapid, consistent Ca^{2+} efflux and a parallel K^+ efflux; it was concluded that progesterone treatment of the oocyte resulted in an increase in internal $[Ca^{2+}]$. Amphibian oocyte membrane resting potentials, in appropriate Ringer's solution (Hagiwara and Jaffe, 1979), are -50 to -70 mV and the potential is

K^+-dependent, although it departs from the Nernst estimate. Oocytes isolated from follicular tissue showed an increased negative potential, which was sensitive to ouabain. Hence, some part of the negative potential in the preparation appeared to originate in electrogenic, active ion transport. Amphibian oocytes treated with progesterone showed a decrease in membrane potential to a final value of -10 to -40 mV, suggesting a progressive increase in ion permeability of the plasma membrane coincident with progressive oocyte maturation.

Action potentials also provide information on the presence of ion channels. In echinoderms, two types of ion channel have been noted in starfish oocytes (Hagiwara and Jaffe, 1979). The properties of one are similar to those of Ca^{2+} channels. The second type differs from the first in electrophysiological properties; however, it appears to be a special form of Ca^{2+} channel, the current being carried by Ca^{2+}. A Ca^{2+}-dependent action potential, probably analogous to the special Ca^{2+} channel of the starfish, is found in the sea urchin egg. With reference to K^+ permeability, three types of K^+ channels have been identified in the oocytes of starfish: the inward K^+ channel, outward K^+ channel, and the fast-inactivating K^+ channel. The electrical properties of these channels appear to vary with species and development state. In the first two types the current is carried by K^+ ions; however, the channels appear to differ in molecular structure, and they respond differently to variations in external conditions. In the latter type, when the membrane potential is held at a level somewhat more positive that the resting potential, the channel is fully inactivated.

B. Teleosts

One is impressed by the paucity of investigative data on teleost oocytes, particularly on the processes that may parallel those in the invertebrates and amphibians. Among the chordates, Masui (1985) lists two fishes in which the stage of meiotic arrest in oocytes (metaphase II) is known: *Catostomus* (Lessman and Huver, 1981) and *Acipenser*. Yanagimachi (1958) recorded the same stage in meiotic arrest in the Pacific herring, *Clupea pallasi*. The resting potential of the *Oryzias latipes* oocyte probably is in the range of -40 mV (Ringer's solution) to -50 mV (0.1 Ringer's) (Hagiwara and Jaffe, 1979). Ion-substitution experiments showed a membrane potential dependent on K^+ and Na^+, but independent of Ca^{2+} or Cl^-. Although Na^+ channels, several types of Ca^{2+} channel, and three types of K^+ channel have been identified, their role in ion transport across the plasma membrane of teleost oocytes is relatively unknown. Hagiwara and Jaffe (1979) speculate that during oocyte and egg development various ion channels may be created, eliminated, or localized; some may be inserted or removed while others may be present continually.

In a study of the brook trout *Salvelinus fontinalis,* Marshall *et al.* (1985) compared the electrophysiological characteristics of oocytes, separated from follicular tissue, with those of ovulated eggs. Control oocytes with the germinal vesicle present were compared with those treated with 17α-20β-dihydroxyprogesterone (DHP) to promote GVBD. The gametes were stored in Cortland's saline and examined in various modifications of Ringer's solution, all having a constant osmolality of 310–320 mOsm/kg. The electrical characteristics at the three stages of maturation are shown in Table I. Compared with oocytes in which the germinal vesicle was present, there was a significant reduction in membrane potential in the ovulated eggs. On the same basis, membrane resistance was significantly higher in the oocyte after GVBD, and was greatly increased after ovulation. Ion conductance before and after GVBD was approximately constant, while K^+ and Na^+ conductances decreased substantially with ovulation. There was no evidence of change in Cl^- conductance. Treatment of control oocytes in low-Na^+ and high-K^+ Ringer's solutions resulted in decreased R_m (membrane resistance) and V_m (membrane potential) values, suggesting that depolarization induces conductive pathways. Significantly, one explanation for this effect would be a voltage-dependent conductance; electrode limitation prevented confirmation of this possibility. The study showed that GVBD and ovulation increased membrane

TABLE I

Electrical Characteristics (Mean ± SE) of Oocytes and Ovulated Eggs of *Salvelinus fontinalis:* Plasma Membrane Potential (V), Membrane Resistance (R), and Conductance (mV per Decade of Change in Ion Concentration, ΔC)[a]

Treatment	V (mV)	R (kΩ cm^2)	Conductance C/R (mV per decade ΔC)		
			K^+	Na^+	Cl^-
Denuded oocytes					
GV present[b]	-46.3 ± 1.2	26.5 ± 1.7	29.1 ± 2.9	13.5 ± 5.7	2.7 ± 0.2
GV absent	-41.5 ± 3.0	40.5 ± 7.0	29.5 ± 11.2	12.4 ± 4.5	2.6 ± 0.7
Ovulated eggs	-38.1 ± 1.8	1640 ± 150	7.1 ± 1.8	0.5 ± 0.2	2.3 ± 0.4

[a] From Marshall *et al.* (1985).
[b] GV, Germinal vesicle.

resistance, decreased membrane current, and that Na^+ and K^+ conductances and membrane depolarization decreased with ovulation. Hence, as a representative salmonid, the mature egg of the brook trout should be relatively impermeable to K^+ and Na^+ ions prior to fertilization.

In general, the presence of the germinal vesicle in late prophase and its breakdown in first metaphase appears to trigger a chain of maturation-inducing events mediated by hormones, and accompanied by ion fluxes. Most of the latter accompany shifts in membrane electrical activity and signal changes in plasma membrane permeability to particular ions. Changes in membrane potentials may not necessarily be instigators of maturation-inducing events: they may be side effects incidental to the process. Nevertheless, membrane potential and permeability changes in oocytes are associated with fluxes of Ca^{2+}, K^+, Na^+, or Cl^-. In amphibians Ca^{2+} ions seem to control the resting potential and Cl^- currents, as well as K^+ and Na^+ permeabilities. Moreau et al. (1985) favor the hypothesis that intracellular Ca^{2+} release is the primary trigger for following events. In the starfish oocyte, hormone application induces Na^+–K^+ electrogenic pump activity, which itself results from Ca^{2+} release, although pump activation is not necessary for restarting meiosis. Many egg membranes in Ringer's solution have K^+-dependent resting potentials (approximately -70 mV); others have smaller potentials and relatively nonspecific ion permeability. Membrane potentials, of course, also reflect conditions in the external medium—the latter being one terminal in the ion gradient. The steady-state current–voltage relation for the starfish (*Mediaster aequalis*) oocyte membrane at the resting potential (0 μA membrane current) (Hagiwara and Jaffe, 1979) was -50 mV in 25 mM K^+ seawater and about -73 mV in normal (10 mM K^+) seawater. In terms of $[H^+]$, Peterson and Martin-Robichaud (1986) found the (well-eyed) eggs of Atlantic salmon (*Salmo salar*) in fresh water to have a membrane potential of -65 mV near pH 6, and -50 mV at pH 3.5. The smaller negative membrane potentials frequently found experimentally (-10 to -20 mV) may be artifacts, resulting from leaks around the impaling microelectrodes. However, Hagiwara and Jaffe (1979) suggest these could be real in certain instances, resulting from the relatively nonselective properties of a membrane in freshwater eggs in the natural state. Peterson and Martin-Robichaud (1986) found that the membrane potential for Atlantic salmon "eggs" in fresh water would be near -70 mV, which would argue against that suggestion.

Ion channels in the plasma membrane may be "gated": that is, they may open and close in response to certain stimuli. Horn (1984) lists such stimuli and related channel properties: (1) transmembrane potential—channels may open more frequently or stay open longer at hyperpolarized voltages; (2) neurotransmitters and drugs may open or close channels; (3) after being activated, channels tend to desensitize and close spontaneously; (4) open time may be influenced by the type of permeant ion present; (5) gating speed is temperature-dependent, involving conformational changes in membrane-bound proteins; and (6) single channels tend to gate very rapidly, the open-to-closed transition being <10 μs.

As alluded to earlier, Hagiwara and Jaffe (1979) suggest that ion channels may be created or eliminated as required during development, and that they may be present in the membrane for future use after oocyte maturation (Hagiwara, 1983). This reasoning may be applied to a current problem in the manipulation of Pacific salmon (*Oncorhynchus*). These anadromous species often are captured on their final approach to fresh water and held in saltwater pens, to be stripped of gametes when ripe, for hatchery culture of the eggs. Normally, the adults would enter fresh water during the last stages of gamete maturation. Held in sea water, the adults may suffer substantial mortality; the gametes of the survivors show highly variable but lower fertilizability. In the female chum salmon (*O. keta*), about to enter fresh water to spawn, the osmolality and [Na$^+$] of the blood plasma and ovarian fluid were very similar, about 340–350 mOsm/kg (Na$^+$ 170–180 mM/l) (Hirano *et al.*, 1978; Lam *et al.*, 1982; Stoss and Fagerlund, 1982). In fresh water, these values declined to 295–300 mOsm/kg (Na$^+$ 140–160 mM/l). With chinook salmon (*O. tshawytscha*) held in seawater, Sower (1981) found that blood plasma of the females reached maxima of 415–430 mOsm/kg (Na$^+$ 200 meq/l) during oocyte maturation, declining to about 390 mOsm/kg (Na$^+$ 175–180 meq/l) at ovulation. In fresh water, blood plasma osmolality and [Na$^+$] declined to 305 mOsm/kg (Na$^+$ 136 meq/l) during oocyte maturation and were 320 mOsm/kg (Na$^+$ 144 meq/l) at ovulation. The comparison suggests that the female Pacific salmon, held in seawater at ovulation, probably carries body-cavity eggs bathed in ovarian fluid about 40–50 mOsm/kg (Na$^+$ 25–30 mM/l) above natural levels. Lam *et al.* (1982) made reciprocal crosses between seawater- and freshwater-held males and females. Fertilization success was higher in crosses between freshwater-held fish; in crosses of seawater-held fish, there was a 30- to 40-fold increase in abnormal embryonic development, and embryonic survival was lower. Sower (1981) concluded that salmon held in seawater were unable to adapt to that medium while undergoing maturation and ovulation. She noted a strong correlation between elevated

[Na$^+$] and osmolality and increased adult mortality. When ovulation was induced in seawater, Sower *et al.* (1982) concluded that hormonal responses were reduced because of osmoregulatory difficulties in the adults. Wertheimer (1984) sums up the problem, suggesting that salmon maturing in seawater seem to be in a race between degeneration of osmoregulatory capacity with resultant loss of reproductive competence or death, and completion of maturation of viable gametes with resultant successful spawning. When the ovulated female is held in seawater, the eggs apparently are unable to cope with their hyperosmotic ovarian fluid medium. Hence, the ovulated egg appears to depend on the adult for osmotic control of its immediate environment, and could be limited in its capacity to regulate ion or water flows across the plasma membrane. The plasma membrane of the mature Atlantic salmon (*Salmo salar*) egg is highly permeable to water. On the other hand, Marshall *et al.* (1985) noted significant reductions in Na$^+$ and K$^+$ conductance in the plasma membrane of the maturing brook trout (*Salvelinus fontinalis*) egg compared to the oocyte after GVBD. Therefore, lower fertilizability and increased abnormal embryonic development in eggs of Pacific salmon (*Oncorhynchus*) females held in seawater could be a result of increased water efflux and ion imbalance from dehydration. The data are limited, but if ion channels are present in the plasma membrane of the mature salmon egg, their role would seem to be one of passive resistance to ion transfer rather than regulation.

Another problem receiving increasing attention is the influence of low pH ("acid rain") on hydromineral ion requirements (Ca^{2+}, Na$^+$, K$^+$) during oocyte and postfertilization egg development in salmonids and other freshwater teleosts. In general, reproductive success in spawning salmonids is lowered at pH levels below 5.5 (Tam and Payson, 1986; Weiner *et al.*, 1986), and plasma ion levels are disturbed in adults and juveniles below pH 5.1–5.5 (Booth *et al.*, 1982; Peterson and Martin-Robichaud, 1986). Salmonid eggs incubating at low pH are more sensitive, showing a threshold for deleterious effects near pH 6.5–6.7 (Menendez, 1976; Craig and Baksi, 1977; Ruby *et al.*, 1977; Rombough and Jensen, 1985). Rudy and Potts (1969) showed that Atlantic salmon (*Salmo salar*) embryos require and take up Na$^+$ from the external medium during embryogenesis. Brown and Lynam (1981) and Brown (1982) found survival to hatch in brown trout (*S. trutta*) at pH 4.5 to be a function of pH-dependent external Na$^+$ and Ca^{2+} levels. Bradley and Rourke (1985) related mortality in juvenile steelhead trout (*S. gairdneri*) to low levels of environmental Na$^+$, postulating that mortality resulted from reduced NH$_4^+$ excretion via a Na$^+$/NH$_4^+$

exchange mechanism. A similar problem is under study in chinook salmon (*O. tshawytscha*), where early alevin mortality appears to involve genetic, osmoregulatory, and pathological components (Alderdice and Harding, 1987). Susceptibility varies among spawning stocks; in a susceptible stock, low [Na^+] and [Ca^{2+}] in the incubation water appear to result in osmotic stress in the alevins, and the debilitated alevins succumb to the effects of an unknown pathogen. Preliminary recommendations in this instance are that eggs and alevins should be incubated under conditions in which pH $\geqslant$ 6.5, [Ca^{2+}] $\geqslant$ 8–10 mg/l, and [Na^+] $\geqslant$ 1 mg/l. The [K^+] seems less critical.

In summary, although oocytes show substantial changes in electrical characteristics and specific ion permeabilities throughout oogenesis, these changes appear to be interdependent processes associated with development and maturation rather than with independent regulation. That is, adjustments in the oocyte and fertilizable egg appear to be passive, and dependent on the homeostatic mechanisms and regulatory capacity of the gravid adult. Although such assumptions seem warranted, it is difficult to extend the argument based on the current, limited knowledge of teleost oocytes. It is evident that much has been learned about ion transport and levels in oocytes of nonteleosts, yet the more modern investigative techniques appear to have been applied only infrequently to teleosts. Although there are likely to be major differences in detail of regulatory mechanisms in coelenterates, echinoderms, and amphibians, compared to teleosts, the groundwork thereby laid for substantial development of a similar knowledge in teleosts appears to be in place. Some of the more obvious possibilities for further investigation of teleost oocytes would include:

The time course of oocyte development during meiosis, including *temperature* (often not stated).

The stage of meiosis attained when the oocyte is mature and ready for insemination.

States of meiotic arrest and the cytochemical chain of events beginning with GVBD and ending with maturation, including intracellular and extracellular hormonal and hydromineral involvements in the maturation process.

Information on oocyte membrane permeability at intervals throughout oogenesis, at least to Ca^{2+}, K^+, Na^+, and Cl^-.

The use of a series of representative, generally available species of differing plasticity: (a) marine eurhyaline (e.g., *Clupea harengus*, *C. pallasi*), (b) marine stenohaline (*Hippoglos-*

sus stenolepis, H. hippoglossus), (c) freshwater (*Cyprinus carpio, Carrasius auratus*), (d) anadromous (*Salmo gairdneri*), and (e) catadromous (*Anguilla rostrata, A. japonica*). Other forms requiring consideration would include *Fundulus* sp. and *Oryzias latipes*.

III. FERTILIZATION

As with oocytes, in the teleosts there is a relative paucity of information on the chronology of events at the molecular level during fertilization. Much of what is known about the process of fertilization has been obtained from studies involving polychaetes, echinoderms, and amphibians. Our knowledge of events during fertilization of teleosts, as with oocytes, is based primarily on the Japanese freshwater medaka (*Oryzias latipes*), first used by Yamamoto (1939) as an appropriate experimental organism. A number of major events occurring at fertilization in the sea-urchin egg (*Lytechinus pictus*) (Whitaker and Steinhardt, 1985) may serve as a guide to the teleosts (Table II). From this may be gained an appreciation of the number and complexity of events triggered in the arrested oocyte by contact with the sperm. The list must be applied with caution to teleosts; between echinoderms and teleosts there are major differences in morphology, likely variations in timing of events, and possible differences in the events themselves.

The teleost egg differs in one major respect from those of the polychaetes, echinoderms, and amphibians, in having a micropyle in the outer egg membrane (zona radiata). The micropyle, a pore whose inner diameter is about 1.5 μm (Ginzburg, 1968), allows sperm to enter and advance into the canal in single file. The first sperm to penetrate the canal and contact the plasma membrane of the egg sets a chain of events in motion that effectively prevents further sperm entry (polyspermy). From the point of sperm contact, the cortical alveoli underlying the oocyte plasma membrane begin discharging their contents (Vacquier, 1975); the reaction proceeds in the medaka egg in association with a pulse of free cytosolic calcium and spreads as a wave over the oocyte membrane (Gilkey, 1983; Gilkey *et al.*, 1978). At the micropyle the everted cortical alveolar material pushes into the micropylar canal to form a plug-like deposit (Nakano, 1969), preventing further sperm entry.

Immediately prior to sperm contact with the plasma membrane, the resting potential of the membrane in freshwater teleosts is about

TABLE II

Estimates of Timing of Major Events (16–18°C) Occurring at and Following
Insemination of the Sea-Urchin Egg (*Lytechinus pictus*)[a]

Event	Time
Membrane potential	
Ca–Na action potential	Before 3 s
Na activation potential	3–120 s
Increases in K conductance (remains at higher levels)	500–3000 s
Intracellular calcium release	40–120 s
Cortical reaction	40–100 s
Activation of NAD kinase	40–120 s
Increases in reduced nicotinamide nucleotides (remains at higher levels)	40–900 s
Acid efflux	1–5 min
Increases in intracellular pH (remains at higher levels)	1–5 min
Increased oxygen consumption (initial burst)	1–3 min
Initiation of protein synthesis	After 5 min
Activation of amino acid transport	After 15 min
Initiation of DNA synthesis	20–40 min
Mitosis	60–80 min
First cleavage	85–95 min

[a] From Whitaker and Steinhardt (1985).

−50 to −90 mV, which, in eggs fertilized externally, will vary in relation
to ion concentration of the external environmental medium (Ito, 1963;
Hagiwara and Jaffe, 1979). In most materials examined, this potential
results from the selective permeability of the egg membrane to K^+
ions and a higher $[K^+]_i$ (inside) than in the external medium.

In the sea-urchin egg, at the instant of sperm contact with the
plasma membrane, evidence supports the occurrence of three consecutive electrical events: an initial depolarization, a Na^+–Ca^{2+} action
potential, and an activation (fertilization) potential that exists until the
egg repolarizes (Whitaker and Steinhardt, 1985). The initial depolarization appears to be sperm-dependent (-gated); it seems to trigger an
action potential dependent on external Na^+ and Ca^{2+}, and suggests
that a calcium-carrying potential-gated membrane channel is primarily responsible. The latter phase of the action potential overlaps the
initial phase of the activation potential, and the latter results from
increased Na^+ conductance by the membrane. The activation potential may also be associated with a Ca^{2+}-dependent change in membrane permeability. Finally, repolarization at the end of the activation
potential appears to be the result of a resequestering of Ca^{2+}, and a

pH-dependent increase in K^+ conductance (Whitaker and Steinhardt, 1985). The total fertilization potential may then involve gated membrane channels and Ca^{2+}, K^+, Na^+, and H^+ ions.

As documented by Hagiwara and Jaffe (1979) and Hagiwara (1983), a very instructive examination of the activation potential and related events is provided by Nuccitelli (1980a,b) for the medaka egg (*Oryzias latipes*). Figure 1A is an example of the electrical activity of the plasma membrane during activation, for which the following *average* values were obtained (20–22°C, in 10% Yamamoto Ringer's).

1. Membrane potential.
   (a) Resting potential, prior to fertilization: -39 ± 9 mV (Yamamoto Ringer's); -47 ± 10 mV (10% Yamamoto Ringer's)
   (b) At activation: small depolarization of 4 ± 3 mV; duration, 20 ± 10 s.
   (c) Hyperpolarization: amplitude, 31 ± 12 mV (e.g., to -66 mV)

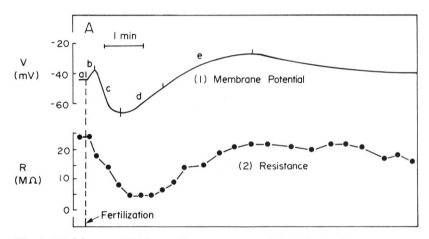

Fig. 1. Medaka egg. (A) Electrical events associated with the fertilization potential. (B, C) Dependence of plasma membrane potential on $[K^+]_o$ and $[Na^+]_o$, respectively. (A1) Membrane potential (V) shows (a) resting potential, (b) small depolarization at fertilization, (c) hyperpolarization phase, (d) fast recovery phase, and (e) slower recovery phase (in 10% Yamamoto's Ringer's solution). (A2) Changes in membrane resistance (R) over the same period (Yamamoto Ringer's). (B, C) Ion dependence of the membrane potential per decade of change (B) in $[K^+]_o$, (C) in $[Na^+]_o$, in (1) unfertilized egg, (2) at hyperpolarization peak (10% Yamamoto Ringer's), and (3) fertilized egg about 8 min later. Some K^+ and Na^+ dependence is shown by the unfertilized egg; conductance at the hyperpolarization peak is strongly K^+-dependent; in the fertilized egg, K^+ dependence falls to near zero while that for Na^+ increases slightly. [Modified from Nuccitelli (1980a).]

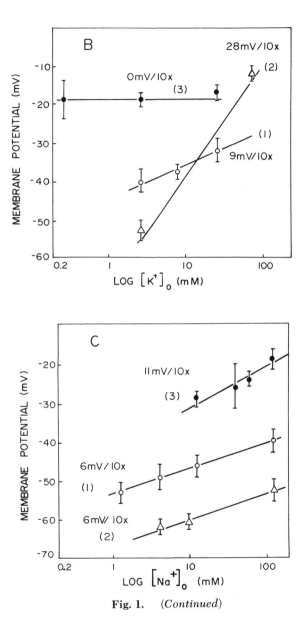

Fig. 1.   (*Continued*)

    (d) Fast phase of recovery from hyperpolarization; duration, 155 ± 18 s.

    (e) Slower phase of recovery, reaching a steady, postactivation resting potential of -19 ± 15 mV by 9.4 ± 1 min after activation.

  2. Membrane resistance. A 10-fold decrease in resistance from 40 to 3 $M\Omega$ occurred over the first 2 min following activation. Thereafter, there was a slow recovery over the next 8 min, resistance ultimately reaching a value 30% greater than the preactivation value.

From electrical events, the time of activation (Fig. 1A) was determined to be about 5 s after a sperm entered the micropyle. Nuccitelli (1980a) also examined the ionic basis for changes in membrane potential. The resting potential showed a limited dependence on external K^+ and Na^+ ions (Figs. 1B, 1C), but independence of external Ca^{2+} and Cl^-. The plasma membrane of the unfertilized egg depolarized 9 mV/decade increase in $[K^+]_o$ (outside) and 6 mV/decade increase in $[Na^+]_o$. Ito (1963) obtained a value of 16 mV/decade increase in $[K^+]_o$ in mature oocytes (eggs) of *Oryzias*. Nuccitelli concluded that the small ion dependencies were due in part to leaks at the membrane–electrode seal and in part to a significant K^+ and Na^+ membrane permeability. Other, preliminary, studies indicated that the small initial depolarization pulse is carried by Ca^{2+} and Na^+ entry as well as K^+ efflux. The strong K^+ dependence (28 mV/decade $[K^+]_o$ (Fig. 1B) and much lower Na^+ response (6 mV/decade $[Na^+]_o$ (Fig. 1C) during hyperpolarization suggest that the increase in conductance during this phase is due mainly to increased K^+ membrane permeability. About 8 min after peak hyperpolarization, K^+ permeability fell to near zero while Na^+ permeability increased slightly (11 mV/decade $[Na^+]_o$). The hyperpolarization and increased K^+ flux are suggested as resulting from two possible mechanisms: (1) an opening of K^+ channels due to a rise in $[Ca^{2+}]_i$ and (2) the addition of K^+ channels, present in the vesicular membranes, that fuse with the plasma membrane by vesicular exocytosis. The transient nature of the K^+ permeability increase may be associated with the concomitant decrease in free $[Ca^{2+}]_i$ after passage of the calcium wave over the cytoplasm at activation, and a corresponding closing of newly inserted K^+ channels in the membrane through endocytosis of the vesicular membrane material (Gilkey, 1981, 1983; Kobayashi, 1985).

Of these several events that coincide with or follow the activation potential immediately, an important one is the very large increase in the internal concentration of free calcium. In the sea-urchin egg this increase likely approaches the 300-fold increase seen in the medaka egg (Jaffe, 1985). Second is the eversion of the cortical alveoli, which moves as a wave over the egg surface from the point of sperm contact. With exocytosis, the cortical alveoli deliver their contents into the presumptive perivitelline space immediately above the plasma membrane. There, the alveoli contents imbibe water from the external medium, raising the fertilization membrane (echinoderms) or zona radiata (teleosts) away from the plasma membrane of the egg proper as the perivitelline fluid forms. Gilkey (1983) injected buffered calcium into unfertilized medaka eggs. He found the level of free calcium required to elicit a transient increase in cytoplasmic free calcium to be between 0.51 and 5.1 mM, depending on internal pH (7.0–7.5). The desequestering of free internal calcium appears to be an autocatalytic Ca^{2+}-stimulated Ca^{2+} release. This threshold, required to elicit the transient calcium wave, is well below the 30 mM of free calcium reached at the peak of the transient. He concluded that the calcium wave is necessary and sufficient to activate the medaka egg. In the sea-urchin egg, after fertilization, the internal calcium content begins to fall; it appears to be pumped out of the cytosol and into the external medium (Azarnia and Chambers, 1976). In some invertebrates (protosomes) egg activation requires calcium, but it appears to enter the cytosol from the external medium through voltage-gated channels in response to a shift in membrane potential (Jaffe *et al.*, 1979).

Internal pH may play a role in egg activation—or in suppression of activation. Medaka eggs with an apparent internal pH of 7.1 (Jaffe, 1985) show a slowing of the calcium wave at pH 6.9 and its acceleration at pH 7.3. Reduction of internal pH from 7.0 to 6.5 in the medaka egg increases threefold the calcium threshold required for initiation of the calcium surge. The ovarian eggs of frogs, taken directly from the oviduct, were shown to be unfertilizable because of their high CO_2 content. Hence, low ovarian pH may act as a brake to activation prior to egg deposition. In sea-urchin eggs there is an apparent doubling of $[OH^-]$ ions shortly after activation; this alkalinization of the cytosol appears, in part, to be the result of an electrically neutral $Na^+–H^+$ ion transfer.

What can be inferred from the available data regarding iono- and osmoregulation in the teleost oocyte and in the egg at fertilization? Few data are available on molecular processes, and the general nature of possible ionic involvements must be cautiously inferred from other animal groups. Basically, various types of ion-transporting channels appear in the plasma membrane of the developing oocyte. At various times these may become functional, may proliferate, or may cease functioning; they may be electrically neutral, electrogenic, or voltage-gated. Some may be blocked by particular pH levels in the surrounding medium. The resting potential of the plasma membrane is dependent on its selective ion permeability and the voltage gradient set up by differences in ion concentrations on either side of the membrane. Generally it becomes more negative as ion concentration decreases in the external (freshwater) medium. Hence, the resting potential may change in relation to external stimuli (such as sperm contact) or internal stimuli (Ca^{2+} release), and to the level of ionic activity in the external medium. As ions move, so too will water move across the membrane in relation to membrane permeability, and to differences in molar concentrations of solute particles on the two sides of the membrane.

Relatively speaking, the oocyte prior to release is subject to the protective regulative mechanisms of the parent. The ovulated egg responds to changes in the ovarian fluid (Sower et al., 1982), the ovarian fluid osmotically is very similar to the blood plasma (Hirano et al., 1978), and the blood plasma is in balance physiologically with the osmotic activity of the external environment (Sower and Schreck, 1982). However, when shed externally at spawning, mature eggs will be subject to major changes in osmotic and ionic composition of the external medium; in general they will be hypotonic to seawater and hypertonic to fresh water. It would seem that the first major regulatory challenge will occur at spawning, adding to the large number of molecular activities set in motion by activation and fertilization. In the newly fertilized egg the only structure available with regulatory capacity would be the plasma membrane. It seems reasonable to assume that the membrane should be well prepared, in terms of permeation, ion-channel function, and electrical activity, to cope with the changes occurring when the eggs are shed and fertilized over a range and variety of external conditions, whose extent would be related to the normal habitat of the species. In terms of tolerance, no doubt these ranges will have limits related both to permeability characteristics of the plasma membrane and to the tolerance of the initial cell and its successors in the newly fertilized egg.

In summary, it appears that an understanding of regulative characteristics of teleost eggs at fertilization will be centered on the plasma membrane and will require, initially, the development of basic, descriptive data on the electrical and ion permeation characteristics of the teleost egg. Further, examples need to be developed for stenoplastic and euryplastic species from the marine and estuarine environments, as well as from fresh water; patterns that may exist would be difficult to identify from the limited information currently available. One may also plead for a greater use of normal external media during such studies to make the physiological results obtained ecologically more meaningful. Areas of interest in specific studies would include:

Resting potentials, ion permeation characteristics, and electrical activity in mature, unfertilized eggs.

The activation potential, and membrane resistance and conductance with respect to Ca^{2+}, K^+, Na^+, and Cl^- in the external medium.

Action potentials, if they occur.

Ion-channel function in preactivation to postfertilization stages.

Internal ion events during activation and fertilization.

Influence of low-ionic-strength fresh water (e.g., low Ca^{2+}, K^+, Na^+) on activation.

The timing of major events during activation and fertilization in relation to temperature and species.

IV. DEVELOPMENT

Of the three levels of potential regulatory capacity mentioned earlier—cellular, tissue, and neurosecretory involvement—it is assumed the oocyte and mature eggs are restricted to the first of these, continuing, in the fertilized egg, for a period extending into very early cell division. It is assumed also that the second level, proliferation of tissues with regulatory capacity, may begin with blastodermal overgrowth of the yolk, and possibly earlier, with an initial level of regulatory capacity achieved by yolk plug closure (Holliday and Jones, 1965). It would seem that neurosecretory involvement would of necessity await the development of specialized tissues, after yolk plug closure, in a process of continuing elaboration through the larval and later stages. From this viewpoint the components of the fertilized egg will be examined that may contribute to regulation. In order of development these would included the plasma membrane, the perivitelline fluid and zona radiata, tissues of the blastoderm, the embryonal epidermis, chloride cells, and, briefly, a consideration of the transition from the embryonic regulative mechanisms to those of the juvenile, including the gills, gut, and kidney. Possible responses to evolving regulatory capacity could involve changes in egg volume, water content, levels of tissue osmolarity, internal ion concentration, hydrostatic pressure, and buoyancy and specific gravity. These will be touched on briefly, but emphasis is placed on regulation, regulative processes, and where and when these might occur.

A. Properties of the Plasma Membrane

The plasma membrane of a cell generally has the form of a lipid bilayer matrix in which islands of protein are interspersed (Korenbrot, 1977). The lipid matrix comprises some 5–10 lipids, including phosphatidyl derivatives, sphingomyelin, cholesterol, neutral lipids, and glycolipids. The islands of protein derive some of their characteristics from the composition of the lipid matrix. They may extend across the entire thickness of the membrane and provide access to reactants at its two surfaces. Most water transport across a membrane appears to occur by a solubility–diffusion process through the hydrophobic bilayer lipid matrix. On the other hand, ions appear to move across the membrane, by active or passive transport, by way of the protein islands— generally referred to as channels. As most cells do not require high water permeability, plasma membrane channels devoted to water movement would seem redundant. Nevertheless, water molecules may penetrate a membrane via channels and do so in relation to channel diameter. In such instances, flows through channels may involve water–water, water–ion, or ion–ion interactions (Levitt, 1984).

Three important functional aspects of a membrane include its water permeability, transfer of ions, and its electrical properties. Various techniques have been used, refined and redefined for measuring the movement of particles across a membrane. There is a potential for confusion because of the various terminologies and measures in the literature, as pointed out by Potts and Parry (1964). The flow of *solute* across a membrane may be measured as a diffusion *constant* (cm^2/s), or as a permeability *constant* (cm/s). Permeability of a membrane to a *solvent* (water) is measured in terms of permeability *coefficients*.

As demonstrated by Kedem and Katchalsky (1961) and Katchalsky and Curran (1965), the movement of particles across a membrane is dependent on the flows and forces operating in the system. A set of phenomenological equations, relating these flows and forces, may be written as

$$J_1 = L_{11} X_1 + L_{12} X_2 + \cdots + L_{1n}X_n$$
$$J_2 = L_{21} X_1 + L_{22} X_2 + \cdots + L_{2n}X_n$$
$$\cdot$$
$$\cdot$$
$$\cdot$$
$$J_n = L_{n1} X_1 + L_{n2} X_2 + \cdots + L_{nn}X_n$$

where the J_i flows are linear functions of the X_k forces; that is,

$$J_i = \sum_{k=1}^{n} L_{ik}X_k \qquad (i = 1, 2, \ldots, n)$$

Hence, the X_i forces will be linear functions of the J_k flows,

$$X_i = \sum_{k=1}^{n} R_{ik}J_k \qquad (i = 1, 2, \ldots, n)$$

The $L_{ik} = (J_i/X_k)_{X_i}$ coefficients are flows per unit force and are characterized as conductances or mobilities. The $R_{ik} = (X_i/J_k)_{J_i}$ coefficients have dimensions of force per unit flow and represent resistances or frictions. In practice, either set of J_i or X_i equations may be converted into the other by standard matrix algebra. Furthermore, the matrices are symmetric, so that

$$L_{ik} = L_{ki} \qquad (i \neq k)$$

Hence, in a two-flow, two-force system (one solute, one solvent), there are three coefficients; in a three by three system, there are six coefficients. In general,

$$L_{ii} \geq 0, \qquad |L| \geq 0, \qquad \text{and} \qquad L_{ii}L_{jj} \geq L_{ij}^2$$

In considering membrane properties, it is advantageous to transform the original (thermodynamic) forces into simpler quantities. For example, it follows, in a binary system, that

$$J_V = L_P \, \Delta P + L_{PD} \, \Delta \pi$$
$$J_D = L_{DP} \, \Delta P + L_D \, \Delta \pi$$

where J_V is the volume flow of solvent (e.g. water); J_D is the velocity of solute relative to solvent, similar to a diffusional flow; ΔP is the hydrostatic pressure difference across the membrane; $\Delta \pi$ is the osmotic pressure difference across the membrane; L_P is the mechanical filtration coefficient, relating J_V to ΔP; L_D is the diffusional mobility coefficient, relating J_D to $\Delta \pi$; and L_{PD} is the coefficient of osmotic flow, which equals L_{DP}, the ultrafiltration coefficient. Several examples may show the value of these multiflow–multiforce relationships.

1. If $\Delta \pi = 0$:

$$J_V = L_P \, \Delta P, \qquad J_D = L_{DP} \, \Delta P$$

Thus, with $\Delta \pi = 0$, hydrostatic pressure will produce both a volume flow and a diffusional flow, the latter by ultrafiltration.

2. If $\Delta P = 0$:

$$J_V = L_{PD} \, \Delta \pi, \qquad J_D = L_D \, \Delta \pi$$

Hence, with $\Delta P = 0$, the osmotic pressure gradient will produce an osmotic flow and a diffusional flow. There also is a relation between osmotic flow and ultrafiltration, as

$$\left(\frac{J_V}{\Delta \pi}\right)_{\Delta P=0} = L_{PD} = L_{DP} = \left(\frac{J_D}{\Delta P}\right)_{\Delta \pi=0}$$

3. If $J_V = 0$:

$$(\Delta P)_{J_{V=0}} = -\frac{L_{PD}}{L_P} \Delta \pi$$

That is, true osmotic equilibrium $[(\Delta P)_{J_{V=0}} = \Delta \pi]$ occurs only if $-L_{PD} = L_P = L_D$ in an ideal semipermeable membrane where the solute flow is zero.

4. Where the membrane permits passage of the solute:

$$-\frac{L_{PD}}{L_P} < 1$$

The ratio $-L_{PD}/L_P$ is called the reflection coefficient σ. When $\sigma = 1$, all of the solute is "reflected" by the membrane (e.g., ideal semipermeable membrane); when $\sigma < 1$, part of the solute permeates the membrane; when $\sigma = 0$, the membrane is nonselective and solute and solvent move with equal velocities.

5. By rearrangement and in terms of σ, the volume flow is

$$J_V = L_P \, \Delta P + L_{PD} \, \Delta \pi = L_P(\Delta P - \sigma \, \Delta \pi)$$

6. In a binary system, the total solute flow is

$$J_S = \bar{C}_S \, (1 - \sigma) \, J_V + \omega \, \Delta \pi$$

where $\bar{C}_S$ is the mean concentration of solute in the membrane and

$$\omega = \bar{C}_S \frac{[L_P L_D - L_{PD}^2]}{L_P}$$

Then, when $J_V = 0$,

$$\omega = \frac{J_S}{\Delta \pi}$$

The term ω is the coefficient of solute permeability at zero volume flow. For an ideal semipermeable membrane, as before, $-L_{PD} = L_P = L_{D,}$, and $\omega = 0$. For a nonselective membrane, where the solute diffuses freely, $L_{PD} = 0$ and $\omega = \bar{C}_S L_D$.

The set of coefficients L_P, σ, and ω is generally accepted as being the most convenient for the description of membrane properties. For a system of n solutes, the equation for volume flow frequently seen is

$$J_V = L_P\left(\Delta P - \sum_{k=1}^{n} \sigma_i \, \Delta \pi_i\right)$$

For a binary system of water and one solute, the equations for volume flow (J_V) and solute flow (J_S) simplify to

$$J_V = L_P(\Delta P - \sigma \, \Delta \pi)$$
$$J_S = \bar{C}_S(1 - \sigma)J_V + \omega \, \Delta \pi$$

or their equivalents

$$J_V = L_P \, (\Delta P - \sigma RT \, \Delta C_S)$$
$$J_S = \bar{C}_S(1 - \sigma)J_V + \omega RT \, \Delta C_S$$

where

$$\pi = RTC_S \quad \text{(atm)}$$
$$R = 0.082\,1 \quad \text{atm } °C^{-1} \text{ mole}^{-1}$$

and where T is the absolute temperature (K), ΔC_S is the molal concentration difference across the membrane, and $\Delta \pi = RT \, \Delta C_S$.

Two widely used permeation coefficients may now be introduced. These often are estimated on the assumption that $\Delta P = 0$ and $\sigma = 1$. Hence we have (1) the filtration permeability coefficient

$$P_f = \frac{L_P}{\bar{V}_w}$$

and (2) the diffusion permeability coefficient

$$P_d = \omega = \left[\frac{J_S}{\Delta \pi}\right]_{J_V=0}$$

The term P_f measures the net flow, the difference between two flows moving in opposite directions across a membrane; P_d measures one of these flows separately, and is used to estimate the unidirectional flow of tritiated water in isotope studies.

When values of the two permeation coefficients are obtained for the same membrane system, it often occurs that $P_f > P_d$. Earlier it was assumed that this inequality resulted from the fact that diffusion was accelerated in the direction of the net flow, and that isotopic methods always measure diffusion against the net flow. Using Poiseuille's law of volume flow, the difference between the two measures provided a means of calculating the width of pores in a membrane through which such flows could occur. The validity of this inequality, however, has been argued for some time (Hansson Mild and Løvtrup, 1985). Although the earlier assumptions now generally are considered unrealistic, comparison of the two coefficients remains valid. Hence, where

$$P_f = \frac{L_P}{\bar{V}_W}, \qquad P_d = \omega, \qquad \Delta P = 0, \qquad \text{and} \qquad \sigma = 1$$

then

$$\frac{P_f}{P_d} = \frac{L_P}{\bar{V}_W}\left(\frac{1}{\omega}\right) = \frac{L_P}{\bar{V}_W \omega}$$

If one wishes to examine the influence of the variables on the ratio, allowing ΔP and σ to vary, where

$$P_f = \frac{L_P}{\bar{V}_W} = \frac{J_V}{(\Delta P - \sigma \, \Delta\pi)\bar{V}_W} \qquad P_d = \omega = \frac{J_s - \{\bar{C}_s(1 - \sigma)J_V\}}{\Delta\pi}$$

then

$$\frac{P_f}{P_d} = \frac{J_V \, \Delta\pi}{(\bar{V}_W \, \Delta P - \bar{V}_W \sigma \, \Delta\pi) \, (J_S - J_V \bar{C}_S + J_V \sigma \bar{C}_S)}$$

$$= \frac{J_V}{\bar{V}_W(\Delta P - \sigma \, \Delta\pi)\omega}$$

Friedman (1986) provides an excellent survey of these and associated relationships, and their utilization in problems of biological transport.

Water permeability coefficients measure the volume of water passing across a unit area of membrane per unit time. Although P_f and P_d usually are reported in centimeters per second, this actually represents a contraction from $cm^3 \, cm^{-2} \, s^{-1}$. The latter ($\times 10^4$) also may be reported as $\mu m^3 \, \mu m^{-2} \, s^{-1}$, equivalent to $\mu m \, s^{-1}$. Other units also have been used (Potts and Parry, 1964). Where possible, the original units used in articles reviewed will be converted to $\mu m^3 \, \mu m^{-2} \, s^{-1}$ ($\mu m \, s^{-1}$).

A long series of most informative investigations has been conducted by Løvtrup, Hansson Mild, and their associates that deserves particular mention; an excellent summary of these studies is found in Løvtrup and Hansson Mild (1981). Using an electromagnetic diver balance, they found that the balance recorded a composite response involving both cytoplasmic diffusion and plasma membrane permeation. That is, the measure of membrane permeability for molecules passing into a cell was influenced by the rate of diffusion of those molecules moving away from the membrane in the cytoplasm. They selected an approach providing an estimate of isotopic water diffusion $(D, cm^2 s^{-1})$ in the cytoplasm[1]; this allowed an independent estimate of the exchange permeation coefficient $(E, cm s^{-1})$ for the plasma membrane. Further, they found that the amphibian oocyte, in contrast to the mature egg, has no measurable barrier to water permeation; that there is a relation between cortical tension, tightness of the plasma membrane, and tonicity of the external medium (low tonicity increases cortical tension and tightness of the membrane); that diffusion in the cytoplasm is a complex function of temperature; and that cell density is a function of water content. Finally, they derived P_f and $P_d (=E, cm s^{-1})$ coefficients for the plasma membrane, and questioned the popular interpretation that $P_f > P_d$ is related to the presence of membrane channels.

Finkelstein (1984) has examined the water permeability coefficients obtained for various bilayer membranes, and finds they range from 2×10^{-1} to $1 \times 10^2 \mu m s^{-1}$. For plasma membranes examined, the range extends from 0.96×10^{-2} to $1.14 \times 10^{-2} \mu m s^{-1}$ for midgastrula eggs of *Fundulus* sp. at 23°C (Dunham *et al.*, 1970)[2] to $2 \times 10^2 \mu m s^{-1}$ for erythrocytes; most values are found to be around $2 \times 10^{-1} \mu m s^{-1}$. Dunham *et al.* (1970) concluded from the low P_d values they obtained for the eggs that water transport across the egg membrane was very slow, suggesting that no special mechanisms are necessary for volume regulation of the *Fundulus* egg. The unfertilized oocyte of the plaice (*Pleuronectes platessa*) shed into seawater (Potts and Eddy, 1973) had an initial P_d of $8.6 \times 10^{-2} \mu m s^{-1}$, reducing to $1.7 \times 10^{-3} \mu m s^{-1}$ after 1 day. Potts and Rudy (1969) estimated P_d for the ovarian egg of the Atlantic salmon (*Salmo salar*), prior to shedding, to be $6 \times 10^{-2} \mu m s^{-1}$, falling to $<4 \times 10^{-3} \mu m s^{-1}$ after water hardening. For fertilized eggs of the same species (gastrula to myomere stage) previously stored in Ringer's solution and examined in that medium at 5.5°C, Loeffler and Løvtrup (1970) obtained a permeability coefficient (E) of $1.0 \times 10^{-2} \mu m s^{-1}$. Loeffler (1971) also noted (his Fig. 3) that ovarian eggs of the zebrafish (*Brachydanio rerio*) had a permeability (E) near $6.4 \times 10^{-1} \mu m s^{-1}$. Prescott and Zeuthen (1953), using mature un-

---

[1] For the derivation of D, see Løvtrup and Hansson Mild (1981); for E, see Løvtrup (1963).

[2] There is a typographical error in Finkelstein (1984); the original record from Dunham *et al.* (1970) places Finkelstein's figure at $1 \times 10^{-6} cm s^{-1} (1 \times 10^{-2} \mu m s^{-1})$, not $1 \times 10^{-16} cm s^{-1}$ as shown in Finkelstein (1984).

fertilized eggs of the same species in Ringer's solution (estimated temperature 20–22°C; Dick, 1959b), obtained an average P_f value of $4.5 \times 10^{-1} \mu m \ s^{-1}$. Prescott (1955) obtained a P_d of $16.8 \times 10^{-1} \mu m \ s^{-1}$ for unactivated, unfertilized mature eggs of the chinook salmon (*Oncorhynchus tshawytscha*) in Ringer's solution at 15°C. Although the range indicated for these permeation estimates of teleost plasma membranes is not extensive, there is evidence that membrane permeability varies in relation to (1) state of embryonic development, (2) composition of the external medium, (3) temperature, (4) surface-to-volume ratio, and (5) presence or absence of membrane channels. These will be considered in turn.

1. STATE OF EMBRYONIC DEVELOPMENT

An examination of permeation characteristics at various stages of embryonic development (Table III) suggests that (1) mature oocytes (ovarian eggs), relatively speaking, have a highly permeable plasma membrane, (2) at fertilization there may be an increase in membrane permeability for a short period, (3) following fertilization, permeability decreases rapidly to a minimum, and (4) subsequently there may be a minor increase in permeability. Potts and Eddy (1973) compared values of the rate constant $(K)^3$ for tritiated water exchange in eggs of the plaice (*Pleuronectes platessa*) (Table III). Their data show that mature eggs shed into isotonic saline were highly permeable initially, retained higher permeability longer, but eventually permeability declined to levels similar to those for eggs shed into seawater. In seawater, the reduction in permeability to a minimum occurred about 1 day after fertilization. Both plaice and Dover sole (*Solea vulgaris*) were shown to have K values of 0.03–0.07 later in development, a small increase from the minimum achieved 1 day following fertilization. Potts and Rudy (1969) noted a similar decline in K values (Table III) for *Salmo salar*, equivalent to a reduction in P_f from $6 \times 10^{-2} \mu m \ s^{-1}$ in freshly stripped eggs to $<4 \times 10^{-3} \mu m \ s^{-1}$ in water-hardened eggs. They found that the permeability of unfertilized eggs declined almost immediately following their placement in water; it continued to decline for 7–8 h to a minimum, remaining at or slightly above the minimum in the following 50 days. The resting value of the unfertilized egg ($K = 0.4 \ h^{-1}$) again appears to rise somewhat after its placement in water. Loeffler (1971) carried out a similar study with fertilized and unfertilized eggs of the pike (*Esox lucius*) (Table III) stored

---

[3] $K = (1/T) \ln[(A - B_1)/(A - B_2)]$, where A is the activity in the external medium and B_1 and B_2 are specific activities at two times, time T apart.

TABLE III

Changes in Water Permeation Rates of Oocytes, and Fertilized Eggs of Three Teleosts at Various Stages of Development

Pleuronectes platessa[a]			Salmo salar[b]			Esox lucius[c]		
Time from fertilization[d]	K (h⁻¹)	Remarks	Time from fertilization[d]	K (h⁻¹)	Remarks	E (μm s⁻¹)	Stage	Remarks
—	1.0	Initial observation in isotonic saline	Oocytes	~0.4	"Resting value"	1.7×10^{-1}	Unfertilized	In 100% Ringer's
			0	>0.5	At fertilization	1.9×10^{-1}	Unhardened	In 100% Ringer's
2½ h	0.5	Unfertilized in isotonic saline	1½ h	~0.10	In river water, water hardened	3.8×10^{-2}	Hardened	In 100% Ringer's
—	0.05	Final observation in isotonic saline	A few hours	0.02–0.03	In river water	1.8×10^{-2}	Fert., early cleavage	In 7.5% Ringer's
			40d	0.05	In river water, "eyed"	2.2×10^{-2}	Blastoderm	In 7.5% Ringer's
1 h	0.2	Fert., in SW[e]	50 d	0.023	In river water, unfertilized	1.9×10^{-2}	Beginning of epiboly	In 7.5% Ringer's
2 h	0.1	Fert., in SW					Advanced	
2½ h	0.05	Fert., in SW				2.2×10^{-2}	myomere	In 7.5% Ringer's
>10 h	0.02–0.04	Fert., in SW						
1 d	0.02	Fert., in SW						
14 d	0.03–0.07	Fert., in SW						

[a] Potts and Eddy (1973), marine.
[b] Potts and Rudy (1969), freshwater.
[c] Loeffler (1971), freshwater.
[d] Time from fertilization in hours (h) and days (d).
[e] Fert., Fertilized; SW, seawater.

in Ringer's and examined at 9.0°C. From a maximum of 1.7×10^{-1} μm s^{-1}, in the unfertilized oocyte, E ($=P_f$) rose slightly, then fell in the fertilized egg to a minimum of 1.8–2.2×10^{-2} μm s^{-1} from early cleavage onward. No subsequent rise in P_f is noted in these data during later development. However, Guggino (1980a) found in the embryos of *Fundulus heteroclitus* following fertilization (25°C) that P_d increased from 4×10^{-3} μm s^{-1} at 4 days to 1×10^{-2} μm s^{-1} at 7 days and 1.6×10^{-2} μm s^{-1} at 10 days. Dunham *et al.* (1970) obtained P_d estimates for midgastrula eggs of *Fundulus* [presumed *heteroclitus* by Guggino (1980a)] at 23°C. For the pregastrula eggs, an earlier developmental stage than Guggino's 4-day embryos, P_d was 9.6×10^{-3} to 1.14×10^{-2} μm s^{-1}. The two lowest values (4×10^{-3}, 9.6×10^{-3} μm s^{-1}) were the lowest values for teleosts found in this review.

Other studies based on differing methodologies show that water uptake by the egg or embryo may vary throughout development. Zotin (1965) distinguished five phases of water uptake by embryos of the sturgeon (*Acipenser güldenstädti colchiens*, beluga (*Huso huso*), and sevruga (*Acipenser stellatus*). Løvtrup (1960) demonstrated complex changes in rate constants (k),[4] diffusion permeabilities and water volumes in eggs of two amphibians (*Siredon mexicanum*, *Rana platyrrhina*) during development. Harvey and Chamberlain (1982) showed that the eggs of the zebrafish (*Brachydanio rerio*) undergo complex density changes throughout development, apparently reflecting changes in water content. Other simple to complex changes in density throughout development have been noted in marine pelagic eggs by Alderdice and Forrester (1968, 1971, 1974) (*Parophrys vetulus*, *Eopsetta jordani*, *Hippoglossoides elassodon*), Forrester and Alderdice (1973) (*Hippoglossus stenolepis*), and Alderdice *et al.* (1987) (*Anoplopoma fimbria*). These changes appear to have survival values in relation to depth distribution and lateral transport of eggs and larvae in the natural environment.

In summary, permeability of the plasma membrane of the mature teleost egg to water appears to be relatively high. In mature amphibian eggs it appears to be so high that the plasma membrane provides no measurable diffusion barrier to the movement of water; that recorded in isotope exchange experiments is now concluded by Løvtrup and Hansson Mild (1981) to be attributable to diffusion in the cytoplasm. When the egg is shed, the permeability of the egg membrane

---

[4] $k = Q D/2.303$, where $Q = 6/d$, d is the diameter of the egg, and D is the diffusion permeability constant (cm s^{-1}).

appears to increase briefly, then drop rapidly to very low values, to increase again, slightly, in later development. The "tightness" of the membrane that develops a few hours after fertilization would seem an appropriate form of "passive" regulation in an embryonic stage obviously lacking regulatory tissue and apparently depending wholly on the plasma membrane to maintain its integrity. There are no other obvious means by which the egg proper could be protected from the hypotonic and hypertonic media into which the freshwater and marine teleost eggs are shed.

One may speculate that the initial period of high water permeability would allow osmotic water flows to occur, assisting in the achievement of a steady state of the internal environment of the newly fertilized egg relative to the ion fluxes that may occur during the activation potential (Nuccitelli, 1980a). Little regulative control would be available to the egg following fertilization if water permeation of the membrane were minimized but substantial ion flows were to continue. It then could be assumed that when water permeation is attenuated, ion flows across the membrane must also reach low values. This appears to be so in the Atlantic salmon (*Salmo salar*) egg (Rudy and Potts, 1969); prior to the "eyed" stage, ^{24}Na uptake was confined to the zona radiata and perivitelline fluid. The latter also was found to concentrate cations when immersed in dilute external solutions, which would tend to reduce the concentration gradient across the plasma membrane, and therefore the potential rate of loss of cations from the yolk. Potts and Eddy (1973) noted a similar pattern in early eggs of *Pleuronectes platessa;* it appears that some uptake of ^{22}Na by the yolk may have occurred in the first five hours after fertilization, but not thereafter.

Zotin (1965) found five periods of differing rates of water influx in the decapsulated eggs of sturgeon, beluga, and sevruga. The first, between fertilization and gastrulation, was characterized by rapid water uptake. In the second period, from gastrulation to yolk plug closure, water uptake virtually ceased. In the third period, from yolk plug closure to the appearance of the heart anlage, water uptake was rapid. In the fourth stage, to heart pulsation, some water was lost. In the fifth period, to hatching, no water uptake occurred. One may speculate that the first period coincides approximately with volume stabilization following fertilization and attenuation of membrane permeability. Exchange may then cease or almost cease until development of the blastoderm and its overgrowth of the yolk at epiboly, tissue suspected to have osmo–ionoregulatory function (Jones *et al.*, 1966).

2. External Medium

Composition of the external medium also influences the permeability of the plasma membrane. Potts and Rudy (1969) stripped mature eggs of *Salmo salar* into various media (3.5°C) and compared their permeation characteristics. Those in isotonic saline had a rate constant (K) of 0.4 h^{-1}, falling to 0.2–0.3 h^{-1} over 2–3 h. In isotonic glucose initial K values were higher (0.9–1.0 h^{-1}), but these fell rapidly to a rather constant level of 0.05–0.1 h^{-1} within about $\frac{1}{2}$ h. In river water (0.45 mM Ca, 0.11 mM Mg/l), initial K values near 0.5 h^{-1} declined to about 0.1 h^{-1} within $1\frac{1}{2}$ h. The presence of Ca^{2+} ions delayed onset of the initial stage of high permeability for a short time; high Na^{+} ion concentration delayed the onset indefinitely. Membrane permeability was highest in solutions containing 0.5–10 mM Ca/l and 0–10 mM Na/l. Loeffler (1971) found that the exchange coefficient $(E = P_f)$ for water in the pike (*Esox lucius*) egg was greater in 100% Ringer's solution (4 × 10^{-2} to 1.0 × 10^{-1} μm s^{-1}) than in 7.5% Ringer's (1.8 × 10^{-2} to 2.2 × 10^{-2} μm s^{-1}), although the comparison is confounded by differences in developmental stages. Berntsson *et al.* (1964) compared the permeability of the plasma membrane of eggs of the amphibian *Siredon mexicanum* after storage for various periods in 100% or 7.5% Ringer's solution (Table IV). Although the data are minimal, it can be seen that eggs stored in 100% Ringer's (isotonic) retain a higher membrane permeability and probably for a longer period than when stored in 7.5% Ringer's. Similar results were obtained by Haglund and Løvtrup (1966) for oocytes and fertilized eggs at a later stage of devel-

TABLE IV

Permeability Coefficients (μm s^{-1}) for the Plasma Membrane of Eggs of the Amphibian *Siredon mexicanum* Stored for Two Periods (>1 day, <1 day) in 100% (Isotonic) or 7.5% (Hypotonic) Ringer's Solution and Examined in 100% or 7.5% Ringer's

Tested in (% Ringer's)	Stored in			
	100% Ringer's		7.5% Ringer's	
	> day	<1 day	>1 day	<1 day
100	1.16 (3)	0.84 (1)	—	0.52 (1)
7.5	—	—	0.49 (2)	0.71 (3)

[a] Figures in the table are averages for $n = 1$–3 trials. Data from Berntsson *et al.* (1964).

opment in the same species. Variations in the original trial values for E for eggs stored or examined in 7.5% Ringer's (0.51–0.94 μm s^{-1}) suggest that the three values on the right side of the table (7.5% Ringer's) are not significantly different. Berntsson *et al.* (1964) also measured cortical stiffness in the eggs of *Siredon*. They attributed the decrease in permeability in hypotonic solution to an increase in stiffness (force required to deform the cell) of the egg cortex and vitelline[5] membrane, and an increase in hydrostatic pressure within the egg that would occur in hypotonic media. One would conclude that the plasma membrane is tighter, and less permeant, in hypotonic media. Hansson Mild and Løvtrup (1974a) determined P_d in mature oocytes of the amphibian *Rana temporaria,* stored in Ringer's solution, and examined in several dilutions of the medium. The P_d values were 1.57, 1.07, and 0.75 μm s^{-1} for oocytes in 100, 25, and 7.5% Ringer's, respectively, over test periods of 2.8–4.6 h. The P_d estimates also were obtained continuously over 8–9 h for oocytes in the same three solutions. In 100% Ringer's, P_d remained constant at 1.5 $\pm$ 0.1 μm s^{-1}. In the dilutions of Ringer's, P_d declined markedly with time more or less proportionate to the degree of dilution; in 7.5% Ringer's, P_d declined from 1.4 to 0.4 μm s^{-1} in 8 h. These authors also computed the tension in the vitelline membrane as a function of oocyte radius and internal hydrostatic pressure. They reasoned that tonicity per se would have no direct influence on permeability of the plasma membrane and that tension in the vitelline membrane is indirect, resulting from the influence of tonicity on imbibition of water, change in egg radius, and internal pressure, all of which influence tension. Their results (Fig. 2) support this contention. They reasoned further that increased vitelline membrane tension is transmitted either directly or proportionately to the plasma membrane, and they concluded that permeability of the plasma membrane (in the amphibian oocyte) varies with the tension in the vitelline membrane surrounding it. It would seem reasonable to conclude that the same relation should apply to the plasma membrane and zona radiata of teleost oocytes and eggs. The internal pressure in an egg may vary considerably with species, state of development, and external conditions, including salinity. Alderdice *et al.* (1984) measured hydrostatic pressure in the eggs of five salmonids throughout development (range when water-hardened, 30–90 mm Hg) and compared the results with those of Kao and Chambers (1954) for *Fundulus* (marine species) eggs (150 mm Hg). In addition, D. F. Alderdice and J. O. T. Jensen (unpublished data) obtained an internal

[5] The outer egg membrane in amphibians, equivalent to the zona radiata of teleosts.

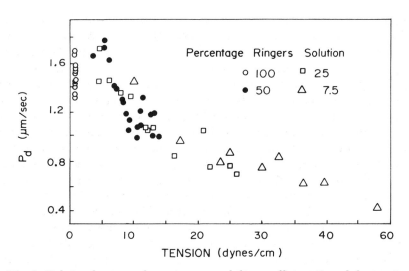

Fig. 2. Relation between the water permeability coefficient P_d and the tension of the vitelline membrane in eggs of the amphibian *Rana temporaria* held in 7.5, 25, 50, and 100% Ringer's solution. Note the inverse relationship: as tension increases, P_d decreases. [Modified from Hansson Mild and Løvtrup (1974a).]

pressure of 162 mm Hg in 72-h-old *Clupea pallasi* eggs (8.5°C, 17‰), with internal pressures increasing in the first 200 h of incubation from 103 to 188 mm Hg. These data show that the tension on the zona radiata may vary between 5.8×10^3 and 1.1×10^4 dyne cm^{-1}, approximately a twofold change in tension. Therefore, following the argument of Hansson Mild and Løvtrup (1974a), plasma membrane permeability could vary considerably and inversely as a function of hydrostatic pressure.

Although plasma membrane permeability reduces to and remains near a minimum during incubation, water exchange across the membrane does not cease. Loeffler and Løvtrup (1970) calculated influx rates for developing eggs of *Salmo salar* to be about 1/300th of the egg volume per day. Over the incubation period (~50 days), the total net influx would be about one-sixth of the egg volume. The salmon egg appears to come into an osmotic steady state with its external environment by a process whereby the concentration difference between the cytoplasm and the external medium is reduced by the sequestering of cations by the perivitelline fluid, and assisted by the increased hydrostatic pressure resulting from water imbibition, sustained by the tension of the zona radiata. In addition, the increased tension in the zona radiata may be directly or proportionally transmitted to the plasma membrane (Hansson Mild and Løvtrup, 1974a) as a further means of limiting water permeation in the plasma membrane. This latter, interesting observation should be examined further.

In summary, water permeation coefficients of the plasma membrane are relatively large in mature oocytes. They remain high at fertilization, and under normal circumstances, permeability declines precipitously thereafter. This period of high permeability may allow the egg to attain a new set of steady-state conditions in relation to the various ion fluxes that may occur across the plasma membrane as a consequence of electrical activity at activation. The absolute value of the permeability coefficient at activation is affected by the nature of the medium into which the egg is shed. In isotonic media, membrane permeability tends to remain higher longer and decrease more slowly. In hypotonic media, membrane permeability decreases at a rate related to the level of tonicity of the medium. In fresh water, the rates of change are altered by ions in the medium; significant effects of Ca^{2+} and Na^+ ions have been demonstrated. However, the permeability coefficient does not vary directly with tonicity of the medium. Rather, there appears to be a direct relation between P_d and tension on the plasma membrane arising through intermediate steps involving imbibition of water, change in egg size, and internal egg pressure. Hence, as the hydrostatic pressure gradient increases, wall tension rises, appears to be transmitted to the plasma membrane, and P_d falls. These differences in permeability must be recognized when eggs of different species or size are compared, or when differences in tonicity of the external medium are involved. It appears that differences in permeation coefficients may be small in isotonic media and large in hypotonic media. However, ecologically meaningful estimates of permeability coefficients require the use of media in which the egg normally is found, a point frequently recognized but one that has seldom been addressed in the long, complex development of permeation theory and methods of measurement. Even then, minor differences in hydromineral content of fresh waters may significantly affect the course and rate of change of permeation characteristics of the plasma membrane.

3. TEMPERATURE

Temperature also is known to influence membrane permeability. Haglund and Løvtrup (1966) obtained exchange permeability coefficients ($E = P_d$) for unfertilized eggs of the amphibian *Siredon mexicanum* in Ringer's solution after pre-adaptation in Ringer's at each test temperature (4–28°C). The values increased from 0.1 μm s^{-1} at 4°C, more or less linearly to ~0.6 μm s^{-1} at 20°C, and to 1.5 μm s^{-1} at 28°C. Using two amphibians with different natural temperature ranges, *Rana temporaria* (10–25°C) and *R. pipiens* (6–17°C), Hansson Mild *et al.* (1974a) and Hansson Mild and Løvtrup (1974b) noted a general trend toward increased plasma membrane permeability at higher temperatures. The P_f values for mature oocytes ("body-cavity eggs") varied between 2.4 ± 0.6 μm s^{-1} in 7.5, 25, and 50% Ringer's at 22–23°C, and 1.8 ± 0.1 μm s^{-1} in 7.5% Ringer's at 19°C. An Arrhenius plot of P_f for *R. temporaria* oocytes ("ovarian eggs") in 50% Ringer's (Hansson Mild *et al.*, 1974a) is complex: there is a local maximum at 16°C, a minimum at 19°C, and a trend to higher permeabilities at higher temperatures (Fig. 3A). In comparison, Fig. 3B shows a similar Arrhenius plot for P_d (Hansson Mild and Løvtrup, 1974b) for mature oocytes ("body-cavity eggs") for both ranid species at isotonic conditions (100% Ringer's). Not only do P_d values vary with temperature, they also vary with tonicity of the medium as discussed in the previous section. It is worthy of note, as pointed out by Hansson Mild and Løvtrup (1974a), that the effect of higher temperature on P can be countered by an increase in plasma membrane tension, as occurs when the egg is exposed to a more hypotonic medium.

In summary, osmotic permeability of the plasma membrane increases at higher temperatures. The nature of this relation may also change with alteration of the tonicity of the external medium. Changes in permeability of the membrane also reflect complex temperature-induced changes in rates of diffusion of water in the cytoplasm, and in the external medium, on each side of the membrane (Hansson Mild and Løvtrup, 1974b); these authors also recall arguments supporting the contention that structural changes in the properties of water may occur with temperature change. Phase transitions frequently are found in the 13–16°C range, suggesting an association with the discontinuities found for cytoplasmic diffusion and plasma membrane permeation at 16°C. Hansson Mild and Løvtrup (1974b) also note the well-known fact that the degree of unsaturation of body lipids tends to be correlated with the general environmental temperature range for a given species (Barańska and Wlodawer, 1969). Therefore, the temperature-associated changes in the plasma membrane of the two ranid species examined, at least in part, may reflect a broad thermal phase transition in the structure or composition of membrane lipids.

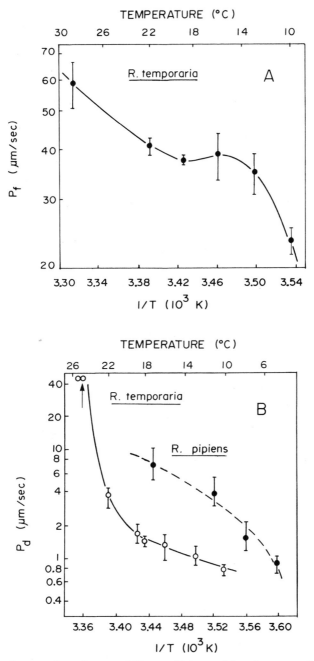

Fig. 3. Arrhenius plots of permeability coefficients of the plasma membrane in two ranid species. (A) P_f estimates for ovarian eggs of *Rana temporaria* in 50% Ringer's. (B) Comparison of P_d estimates for body-cavity eggs (mature oocytes) of *R. temporaria* and *R. pipiens* in 100% Ringer's. Note that the permeability of the oocytes is an order of magnitude higher than that of the mature eggs. [Modified from Hansson Mild *et al.* (1974a) and Hansson Mild and Løvtrup (1974b).]

4. EGG VOLUME

One might assume that osmoregulation would pose a greater problem in smaller eggs, which have a higher surface-to-volume ratio than larger eggs. That is, given equal rates of water permeation across the plasma membrane, smaller eggs with their smaller volume would tend to come into a steady state with the external medium more rapidly than larger eggs. Dick (1959b) examined the question earlier; he found, initially, an inverse correlation between cell size and permeability coefficient in a variety of cell types and sizes. This could be expressed as a direct correlation between the permeability coefficient and the surface/volume (S/V) ratio (for spheres) for groups of cells of similar phylogenetic or histological type. When the same parameters were examined for all cells without grouping, the increased scatter in the pattern was suggestive of several confounded relationships. The confounding was removed when the diffusion coefficients for water movement in the cytoplasm were examined; an inverse relation was found with S/V, the lower S/V ratio of the larger eggs being associated with higher measured values for diffusion of water in the cytoplasm. Dick (1959b) suggested an attractive interpretation of the $S/V-P_f$ correlation: the apparent decrease in permeability in the plasma membrane of larger cells (with smaller S/V ratios) is due to the greater length of time water takes to diffuse through the larger volume of internal cytoplasm. That is, measures of membrane permeability (P) and cytoplasmic diffusion (D) were confounded and interactive, each influencing the measured value of the other. The development of this problem and its resolution is well described by Løvtrup (1963), Hansson Mild (1971, 1972), Hansson Mild et al. (1974b), and Løvtrup and Hansson Mild (1981). More recent determinations of membrane permeability take into account the rate of diffusion of water in the cytoplasm (D_1) and in the external medium (D_2) (Hansson Mild and Løvtrup, 1974b).

Løvtrup (1963) confirmed Dick's argument, and Hansson Mild and Løvtrup (1985) describe how permeability of a cell membrane (P_d) is a function of both the rate of cytoplasmic diffusion and cell size. For example, for a spherical egg with radius (r) of 100 μm and $P_d = 10^1$, 10^0, or 10^{-1} μm s^{-1}, cytoplasmic diffusion (D_1) must be below about $10^{3.5}$, $10^{2.5}$, or $10^{1.5}$ μm^2 s^{-1}, respectively, before D_1 has a significant effect on water exchange. In terms of cell size and P_d value of 10^1 μm s^{-1}, cytoplasmic diffusion cannot be neglected in eggs of $r=100$, 10, or 1 μm when D_1 values are below about $10^{3.5}$, $10^{2.5}$, or $10^{1.5}$ μm^2 s^{-1}, respectively.

In summary, the direct correlation between the permeability coefficient of a cell membrane and its S/V ratio discussed in the earlier literature is now considered spurious. There is an inverse curvilinear relation between P_d and D_1 such that P_d tends to be overestimated as a result of cytoplasmic diffusion at lower values of D_1. The influence of D_1 on P_d is greater in large cells (with a smaller S/V ratio) when the membrane is tighter. At a given cell size, D_1 must decrease in order not to inflate the real value of P_d. A further correction to earlier estimates takes into account the presence of stirred or unstirred layers external to the membrane (D_2). Hence, earlier estimates of the permeability coefficient P were biased by the effect of D_1 and D_2 on permeation influenced by cell size. Yet, the bias may lie in how one interprets the information. If one wishes to estimate the *ultimate capacity* for transfer of water through a plasma membrane, then the peripheral effects of cytoplasmic diffusion, and of unstirred layers external to the membrane, must be removed. On the other hand, if one wishes to know how a given membrane capacity is *utilized* under natural conditions, then the "biased" estimate, at least relative to D_1, probably is more meaningful. As an example of the differences involved, Prescott and Zeuthen (1953) obtained an uncorrected estimate for P_d of 0.75 μm s^{-1} for mature oocytes (body-cavity eggs) of *Rana temporaria* in Ringer's (22°C). Corrected for D_1, this rises to 1.6 μm s^{-1}; when corrected for D_1 and D_2 its value is 2.6 μm s^{-1} (Hansson Mild and Løvtrup, 1974b). A reexamination and adjustment of earlier uncorrected estimates, where possible, would seem highly appropriate. However, the question of interpretation of "bias" remains. The corrected value of P may provide a measure of ultimate capacity of the membrane; the value, at least uncorrected for D_1, may be more meaningful in terms of how a given capacity is utilized by the living cell.

5. P_f/P_d RATIOS

It is recognized that water molecules may cross plasma membranes in two ways: through the matrix of the membrane by a solubility–diffusion process, or by passage through transmembrane channels. Finkelstein (1984) gives examples of special materials that "sieve" nonelectrolyte molecules, having P_f/P_d ratios of 3 to 5. Their characteristics indicate that they would pass water molecules through channels in single file array. The question is discussed by Friedman (1986). When solute molecules pass through a pore in single file, the factors that determine permeation rate differ from those governing diffusion in a larger pore, and the probability of higher solute–solute interactions is greater. In the previously described relation, $P_f/P_d = L_P/\bar{V}_n\omega$, where water molecules pass through a pore in single file and n molecules fill the pore, Lea (1963) shows that osmosis is $n + 1$ times more "efficient" than tracer diffusion; that is,

$$\frac{P_f}{P_d} = n + 1$$

Hence the number of molecules needed to fill the pore is of less consequence. If P_f/P_d is large, transport takes place through large pores. In smaller pores the ratio will be closer to unity. When pores are very small or absent and solvent cannot pass through, water will cross the membrane by diffusion only, so that $P_f \simeq P_d$. Of importance is the case when $P_f/P_d > 1$, and a pore transport mechanism is implied. Some examples are examined from this viewpoint.

Prescott and Zeuthen (1953) compared P_f and P_d values for oocytes of the zebrafish (*Brachydanio rerio*) (a) teased from the follicular membranes, and (b) as eggs normally shed, without manipulation. In the former, $P_f/P_d = 29.3/0.68$ (μm s^{-1}) gives a ratio of 43; in the latter, $P_f/P_d = 0.45/0.36$ (μm s^{-1}), a ratio of 1.3. At first glance it appears that water permeation of the membrane of the teased (ovarian) oocytes ($P_f/P_d = 43$) would involve transmembrane channels, which are absent in the mature (body cavity) egg ($P_f/P_d \simeq 1$). Dick (1959b) and others have questioned the validity of this interpretation, suggesting that it reflects injury to the plasma membrane resulting from dissection of the oocytes from the ovarian tissue. Potts and Rudy (1969) calculated the filtration and diffusional water fluxes across the plasma membrane in the egg of Atlantic salmon (*Salmo salar*). The ratio of these inflows was about 4. The authors noted that correction for unstirred layers was not made, nor apparently was a correction made for diffusion in the cytoplasm. The authors concluded that the plasma membrane was an apparent barrier to diffusion. If the order of magnitude of the difference between corrected and uncorrected values of P_d should be similar to that for *Rana temporaria* (Hansson Mild and Løvtrup, 1974b), discussed in the previous section (2.6/0.75=3.5), then the ratio found by Potts and Rudy (1969) for *S. salar* would reduce from near 4 to about 1.1; on this basis one might conclude that channels were not involved in transport of water across the plasma membrane of the *S. salar* egg. Hansson Mild *et al.* (1974a) also compared P_f and P_d estimates, using oocytes of the amphibian *R. temporaria*. They obtained (μm s^{-1})

$$\frac{P_f}{P_d} = \frac{2.4 \pm 0.6}{2.9 \pm 0.5} = 0.83 \qquad \text{at} \quad 22°C$$

and

$$\frac{1.8 \pm 0.3}{1.7 \pm 0.2} = 1.06 \qquad \text{at} \quad 19°C$$

In this particular instance experimental error is taken into consideration, and there seems to be no justification in assuming $P_f > P_d$; in each case $P_f/P_d \simeq 1$.

In summary, there are few instances in the literature where P_f/P_d ratios are capable of statistical comparison. It is possible that some eggs may possess transmembrane channels; further data on teleosts would help clarify the question. Moreover, Finkelstein (1984) concludes that there are too few channels in most plasma membranes to provide a significant path for water transport, and that the major route across most plasma membranes must be through their lipid bilayers. He also points out that most cells do not require high water permeabilities, and that one therefore would not expect to find channels devoted to water transport in such membranes. It is of interest to note that teleost egg plasma membranes are highly permeable to water when imbibition of water and processes related to perivitelline fluid formation occur naturally, shutting down thereafter when the internal hydrostatic pressure gradient is higher and the water permeability of the plasma membrane is presumed to be at a minimum. Although there is little further evidence to add regarding teleosts, the current impression that there are no special mechanisms devoted to transmembrane water movement does not appear to contradict Finkelstein's (1984) conclusions.

B. Influence of the Perivitelline Fluid and Zona Radiata

1. FORMATION

The outermost nongelatinous membrane in teleost eggs has been called variously the zona radiata, chorion, or capsule. The egg proper, bounded by its plasma membrane, is bathed by the perivitelline fluid after activation and imbibition, and both are enclosed by the outermost membrane. Arguments regarding the naming of the outermost membrane are based primarily on its origin (Anderson, 1967). Here the outermost coat is called the zona radiata. The term "vitelline membrane" also is used in the nonteleost literature for the outermost egg membrane. To avoid confusion, the lipid membrane here has been called the plasma membrane. The structure of the zona radiata varies substantially, particularly between marine and freshwater teleosts. Among the anadromous salmonids (*Oncorhynchus, Salmo*), the zona radiata appears to consist of two main layers, the internus and externus; in some species the internus may be further subdivided (Groot and Alderdice, 1985). These authors believe the very thin externus (0.1–0.3 μm) primarily is responsible for the permeation characteristics of the zona radiata, the internus and the subinternus (30–60 μm), providing structural integrity to the zona radiata. Many marine teleost eggs are smaller and possess a proportionately thinner zona radiata (Lönning, 1972; Stehr and Hawkes, 1979; Grierson and Neville, 1981).

At fertilization, and as a consequence of the changes occurring in the plasma membrane resting potential and correlated cytoplasmic calcium surge, cortical alveolar exocytosis occurs, the contents of the alveoli underlying the plasma membrane being everted into the presumptive perivitelline space between the plasma membrane and zona radiata. There the alveolar contents cause imbibition of water across the zona radiata from the external medium into the presumptive perivitelline space. The zona radiata lifts away from the plasma membrane by displacement, and some shrinkage occurs of the egg proper. The initially flaccid egg swells and becomes turgid under the action of the colloid osmotic pressure of the forming perivitelline fluid. Swelling continues as water and ions enter the forming perivitelline fluid, until a steady state ensues between the increasing tension of the zona radiata and the increasing hydrostatic pressure resulting from the osmotic pressure difference between the perivitelline fluid and the external medium. Alderdice *et al.* (1984) measured hydrostatic pressures in eggs of five salmonids (*Oncorhynchus, Salmo*), which varied with species and stage of development. In eggs of *S. gairdneri* an effective filtration pressure (Starling, 1895; Florey, 1966) of 62 mm Hg was calculated, which would result in an efflux of water and ions from the perivitelline fluid to the external medium against the established osmotic gradient. The expected rapid turnover of water and ions associated with this effective filtration pressure is supported by evidence from a number of sources (Potts and Rudy, 1969; Rudy and Potts, 1969; Loeffler and Løvtrup, 1970; Loeffler, 1971; Eddy, 1974). The tension of the zona radiata, resulting from the internal hydrostatic pressure created by the osmotic properties of the perivitelline fluid, produces a cushioned environment for the embryo against external deformation and, as indicated earlier, a decrease in the permeability of the plasma membrane.

2. PERIVITELLINE FLUID INCLUSIONS

Eddy (1974) found the perivitelline fluid of the Atlantic salmon (*S. salar*) egg to consist of 58% water, 25% protein, 12% lipid, and 1.7% carbohydrate. He suggested that if the colloidal osmotic pressure were due to a single molecular species, the molecular weight of the organic fraction of the perivitelline fluid would be about 300,000. Schuel (1985) noted that sea-urchin cortical granules prior to alveolar eversion contain membrane-bound calcium, β-1,3-glucanase, protease, peroxidase, acid mucopolysaccharides, structural proteins that contribute to the hardening of the fertilization envelope, and hyalin. During and following imbibition, the zona radiata is impermeable, or almost so, to higher-molecular-weight substances such as dextran (Eddy, 1974). Hence, the mucopolysaccharides of the perivitelline fluid, assumed to be largely responsible for the fluid's colloid osmotic pressure, are confined to the perivitelline fluid by the zona radiata. At normal internal and external pH conditions, the macromolecular inclusions likely will be above their isoelectric point and will be predominately acidic and negatively charged. Hence there will be a tendency for the perivitelline fluid to accumulate cations in excess of their concentrations in the external medium. Because the perivitelline fluid contains diffusible as well as nondiffusible ions, a Donnan type equilibrium will be established resulting in an electrical potential difference, the perivitelline fluid in freshwater eggs being negative relative to the external medium. In most marine teleost eggs the perivitelline potential is likely to be neutral to positive (Peterson and Martin-Robichaud, 1986). Rudy and Potts (1969) obtained for Na^+ a perivitelline fluid-to-external medium ratio of 1 : 1 for 150 mM NaCl/l in the external medium. At 0.1 mM NaCl/l the ratio was about 10 : 1. In the more dilute condition an appreciable number of the available cations is assumed to associate with the negative charges on the perivitelline fluid colloids, thereby reducing the ionic excess (Peterson, 1984). Under normal circumstances Na^+, K^+, Ca^{2+}, Mg^{2+}, and H^+ probably are involved in establishment of these equilibria.

The perivitelline fluid also appears to change in composition as embryonic development proceeds. In newly fertilized eggs of *Clupea pallasi*, the perivitelline fluid is watery in appearance; as development proceeds it becomes increasingly viscous, suggesting that water content is decreasing or that larger molecule metabolites are accumulating. Yarzhombek and Maslennikova (1971) listed 18 nitrogenous metabolites that leached from the perivitelline fluid of S. *trutta, S. gairdneri, Acipenser nudiventris*, and *Huso huso*, of which urea, ammonia, and ornithine were the major components. Among the remainder were 14 amino acids. These authors also reported previously having found creatine and creatinine in the perivitelline fluid of eggs of far eastern salmon, presumably *Oncorhynchus*.

3. Perivitelline Potential

Peterson (1984) examined the perivitelline potential (PVP) of the Atlantic salmon (S. *salar*) egg using microelectrodes filled with 4 M NaCl in 1% agar. The PVP is the voltage gradient between the perivitelline fluid and the external medium, expressed by the Nernst equation as a function of the cation concentration difference across the zona radiata. Varying [H^+], Peterson obtained for fertilized, water-hardened eggs prior to the "eyed" stage a value very close to Nernst equation prediction (-56.5 mV/decade change in [H^+]). Reversal of potential occurred at pH$=4.0$, and maximum and minimum potentials of -80 and $+50$ mV were obtained at [H^+] equivalent to pH 5.3 and pH 2.0, respectively. In deionized water the change in potential for [H^+], [K^+], and [Na^+] was 56, 40, and 30 mV per decade change in concentration, respectively. Peterson suggests that these differences in response may reflect differences in mobility of the ions in aqueous solution, resulting in differing permeabilities through the zona radiata. For Ca^{2+} and Mg^{2+} the change in potential followed a Nernst slope of 26–28 mV decade^{-1}; a high selectivity coefficient suggested that Ca^{2+} was being adsorbed at binding sites in the zona radiata. Departures from Nernst slopes were also noted for Na^+ and Ca^{2+} in more acid media, indicative of cation interference where the measuring electrode is differentially more sensitive to H^+, relative to Na^+ and Ca^{2+}, at more acid pH conditions. In normal tapwater (pH 6.8, Na^+ 0.1–0.2 mM, Ca^{2+} 50 μM, Mg^{2+} 24 μM) the mean PVP was -44 mV. The difference between the low value obtained in tapwater (-44 mV) and the maximum obtained in deionized water at pH 5.8 (-80 mV) could largely be explained by the residual levels of [Na^+] and [Ca^{2+}] occurring in the tapwater. The data indicate the presence of a net negative charge of the perivitelline colloids at normal (near 7.0) pH, and a tendency for the perivitelline fluid to concentrate cations. This net negative charge would reduce to zero at a pH of 4.0. At pH 2.0 the system would be saturated with H^+ ions; below pH 4.0 there could be a tendency for the perivitelline fluid to concentrate anions. For developing salmonid eggs, pH levels of 6.0–6.5 or lower are of increasing environmental concern with the growing acidification of natural waters. Rombough and Jensen (1985) found water uptake and resistance to deformation in eggs of S. *gairdneri* to be significancly reduced at pH 5.5 and 6.0 respectively, although the trends in each case, after 24 h of exposure to various [H$^+$], appear to begin at a threshold near pH 6.5. Hence, in fresh waters inhabited by salmonids, the PVP likely could range between −45 and −80 mV, depending on the ionic strength of an ion species present in the water involved, with Na$^+$ and Ca^{2+} being the most important ions. Peterson and Martin-Robichaud (1986) also examined PVPs in eggs of *Gadus morhua*, the white sucker *Catostomus commersoni*, and Atlantic salmon S. *salar*, using microelectrodes filled with 3 M KCl. The authors note that the smaller PVPs obtained with KCl electrodes likely are a function of the greater similarity in mobilities of K$^+$ and Cl$^-$ ions, compared with Na$^+$ and Cl$^-$ ions. With KCl electrodes they obtained, respectively, for H$^+$, K$^+$, Na$^+$ and Ca^{2+}, perivitelline potentials of 40, 27, 19, and 15 mV per decade change in ion concentration for S. *salar*. The PVP values for *Catostomus* and *Salmo* were very similar (15 mV decade^{-1}) for [Cd^{2+}] between 10^{-3} and 10^{-5} M, zero slopes occurring near −50 and −10 mV (10^{-6} and 10^{-2} M Cd^{2+}), respectively. The PVP for *Gadus* became more positive as the external medium (30‰ sea water) was diluted; the response was attributed to diffusion of salts across the zona radiata. For the cod, PVPs were approximately 0 to +40 mV at [Na$^+$] of 1 to 10^{-4} M, respectively. Plasma membrane potentials also were measured for eggs of S. *salar*. At "normal" conditions of high ambient pH and low ambient ion levels the yolk was 25–30 mV negative to the perivitelline fluid. The plasma membrane potential so measured was less responsive to change in [H$^+$] (10 mV decade^{-1}) compared with the PVP (40 mV decade^{-1}); the two measures would appear (Peterson and Martin-Robichaud, 1986, their Fig. 5) to converge at a potential near −70 mV at a pH near neutrality (pH ~6.7). At pH 3.5 the plasma membrane potential was −60 mV relative to the perivitelline fluid and −40 mV relative to the external medium of deionized water. Hence, the potential difference between the perivitelline fluid and external medium at pH 3.5 should be about +20 mV, a value obtained by Peterson and Martin-Robichaud (their Fig. 5) and one close to the Nernst estimate (Peterson, 1984, his Fig. 1). These results suggest that in the developing salmonid egg the plasma membrane has a low permeability and the plasma membrane–perivitelline fluid potential is much less responsive to ambient pH compared with the potential between the perivitelline fluid and the external medium. The perivitelline fluid should have a substantial capacity to accumulate K^+ and Na^+ ions in natural fresh waters. As the salmonid embryo may begin to take up Na^+ from the perivitelline fluid at the "eyed" stage (Rudy and Potts, 1969), this accumulatory capacity could be of distinct advantage to eyed salmonid eggs and the resulting larvae, particularly in waters of very low hydromineral content involving Ca^{2+}, Na^+, and K^+ in particular, (Alderdice and Harding, 1987). Peterson and Martin-Robichaud (1986) question whether the PVP of the cod egg is the result of a Donnan mechanism or of an ion selectivity of the zona radiata. The substantial movement of K^+, Na^+, and Cl^- ions across the zona radiata of *G. morhua* eggs held in 15‰ salinity and transferred to 15 and 35‰ (Kändler and Tan, 1965) would seem to negate the latter suggestion.

From the results of Peterson (1984) and Peterson and Martin-Robichaud (1986), several general comments may be made. (1) Over a pH range of 4.0–6.7 the plasma membrane of the freshwater teleost egg maintains a substantial negative potential relative to the external medium. It appears to be a tight membrane with a low permeability to H^+, as well as to water (Table II) and Na^+ (Rudy and Potts, 1969). (2) The perivitelline fluid is very responsive to H^+ in the external medium; it appears to act as a buffer, minimizing changes in plasma membrane potential with reference to large variations in $[H^+]$ in the external medium. (3) The ability of the perivitelline fluid to accumulate hydromineral ions is maximized, relative to PVP, at neutral pH in fresh waters of low ionic strength, and minimized in acid medium. The former could be advantageous to the developing salmonid embryo, which may incubate in natural waters neutral to slightly acid and of low ionic activity. Under these conditions the sequestering of physiologically important ions by the perivitelline fluid may be essential for continued embryonic development beyond the eyed stage. Conversely, the acidification of low-ionic-strength fresh waters ("acid rain") may deny the accumulation of hydromineral ions at a level sufficient for post-eyed-stage development. (4) Relative to freshwater incubation, where the PVP is substantially negative, that of the marine egg perivitelline fluid may be near zero or positive in estuarine waters. Hence, the marine egg perivitelline fluid may sequester anions, such as Cl^-. (5) There remains some uncertainty in the interpretation of the role of K^+, Na^+, Ca^{2+}, and Mg^{2+} ions and their necessary levels during freshwater incubation. Brown and Lynam (1981) show that *S. trutta* eggs incubating at low pH (4.5) were most successful in waters containing calcium at 10 mg/l, independent of $[Na^+]$ (0, 1, 10 mg/l). After hatching, however, survival of alevins was high in water containing sodium (1 or 10 mg/l). Eyed eggs of *Oncorhynchus tsha-*

wytscha exposed to acid pH conditions (Swinehart and Cheney, 1984) showed losses of Ca^{2+} and Mg^{2+} at pH 5.0, exceeding those losses that occurred in distilled water. Leaching of primary amines at pH 5.0 did not exceed the rate in distilled water; some losses of amines could be expected under normal conditions (Yarzhombek and Maslennikova, 1971). The fact that Swinehart and Cheney (1984) noted divalent ion losses within minutes of exposure to acid water would suggest the source of this loss to be the perivitelline fluid or desorption from zone radiata binding sites. Bradley and Rourke (1985) found a correlation between elevated NH_4^+ and reduced Na^+ plasma levels, and mortality, in juvenile steelhead trout (*Salmo gairdneri*) cultured in low-ionic-strength fresh water. They suggested that low environmental $[Na^+]$ likely was inhibiting NH_4^+ excretion occurring normally by an NH_4^+/Na^+ exchange mechanism. Before the gills are functional, the blockage of transfer of NH_4^+ to the exterior via H^+/Na^+ or NH_4^+/Na^+ ion exchange mechanisms (Heisler, 1982) presumably could lead to acidosis, ammonia toxicity, and loss of regulatory capacity. Alderdice and Harding (1987) found the addition of Ca^{2+} and Na^+ to hatchery water of low ionic activity resulted in a significant reduction in mortality of chinook salmon (*O. tshawytscha*) alevins. In this instance, however, the problem is complicated by genetic and pathological components (Alderdice and Harding, 1987).

The point to be made is that pH and ionic activity of fresh waters in which teleost eggs incubate may have a much more important bearing on incubation success than is currently realized. In this respect, more work on the minimum requirements of hydromineral ions during incubation in fresh water would seem highly appropriate. The zero to positive perivitelline potentials obtained for eggs of marine species by Peterson (1984) and Peterson and Martin-Robichaud (1986) indicate that major differences exist in the electrochemistry of the various compartments, compared to fresh water eggs. Yet, concentration differences between the embryo and the external medium in cod eggs (*Gadus morhua*) obtained by Leivestad (1971) suggest a plasma membrane potential near −61 mV, not unlike that found in freshwater eggs.

In summary, a number of questions emerge whose further study could lead to an improved understanding of the role of the zona radiata and perivitelline fluid on the function of the plasma membrane during early cell division:

How does the zona radiata inhibit the transzonal passage of macromolecules, including those of the perivitelline fluid? The microvillar canals appear to be too large to inhibit such movements. Is the thin externus layer responsible for the semipermeable characteristics of the zona radiata?

The relation between plasma membrane tension and permeability as influenced by internal hydrostatic pressure, first noted by Hansson Mild and Løvtrup (1974a), is unexplored. Low plasma membrane permeability during early cell division appears to be the major source of ionic and osmotic control. Factors that influence hydrostatic pressure, such as activity of the external medium, could lower hydrostatic pressure and increase plasma membrane permeability.

What is the influence of H^+, K^+, Na^+, Ca^{2+}, and Mg^{2+} on the electrical gradients between the yolk and perivitelline fluid, the yolk and the external medium, and the perivitelline fluid and external medium in freshwater eggs?

What ions are involved in the foregoing relationships in marine eggs?

It appears that the perivitelline fluid buffers $[H^+]$ in the external medium, thereby preventing large variations in the plasma membrane potential of eggs in fresh water.

The perivitelline fluid accumulates cations in fresh waters of low ionic strength, and salmonid embryos begin taking up Na^+ beginning at the eyed stage. Is the sequestering of cations essential to the later development of the embryo? If so, what are the characteristics of a water supply that would make these minimum requirements available? In other words, what are the hydromineral requirements in the external medium that are necessary for continued embryonic development? How are these influenced by $[H^+]$?

C. First Cell Division to Beginning of Epiboly

To recapitulate, during development in the ovary the oocyte appears to be relatively permeable to water up to the time of activation. During that formative period the transfer of nutrients and ions across the oocyte membrane probably occurs mainly through the contacting oocyte and follicular cell microvilli. Hence ionic and osmotic control likely would be a function of the parental regulatory system. At ovulation the mature egg is removed from intimate contact with the follicular cells, and becomes free in the ovary or body cavity of the adult.

There the ovarian fluid, under control of the adult regulatory system, maintains the egg in a medium very similar to that of the adult blood (Lam *et al.*, 1982). Changes in ion permeability of the plasma membrane prior to gamete maturation do occur, but they appear to be short-term changes associated with transitory events during meiosis. During this period the oocyte appears to be relatively permeable to K^+ and Na^+, but independent of Ca^{2+} and Cl^- in the external medium. At activation, and concurrent with increase in free $[Ca^{2+}]_i$ associated with the calcium wave, is an increase in K^+ permeability of the plasma membrane, both events being transitory. During imbibition plasma membrane water permeability reaches a maximum, then begins to shut down, reaching a state of very low permeability a few hours following activation. The presence of ion channels and their function in the plasma membrane remains enigmatic. K^+ channels appear to be ubiquitous in their phylogenetic distribution, which includes teleosts.

To provide a basis for further examination of potential regulatory capacity during development, three examples are illustrated in Fig. 4 of yolk osmoconcentration during incubation of the plaice (*Pleuronectes platessa*) (Holliday and Jones, 1967), Atlantic herring (*Clupea harengus*) (Holliday and Jones, 1965), and Pacific herring (*C. pallasi*) (Alderdice *et al.*, 1979). The plaice eggs (Fig. 4A), fertilized and incubated in four salinities (5, 17.5, 3.5, 50‰), show a remarkable consistency in yolk osmotic concentration from about 12 h onward. For the first 132 h yolk osmotic concentration is highly uniform; from day 7 to day 13 it tends to rise, particularly at the extreme salinities (5, 50‰), yet recovery appears to occur thereafter (day 13–17). The authors suggest that regulation occurs within relatively narrow limits from fertilization to hatching. Early mortality occurred in salinities lower than 17.5‰ and was highest during gastrulation and before yolk plug closure. Following hatching, yolk-sac larvae tolerated salinities ranging from 5 to 65‰ for 24 h, 10 to 60‰ for 48 h, and 15 to 60‰ for 1 week. Between yolk-sac absorption and metamorphosis, limits of salinity tolerance shifted downward to 1–50‰ for 24 h and 2.5–45‰ for 1 week of exposure. Of major interest in this example is the constancy of yolk osmotic concentration from 12 h onward. It appears from Holliday and Jones (1967) that a temperature of 7°C could be applied to these incubations: therefore egg development at 12 h probably would not have proceeded beyond the early blastula (Fullarton, 1890). The osmotically equivalent salinity of the yolk of the unfertilized oocyte was about 10.5‰ and of the activated egg at fertilization about 13‰. From examination of rate constants for water permeation across the plasma membrane in *P. platessa* (Table II) there is no indication of a

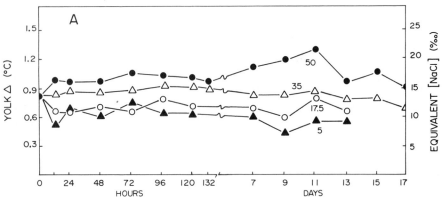

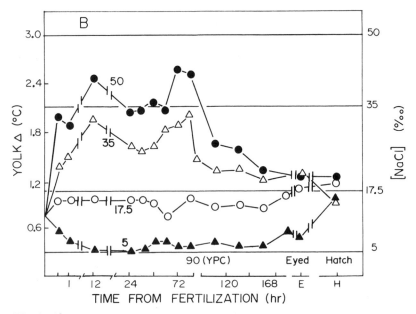

Fig. 4. Changes in osmoticity of the yolk (A, B, Δ°C; C, mOsm) in embryos of three teleosts throughout the incubation period. (A) *Pleuronectes plattessa* eggs fertilized and incubated in each of four salinities: 5, 17.5, 35, and 50‰. (B) *Clupea harengus* eggs fertilized in 35‰ and transferred to 5, 17.5, 35, and 50‰, 30 min after fertilization. (C) *Clupea pallasi* eggs fertilized in 20‰ and transferred to 5, 20, and 35‰, 10 min after fertilization. Temperatures: (A) between 6 and 12°C; (B) between 8 and 11°C; (C) 10 ± 0.1°C. Symbols: GRE, germ ring at equator; YPC, yolk plug closure; E, eyed stage; H, hatching. [Redrawn from (A) Holliday and Jones (1967); (B) Holliday and Jones (1965); (C) Alderdice *et al.* (1979).]

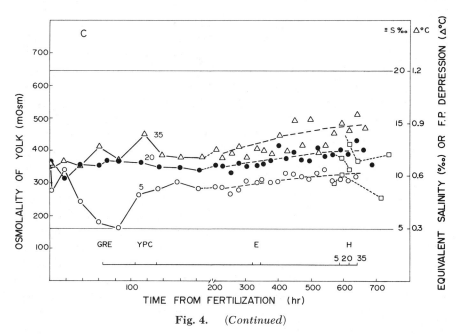

Fig. 4. (*Continued*)

more rapid than usual shutdown in membrane permeability. From the results of Guggino (1980b) for *Fundulus* and Peterson and Martin-Robichaud (1986) for *G. morhua* one might expect a yolk-to-perivitelline fluid potential of +40 to +50 mV in *P. platessa*, and an increase in [Na$^+$] and a decrease in [K$^+$] over the first few days of development. Such evidence would indicate rather constant osmotic concentration, similar to that in *P. platessa* (Fig. 4A). Although the argument is speculative, it appears that *P. platessa* maintains osmotic equilibrium through compensating ion fluxes and low water permeation rates, at least in the period between fertilization and the early blastula stage.

In comparison, the Atlantic herring (*C. harengus*) egg (Holliday and Jones, 1965) shows major differences in osmolarity during development (Fig. 4b). Major differences in the osmotic concentration pattern in this species involve the rapid change in yolk osmoconcentration in the 12 h following fertilization, particularly at higher salinities, as well as the reaching of maximum values between 12 and 96 h, and the subsequent recovery to a rather narrow, apparently regulated level at an osmotic concentration equivalent to 17.5‰ salinity. In this species there appears to be little or no regulation of the yolk in the 12-h period following fertilization. The equivalent salinity of the unfertilized egg was 12.5‰. In the interval between 12 and 96 h the yolk becomes approximately isosmotic in relation to the external medium, except for the suggestion of some regulation occurring in the higher salinities (35, 50‰). At the apparent incubation temperature of 7°C, this period would include development of the blastula and gastrula, epiboly, and growth to the stage of yolk plug nearing closure at about 84 h. Between 84 and 96 h (yolk plug closure, 90 h), corrective trends begin and are well established by 120 h, at which time the embryo is well defined. In the initial 12-h period there must be a major ion influx into or water efflux from the yolk, followed between 84 and 96 h by the beginning of regulative function. Following hatching, herring larvae survived salinities of 1.4–60.1‰ for 24 h, and 2.5–52.5‰ for 1 week. Compared with *P. platessa,* one would conclude that the plasma membrane of the *C. harengus* egg must be considerably more permeable to influx of ions up to the development of the blastula. It seems appropriate to suspect the development of some regulative capacity in the cells of the blastoderm, particularly at higher salinities. Holliday and Jones (1965) suggested that this capacity could relate to the intucking of blastula cells with the commencement of epiboly. From Guggino (1980b) it would appear that regulative cells may be found in the cell sheet established with germ ring overgrowth of the yolk, as differentiated chloride cells are to be found in the resulting epithelium of the yolk sac of the later embryo. Dépêche (1973) found chloride cells in the superficial ectoderm of the blastodermal sheet during epiboly in *Poecilia reticulata.* The work of Holliday and Jones (1965) supports the contention that these cells become functional prior to, and become an effective regulatory mechanism following, yolk plug closure. Jones *et al.* (1966) examined the epidermis of *C. harengus* larvae 12–24 h after hatching in a search for cells with potential ion regulatory function. They examined the outer membranes and buccal membranes of the head, the lateral and ventral yolk sac, and the tail, in larvae incubated and hatched in salinities of 5, 17.5, 35, and 50‰. In spite of their extensive examination of fine structure of the epidermis, no special cells were found. Guggino (1980b) found chloride cells in close association with the vitelline blood vessels in *F. heteroclitus.* He first recorded their presence in larvae 4 days after fertilization (25°C), a stage about 1–2 days after yolk plug closure and shortly before the pronephros becomes functional (Armstrong and Child, 1965). The removal of excess ions from the established yolk-sac circulation would seem to be a natural and effective approach to ion regulation. Is there such an association between chloride cells and subepithelial circulatory vessels, such as those on the yolk sac? Might not these be present in herring embryos as well as marine teleost embryos in general? However, if the chloride cells do appear after yolk plug closure, they cannot be the source of regulation seen in Atlantic herring embryos in the period between 12 and 84 h postfertilization (Fig. 4b). One seems forced to reexamine the function of the blastoderm, which initially forms a two-layered cell sheet enveloping the yolk during epiboly.

The third species compared is the Pacific herring (*C. pallasi*) (Fig. 4C). Alderdice *et al.* (1979) fertilized Pacific herring eggs in sea water (20‰, 5°C) and incubated groups of these eggs at 5, 20, and 35‰, the transfer occurring 10 min after fertilization. Compared with Atlantic herring (Fig. 4b), a greater similarity in response might have been anticipated for two forms whose similarities and taxonomic relations (species or subspecies) are still being debated. Alderdice *et al.* (1979) conclude that Atlantic herring eggs are more tolerant of higher salinities; maximum hatch of Atlantic herring eggs occurs near 25‰, while that in Pacific herring occurs near 17‰. Under these circumstances yolk osmotic changes should be near minimal throughout development in both species at an external salinity near 20‰ (Fig. 4b, 17.5‰; Fig. 4C, 20‰), and they are. At 5 and 35‰, respectively, initial undershoot and overshoot of yolk osmolality occur more rapidly and reach a higher final level of osmotic concentration in 35‰ in the Atlantic herring. In both species stabilization of yolk osmoconcentration occurs at or shortly after yolk plug closure. The major difference in response to external salinity occurs at this time. In the Atlantic herring there is a return from undershoots and overshoots toward a medium osmotic value near the equivalent of 17.5‰; in the Pacific herring egg in the same period (yolk plug closure to hatching) there is a slow increase in yolk osmolality and the egg appears to achieve a steady state at the equivalent external salinity of 12‰. An inference to be drawn from the comparison is that the Atlantic herring egg plasma membrane is more permeable and that the cells of the early blastoderm are more tolerant of higher salinity than in the Pacific herring egg.

From the comparison of the eggs of the three species, the following inferences are drawn, which agree with Holliday and Jones (1967) that survival is based on a combination of tissue tolerance and regulation.

> *Plaice.* When plaice eggs are transferred between salinities, measures of osmoconcentration of the yolk prior to gastrulation (Holliday and Jones, 1967), and of egg, blastodermal cap, and blastula cell size (Holliday, 1965), suggest the plasma

membrane is relatively impermeable to water and ions from fertilization to yolk plug closure. The individual blastula cells therefore may be less tolerant of internal ion concentration changes, particularly at low salinities.

Atlantic herring. In similar transfers and from measures of yolk osmoconcentration (Holliday and Jones, 1965) and egg size in various salinities (Holliday and Blaxter, 1960), the Atlantic herring egg appears to be more poikilosmotic, the plasma membrane being more permeable to water and ions. The cells of the blastula therefore are presumed to be more tolerant of internal ion concentration changes.

Pacific herring. Eggs of the Pacific herring appear to be intermediate in response to salinity change, being less poikilosmotic than the Atlantic herring and less homoiosmotic than the plaice. However, the individual blastula cells are assumed to be only slightly less tolerant of internal ion concentration changes than those of the Atlantic herring (Alderdice *et al.*, 1979).

From the assumptions made in the foregoing comparison, the following argument will be examined. The first passive stage of "regulation" following fertilization may more properly be considered one of resistive maintenance of the integrity of the egg proper, achieved through water and ion permeation characteristics of the plasma membrane. The second stage, and the first involving embryonic "tissue," is assumed to commence with development of the blastoderm, a tissue whose transitory regulatory function would be an assumption of the role of the diminishing area of the plasma membrane, and provision of greater regulative capacity. The third stage, beginning with gastrulation, would seem to be an increasing restriction of water and ion transfer across the developing ectodermal layer of the blastoderm as it spreads to cover the yolk sac and pericardial region of the embryo, ending with the appearance of chloride cells near or following yolk plug closure.

The fact that chloride cells have been found in some embryos at epiboly (*Poecilia*), in later embryos (*Fundulus*), and not at all in others (*C. harengus*) suggests natural flexibility in the regulatory process. If the three stages of embryonic regulatory development—involving successively the plasma membrane, the blastoderm, and chloride cells—provide a regulative function of greater and growing effectiveness, it seems reasonable to assume that the basic ground plan would not necessarily be followed by all teleost embryos in an identical manner. If, as suggested, the plaice blastula cells were relatively intolerant of ion concentration extremes, the egg might compensate by providing greater control over water and ion fluxes across the limiting cell membrane. Conversely, if herring blastula cells are relatively more tolerant of internal ionic change in concentration, there would be less need to maintain a homoiosmotic environment around the cells, as suggested by the larger variations in yolk osmolarity in *C. harengus* and to a lesser extent in *C. pallasi*. The same argument would support the apparent lack of chloride cells in the yolk sac and pericardial epidermis of *C. harengus;* it also would suggest those cells should be present in the plaice embryo.

Based on Guggino's (1980b) findings in *Fundulus,* it seems highly appropriate that the chloride cells should be found in close association with the blood vessels of the yolk sac. Their juxtaposition would support removal of Na^+ and Cl^- from the circulating plasma. Following this pattern, the fourth and final stage of regulatory development, probably following hatching, would see development of chloride cells in the branchial epithelium, the first appearance of this component of the ultimate regulatory array (gills, gut, kidney) of the adult teleost.

Much of the foregoing is circumstantial, based on incomplete evidence, yet it serves to pose the following questions: (1) what are the electrical characteristics of the limiting membrane (the earlier plasma membrane) of the blastula cells and how do they vary near the boundaries of the egg's salinity tolerance; (2) do the undifferentiated blastula cells have an iono- or osmoregulatory function; (3) what specific ion channels and cell-to-cell channels may occur in the blastoderm; (4) when do the chloride cells first appear in embryonic development; (5) do the yolk sac and pericardial epithelia and their chloride cells form the primary embryonic ion transport system; (6) are chloride cells absent or less developed in forms where the individual cells have a high tolerance to salinity change; and (7) do forms with a low tolerance to salinity change compensate by having a tighter limiting membrane? These questions are addressed in the following sections.

1. PLASMA (LIMITING) MEMBRANE

Activation of the teleost egg increases internal free calcium, possibly modified by membrane potential-dependent changes in Ca^{2+} influx (Nuccitelli, 1980a). A temporary incorporation of K channels into the plasma membrane also occurs through condensation of cortical alveolar vesicular membrane (Hagiwara and Jaffe, 1979; Gilkey, 1981). In addition, transmembrane movement of Ca^{2+}, K^+, and Na^+

may occur in relation to changes in the membrane potential. The K^+ channels added to the membrane in *Oryzias latipes* eggs appear during hyperpolarization, reaching a maximum within about 1 min following sperm contact. The fast and slow recovery phases following hyperpolarization are coincident with removal of the K^+ channels, and hence K^+ permeability of the membrane returns to an undetectable level about 8 min after the conductance peak (Nuccitelli, 1980a). Nuccitelli (1980b) also noted that the duration of the phase of increased K^+ conductance was dependent on voltage; the greater the level of polarization in voltage-clamped eggs, the more rapidly K^+ conductance diminished. Conversely, eggs clamped at positive voltage levels retained a higher conductance, suggesting that K^+ channels remained open. In addition, Na^+ permeability of the membrane increased slightly as K^+ permeability fell during posthyperpolarization recovery.

Nuccitelli (1980a,b) indicated that increased ion fluxes could occur between the yolk and cytoplasm and the external medium during activation. Yet these adjustments may be minimal in view of the transient nature of the electrical events. Further adjustments of a longer but still transient nature must occur in the period of higher and attenuating water permeability at and immediately after (12–24 h) activation. Although there is evidence of increased Na^+ and K^+ exchange from the eyed egg stage until hatching (Rudy and Potts, 1969; Shen and Leatherland, 1978a), there is little on which to judge the chronology of possible changes in limiting membrane permeability or ion channel function between the recovery phase after hyperpolarization and yolk plug closure. Apparently there is no evidence of action potentials during this period (Hagiwara and Jaffe, 1979). Since the resting potential of the limiting membrane relative to the external medium is a function of the concentration gradient, the lower the ionic activity of a freshwater medium, the more negative the resting potential should be with respect to external hydromineral ion concentration. In *S. salar* this negative potential appears to reach an asymptote at a pH of 6.7–7.0 (Peterson and Martin-Robichaud, 1986). Is it possible, at such levels (-50 to -80 mV), that channels in the limiting membrane cells might be activated to open in response to physiological requirements for Ca^{2+}, K^+, or Na^+? On the other hand, the period from shortly after fertilization until the end of the blastula stage may be one of very limited regulatory capability, where a tight limiting membrane would provide a major defence of internal stability.

In acidified fresh water, or fresh water with a low hydromineral content—and low buffering capacity—egg mortality can be rapid and substantial in the early egg stage. Over the first 26 days of development, mortality of *S. trutta* eggs at 8°C and pH 4.5 was 90–100% in media containing $CaCl_2$ at 0 or 1 mg/l, and 10–33% mortality at 10 mg/l, whether Na^+ was added or not (0, 1, 10 mg/l of NaCl) (Brown and Lynam, 1981). A control with 10 and 18 mg/l of NaCl and $CaCl_2$, respectively, resulted in 12% mortality. Although NaCl addition had little influence on early egg survival, later survival of alevins was maximal at 10 mg $CaCl_2$ and 1–10 mg NaCl/l. These results suggest that some minimum level of Ca^{2+} in the incubation water is of prime importance, while Na^+ requirements probably are met from internal stores or from perivitelline fluid accumulation up to the eyed stage. Peterson's (1984) study of perivitelline potentials in *S. salar*, relative to the external medium, indicates that lowering external pH and the attendant reduction in potential across the zona radiata can result in loss of cations from the perivitelline fluid. Yet under these conditions the yolk-to-external medium potential appears to remain substantially negative (Peterson and Martin-Robichaud, 1986). Hence, the limiting membrane appears to remain tight, even when the perivitelline fluid is less effective in acid media as an "ion trap" (Eddy and Talbot, 1985) in binding Na^+ and other ions as they diffuse from the embryo. Eddy and Talbot (1985) also found the Na^+ level in the yolk of newly stripped eggs of *S. salar* to be higher than in the eyed stage, indicating a low net loss (influx versus efflux) during early development. This loss was very small and unaffected by the presence of H^+; however, similar conditions at and after the eyed stage resulted in marked loss of Na^+ from the embryo.

In summary, the period between activation and the end of the blastula stage appears to be transitional in the teleost egg. The oocyte prior to spawning is under regulatory control of the adult through the ovarian fluid; beginning with gastrulation, the embryonic tissue begins to assume this control (Fig. 4). For a short period during and following activation there could be movement of Ca^{2+}, K^+, and Na^+ ions across the plasma membrane concurrent with a high initial water permeability attenuating within 12–24 h. From that time until the end of the blastula stage it appears that the egg probably depends, for its internal regulation, on three interrelated factors: (1) control of water permeation at a low rate, (2) tightness of the plasma (limiting) membrane to ion flows, and (3) compensatory control of these water and ion permeation rates relative to the innate tolerance of the blastula cells to osmotic and ionic variation. During this period external Ca^{2+} seems essential for normal development, while internal stores of Na^+ seem sufficient for development to the eyed stage. The role of K^+ during this period is less clear. An examination of the electrical properties of the plasma (limiting) membrane during the blastula stage would be instructive, particularly regarding the presence and function of specific ion channels and the influence of hydromineral ions in the external medium. Since ion concentration gradients may differ between the yolk and perivitelline fluid, and between the perivitelline fluid and the external medium, and the associated potentials are influenced by pH, a comparison of intact and dechorionated eggs would also be useful.

2. BLASTODERM

Limited evidence suggests the blastula stage is one in which there is a major limitation in exchange of water and ions between the egg and the external environment, a holding action awaiting the development of new tissue with regulatory capacity. There is growing evidence (Bennett *et al.*, 1981; Caveney, 1985) that the individual blastomeres begin very early to acquire properties associated with homeostasis, growth control, pattern formation, and tissue differentiation through electrical coupling of adjacent cells. Electrical coupling assumes the presence of special pathways for current flow where the resistance between adjacent cells is known to be low. The sharing of electrical or chemical information by an array of cells is suggestive of the function of a primitive tissue. In initial cell division in an egg this connection may occur over cytoplasmic bridges, where cell division is incomplete. In most instances, however, primary control of intercellular communication appears to be associated with "gap junctions" (Revel *et al.*, 1985) between adjacent cells. Gap junctions form cell-to-cell channels through adjacent plasma membranes. They are permeable to inorganic ions and small organic molecules with a diameter up to 1.2 nm (molecular weight about 450–1500) (Bennett *et al.*, 1981). Loewenstein (1984) estimates these channel diameters, using fluorescent labeled molecular probes, to be 16–20 Å (1.6–2.0 nm). The channel behaves as though it has a fixed or induced charge, and it selectively discriminates against negatively charged molecules. These gap junctions (Unwin and Zampighi, 1980) or cell-to-cell channels (Loewenstein, 1984) have been convincingly demonstrated by dye tracer studies, electron microscopy, and X-ray diffraction techniques (Fig. 5). Gap junctions are found in most metazoan tissues, at least from annelids to mammals (Bennett *et al.*, 1981). Four-cell mouse embryos are coupled in pairs by cytoplasmic bridges; coupling is general and gap junctions are present in the eight-cell stage (Ducibella *et*

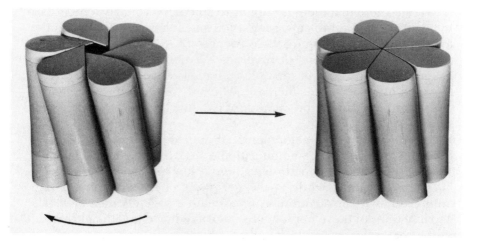

Fig. 5. Model of connexons, units of gap junction, open (left) and closed (right). Organized as a hexagonal lattice and presumed constructed of protein oligomers, the unit is about 75 Å long, 25 Å in diameter, and bridges a gap of 30–40 Å of extracellular space between the plasma membranes of juxtapositioned cells. The central opening widens to a diameter of about 20 Å at the cytoplasmic face. Closed, the radial displacement of each subunit (cytoplasmic side) would be about 6 Å. The unit acts as a three-dimensional iris diaphragm with two apparent states, open and closed. [Reprinted by permission from *Nature*, Unwin and Zampighi (1980), copyright 1980, Macmillan Journals Limited.]

al., 1975; Lo and Gilula, 1979). Early *Fundulus* blastomeres re-form junctions within minutes of cell separation and reaggregation (Ne'eman *et al.*, 1980). Hagiwara (1983) noted that in the tunicate *Halocynthia roretzi* all blastomeres are electrically coupled by the 16- to 64-cell stage. In the limpet (*Patella*), gap junctions appear at the two-cell stage but junctional communication may not occur until the 32-cell stage (Caveney, 1985). Blastoderm cells of *Fundulus* show a rapid distribution of injected fluorescent dye, indicative of cell-to-cell communication (Bennett *et al.*, 1978). Increase of $[H^+]_i$ reduces the conductance of treated cells, as does an increase in either internal or external $[Ca^{2+}]$; the sensitivity of gap junctions to Ca^{2+}, however, is much lower than that to H^+. Blastomeres may be uncoupled by application of polarizing current and channels may close with membrane depolarization. Gap junctions are stable in the open state (Loewenstein, 1984), whereas open time in other types of channel is of short duration. Dye molecules that cross high conductance junctions do not cross when the junctions are held at low conductance by transjunctional voltage. In a comparison with amphibian eggs, Bennett *et al.*

(1981) also note that there is very little junctional voltage dependence in teleost eggs. Cell-to-cell recognition and adhesion prior to junction formation appear to require the presence of Ca^{2+} or Mg^{2+} and a glycoprotein on the membrane surface. The junctional permeability of various mammalian cell types in culture increases when [cyclic AMP] is elevated internally. This increase, which depends on protein synthesis, is correlated with the number of gap junctions present (Loewenstein, 1984).

At later stages of development, patterns of junctional communication appear in which "communication compartments" form (Caveney, 1985). Within such compartments there is low resistance to intercellular communication, while at the periphery of a compartment, communication with an adjoining compartment is severely restricted. The gap junctions of these border cells have a reduced junctional permeability; they effectively block cell-to-cell movement of tracer dyes, although ionic coupling is unaffected (Caveney, 1985). The *outer* membranes of border cells that form enveloping layers may be of very high resistance, and be joined at their apical surfaces by "tight junctions" (Ne'eman *et al.*, 1980), which prevent ion flow as well as water permeation (Bennett *et al.*, 1981). Loss of cell coupling through gap junctions also appears to occur during final stages of cell differentiation that lead to normal physiological function in mature tissues (Caveney, 1985).

The function of specific ion channels and gap junctions in the blastomeres of teleosts is not yet well understood. Gap junctions in the blastomeres and blastoderm may be only one of various components contributing to the later development of homeostasis. There appears to be little or no limitation of ion entry into the yolk of the Atlantic herring (*C. harengus*) egg during initial cell division and at least to the end of the blastula stage (Fig. 4b); evidence suggests gap junctions could be functional during this period.

Based on available evidence, the following inferences are drawn regarding regulatory developments from first cell division to the end of the blastula stage. Presumably gap junctions between blastomeres may appear as early as the third cell division, and the junctions appear to provide for the establishment of ionic and osmotic stabilization within the compartments of connected cells. The outer enveloping border cells form tight junctions, which would limit ion and water transport between the cell compartment interior and the perivitelline fluid. Perhaps the differences in particle transport across the apical cell surfaces of the developing compartments of the three species illustrated in Fig. 4 is related to species variation in the function of the tight junctions between enveloping layer cells and the permeability of their apical membranes on the one hand, and tolerance of the cell cytoplasm to salinity (osmotic and ionic) variation on the other.

As suggested by Holliday and Jones (1965), the in-tucking of blastula cells at the beginning of gastrulation produces a cell sheet of two layers, the inner layer changing polarity in the process. The cells at the margin of the blastoderm uncouple selectively from the yolk in *Fundulus* (Caveney, 1985) in the late blastula, just prior to gastrulation. The uncoupling involves closure of gap junction channels and would appear to create two communication domains, the blastoderm and the syncitial periblast. Recent evidence (Balinsky, 1975) suggests that the "in-tucked" cells of the endomesoderm actually are formed by a rearrangement of deep-lying cells within the blastoderm. Hence, with gastrulation, compartmentalization could result in the establishment of three major communication domains—the outer "firmly adherent" covering layer of the blastoderm (Balinsky, 1975), the deeper blastoderm, and the intermediate endomesoderm. Evidence indicates that the outer blastodermal layer would be relatively impermeant. The periblast would be involved with yolk metabolism. Hence, any regulatory activity would seem of necessity to be restricted to the deeper blastoderm or the endomesoderm. Positionally, the endomesoderm would seem to be unfavorably located if it were to have a regulatory role. In *F. heteroclitus* the cells of the enveloping layer have typical gap junctions in their lateral and basal membranes. The deeper blastodermal cells also have gap junctions, both in the blastula and gastrula, but in contrast to the enveloping layer, there are no tight junctions (Ne'eman *et al.*, 1980). In the amphibian *Xenopus*, the blastomeres during early cell division show electrogenic sodium pumping (Turin, 1984); there is little evidence of active ion transport in early teleost embryos prior to gastrulation.

In summary, it seems that embryonic regulation per se does not begin in the blastula, although compartmentalization probably signals the beginning of tissue formation. As suggested earlier, unmodulated "regulation" may occur from the eight-cell stage to the end of the blastula stage through limitations on particle transfer provided by the high resistance and tight junctions of the cells of the enveloping layer of the outer blastoderm. The establishment of osmotic regulation beginning after gastrulation, and in place by yolk plug closure (Figs. 4b, 4c), indicates that embryonic regulation begins with epiboly, the stage at which chloride cells have first been observed (Dépêche, 1973).

D. Epiboly to Hatching

From the large number of investigations conducted on aspects of iono- and osmoregulation in juvenile and adult teleosts (Evans, 1979, 1980), it is well recognized that the renal complex, the gut, and the branchial epithelium are primary sites of regulation. The evolving integument would be included among these primary sites of regulation (Marshall and Nishioka, 1980; Marshall, 1985). All of these systems are as yet undeveloped at gastrulation. From the rather impermeant limiting membrane of the blastula and gastrula (Bennett *et al.*, 1981) arises the next stage in the development of regulatory capacity—the appearance of chloride cells in the integument, likely centered in the yolk-sac epithelium, and becoming functional at the earliest in the period from beginning of epiboly until shortly after yolk plug closure.

1. THE CHLORIDE CELL

Shelbourne (1957) was the first to show that the integument of the teleost larva (plaice, *Pleuronectes platessa*) was the probable site of iono- and osmoregulation in teleosts. He found activity in the whole integument, suggestive of chloride cells, but particularly in the integument of the yolk sac. Since then chloride cells have been found in the integument of a number of embryonic and larval forms, summarized by Hwang and Hirano (1985) to include (a) saltwater species— plaice (*P. platessa*), puffer (*Fugu niphobles*), red seabream (*Pargues major*), northern anchovy (*Engraulis mordax*), sardine (*Sardinops caerulea*) [*S. sajax*], and flounder (*Kareius bicoloratus*); (b) freshwater species—carp (*Cyprinus carpio*); (c) anadromous or estuarine forms— rainbow trout (*Salmo gairdneri*), killfish (*Fundulus heteroclitus*), molly (*Poecilia reticulata*), and ayu (*Plecoglossus altivelis*). Chloride cells were not found in the embryonic or larval integument of the Atlantic herring (*Clupea harengus*), nor in the anadromous coho salmon (*Oncorhynchus kisutch*). In many of these studies embryos were examined at intervals and it is not possible to determine the developmental stage at which chloride cells first appeared. Nor is the incubation temperature always given, from which one might estimate the developmental stage of interest. In such descriptive studies, incubation temperatures and descriptions of embryonic stages allow confirmation or comparison and should always be given.

The function of "chloride-secreting cells," first proposed by Keys and Willmer (1932), has until recently been the subject of continuing controversy. Marshall and Nishioka (1980) showed that physiologically important chloride current in teleost (*Gillichthys mirabilis*) skin varies directly with the density of chloride cells in the epithelium, as detected by fluorescence microscopy. Foskett and Scheffey (1982), using a technique that localized conductance and chloride current with reference to individual cells, now have shown rather conclusively in the tilapia (*Oreochromis mossambicus*) [*Sarotherodon mossambicus*] [*Tilapia mossambica*] that the so-called chloride or mitochondria-rich cells indeed are the site of salt transport.

In the teleost species studied, there appears to be considerable variation in the location and time of appearance of embryonic and larval chloride cells. They have been found primarily in the integument covering the epicardial region of the anterior yolk sac, on the yolk sac, and in the tail region of the trunk. O'Connell (1981) notes the close association of the yolk-sac blood vessels with the yolk syncytium and the yolk-sac ectoderm in the anterior portion of the yolk sac.

In summary, chloride-cell density varies between integumental regions and between species. The cells may develop or degenerate at various ages, or following transfer between saltwater and fresh water. They may appear during epiboly, at yolk plug closure, at the eyed stage, or shortly before hatching, and they may disappear in the post-yolk-sac stage, presumably with concurrent development of branchial and other adult-type regulatory tissues. As yet the data are too fragmentary to allow close definition of the chronology of chloride-cell development in extra-branchial tissue. The interesting point is the fact that these cells provide continuity, with increasing complexity, between the cellular property of very limited transmembrane particle exchange in the blastomeres and blastula, and the final function of the tissues constituting the regulatory apparatus of the juvenile and adult.

2. STRUCTURE AND FUNCTION IN SALT WATER AND FRESH WATER

Zadunaisky (1984) recently reviewed the structure and function of the chloride cells in the teleost branchial epithelium. The morphology of extrabranchial chloride cells (Dépêche, 1973; Guggino, 1980b) is less well documented. On the other hand, Hwang and Hirano (1985) conclude that the same general cell types occur in both integumental and branchial tissues. On that basis the general morphology of the chloride cell may be illustrated from the studies of Sardet *et al.* (1979)

and Sardet (1980) (Fig. 6) on mullet (*Mugil capito*), eel (*Anguilla anguilla*), trout (*Salmo gairdneri*), killfish, (*F. heteroclitus*), and the molly *Peocilia reticulata* [*Lebistes reticulatus*].

In the integument, the chloride cell underlies and protrudes through an outer enveloping layer of epidermal cells. Discontinuity in the surface epithelial cells surrounding a chloride cell forms an orifice or "pore" over the chloride cell apical surface. The apical surface

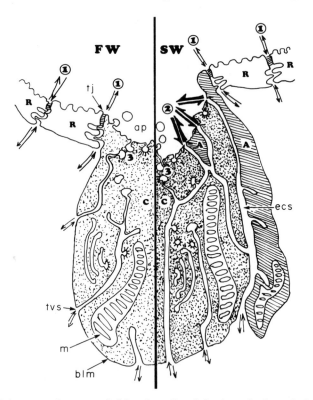

Fig. 6. Schematic drawing of chloride cells of the branchial epithelium in fresh water (FW) and saltwater (SW). With some minor differences the sketch is representative of chloride cells in the embryonic integument. C, Chloride cell; A, adjacent, accessory, or small chloride cell; R, epithelial cell (in the integument); TVS, tubulo–vesicular system; M, mitochondrion; BLM, basolateral membrane (serosal side); AP, apical pit of the chloride cell (mucosal side); (1) tight junction between chloride cell and adjacent epithelial cell; (2) leaky junction between chloride cell and adjacent or smaller chloride cell. The interdigitations between these cells provide channels for the TVS to communicate via the extracellular spaces (ECS) with the apical pit. It is also assumed that vesicles of the TVS may communicate directly with the apical membrane in both fresh water and saltwater (3). [Reproduced (slightly modified) from the *Journal of Cell Biology*, Sardet *et al.* (1979), by copyright permission of the Rockefeller University Press.]

forms a pit or crypt, most obvious in seawater-adapted cells of older individuals; there may be little or no concavity to the crypt in freshwater-adapted cells or in seawater-adapted larvae. The body of the cell is rich in mitochondria and tubules of the tubulovesicular system (Fig. 6). The latter is an extension of the bastolateral border (Sardet *et al.*, 1979) and extends throughout the body of the cell as far as the region below the apical pit, which is relatively free of mitochondria and tubules. The tubulovesicular system contains many ouabain-binding sites. Sardet *et al.* (1979) speculated that small projections on the tubule walls may be Na^+,K^+-dependent ATPase pumping units, with the Na^+-coupling site on the cytoplasmic side and the ouabain-sensitive K^+-coupling site toward the lumen of the tubule. Recent autoradiographic studies support this view (Karnaky, 1986). The tubular system appears to terminate in small vesicles below the apical crypt; these vesicles could be secretory in function. Sardet *et al.* (1979) also suggested that the vesicles could be a means of transport of particles between the tubular reticuli and the apical membrane, across the "vesicular–tubular" space. Larger molecules do not penetrate the space; the function of the vesicles and the vesicular–tubular space remains somewhat uncertain, although there is little doubt that chloride efflux occurs across the apical membrane in seawater-adapted fishes (Degnan, 1986; Karnaky, 1986).

In both freshwater- and seawater-adapted individuals the epithelial cells of the enveloping layer form tight junctions between themselves, and with adjacent chloride cells at the apical surface. The tight junctions appear to form a barrier to ion or water fluxes, limiting or preventing flows between paracellular pathways (Fig. 6, ECS) and the mucosal side of the epithelium. The junctions are deep (0.2–0.4 μm), consist of five to nine anastomosing strands, and are presumed to provide the high electrical resistance characteristic of other tight, impermeant epithelia.

In freshwater-adapted teleosts the apical pits generally are associated with single chloride cells. In seawater adaptation the accessory cells adjacent to the chloride cells differentiate to form "small chloride cells"; these are sheet-like in cross section. Such accessory cells have not been reported in freshwater forms. The pluricellular chloride cell complexes form interdigitations in the apical region, and extracellular (paracellular) channels (Fischbarg *et al.*, 1977; Degnan and Zadunaisky, 1980; Lewis *et al.*, 1984) form between the adjoining chloride cells, pathways not present in freshwater-adapted cells. In seawater-adapted cells the extracellular channels appear to terminate at leaky junctions in the apical crypt (Fig. 6).

In seawater-adapted cells there are large internal and external fluxes of NaCl. Karnaky (1986) suggests the following model for the chloride cell. The primary driving force is the tubular system Na^+,K^+-ATPase (primary active transport), which creates a large Na^+ gradient with high concentrations in the tubular system lumen and low concentrations in the cell cytoplasm. Na^+ entering the cell is recycled for K^+ via the Na^+,K^+-ATPase. This gradient drives a (secondary active transport) NaCl carrier, also located in the tubular membrane; Cl^- enters the cell via the carrier and diffuses to the mucosal side through the apical membrane into the apical crypt. The blood (serosal side) is electropositive relative to sea water, and Na^+ ions concentrating in the tubular reticulum would diffuse down the concentration gradient to the paracellular pathways, to exit at the mucosal side through the leaky junctions between the chloride cell and accessory cell interdigitations into the apical crypt. Foskett *et al.* (1982) note that chloride secretion by the chloride cell can be modified by two different hormonally controlled pathways: (1) slow changes in the number and differentiation of the cells, as described by Hwang and Hirano (1985), and (2) rapid changes by direct effects on the cell transport mechanisms. Numerous examples of the latter are given by Zadunaisky (1984), Foskett *et al.* (1982), and Degnan (1986), including epinephrine, somatostatin, cortisol, prolactin, urotensin I and II, and phosphodiesterase inhibitors.

In freshwater and euryhaline species the situation remains unclear. However, Karnaky (1986) made an interesting observation in *F. heteroclitus*, adapted for several weeks to 1% seawater. The chloride cells remained in pluricellular complexes and shared apical crypts. He noted a new type of deep, tight junction between the cells. In such epithelia, placed in normal Ringer's solution, these deep tight junctions disassembled within 30–50 min to the shallow junctions characterizing chloride cells from seawater-adapted fish.

Laurent and Dunel (1980) provided some provocative information on chloride cell function in the gills of teleosts (*Anguilla anguilla, S. gairdneri, Mugil cephalus, F. heteroclitus*) in fresh water. They noted that in gill development there are two major tissues: primary and secondary epithelium. The former covers the primary lamellae and the interlamellar region, and the latter the free part of the secondary lamellae. Chloride cells are associated largely with the primary epithelium. In *S. gairdneri* larvae, at hatching, the embryonic primary lamellae are sheathed in pavement cells, and interspersed among them are fully developed chloride cells. Chloride cells are not incorporated into the epithelium of the secondary lamellae. When transferred to seawater, the fish may show an accelerated development of "fast renewal cells" in seawater acclimation, and these cells originate from the inner primary epithelium. Fish transferred to *deionized* water show a proliferation of chloride cells in the *secondary* epithelium (*S. gairdneri, A. anguilla*). On their return to 10% seawater, many of the chloride cells showed signs of degeneration. Trout (*S. gairdneri*) moved to deionized water show an intensive ion loss and a proliferation of chloride cells, apparently in the secondary lamellae ("freshwater type"). Eels transferred to seawater show a proliferation of chloride cells in the primary epithelium, but not in the secondary epithelium. Transfer of fishes to deionized water stimulates ion flux by Ca^{2+} removal. Therefore the acute development of secondary lamellae chloride cells could provide a compensatory mechanism for increasing net ion uptake in fresh water. It would be of interest to know whether external acid pH conditions (acid rain) might also provoke the acute development of secondary lamellar chloride cells. In that regard, Perry and Wood (1985) showed that calcium uptake in *S. gairdneri* in freshwater (Ca^{2+} 36.07 mg/l, Na^+ 46.75 mg/l) occurred at equal rates through the gills and general body surface. Exposure of the fish to low Ca^{2+} (1 mg/l) increased the Ca^{2+} flux, which was correlated with a proliferation of lamellar chloride cells. Plasma [Ca^{2+}] remained constant regardless of the external [Ca^{2+}]. However, Na^+ and Cl^- initially were reduced internally, but recovered in 7 days. Hence, while chloride cells in saltwater-adapted fish serve to excrete NaCl, it would seem that in freshwater-adapted fish similar chloride cells may promote the net influx of Na^+ and Cl^-. The differences between seawater and freshwater adaptation, in relation to chloride-cell function, may require a close examination of the origin and location of branchial (and extrabranchial) chloride cells and their relation to particular compartments of the circulatory system, as hypothesized by Laurent and Dunel (1980).

Hwang and Hirano (1985) compared survival with morphology of chloride cells in the skin and branchial epithelium of the freshwater ayu (*Plecoglossus altivelis*), carp (*Cyprinus carpio*), and marine flounder (*Kareius bicoloratus*) in seawater (30‰), fresh water, and in transfers between the two media. In transfers of larval ayu to seawater or larval flounder to fresh water, these authors found that interdigitations and leaky junctions may form or be lost rapidly (3 h). Transfer of seawater-adapted individuals to fresh water generally, but not always, resulted in the degeneration of chloride cells or a reversion to the typical freshwater cell structure within a few days. No interdigitations or leaky junctions were found in the chloride cells of larval ayu incubated in fresh water. Ayu[6] eggs, transferred to seawater at intervals up to 120 h following fertilization, showed no survival to hatching (17–20°C). When transferred at or after 144 h of incubation, survival to hatching rose to about 21–28%. Apparently these embryos formed interdigitations and leaky junctions within 24 h after transfer as 144-h embryos. Freshwater-incubated ayu, challenged at hatching, appeared to begin developing interdigitations and leaky junctions within 3 h; after 24 h these chloride cells showed the morphological characteristics of seawater-adapted fish. Juvenile (60-day) ayu in fresh water showed no interdigitations or leaky junctions; transferred to seawater (30‰), they died within 6 h and showed no modification of the epithelium. However, transfer to salinities below 20‰ resulted in limited formation of interdigitations and leaky junctions. Larval flounder incubated in seawater possessed interdigitations and leaky junctions between integumental chloride cells. Newly hatched flounder larvae transferred to fresh water or mild seawater (<10‰) were unable to survive; chloride cells showed either abnormalities or no morphological change, while the integument was severely disrupted. Juvenile (60-day) flounder showed typical interdigitations and leaky junctions in seawater. Transferred to lower salinities or fresh water the juveniles survived and the interdigitations and leaky junctions began to degenerate within 3 h; after 3 days, chloride-cell morphology was typical of freshwater cells. Finally, juvenile carp showed no interdigitations or leaky junctions in fresh water. When juveniles were transferred to dilute saltwater these structures did not form, there was extensive damage to the skin and branchial epithelium, and the juveniles did not survive beyond 12 h in 15‰. In the three species, survival was correlated with the elaboration or degeneration of chloride cells following transfer between external media. In this regard, the juvenile carp tolerated only fresh water. Larval or embryonic ayu generated chloride cells, but that adaptive ability seems to be lost in the juvenile. Larval flounder could not tolerate fresh water but the juveniles were relatively euryhaline. The results of Hwang and Hirano (1985) parallel those of Sardet et al. (1979) and Sardet (1980). Interdigitations and leaky junctions occurred in those species normally found in seawater, and not in those restricted to fresh water. In species

---

[6] Kashiwagi et al. (1986) have reported on the relation between temperature and development time to 50% hatch in the ayu egg. The Schnute (1981) general growth model provides the following relation between their data (their Table 5) and temperatures of 11–26°C:

$$Y \text{ (days)} = [9.2092 - [6.3693(1 - e^{-0.1248X°C + 1.3728}]]^{1.4556}.$$

showing evidence of eurhalinity these structures either formed or degenerated after transfer. Although differences in adaptability varied with species, age, and embryonic salinity experience (ayu), there seems little doubt that chloride cells, either integumental or branchial, were involved with regulation and survival.

The observations of Hwang and Hirano (1985) regarding morphological changes in the chloride cells, after adaptation of the organism to saltwater or fresh water, are convincing. However, the assumption that morphologically shallow (leaky, one-to-four-junctional strand) and deep (tight, five-to-nine-junctional strand) junctions are fully correlated functionally with high and low permeabilities may be too simplistic. Where there are leaky junctions between the cells, paracellular conductances are greater than transcellular conductances. With tight junctions, the paracellular pathway is more resistive. In spite of the distinctive differences between the two types of junction, attempts to correlate junctional resistance with the number of junctional strands remain arguable (Friedman, 1986).

Dépêche (1973) found the yolk sac and pericardial epithelium were involved in embryonic homeostasis in the live-bearer molly, *Poecilia reticulata*. Chloride cells occurred at a density of 80–100/mm^2 on the yolk sac, and 60–80/mm^2 on the pericardial epithelium in 3.5-mm embryos. Embryos exhibited osmoregulative capacity after the seventeenth day of development. Guggino (1980b), who found chloride cells in close association with the vitelline vessels in the yolk sac and pericardial epithelium in *F. heteroclitus* embryos, found no chloride cells in the branchial epithelium before hatching. In *F. heteroclitus* dechorionated (7-, 10-day) embryos and in hatchlings, $[Na^+]_i$ tended to increase while $[K^+]_i$ and $[Cl^-]_i$ tended to decrease toward hatching. In intact eggs the $[Na^+]$ and $[K^+]$ of the perivitelline fluid were similar to that of the seawater medium; that in the pericardial cavity resembled the seawater-acclimated blood plasma of the adult. From transmembrane potentials and ion replacement trials, Guggino concluded that the embryos were permeable to K^+ and Na^+, the former more so than the latter, and that the epithelium was cation-selective. The transepithelial potential could not be explained by simple diffusion, as expressed by the Goldman equation, suggesting strongly that while Na^+ efflux is diffusional, Cl^- is moved out of the embryo by electrogenic pumping. Guggino concluded that the embryo maintains osmotic and ionic balance by Na^+ and Cl^- excretion across the epithelium of the yolk sac and embryonic cavity, the site of ion transport moving to the gills later, after hatching.

Guggino (1980a) made a further observation of major interest, regarding water balance in the marine embryo. If the osmolality of an embryonic tissue fluid, for example, is 350 mOsm/kg and that of the perivitelline fluid is 600 mOsm/kg (e.g., Fig. 4C), then the tissue fluid is hypotonic relative to the perivitelline fluid and the gradient of 250 mOsm/kg between them will tend to remove water from the tissues. If the embryonic epithelium is not completely impermeant, which is a reasonable generalization, there will be a tendency for the embryo to lose water. Guggino estimates these losses in *F. heteroclitus* at 2.6 and 12% per day for 4.7- and 10-day-old embryos, respectively. Juvenile and adult teleosts tend to balance such water losses by drinking or presumably by using the fluid intake provided by their food, removing the ions so gained and eliminating them through the gills, kidney, and gut. In the head region of the 10-day *F. heteroclitus* embryo (25°C), Guggino found lateroventral pores, anterior to the pectoral fins. These were open to the perivitelline fluid and connected via paired branchial chambers and embryonic gill arches with an open pharynx, which extended forward to a position posterior to the eyes. These structures, in rudimentary form, are present in the 4-day embryo (assumed 1–2 days after yolk plug closure), and the pharynx anteriorly and the gut posteriorly are undeveloped. In the 10-day embryo the mouth is still nonfunctional; the gut is formed, but it is a blind sac, closed at the cloaca. Hence, in the embryo there is a channel available for the perivitelline fluid to enter the pharynx and gut. There water presumably is absorbed, from which ions would be returned via the integumental chloride cells to the perivitelline fluid. One must surely question whether a similar water-conservation mechanism may be found in larvae of other marine teleosts.

Development of salinity tolerance in eggs of the freshwater ayu (Hwang and Hirano, 1985) provides a possible interpretation for similar observations in embryonic development of Pacific herring (*Clupea pallasi*). Dueñas (1981) incubated Pacific herring eggs at two temperatures (6, 12°C) and three salinities (13, 21, 29‰), and cultured the resulting larvae at three temperatures (6, 9, 12°C). Estimates of salinity tolerance (72-h ED_{50}, ‰) were obtained for samples of newly hatched larvae for each of the 18 incubation–culture conditions. Incubation salinity (IS), incubation temperature (IT), and the IS × IT interaction had a significant effect on salinity tolerance of the larvae. Larvae from eggs incubated at 13, 21, and 29‰ had median tolerance estimates of 41.7, 43.6, and 48.2‰, respectively. Those from the 6 and 12°C incubations had median tolerance values of 48.5 and 40.4‰, respectively. With reference to the interaction term, larvae were more tolerant of higher salinities when incubated as eggs at the lower temperature. Culture temperature (6, 9, 12°C) had a small but insignificant effect on salinity tolerance. These data are examined in greater detail by Alderdice and Hourston (1985).

In relation to the results of Dueñas (1981), Hwang and Hirano (1985) show that the extent of chloride-cell development can be related to the extent of salinity change involved in a challenge; that is, the development of chloride cells is not an "all-or-nothing" response. This would suggest that the greater larval salinity tolerance associated with higher incubation salinities could result from a greater differentiation of regulative tissue—such as an increased density of chloride cells. If such cells occur, are they in the yolk-sac integument? In the Atlantic herring (*C. harengus*), functional gill filaments do not occur until some weeks after hatching (Blaxter, 1969). One might assume, from the regulatory capabilities of the embryos of both Atlantic and Pacific herring, that integumental chloride cells would be found in both forms. They were not found in the Atlantic herring larva (Jones *et al.*, 1966), yet a reexamination of their results (e.g., their plate 1A2) suggests these cells may have been present but unrecognized. The inference is that chloride cells *may* be found in the yolk-sac integument of embryos of both species, a suggestion in need of further examination. As an alternative explanation, a greater basic cell tolerance to salinity might account for the greater tolerance of Dueñas' (1981) Pacific herring larvae hatched from higher incubation salinities, although this would seem less probable.

In summary, there is considerable evidence that true osmoregulation in teleost embryos begins following gastrulation and is functional by yolk plug closure. Although the timing is variable, regulation appears to be correlated with the development of chloride cells in the integument of the yolk sac, particularly over the pericardial region, and on the trunk. In freshwater forms the chloride cells are sparse and are joined to surrounding epithelial cells by impermeant tight junctions at their apical margins. In marine forms there is an elaboration of chloride cells and accessory chloride cells, whose interdigitations provide paracellular channels connected with the tubular vesicular system of the chloride cell interior and with the apical crypt via leaky junctions at the mucosal ends of the paracellular channels. It appears that Na^+ may be concentrated electrogenically in the tubular vesicular system, an extension of the basolateral membrane, where it diffuses down the concentration gradient, via the paracellular channels, to the apical crypt. Cl^- ions appear to diffuse passively across the cytoplasm to the apical crypt, although Guggino's (1980b) results indicate Cl^- efflux may involve active transport. As in freshwater forms, tight junctions occur between the chloride cells and adjoining epithelial cells. Transfer of eggs or larvae between seawater and fresh water can produce rapid changes in chloride-cell morphology; elaboration or degeneration of chloride cells may begin in as little as 3 h, and be complete 1–3 days after transfer. Some early larvae respond in this manner to transfer, while later juveniles cannot. Alternatively, in some species early larvae do not respond while later juveniles do. Others show no morphological change in chloride cells, and cannot tolerate transfer. Marine embryos in hypertonic (seawater) media tend to lose tissue water at a slow rate; juveniles and adults balance water loss by "drinking." The *Fundulus* embryo appears to accomplish "drinking" by a mechanism involving external pores communicating with the pharynx. Perivitelline fluid apparently is taken into the pharynx, water is absorbed, and the excess ions are returned to the perivitelline fluid. Such a system, balancing water loss, could occur in other marine embryos. Evidence also indicates that *Clupea pallasi* embryos are more salt-tolerant when incubated in higher salinities, suggesting that the extent of regulative tissue development may correlate with the salinity of the external medium. Although chloride cells were not found earlier in *C. harengus*, growing evidence suggests both species be examined for the presence of chloride cells.

E. Transition to the Adult Mechanism for Regulation

Marine teleosts function in a medium where there is a tendency for osmotic loss of water and diffusional gain of Na^+ and Cl^-. In fresh water the reverse occurs: there is an osmotic gain of water and a diffusional loss of Na^+ and Cl^-. The preceding section indicated that many marine teleosts may develop extrabranchial chloride cells in embryonic (epiboly and later) or postembryonic stages, with these cells functioning to eliminate excess ions and stabilize extracellular fluid osmoconcentration. In the postembryonic stage, regulation shifts to adult-type tissues (kidney, gills, and gut), and water uptake to offset osmotic loss in sea water is considered to occur largely by drinking. "Drinking" to offset osmotic water loss appears to have been anticipated by the apparent pharyngeal and gut absorption of water from the perivitelline fluid as documented by Guggino (1980a) in *F. heteroclitus* and *F. bermudae* embryos. However, in marine teleosts, absorbed water carries with it ions that, if they remained, would defeat the purpose of "drinking." In adult-type regulation in marine teleosts most divalent ions remain in the gut and are excreted via the anus;

some are absorbed through the epithelium of the gut and esophagus and are excreted by the kidney. Monovalent ions are absorbed by the gut and excreted either by the kidney or by the branchial epithelium. In fresh water, osmotic influx of water is balanced by renal production of copious volumes of hypoosmotic urine. Na^+ and Cl^- ions also are resorbed from the urine by the kidney, while divalent ions are excreted. In those freshwater teleosts possessing a urinary bladder, further Na^+ and Cl^- are actively resorbed through an Na^+,K^+-activated ATPase mediating an Na^+-K^+ exchange system resulting in sodium uptake (Evans, 1980). Absorption of Na^+ and Cl^- by the renal system is not total; if the kidney were the only strucutre involved, there would be a net loss of NaCl. This potential net loss by the kidney appears to be offset through active uptake of Na^+ and Cl^- by the branchial epithelium. Various models of the chloride cell in marine teleosts have been proposed, based on growing detailed evidence of their salt-secreting function (Sardet *et al.*, 1979, Zadunaisky, 1984; Karnaky, 1986). Heisler (1980) and Aronson (1985) have proposed models for salt balance and pH control in marine and freshwater teleosts. To these systems may be added the probability that the resistance of individual cells to salinity change may vary between species, as suggested by Weisbart (1968) for the five species of North American Pacific salmon (*Oncorhynchus*). On the other hand, Shen and Leatherland (1978b) found very few "mitochondria-rich" chloride-type cells in the yolksac epithelium of the salmonid *S. gairdneri*. They also found a small number of chloride-type cells at the base of the gill filaments in the branchial epithelium 1 week before hatching. They concluded that there did not appear to be a significant number of chloride-type cells and that their role in regulation during the early alevin period might be a limited one.

The chronology of chloride-cell appearance, density, location, and transfer of regulative function from the integumental system to the postembryonic renal–branchial–gut system is not well documented. From the scattered evidence available, it appears that the timing of appearance and properties of integumental chloride cells and the transfer of regulatory function to the adult-type regulatory system in the larvae or juvenile may be highly variable. Chronology may vary in relation to the ecology of the species—whether it is obligatory marine and stenohaline, obligatory fresh water, euryhaline with episodes of marine, estuarine, or freshwater experience in its life history, and anadromous or catadromous.

In summary, the timing of the transition from integumental regulation to the adult type is not well documented, but it would seem to be restricted mainly to the period from just prior to hatching until development of the branchial epithelium occurs some days after hatching. As the adult-type of regulative system involves the renal complex and gut, as well as the branchial epithelium, the inference is compelling that all of these structures should be examined concurrently to obtain a full appreciation of regulative function. The question of whether and how chloride cells function in freshwater teleosts, such as the salmonids, remains obscure. To be noted is the fact that in fresh waters of acid pH or very low ion content, sequestering of Na^+ by the embryo from the eyed stage onward may be subnormal, levels of Na^+ and Ca^{2+} in the resulting hatched alevins may be low, and mortality may occur in these fish prior to their reaching the first feeding stage when exogenous sources of hydrominerals would become available in the food. If chloride cells are present under such circumstances, it would seem reasonable that they would be involved, at least in the uptake of Na^+ ions from the medium.

V. CONCLUSIONS

Three questions were posed earlier regarding regulatory mechanisms, patterns of regulatory activity, and the developmental states at which regulatory processes become functional. A wealth of information is accumulating that addresses these questions in particular stages and species of invertebrates, lower chordates, and vertebrates. However, its consolidation and integration into cause–effect associations and developmental chronologies has yet to be achieved, particularly in teleosts. Major phylogenetic differences in cell function and regulatory behavior in different groups of organisms are such that a sequence of events determined for an echinoderm or amphibian may not apply among teleosts. Even among the teleosts a surprising number of gaps occur in our understanding of general as well as specific aspects of regulatory development.

A. Oocytes

The oocyte appears to depend largely on its adult female carrier for osmo- and ionoregulation. Teleosts, being homoiosmotic regulators, maintain a gradient between the external medium and the oocyte via the circulatory system, which serves to regulate the tissues including the follicular cells of the ovary. Following ovulation, intimate cellular contact is replaced by the ovarian fluid, which also follows the gradient. To what extent might the oocyte take part in these regulatory processes? The oocyte plasma membrane would seem to be the only structure that could partake in regulatory activities. Yet changes that do occur in the properties of the membrane appear to be no more than transient adjustments associated with episodes of development within the bounds of a regulated ovarian or coelomic environment. A number of changes in the properties of the membrane occur between GVBD and ovulation. The membrane potential of the teleost oocyte is dependent largely on K^+ and Na^+ concentrations, and developmental episodes may involve the freeing of bound Ca^{2+} and its movement across the membrane. The ovulated oocyte likely will show a lower membrane potential, increased electrical resistance, a decrease in K^+ and Na^+ conductance, and high water permeability compared with the immature oocyte.

B. Fertilization

Egg extrusion and fertilization present the first major regulatory challenge to the teleost gamete. On leaving the protective internal environment of the adult body cavity, the fertilized egg is immediately bathed in a hypertonic (marine) or hypotonic (freshwater) external medium. The resting potential of the extruded egg, just prior to activation, will probably be around -50 to -90 mV in fresh water, the potential being the result of a limited, selective permeability to K^+ and Na^+, but independent of Ca^{2+} or Cl^-. The membrane potential in general is a function of sperm contact, internal factors such as calcium release at activation, and the ions present and their concentrations in the external medium. At the instant of sperm contact with the plasma membrane, a number of molecular and electrical events are set in motion. As shown in the medaka egg, the activation potential consists consecutively of an initial small depolarization, a major hyperpolarization, followed by phases of fast and slow recovery, the whole process lasting about 10 min. During this period, membrane resistance shows a marked decrease for the first few minutes, followed by recovery to a final resistance value somewhat greater than the initial level. The initial small depolarization apparently may be carried by Na^+ and Ca^{2+} influx and K^+ efflux. In the following hyperpolarization phase, cortical aleveolar exocytosis occurs with a concurrent large increase in cytosolic free calcium; this occurs as a wave, spreading from the point of sperm contact to encompass the egg and terminate at the vegetal pole. During hyperpolarization the plasma membrane shows an increased conductance due mainly to increased K^+ permeability. This may result from the opening of K^+ channels in the membrane, or from the transient addition of vesicular membrane, and its K^+ channels, to the plasma membrane. After peak hyperpolarization the vesicular membrane material appears to be removed from the plasma membrane: K^+ permeability of the membrane falls rapidly, while Na^+ permeability increases slightly. After the main Ca^{2+} surge, internal Ca^{2+} is pumped out into the external medium, the net internal loss being about one-third of the initial free calcium concentration. The Ca^{2+} wave also is pH-dependent; acid pH slows the Ca^{2+} wave and increases the $[Ca^{2+}]$ required for initiation of the Ca surge.

What do these changing rates of permeation and ion concentration mean to the activated egg? Rates of water permeation are very high at activation, reducing to a minimum within approximately 24 h or less. Water fluxes are certain to occur in response to changes in ion concentrations in the cytoplasm, presumably involving Na^+, K^+, and Ca^{2+}. Hence, for a limited period following activation the plasma membrane may be selectively permeable and the concentration gradient between the cytoplasm and the new external medium may be of particular importance in the establishment of a new steady state between the yolk and the external medium.

Before turning to the first stage of cell division several factors may be mentioned that influence following events. They comprise conditions influenced by properties of the plasma membrane and perivitelline fluid. The available evidence rather definitely shows that plasma membrane permeability, high at activation, decreases rapidly within the first hour or so, to reach a minimum within 24 h. This tightening of the plasma membrane appears to restrict transmembrane flows of both water and ions. The reduction in membrane permeability, starting with activation, is correlated with increasing tension in the membrane, resulting from the increase in hydrostatic (internal) pressure occurring as a consequence of water imbibition during formation of the perivitelline fluid. Restricted permeability of the membrane initially, and of its derivative limiting membrane in later multicellular stages, is retained at least to the eyed embryo stage. Following that, there may be a small increase in permeability of the embryo to both water and external ions.

Membrane permeability tends to be lowest in media of very low ionic activity. Many studies of permeability have been conducted using eggs bathed in Ringer's solution or dilutions of Ringer's, with the result that permeation characteristics tend to reflect the level of ionic activity of the external medium used. Under such conditions standard measures may be obtained, but it is then difficult to appreciate the ecological significance of permeation characteristics in natural media. The same argument applies in the estimation of membrane potentials. Hydromineral ion concentrations in freshwater media also influence permeability; Ca^{2+} and Na^+ appear important in this regard for continued successful embryonic development. While Ca^{2+}, Na^+, K^+, Mg^{2+}, and H^+ in the external medium may be of critical importance during development, it is surprising that few hydromineral ion criteria are available for incubation of teleost eggs in fresh water, particularly where Na^+/H^+ and Na^+/NH_4^+ exchange may be of critical concern in low ionic strength natural waters with a pH of 6.5–6.6 or lower. The impermeant colloids of the perivitelline fluid are negatively charged in freshwater eggs, and the excess negative charge tends to sequester cations from the external medium. In fresh water, the perivitelline potential (PVP, between the perivitelline fluid and the external medium) appears explainable in terms of $[Na^+]$ and $[Ca^{2+}]$ in the medium. In fresh water the PVP may be -40 to -80 mV; in marine eggs the PVP varies between about 0 and $+40$ mV, and there is a substantial movement of Na^+, K^+, and Cl^- across the zona radiata.

The influence of temperature on permeability is complex. It is argued that the properties of water near interfaces are different from those of bulk water, and that phase transitions may be expected to occur as temperature changes (Hansson Mild and Løvtrup, 1974b). These transitions frequently occur at 13–16°C. They appear to influence the rate of diffusion of water in the cytoplasm, tension in the plasma membrane, measures of osmotic water permeability of the membrane, and internal egg pressure. In ranid species the permeability coefficient increases with higher temperature; this can be counteracted by reducing the tonicity of the external medium—presumably through increased membrane tension resulting from increased hydrostatic pressure occurring because of further water uptake in a medium of lower ionic activity. Differences in temperature dependence between species also have long been assumed related to the fatty acid composition of membrane lipids. Hence differences in the temperature distribution of permeability coefficients may also reflect differences in thermal phase transitions associated with variations in membrane lipid composition between species.

In the earlier literature much discussion centered on an observed inverse relation between the size of a cell and its permeability coefficient, equivalent to a direct relation between the surface-to-volume ratio (S/V) of a cell and its membrane permeability. That is, larger eggs with a smaller S/V ratio appeared less permeable than smaller eggs with a greater S/V ratio. Dick (1959b) showed that there is a significant inverse relation between S/V ratio and the coefficient for diffusion in the cytoplasm. He reasoned that the relative effect of plasma membrane resistance in reducing the rate of diffusion, in comparison with the effect of resistance of the cytoplasm, would be greater in smaller cells (with a greater S/V ratio) than in larger cells (with a smaller S/V ratio). Hansson Mild and Løvtrup (1985) appear to have a final explanation. Plasma membrane permeation, cytoplasmic diffusion and cell size are interrelated. The value of P_d as a function of D_1 (diffusion in the cytoplasm) varies with cell size when the effect of cytoplasmic diffusion is taken into account.

Another factor possibly involved in this relationship is hydrostatic pressure. Although exceptions occur, smaller eggs tend to have higher hydrostatic pressures. Hence smaller eggs (greater S/V ratio), with a greater tension in the plasma membrane, may have lower permeability coefficients. Interestingly, the unbiased estimate of P should provide a measure of the permeation *capacity* of the plasma membrane. However, the manner in which that capacity is *utilized,* in an ecological sense, may be best provided by the biased estimate, at least uncorrected for D_1.

C. First Cell Division to Beginning of Epiboly

The first stage of "regulation" in the teleost egg, following activation, appears to be one of resistive maintenance of the integrity of the egg proper, achieved through the presence of a tight plasma membrane and limited transmembrane water and ion fluxes. Natural variations in this pattern may involve differences in membrane tightness, together with differences in the tolerance of the dividing cells of the blastula to salinity change. Hence, it is suggested that a species with a low tolerance to salinity extremes may compensate by shielding the blastomeres with a tight membrane; conversely, a species with a broad cellular salinity tolerance may demonstrate less limited transmembrane fluxes. The fact that compartmentalization of the cells begins rather early in cell division suggests that such differentiation may signal the initial stages in development of real regulatory capacity, particularly where the enveloping layer cells show high membrane resistance and tight intercellular junctions. The fact that cells within compartments share communication and ion fluxes via gap junctions suggests the emerging development of protected prototissues among whose new functions may be those of osmo- and ionoregulation.

D. Epiboly to Hatching

The development of extrabranchial chloride cells appears to signal the beginning of modulated and selective ion regulation in the teleost embryo. Prior to epiboly there was a suggestion of some regulatory capacity. However, this capacity seems to result from low transmembrane fluxes based on tight membranes of enveloping cell layers, and as such appears to be neither modulated nor selective. The chronology of development and first appearance of chloride cells is not yet well defined. These cells are recorded in the yolk-sac integument as early as during epiboly, although in various species they may develop from that time until shortly before hatching. The transport of Na^+ and Cl^- by chloride cells has now been clearly demonstrated. The morphology of the chloride and epithelial cells and their tight and leaky junctions leads to various models of ion transport in a saltwater medium. Although the picture is less clear in fresh water, it would seem that the simplest, most efficient process for ionoregulation would be that involving modification of one tissue to serve as an intake device in fresh water or an output device in saltwater. Based on available data, one would expect to see considerable variation in the time of appearance, cell density, and location of extrabranchial chloride cells. One might expect to find them in embryos of marine, estuarine, and catadromous species. They may be absent in strictly freshwater species, and they might be induced to form in some freshwater forms, particularly those with anadromous life histories. The quantitative development of chloride cells in relation to the level of salinity experienced during incubation is not as yet fully substantiated, but is supported by the fact that higher incubation salinity has been shown to increase salinity tolerance in resulting larvae. Guggino's (1980a) observation of "pharangeal pores" in *Fundulus* opens up another avenue of enquiry into embryonic osmoregulation, particularly regarding water uptake by hypoosmotic embryos in seawater. In general, true regulation appears to begin in the embryo with the development of extrabranchial chloride cells, and the apparent capacity to provide water balance by the pharangeal complex. These inferences, however, require considerable experimental confirmation in view of the limited number of species examined and the variation in the results obtained.

E. Transition to the Adult Mechanism for Regulation

In teleosts, as well as in other organisms, two major aspects of regulation are certain to be water and salt balance. Marine teleosts tend to lose water osmotically and gain salts diffusionally. Freshwater teleosts tend to gain water and lose salts by the same processes. In juvenile and older stages of marine teleosts, water loss is made up by "drinking," the ions absorbed in the process being excreted by the renal complex and branchial epithelium (monovalent cations) and gut (mainly divalent cations). In juvenile and older freshwater teleosts a copious volume of hypoosmotic urine is produced from which Na^+ and Cl^- are resorbed by the renal complex and, where present, by the urinary bladder. Some Na^+ and Cl^- may be taken up from the external medium by the branchial epithelium.

One might anticipate that the same regulatory strategies would apply in the late embryo, although the tactics employed may vary with developmental stage and species. The chronology of such development and the timing of the transition between embryonic extrabranchial chloride-cell function and juvenile branchial function largely remain to be documented. Perhaps our fascination with chloride-cell function has diverted attention from the renal complex and gut, which may be more involved in embryonic regulation than has been appreciated. Hence the search for further understanding of patterns of regulation in teleost embryos seems to suggest, both for marine and freshwater forms, that the chronology of tissue development be examined more closely and concurrently in the gut and pharynx, the renal complex, and the extrabranchial and branchial apparatus of the embryo before, during, and after its transition to the first juvenile form.

REFERENCES

Alderdice, D. F., and Forrester, C. R. (1968). Some effects of salinity and temperature on early development of the English sole (*Parophrys vetulus*). *J. Fish. Res. Board Can.* **25**, 495–521.

Alderdice, D. F., and Forrester, C. R. (1971). Effects of salinity and temperature on embryonic development of the petrale sole (*Eopsetta jordani*). *J. Fish. Res. Board Can.* **28**, 727–744.

Alderdice, D. F., and Forrester, C. R. (1974). Early development and distribution of the flathead sole (*Hippoglossoides elassodon*). *J. Fish. Res. Board Can.* **31**, 1899–1918.

Alderdice, D. F., and Harding, D. R. (Eds.) (1988). Studies to determine the cause of mortality in alevins of chinook salmon (*Oncorhynchus tshawytscha*) at Chehalis Hatchery, British Columbia. *Can. Tech. Rep. Fish. Aquat. Sci.* in press.

Alderdice, D. F., and Hourston, A. S. (1985). Factors influencing development and survival of Pacific herring (*Clupea harengus pallasi*) eggs and larvae to beginning of exogenous feeding. *Can. J. Fish. Aquat. Sci.* **42**, 56–68.

Alderdice, D. F., Rao, T. R., and Rosenthal, H. (1979). Osmotic responses of eggs and larvae of the Pacific herring to salinity and cadmium. *Helgol. Wiss. Meeresunters.* **32**, 508–538.

Alderdice, D. F., Jensen, J. O. T., and Velsen, F. P. J. (1984). Measurement of hydrostatic pressure in salmonid eggs. *Can. J. Zool.* **62**, 1977–1987.

Alderdice, D. F., Jensen, J. O. T., and Velsen, F. P. J. (1987). "Preliminary Trials on Incubation of Sablefish Eggs (*Anoplopoma fimbria*)." Pac. Biol. Stn., Nanaimo, B.C.

Anderson, E. (1967). The formation of the primary envelope during oocyte differentiation in teleosts. *J. Cell Biol.* **35**, 193–212.

Armstrong, P. B., and Child, J. S. (1965). Stages in the normal development of *Fundulus heteroclitus*. *Biol. Bull. (Woods Hole, Mass.)* **128**, 143–168.

Aronson, P. S. (1985). Kinetic properties of the plasma membrane Na^+-H^+ exchanger. *Annu. Rev. Physiol.* **47**, 545–560.

Azarnia, R., and Chambers, E. L. (1976). The role of divalent cations in activation of the sea urchin egg. I. Effect of fertilization on divalent cation content. *J. Exp. Zool.* **198**, 65–77.

Balinsky, B. I. (1975). "An Introduction to Embryology." Saunders, Philadelphia, Pennsylvania.

Barańska, J., and Wlodawer, P. (1969). Influence of temperature on the composition of fatty acids and on lipogenesis in frog tissues. *Comp. Biochem. Physiol.* **28**, 553–570.

Barry, H., and Gage, P. W. (1984). Ionic selectivity of channels at the end plate. *Curr. Top. Membr. Transp.* **21**, 1–51.

Bennett, M. V. L., Spira, M. E., and Spray, D. C. (1978). Permeability of gap junctions between embryonic cells of *Fundulus*: A reevaluation. *Dev. Biol.* **65**, 114–125.

Bennett, M. V. L., Spray, D. C., and Harris, A. L. (1981). Electrical coupling in development. *Am. Zool.* **21**, 413–427.

Berntsson, K.-E., Haglund, B., and Løvtrup, S. (1964). Water permeation at different tonicities in the amphibian egg. *J. Exp. Zool.* **155**, 317–324.

Blaxter, J. H. S. (1969). Development: eggs and larvae. In "Fish Physiology" (W. S. Hoar and D. J. Randall, eds.), Vol. III, pp. 177–252. Academic Press, New York.

Booth, J. H., Jansz, G. F., and Holeton, G. F. (1982). Cl^-, K^+ and acid–base balance in rainbow-trout during exposure to and recovery from, sublethal environmental acidification. *Can. J. Zool.* **60**, 1123–1130.

Bradley, T. M., and Rourke, A. W. (1985). The influences of addition of minerals to rearing water and smoltification on selected blood parameters of juvenile steelhead trout, *Salmo gairdneri* Richardson. *Physiol. Zool.* **58**, 312–319.

Brown, D. J. A. (1982). Influence of calcium on the survival of eggs and fry of brown trout (*Salmo trutta*) at pH 4.5. *Bull. Environ. Contam. Toxicol.* **28**, 664–668.

Brown, D. J. A., and Lynam, S. (1981). The effect of sodium and calcium concentrations on the hatching of eggs and the survival of the yolk sac fry of brown trout, *Salmo trutta* L. at low pH. *J. Fish Biol.* **19**, 205–211.

Caveney, S. (1985). The role of gap junctions in development. *Annu. Rev. Physiol.* **47**, 319–335.

Craig, G. R., and Baksi, W. F. (1977). The effects of depressed pH on flagfish reproduction, growth and survival. *Water Res.* **11**, 621–626.

Dainty, J. (1965). Osmotic flow. *Symp. Soc. Exp. Biol.* **19**, 75–85.

Degnan, K. J. (1986). Cyclic AMP stimulation of Cl⁻ secretion by the opercular epithelium: The apical membrane chloride conductance. *J. Exp. Zool.* **238**, 141–146.

Degnan, K. J., and Zadunaisky, J. A. (1980). Passive sodium movements across the opercular epithelium: The paracellular shunt pathway and ionic conductance. *J. Membr. Biol.* **55**, 175–185.

Dépêche, J. (1973). Infrastructure superficielle de la vésicule vitelline et du sac péricardique de l'embryon de *Poecilia reticulata* (poisson téléostéen). *Z. Zellforsch. Mikrosk. Anat.* **141**, 235–253.

Dick, D. A. T. (1959a). Osmotic properties of living cells. *Int. Rev. Cytol.* **8**, 387–448.

Dick, D. A. T. (1959b). The rate of diffusion of water in the protoplasm of living cells. *Exp. Cell Res.* **17**, 5–12.

Dorée, M. (1981). Hormonal control of meiosis in starfish oocytes. Calcium release induced by 1-methyladenine increases permeability to sodium ions. *Exp. Cell Res.* **131**, 115–120.

Ducibella, T., Albertini, D. F., Anderson, E., and Biggers, J. D. (1975). The preimplantation mammalian embryo: Characterization of intercellular junctions and their appearance during development. *Dev. Biol.* **45**, 231–250.

Dueñas, C. E. (1981). Influence of incubation salinity and temperature and post-hatching temperature on salinity tolerance of Pacific herring (*Clupea pallasi* Valenciennes) larvae. M. Sc. Thesis, Department of Zoology, University of British Columbia.

Dunham, P. B., Cass, A., Trinkaus, J. P., and Bennett, M. V. L. (1970). Water permeability of *Fundulus* eggs. *Biol. Bull. (Woods Hole, Mass.)* **139**, 420–421.

Eddy, F. B. (1974). Osmotic properties of the perivitelline fluid and some properties of the chorion of Atlantic salmon eggs (*Salmo salar*). *J. Zool.* **174**, 237–243.

Eddy, F. B. (1982). Osmotic and ionic regulation in captive fish with particular reference to salmonids. *Comp. Biochem. Physiol. B* **73B**, 125–141.

Eddy, F. B., and Talbot, C. (1985). Sodium balance in eggs and dechorinated embryos of the Atlantic salmon *Salmo salar* L. exposed to zinc, aluminum and acid waters. *Comp. Biochem. Physiol. C* **81C**, 259–266.

Evans, D. H. (1979). Fish. *In* "Comparative Physiology of Osmoregulation in Animals" (G. M. O. Maloiy, ed.), Vol. I, pp. 305–390. Academic Press, London.

Evans, D. H. (1980). Osmotic and ionic regulation by freshwater and marine fishes. *In* "Environmental Physiology of Fishes" (M. A. Ali, ed.), pp. 93–122. Plenum, New York.

Evans, D. H. (1984a). Gill Na⁺/H⁻ and Cl⁻/HCO₃⁻ exchange systems evolved before the vertebrates entered fresh water. *J. Exp. Biol.* **113**, 465–469.

Evans, D. H. (1984b). The roles of gill permeability and transport mechanisms in euryhalinity. *In* "Fish Physiology" (W. S. Hoar and D. J. Randall, eds.), Vol. 10, Part B, pp. 239–283. Academic Press, New York.

Evans, D. H., Claiborne, C. B., Farmer, L., Mallery, C., and Krasny, E. J., Jr. (1982). Fish gill ionic transport: Methods and models. *Biol. Bull. (Woods Hole, Mass.)* **163**, 108–130.

Finkelstein, A. (1984). Water movement through membrane channels. *Curr. Top. Membr. Transp.* **21**, 295–308.

Fischbarg, J., Warshavsky, C. R., and Lim, J. J. (1977). Pathways for hydraulically and osmotically induced water flows across epithelia. *Nature (London)* **266**, 71–74.

Florey, E. (1966). "An Introduction to General and Comparative Animal Physiology." Saunders, Philadelphia, Pennsylvania.

Folmar, L. C., and Dickhoff, W. W. (1980). The parr–smolt transformation (smoltification) and sea water adaptation in salmonids. *Aquaculture* **21**, 1–37.

Forrester, C. R., and Alderdice, D. F. (1973). Laboratory observations on early development of the Pacific halibut. *Int. Pac. Halibut Comm., Tech. Rep.* **9**, 1–13.

Foskett, J. K., and Scheffey, C. (1982). The chloride cell: Definitive identification as the salt-secretory cell in teleosts. *Science* **215**, 164–166.

Foskett, J. K., Hubbard, G. M., Machen, T. E., and Bern, H. A. (1982). Effects of epinephrine, glucagon and vasoactive intestinal polypeptide on chloride secretion by teleost opercular membrane. *J. Comp. Physiol.* **146**, 27–34.

Friedman, M. H. (1986). "Principles and Models of Biological Transport." Springer-Verlag, Berlin and New York.

Fullarton, J. H. (1890). On the development of the plaice. *Fish. Board Scotland, 9th Annu. Rep., Part III: Sci. Invest.*, pp. 311–316.

Gilkey, J. C. (1981). Mechanisms of fertilization in fishes. *Am. Zool.* **21**, 359–375.

Gilkey, J. C. (1983). The roles of calcium and pH in activation of eggs of the medaka fish, *Oryzias latipes. J. Cell Biol.* **97**, 669–678.

Gilkey, J. C., Jaffe, L. F., Ridgeway, E. B., and Reynolds, G. T. (1978). A free calcium wave traverses the activating egg of the medaka, *Oryzias latipes. J. Cell Biol.* **76**, 448–466.

Ginzburg, A. S. (1968). "Fertilization in Fishes and the Problem of Polyspermy," Acad. Sci. USSR (Israel Program for Scientific Translations, Cat. No. 600418. U.S. Dept. of Commerce, Natl. Tech. Inf. Serv., Springfield, Virginia).

Grierson, J. P., and Neville, A. C. (1981). Helicoidal architecture of fish eggshell. *Tissue & Cell* **13**, 819–830.

Griffiths, R. W. (1974). Environment and salinity tolerance in the genus *Fundulus. Copeia*, pp. 319–331.

Groot, E. P., and Alderdice, D. F. (1985). Fine structures of the external egg membrane of five species of Pacific salmon and steelhead trout. *Can. J. Zool.* **63**, 552–566.

Guggino, W. B. (1980a), Water balance in embryos of *Fundulus heteroclitus* and *F. bermudae* in seawater. *Am. J. Physiol.* **238**, R36–R41.

Guggino, W. B. (1980b), Salt balance in embryos of *Fundulus heteroclitus* and *F. bermudae* adapted to seawater. *Am. J. Physiol.* **238**, R42–R49.

Hagiwara, S. (1983). "Membrane Potential-dependent Ion Channels in Cell Membranes: Phylogenetic and Developmental Approaches," Distinguished Lect. Ser. Soc. Gen. Physiol., Vol. 3. Raven Press, New York.

Hagiwara, S., and Jaffe, L. A. (1979). Electrical properties of egg cell membranes. *Annu. Rev. Biophys. Bioeng.* **8**, 385–416.

Haglund, B., and Løvtrup, S. (1966). The influence of temperature on the water exchange in amphibian eggs and embryos. *J. Cell. Physiol.* **67**, 355–360.

Hansson Mild, K. (1971). The kinetics of diffusion between a spherical cell and a surrounding medium with different diffusion properties. *Bull. Math. Biophys.* **33**, 19–26.

Hansson Mild, K. (1972). Diffusion exchange between a membrane-bounded sphere and its surroundings. *Bull. Math. Biophys.* **34**, 93–102.

Hansson Mild, K., and Løvtrup, S. (1974a). Diffusion and permeation of water in the frog egg: The effect of tension and tonicity. *J. Exp. Biol.* **61**, 697–703.

Hansson Mild, K., and Løvtrup, S. (1974b). Diffusion and permeation of water in the frog egg: The effect of temperature. *Biochim. Biophys. Acta* **373**, 383–396.

Hannson Mild, K., and Løvtrup, S. (1985). Movement and structure of water in animal cells. Ideas and experiments. *Biochim. Biophys. Acta* **822**, 155–167.

Hansson Mild, K., Carlson, L., and Løvtrup, S. (1974a). The identity of filtration and diffusion permeability coefficients from frog egg membrane. *J. Membr. Biol.* **19**, 221–228.

Hansson Mild, K., Løvtrup, S., and Bergfors, T. (1974b). On the mechanical properties of the vitelline membrane of the frog egg. *J. Exp. Biol.* **60**, 807–820.

Harvey, B., and Chamberlain, J. B. (1982). Water permeability in the developing embryo of the zebrafish, *Brachydanio rerio. Can. J. Zool.* **60**, 268–270.

Heinz, E., and Grassl, S. M. (1984). Electric aspects of co- and counter-transport. *Soc. Gen. Physiol. Ser.* **38**, 93–104.

Heisler, N. (1980). Regulation of the acid–base status in fishes. *In* "Environmental Physiology of Fishes" (M. A. Ali, ed.), pp. 123–162. Plenum, New York.

Heisler, N. (1982). Transepithelial ion transfer processes and mechanisms for fish acid–base regulation in hypercapnia and lactacidosis. *Can. J. Zool.* **60**, 1108–1122.

Hirano, T., Morisawa, M., and Susuki, K. (1978). Changes in plasma and coelomic fluid composition of the mature salmon (*Oncorhynchus keta*) during freshwater adaptation. *Comp. Biochem. Physiol. A* **61A**, 5–8.

Hoar, W. S., and Randall, D. J. (eds). (1984). "Fish Physiology," Vol. 10, Part B. Academic Press, New York.

Holliday, F. G. T. (1965). Osmoregulation in marine teleost eggs and larvae. *Calif. Coop. Oceanic Fish. Invest.* **10**, 89–95.

Holliday, F. G. T. (1969). The effects of salinity on the eggs and larvae of teleosts. *In* "Fish Physiology" (W. S. Hoar and D. J. Randall, eds.), Vol. 1, pp. 293–311. Academic Press, New York.

Holliday, F. G. T., and Blaxter, J. H. S. (1960). The effects of salinity on the developing eggs and larvae of the herring. *J. Mar. Biol. Assoc. U.K.* **39**, 591–603.

Holliday, F. G. T., and Jones, M. P. (1965). Osmotic regulation in the embryo of the herring (*Clupea harengus*). *J. Mar. Biol. Assoc. U.K.* **45**, 305–311.

Holliday, F. G. T., and Jones, M. P. (1967). Some effects of salinity on the developing eggs and larvae of the plaice (*Pleuronectes platessa*). *J. Mar. Biol. Assoc. U.K.* **47**, 39–48.

Horn, R. (1984). Gating of channels in nerve and muscle: A stochastic approach. *Curr. Top. Membr. Transp.* **21**, 53–97.

Hwang, P. P., and Hirano, R. (1985). Effects of environmental salinity on intercellular organization and junctional structure of chloride cells in early stages of teleost development. *J. Exp. Zool.* **236**, 115–126.

Ito, S. (1963). Resting potential and activation potential of the *Oryzias* egg. III. The effect of monovalent ions. *Embryologia* **7**, 344–354.

Jaffe, L. A., Gould-Somero, M., and Holland, L. (1979). Ionic mechanism of the fertilization potential of the marine worm, *Urechis caupo* (Echiura). *J. Gen. Physiol.* **73**, 469–492.

Jaffe, L. F. (1985). The role of calcium explosions, waves, and pulses in activating eggs. *In* "Biology of Fertilization" (C. B. Metz and A. Monroy, eds.), Vol. 3, pp. 127–165. Academic Press, Orlando, Florida.

Jones, M. P., Holliday, F. G. T., and Dunn, A. E. G. (1966). The ultra-structure of the epidermis of larvae of the herring (*Clupea harengus*) in relation to the rearing salinity. *J. Mar. Biol. Assoc. U.K.* **46**, 235–239.

Kanatani, H. (1985). Oocyte growth and maturation in starfish. *In* "Biology of Fertilization" (C. B. Metz and A. Monroy, eds.), Vol. 1, pp. 119–140. Academic Press, Orlando, Florida.

Kanatani, H., and Nagahama, Y. (1980). Mediators of oocyte maturation. *Biomed. Res.* **1**, 273–291.

Kändler, R., and Tan, E. O. (1965). Investigations on the osmoregulation in pelagic eggs of gadoid and flatfishes in the Baltic. II. Changes in chemical composition at different salinities. *Int. Counc. Explor. Sea, Counc. Meet.* **CM65**, (Baltic-Belt Seas Committee), No. 44, 1–3.

Kao, C.-Y., and Chambers, R. (1954). Internal hydrostatic pressure of the *Fundulus* egg. I. The activated egg. *J. Exp. Biol.* **31**, 139–149.

Karnaky, K. J., Jr. (1986). Structure and function of the chloride cell of *Fundulus heteroclitus* and other teleosts. *Am. Zool.* **26**, 209–224.

Kashiwagi, M., Iwai, T., Yamamoto, H., and Sokabe, Y. (1986). Effects of temperature and salinity on egg hatch of the ayu *Plecoglossus altivelis. Bull. Fac. Fish., Mie Univ.* **13**, 17–24.

Katchalsky, A., and Curran, P. F. (1965). "Nonequilibrium Thermodynamics in Biophysics." Harvard Univ. Press, Cambridge, Massachusetts.

Kedem, O., and Katchalsky, A. (1961). A physical interpretation of the phenomenological coefficients of membrane permeability. *J. Gen. Physiol.* **45**, 143–179.

Keys, A. B., and Willmer, E. N. (1932). "Chloride secreting cells" in the gills of fishes, with special reference to the common eel. *J. Physiol. (London)* **76**, 368–381.

Kinne, O. (1964). The effects of temperature and salinity on marine and brackish water animals. II. Salinity and temperature salinity combinations. *Oceanogr. Mar. Biol.* **2**, 281–339.

Kobayashi, W. (1985). Electron microscopic observation of the breakdown of cortical vesicles in the chum salmon egg. *J. Fac. Sci. Hokkaido Univ., Ser. 6* **24**, 87–102.

Kofoid, E. C., Knauber, D. C., and Allende, J. E. (1979). Induction of amphibian oocyte maturation by polyvalent cations and alkaline pH in the absence of potassium ions. *Dev. Biol.* **72**, 374–380.

Korenbrot, J. I. (1977). Ion transport in membranes. *Annu. Rev. Physiol.* **39**, 19–49.

Lam, C. N. H., Jensen, J. O. T., and Alderdice, D. F. (1982). Preliminary study of low gamete viability in adult chum salmon (*Oncorhynchus keta*) held in sea pens at Deserted Creek, Hisnit Inlet, B.C. *Can. Tech. Rep. Fish. Aquat. Sci.* **1133**, 1–49.

Laurent, P., and Dunel, S. (1980). Morphology of gill epithelia in fish. *Am. J. Physiol.* **238**, R147–R159.

Lea, E. J. A. (1963). Permeation through long narrow pores. *J. Theor. Biol.* **5**, 102–107.

Leivestad, H. (1971). Osmotic and ionic conditions in pelagic teleost eggs. *Acta Physiol. Scand.* **82**, 12A.

Lessman, C. A., and Huver, C. W. (1981). Quantification of fertilization-induced gamete changes and sperm entry without egg activation in a teleost egg. *Dev. Biol.* **84**, 218–224.

Lessman, C. A., and Marshall, W. S. (1984). Electrophysiology of in vitro insulin- and progesterone-induced reinitiation of oocyte meiosis in *Rana pipiens. J. Exp. Zool.* **231**, 257–266.

Levitt, D. G. (1984). Kinetics of movement in narrow channels. *Curr. Top. Membr. Transp.* **21**, 181–197.

Lewis, S. A., Hanrahan, J. W., and Van Driessche, W. (1984). Channels across epithelial cell layers. *Curr. Top. Membr. Transp.* **21**, 253–293.

Lo, C. W., and Gilula, N. B. (1979). Gap junctional communication in the preimplantation mouse embryo. *Cell (Cambridge, Mass.)* **18**, 399–409.

D. F. ALDERDICE

Loeffler, C. A. (1971). Water balance in the pike egg. *J. Exp. Biol.* **55**, 797–811.
Loeffler, C. A., and Løvtrup, S. (1970). Water balance in the salmon egg. *J. Exp. Biol.* **52**, 291–298.
Loewenstein, W. R. (1984). Channels in the junctions between cells. *Curr. Top. Membr. Transp.* **21**, 221–252.
Lönning, S. (1972). Comparative electron microscopic studies of teleostean eggs with special reference to the chorion. *Sarsia* **49**, 41–48.
Løvtrup, S. (1960). Water permeation in the amphibian embryo. *J. Exp. Zool.* **145**, 139–149.
Løvtrup, S. (1963). On the rate of water exchange across the surface of animal cells. *J. Theor. Biol.* **5**, 341–359.
Løvtrup, S., and Hansson Mild, K. (1981). Permeation, diffusion, and structure of water in living cells. *In* "International Cell Biology 1980–1981" (H. G. Schweiger, ed.), pp. 889–903. Springer-Verlag, Berlin and New York.
Marshall, W. S. (1985). Paracellular ion transport in trout opercular epithelium models osmoregulatory effects of acid precipitation. *Can. J. Zool.* **63**, 1816–1822.
Marshall, W. S., and Nishioka, R. S. (1980). Relation of mitochondria-rich chloride cells to active chloride transport in the skin of a marine teleost. *J. Exp. Zool.* **214**, 147–156.
Marshall, W. S., Habibi, H. R., and Lessman, C. A. (1985). Electrophysiology of oocytes during meiotic maturation and after ovulation in brook trout (*Salvelinus fontinalis*). *Can. J. Zool.* **63**, 1904–1908.
Masui, Y. (1985). Meiotic arrest in animal oocytes. *In* "Biology of Fertilization" (C. B. Metz and A. Monroy, eds.), Vol. 1, pp. 189–219. Academic Press, Orlando, Florida.
Menendez, R. (1976). Chronic effects of reduced pH on brook trout (*Salvelinus fontinalis*). *J. Fish. Res. Board Can.* **33**, 118–123.
Metz, C. B., and Monroy, A., eds. (1985). "Biology of Fertilization," Vols. 1–3. Academic Press, Orlando, Florida.
Miller, C., Bell, J. E., and Garcia, A. M. (1984). The potassium channel of sarcoplasmic reticulum. *Curr. Top. Membr. Transp.* **21**, 99–132.
Moreau, M., Guerrier, P., and Vilain, J. P. (1985). Ionic regulation of oocyte maturation. *In* "Biology of Fertilization" (C. B. Metz and A. Monroy, eds.), Vol. 1, pp. 299–345. Academic Press, Orlando, Florida.
Nakano, E. (1969). Fishes. *In* "Fertilization: Comparative Morphology, Biochemistry and Immunology" (C. B. Metz and A. Monroy, eds.), Vol. 2, pp. 295–324. Academic Press, New York.
Ne'eman, Z., Spira, M. E., and Bennett, M. V. L. (1980). Formation of gap and tight junctions between reaggregated blastomeres of the killifish, *Fundulus*. *Am. J. Anat.* **158**, 251–262.
Nuccitelli, R. (1980a). The electrical changes accompanying fertilization and cortical vesicle secretion in the medaka egg. *Dev. Biol.* **76**, 483–498.
Nuccitelli, R. (1980b). The fertilization potential is not necessary for the block to polyspermy or the activation of development in the medaka egg. *Dev. Biol.* **76**, 499–504.
O'Connell, C. P. (1981). Development of organ systems in the northern anchovy, *Engraulis mordax*, and other teleosts. *Am. Zool.* **21**, 429–446.
O'Connor, C., Robinson, K. R., and Smith, L. D. (1977). Calcium, potassium, and sodium exchange by full-grown and mature *Xenopus laevis* oocytes. *Dev. Biol.* **61**, 28–40.
Parry, G. (1966). Osmotic adaptation in fishes. *Biol. Rev. Cambridge Philos. Soc.* **41**, 392–444.

Perry, S. F., and Wood, C. M. (1985). Kinetics of branchial calcium uptake in the rainbow trout: Effects of acclimation to various external calcium levels. *J. Exp. Biol.* **116**, 411–433.

Peterson, R. H. (1984). Influence of varying pH and some inorganic ions on the perivitelline potential of eggs of Atlantic salmon (*Salmo salar*). *Can. J. Fish. Aquat. Sci.* **41**, 1066–1069.

Peterson, R. H., and Martin-Robichaud, D. J. (1986). Perivitelline and vitelline potentials in teleost eggs as influenced by ambient strength, natal salinity, and electrode electrolyte; and the influence of these potentials on cadmium dynamics within the egg. *Can. J. Fish. Aquat. Sci.* **43**, 1445–1450.

Potts, W. T. W. (1968). Osmotic and ionic regulation. *Annu. Rev. Physiol.* **30**, 73–104.

Potts, W. T. W. (1976). Ion transport and osmoregulation in marine fish. *In* "Perspectives in Experimental Biology" (P. Spencer-Davies, ed.), Vol. 1, pp. 65–75. Pergamon, Oxford.

Potts, W. T. W. (1977). Fish gills. *In* "Transport of Ions and Water in Animals" (B. L. Gupta, J. C. Oschman, and B. L. Wall, eds.), pp. 453–480. Academic Press, London.

Potts, W. T. W., and Eddy, F. B. (1973). The permeability to water of the eggs of certain marine teleosts. *J. Comp. Physiol.* **82**, 305–315.

Potts, W. T. W., and Parry, G. (1964). "Osmotic and Ionic Regulation in Animals." Macmillan, New York.

Potts, W. T. W., and Rudy, P. P., Jr. (1969). Water balance in the eggs of the Atlantic salmon *Salmo salar*. *J. Exp. Biol.* **50**, 223–237.

Prescott, D. M. (1955). Effect of activation on the water permeability of salmon eggs. *J. Cell. Comp. Physiol.* **45**, 1–12.

Prescott, D. M., and Zuethen, E. (1953). Comparisons of water diffusion and filtration across cell surfaces. *Acta Physiol. Scand.* **28**, 77–94.

Prosser, C. L. (1973). "Comparative Animal Physiology." Saunders, Philadelphia, Pennsylvania.

Revel, J.-P., Nicholson, B. J., and Yancy, S. B. (1985). Chemistry of gap junctions. *Annu. Rev. Physiol.* **47**, 263–279.

Robinson, K. R. (1979). Electrical currents through full-grown and maturing *Xenopus* oocytes. *Proc. Natl. Acad. Sci. U.S.A.* **76**, 837–841.

Rombough, P. J., and Jensen, J. O. T. (1985). Reduced water uptake and resistance to deformation in acid-exposed eggs of steelhead *Salmo gairdneri*. *Trans. Am. Fish. Soc.* **114**, 571–576.

Ruby, S. M., Aczel, J., and Craig, G. R. (1977). The effects of depressed pH on oogenesis in flagfish *Jordanella floridae*. *Water Res.* **11**, 757–762.

Rudy, P. P., Jr., and Potts, W. T. W. (1969). Sodium balance in the eggs of the Atlantic salmon, *Salmo salar*. *J. Exp. Biol.* **50**, 239–246.

Sardet, C. (1980). Freeze fracture of the gill epithelium of euryhaline teleost fish. *Am. J. Physiol.* **238**, R207–R212.

Sardet, C., Pisam, M., and Maetz, J. (1979). The surface epithelium of teleostean fish gills: Cellular and junctional adaptations of the chloride cell in relation to salt adaptation. *J. Cell Biol.* **80**, 96–117.

Schnute, J. (1981). A versatile growth model with statistically stable parameters. *Can. J. Fish. Aquat. Sci.* **38**, 1128–1140.

Schuel, H. (1985). Function of egg cortical granules. *In* "Biology of Fertilization" (C. B. Metz and A. Monroy, eds.), Vol. 3, Academic Press, New York.

Shelbourne, J. E. (1957). Site of chloride regulation in marine fish larvae. *Nature (London)* **180**, 920–922.

Shen, A. C. Y., and Leatherland, J. F. (1978a). Effect of ambient salinity on ionic and osmotic regulation of eggs, larvae, and alevins of rainbow trout (*Salmo gairdneri*). *Can. J. Zool.* **56**, 571–577.

Shen, A. C. Y., and Leatherland, J. F. (1978b). Structure of the yolk sac epithelium and gills in the early developmental stages of rainbow trout (*Salmo gairdneri*) maintained in different ambient salinities. *Environ. Biol. Fishes* **3**, 345–354.

Sower, S. A. (1981). Sexual maturation of coho salmon (*Oncorhynchus kisutch*): Induced ovulation, in vitro induction of final maturation and ovulation, and serum hormone and ion levels of salmon in seawater and fresh water. Ph.D. Thesis, Oregon State University, Corvallis.

Sower, S. A., and Schreck, C. B. (1982). Sexual maturation of coho salmon (*Oncorhynchus kisutch*): Accelerated ovulation and circulating steroid hormone and ion levels of salmon in freshwater and seawater. *Proc. N. Pac. Aquacult. Symp., 1980*, pp. 226–235.

Sower, S. A., Schreck, C. B., and Donaldson, E. M. (1982). Hormone-induced ovulation of coho salmon (*Oncorhynchus kisutch*) held in sea water and fresh water. *Can. J. Fish. Aquat. Sci.* **39**, 627–632.

Starling, E. H. (1895). On the absorption of fluids from the connective tissue spaces. *J. Physiol. (London)* **19**, 312–326.

Stehr, C. M., and Hawkes, J. W. (1979). The comparative ultrastructure of the egg membrane and associated pore structures in the starry flounder, *Platichthys stellatus* (Pallas) and pink salmon, *Oncorhynchus gorbuscha* (Walbaum). *Cell Tissue Res.* **202**, 347–356.

Stoss, J., and Fagerlund, U. H. M. (1982). Influence of saltwater spawning and stress on quality of sex products of chum salmon (*Oncorhynchus keta*) *Proc. Int. Symp. Reprod. Physiol. Fish, 1982*.

Swinehart, J. H., and Cheney, M. A. (1984). The effect of acid water on the loss of divalent cations and primary amines from natural membranes. *Comp. Biochem. Physiol. C* **77C**, 327–330.

Tam, W. H., and Payson, P. D. (1986). Effects of chronic exposure to sublethal pH on growth, egg production and ovulation in brook trout, *Salvelinus fontinalis*. *Can. J. Fish. Aquat. Sci.* **43**, 275–280.

Turin, L. (1984). Electrogenic sodium pumping in *Xenopus* blastomeres: Apparent pump conductance and reversal potential. *Soc. Gen. Physiol. Ser.* **38**, 345–351.

Unwin, P. N. T., and Zampighi, G. (1980). Structure of the junction between communicating cells. *Nature (London)* **283**, 545–549.

Vacquier, V. D. (1975). The isolation of intact cortical granules from sea urchin eggs: Calcium ions trigger granule discharge. *Dev. Biol.* **43**, 62–74.

Vitto, A., Jr., and Wallace, R. A. (1976). Maturation of *Xenopus* oocytes. I. Facilitation by ouabain. *Exp. Cell Res.* **97**, 56–62.

Wallace, R. A., and Steinhardt, R. A. (1977). Maturation of *Xenopus* oocytes. II. Observations on membrane potential. *Dev. Biol.* **57**, 305–316.

Weiner, G. S., Schreck, C. B., Li, H. W. (1986). Effects of low pH on rainbow trout. *Trans. Am. Fish. Soc.* **115**, 75–82.

Weinstein, S. P., Kostellow, A. B., Ziegler, D. H., and Morrill, G. A. (1982). Progesterone-induced down-regulation of an electrogenic Na^+,K^+-ATPase during the first meiotic division in amphibian oocytes. *J. Membr. Biol.* **69**, 41–48.

Weisbart, M. (1968). Osmotic and ionic regulation in embryos, alevins, and fry of the five species of Pacific salmon. *Can. J. Zool.* **46**, 385–397.

Wertheimer, A. C. (1984). Maturation success of pink salmon (*Oncorhynchus gorbuscha*) and coho salmon (*O. kisutch*) held under three salinity regimes. *Aquaculture* **43**, 195–212.

Whitaker, M. J., and Steinhardt, R. (1985). Ionic signalling in the sea urchin egg at fertilization. *In* "Biology of Fertilization" (C. B. Metz and A. Monroy, eds.), Vol. 3, pp. 167–221.

Yamamoto, T. (1939). Changes of the cortical layer of the egg of *Oryzias latipes* at the time of fertilization. *Proc. Imp. Acad. (Tokyo)* **15**, 269–271.

Yanagimachi, R. (1958). Studies of fertilization in *Clupea pallasi*. VIII. On the fertilization reaction of the under-ripe eggs. *Jpn. J. Ichthyol.* **7**, 61–66.

Yarzhombek, A. A., and Maslennikova, N. V. (1971). Nitrogenous metabolites of the eggs and larvae of various fishes. *J. Ichthyol. (Engl. Transl.)* **11**, 276–281.

Zadunaisky, J. A. (1984). The chloride cell: The active transport of chloride and the paracellular pathways. *In* "Fish Physiology" (W. S. Hoar and D. J. Randall, eds.), Vol. 10, Part B, pp. 129–176. Academic Press, New York.

Zotin, A. I. (1965). The uptake and movement of water in embryos. *Symp. Soc. Exp. Biol.* **19**, 365–384.

SUBLETHAL EFFECTS OF POLLUTANTS ON FISH EGGS AND LARVAE

H. von WESTERNHAGEN

Biologische Anstalt Helgoland (Zentrale)
D-2000 Hamburg 52, Federal Republic of Germany

I. INTRODUCTION

The developing fish embryo or larva is generally considered the most sensitive stage in the life cycle of a teleost, being particularly sensitive to all kinds of low-level environmental changes to which it might be exposed. Available data suggest that certain stages in the life

253

cycle of marine and freshwater fishes are more susceptible to environmental and pollutional stress than others (von Westernhagen, 1968,
1970; Rosenthal and Alderdice, 1976). Differences in susceptibility
are particularly known to exist between the early developmental
stages, that is, embryos and larvae. The younger embryonic stages
(before gastrulation) are more vulnerable than those that have completed gastrulation.

This has been shown by Kühnhold (1972) for the reaction of cod
(*Gadus morhua*), herring (*Clupea harengus*), and plaice (*Pleuronectes
platessa*) embryos when exposed to crude oil extracts and could be
confirmed for cod by Davenport *et al.* (1979), Black Sea flounder
(*Platichthys flesus luscus*) by Mazmanidi and Bazhasvili (1975), northern pike (*Esox lucius*) by Häkkilä and Niemi (1973), and several other
marine species (Wilson, 1972).

The decrease of sensitivity of cod *G. morhua* embryos to methyl-
naphthalene during development is similar. Newly fertilized eggs
are more sensitive than after completion of gastrulation (Stene and
Lönning, 1984). The effects of heavy metals on embryonic development are likewise particularly pronounced during early embryonic
stages as documented experimentally by P. Weis and Weis (1977) and
Sharp and Neff (1980, 1982) for the effects of methylmercury on embryos of the killifish (*Fundulus heteroclitus*) and Akiyama (1970) and
Dial (1978) on *Oryzias latipes*. Another toxicant, such as the lamprey
larvicide TFM, was also found to exert its most detrimental effects on
rainbow trout embryos immediately following fertilization (Niblett
and McKeown, 1980).

Even though the relatively high susceptibility of the early embryonic stages has been well documented, the yolk-sac or alevin stage is
known to be the most sensitive stage in the teleost life cycle. Linden
(1974) reports a factor of 10 for the higher sensitivity of Baltic herring
(*C. harengus membras*) larvae toward water-soluble crude oil components, while the larvae of northern pike (*E. lucius*) are 100 times more
sensitive to oil than the embryos (Häkkilä and Niemi, 1973). The same
can be observed in the alevins of pink salmon (*Oncorhynchus gorbuscha*) (Rice *et al.*, 1975).

Likewise, heavy-metal toxicity to early life stages of fish expresses
itself most sensitively in larvae and alevins, with significant inhibition
of growth in Atlantic salmon (*Salmo salar*) upon exposure to low (0.47
μg Cd/l) levels of cadmium, while a significant reduction in viable
hatch is noticed only between 300 and 800 μg Cd/l (Rombough and
Garside, 1982). Similar effects of copper on the larvae of eight species
of freshwater fish are known from McKim *et al.* (1978). Low concentra-

tions of methylmercuric (0.075 μg/l), cadmium (0.55 μg/l), and lead (0.7 μg/l) ions appear to cause little "biochemical stress" on brook trout (*Salvelinus fontinalis*) embryos, but cause definite changes in alevins. The higher susceptibility of larvae compared to embryos has been also demonstrated for organophosphate insecticides (Paflits-chek, 1979) as well as chlorinated hydrocarbons [Schimmel *et al.* (1974), Aroclor 1254; Dethlefsen (1977), DDT, DDE].

Thus, within a variety of fluctuating natural environmental factors such as temperature, salinity, oxygen content, etc., human-made activities have added a series of new parameters that may exert considerable impact on the developing fish. It is during early life that fish seem to be particularly sensitive to substances, many of which are commonly called pollutants; during early life the yolk-sac stage may be considered the most sensitive one, followed by the embryonic stage prior to completion of gastrulation.

The effects caused during these and other stages of development by pollutants may be very subtle and go unrecognized at the individual level. Yet these effects may proceed until they develop into very conspicious lethal effects. Ultimately their significance may be recognized at the population level.

The urgent need for the determination of legally applicable measures of pollutants and their effects on the aquatic biota has recently triggered a multitude of investigations on the effects of hitherto unrecognized pollutants and the determination of their LC_{50} (concentration at which 50% of the experimental animals died within a given time span, usually 48 h) or their MATC (maximal acceptable toxicant concentration). The application of the LC_{50} has found widespread entry into the methods for determination of water quality by means of test organisms (Anonymous, 1972a,b). In combination with application factors, the LC_{50} is used to determine specific pollutant toxicity (Sprague, 1969; Anonymous, 1976). Among the aquatic organisms, fish (Stephan and Mount, 1973) as well as their eggs and larvae have attracted considerable interest for toxicity testing, particularly so because an early life stage test with fish eggs can be considered equivalent to life-cycle tests, as was shown by McKim (1985) when comparing 72 life-cycle tests in four freshwater fish with early life-cycle tests of these species.

Relatively few works have been devoted to the description of sublethal effects; with the exception of the paper by Rosenthal and Alderdice (1976) on sublethal effects of environmental stressors, relatively little synthesis of data appears to have occurred with regard to sublethal effects of pollutants on fish eggs and larvae. One of the reasons for the lack of work on sublethal effects may be the need of decision makers for clear-cut answers regarding damage to a test organism. Since death of the individual is easily recognized in most cases, LC_{50} investigations are preferred for reasons of simplicity of interpretation (see Rand and Petrocelli, 1985). A sublethal response requires more subtlety in its interpretation if it is to be well understood.

The term "sublethal" is not easily defined. According to Rosenthal and Alderdice (1976), sublethal effects may be defined as "those responses to environmental changes—histological, morphological, physiological, or ethological—that may be induced in one stage of development but be expressed at a later stage of organization or development in terms of reduced survival potential." This implies that a "sublethal" effect on an embryo may give rise to a lethally damaged larva. Thus the term "sublethal" for developing fish eggs applies to the different stages of development in the early life history, which may begin as early as in the ovarian, unfertilized egg; even impairment of fertilization would then be considered as a sublethal effect on an egg.

When trying to define "sublethal" in terms of effects on groups of fish eggs and larvae rather than on individuals, the connotation may change significantly. An individual's sublethal response always excludes immediate death, though its life expectancy may be shortened and it might die earlier than usual, while the sublethal response of, say, a batch of eggs toward environmental factors accepts the death of a few individuals as normal (such as found in the controls of most incubation experiments with fish eggs); thus even an increasing death rate in any given experiment may still be considered "sublethal" as long as the group concerned is not lethally damaged. That is, before we observe a mortality rate that is so high that no new recruitment can be assured, we may still call an observed effect "sublethal."

Thus the term "sublethal" in the context of this chapter will describe a rather subjective situation, which is strongly dependent on the biological end point considered and, as indicated above, allows a stochastic approach (which applies for dealing with sublethal effects on the population level) as well as a more deterministic interpretation, which may be referred to when discussing responses on the individual level. I shall thus discuss sublethal responses elicited in one ontogenetic stage (the earliest of which may be the ovarian egg), and its significance for that ontogenetic stage or ensuing stages of development in terms of consequences to the afflicted individual or group (reduced survival percentage).

As one looks at today's knowledge of sublethal effects of pollutants on fish eggs and larvae, it would be preferable to use the term "not acutely lethal" rather than "sublethal," since most of the described phenomena are defined for an arbitrarily chosen biological end point. Many will be lethal for the individual at a later stage of development. The significance of the effects for the survival of the population will be discussed later. In this review, due to lack of information, impact on the genetic information of the population via selection pressure must be totally neglected. The only available data of mutagenic effects of aquatic pollutants on the reproduction of fish are the experiments conducted by Prein *et al.* (1978) and Alink *et al.* (1980), exposing the eastern mud minnow *Umbra pygmaea* to Rhine water. The mutation rate in the testes of exposed fish (measured as sister chromatid exchange) increased by the factor 2 and 3, respectively, when exposing the specimens for 3 or 11 days. The water contained, among other substances, 16 phenolic compounds, 28 aromatic hydrocarbons, and 16 aromatic bases, of which the fraction of aromatic compounds was found to have a mutagenic potential (Prein *et al.*, 1978). The authors speculated that the common known mutagenic compounds responsible for increased mutation rate [i.e., some chlorophenols, chlorobenzenes, nitrobenzene, bis(chloroisopropyl)ether] that were detected in the water represented only a fraction of the total mutagenic load of the Rhine water, since many pollutants that may be biologically active are present in concentrations below detection levels.

This review is limited to sublethal effects defined by Rosenthal and Alderdice (1976) as histological, morphological, physiological, and ethological. From the available literature, it appears that within the framework of the review there are four main types of pollutants that are likely to produce sublethal effects in the early stages of fish either experimentally or in the field. These are heavy metals, petroleum hydrocarbons, chlorinated hydrocarbons, and acidifying substances (pH). This short list is not quite consistent in itself (for a more complete enumeration, see McKim *et al.*, 1975; McKim, 1985; Nimmo, 1985; Russo, 1985), since with the first three pollutant types we are referring to substances while the fourth in the literature is referred to rather as a chemical state—the hydrogen ion concentration (pH) in the water—and not the substances causing a possible shift in the H^+/OH^- equilibrium. Thus this chapter will be devoted to the above mentioned "substances" and their effects on gonadal tissue and ovarian eggs, eggs and sperm, embryos, and larvae, up to metamorphosis.

II. SUBLETHAL EFFECTS DURING DEVELOPMENT

Sublethal effects of pollutants on early developmental stages may be caused in two different ways. The first is by exposure of the parent fish and ensuing reduction in eggs deposited. This may be the case when fish are chronically exposed to low levels of metals or pesticides, and cuts down egg production by up to 80%; short-term exposure to cyclic hydrocarbons has similar effects. The second is by exposure of the extruded egg. Then exposure of the unfertilized egg may lead to an impairment of the fertilization process and the hardening of the egg shell. Exposure of the fertilized egg causes disturbances of early cleavage patterns or morphological aberrations, particularly in axis formation and head and eye development. Morphological aberrations are not pollutant-specific and may be caused also by natural stressors. Other, nonmorphological, aberrations from the normal pattern of development include depression of embryo activity (metals, hydrocarbons), inhibition of hatching enzyme (low pH), and alteration in incubation time (metals, hydrocarbons). All of the pollutants dealt with reduce hatchability. The parameter "viable hatch"—that is, the production of viable larvae—is a more sensitive indicator for sublethal effects than hatchability.

A. Ovarian Eggs and Egg Deposition

Exposure of mature fish to low (micrograms per liter) levels of zinc, cadmium, copper, mercury, hydrocarbons, or pesticides may lead to an 80% reduction of eggs produced. Low pH (4.5–6.0) exerts similar effects on some freshwater species.

The requisite for successful reproduction is the production of enough eggs by the parental generation to preserve the species. This first and basic requirement may be considerably impeded by the action of the above-mentioned pollutants. Thus, copper administered at concentrations of 18–32 μg/l to aquaria inhabited by fathead minnows (*Pimephales promelas*) totally prevents egg deposition in this species (Mount, 1968; Mount and Stephan, 1969). Although the complete suppression of ovarian egg development may be a rare case, a reduction in number of eggs produced upon exposure to pollutants is relatively common and has been reported by Brungs (1969) for the effects of low levels of zinc (0.18 mg/l). The number of eggs produced by female *Pimephales promelas* was only 17% of the eggs produced in the controls. In addition to zinc, cadmium and copper at low concentrations (Cu, 3.7–31 μg/l; Cd, 0.6–60 μg/l) caused progressively decreasing spawning activity and egg numbers spawned per female (Eaton, 1973). A reduction in the number of eggs spawned (up to 21%) after exposure to zinc at 0.13 and 0.2 mg/l is known to occur in the minnow *Phoxinus phoxinus* (Bengtsson, 1974); the number of eggs spawned is also reduced in the zebrafish *Brachydanio rerio*, when exposed to 5 mg Zn/l for a 9-day period during the time of gamete maturation (Speranza *et al.*, 1977). Effective zinc and cadmium concentrations appear to be fairly low, as documented above and also by Spehar *et al.* (1978) in experiments with flagfish *Jordanella floridae*, even in the parts per billion (ppb) range (micrograms per liter). This is true also in the live-bearing guppy *Poecilia reticulata*, where the clutch size of young suffers a reduction down to 50% upon exposure to low levels of zinc (0.36–17 μg/l) (Uviovo and Beatty, 1979). Aside from the above-mentioned metals, phenylmercuric acetate is known to exert detrimental effects on the number of eggs spawned by zebrafish *B. rerio*, at concentrations of 1 μg/l or less (Kihlström *et al.*, 1971). However, between this concentration and the controls, the authors observed a beneficial effect ("hormesis") of 0.2 μ/l of phenylmercuric acetate on egg production, a phenomenon that one encounters frequently in pollution research.

Besides metals, other substances may reduce the number of viable ovarian eggs. Among the hydrocarbons, benzene, a toxic component of petroleum, is known to be very active. When female Pacific herring *C. pallasi* are exposed to low (ppb) levels of benzene for 48 h just prior to spawning, a significant reduction in survival of ovarian eggs is recorded in the range of 10–25% (Struhsaker, 1977).

Effects on the number of eggs produced or spawned are also known for pesticides and chlorinated hydrocarbons in general. Again, minnows *Pimephales promelas*, *Phoxinus phoxinus*, and *Cyprinodon variegatus* have been used for the oral or external administration of substances. In experiments with *Pimephales promelas* and the carbamate insecticide carbaryl lasting for 9 months, 0.68 μg/l of the insecticide reduced the number of eggs produced per female significantly (Carlson, 1971). Also, diazinon, an organophosphate insecticide, employed in continuous flow-through tests with sheepshead (*Cyprinodon variegatus*) minnows caused a reduction of up to 55% in the number of eggs produced per female at concentrations higher than 0.47 μg/l in seawater. Similar effects are known to be caused by PCBs (polychlorobiphenyl) when applied orally (*Phoxinus phoxinus;* Bengtsson, 1980) or in solution (*Pimephales promelas;* Nebeker *et al.*, 1974) at low (1.8 μg/l) doses.

Impairment of egg production in some of the above-mentioned species is also known to be caused by depressed pH of the holding water. Thus fathead minnow (*P. promelas*) (Mount, 1973) and flagfish (*J. floridae*) when kept in water at pH levels between 4.5 and 6.0 produce significantly fewer eggs than in an environment with higher pH (pH 6.8) (Craig and Baksi, 1977). In fact, not only is the number of eggs laid reduced, but the production of fully mature (stage 6) oocytes capable of being fertilized is reduced to 20.7, 15.8, or 8.2% at pH 6, 5.5, or 4.5, respectively (Ruby *et al.*, 1977). At the lowest pH, mature oocytes soon undergo resorption and become atretic, while there is a preponderance of early oocyte stages in the ovary. In brook trout (*Salvelinus fontinalis*), low pH delays ovulation considerably (Tam and Payson, 1986). Beamish *et al.* (1975) suggested that due to altered pH, normal calcium metabolism required for successful ovarian maturation (Urist and Schjeide, 1961) is altered, resulting in abnormally low calcium concentration in female serum, which in turn affects the female reproductive physiology.

B. Fertilization, Water Uptake, and Water Hardening

While heavy metals have only minor effects on the process of fertilization, low pH through reduced sperm activity may lower fertilization rate in freshwater fish considerably. Petroleum hydrocarbons disturb formation of a fertilized egg, probably via effects on cleavage membrane formation, while chlorinated hydrocarbons inside the egg are responsible for a considerable reduction of fertilization rate due to cytogenetic damage. Impairment of osmotic activity of perivitelline colloids reduces water uptake in eggs reared in water of low pH or high metal levels.

An important step toward successful reproduction in fish is the fertilization of the extruded egg and the water-hardening process, two events that may very well be reversed in sequence depending on the fish species involved. Thus, in the pelagic eggs of many marine fish, the process of water hardening and water uptake may preceed the actual event of fertilization (Kjörsvik and Lönning, 1983; Lönning *et al.*, 1984); fertilization may not be necessary for water hardening (Potts and Eddy, 1973), or fertilization is required for the hardening process as in most demersal eggs, such as in herring and salmonids.

There are few reports of heavy metals exerting detrimental effects on the fertilization rate. In herring (*C. harengus*), Ojaveer *et al.* (1980) observed reduced fertilization when this process was conducted in copper contaminated water at an ambient copper concentration of 0.01 mg/l. Yet Blaxter (1977) was able to fertilize herring eggs at copper concentrations of 0.9 mg/l, obtaining a fertilization rate of 22%. These results fall in line with information provided by Rosenthal and Alderdice (1976), who reported that fertilization of Pacific herring eggs is virtually unaffected by exposure of up to 10 mg/l cadmium/l prior to fertilization.

In contrast to the action of metals, the pH of the incubation and fertilization media seems to have a delicate bearing on fertilization rate as shown by Petit *et al.* (1973) through work with rainbow trout (*Salmo gairdneri*). At optimum sperm concentration, fertilization rate drops from about 90% at pH 9.5 to only 50% at pH 7.3, although it is not clear whether this is an effect on the eggs or the sperm or both. Earlier experiments with sperm of trout (Inaba *et al.*, 1958) confirm that sperm mobility increases with rising pH; thus the reduced fertilization rate described above may be an effect of the lowered sperm activity. Reduced sperm activity with resulting lowered fertilization success occurs also after treatment of herring sperm with oil dispersants (Wilson, 1976) as well as after the exposure of trout (*S. gairdneri*) sperm to methylmercury at concentrations >1 mg/l (McIntyre, 1973). While the latter author used seawater of higher salinity for the incubation of herring eggs, the experiments of Ojaveer *et al.* (1980) were conducted in seawater of only 5.6–5.8‰ salinity. In these experiments, not only copper but also cadmium at concentrations higher than 0.005 mg/l affected fertilization negatively, ultimately yielding only 62% fertilized eggs at 1.0 mg copper/l, and 60% at 0.5 mg cadmium/l. Due to the fact that the two experimental series were conducted in different salinities, they are difficult to compare in terms of metal effects on fertilization, since we know that particularly the effects of cadmium are salinity-influenced (von Westernhagen *et al.*, 1974). The origin of this sublethal response is not clear. It might be caused by the interference of cadmium with the jelly coat of the egg, thus altering the site of the penetration of the sperm into the egg, the micropyle.

A direct influence on the formation of the zygote can probably also be excluded as the mode of action of pollutants such as aromatic hydrocarbons (xylene), which prevent the formation of a fertilized egg and early cleavage stages at concentrations higher than 10 mg/l in cod eggs (Kjörsvik *et al.*, 1982). The action of aromatic hydrocarbons such as *para*-xylene on fertilization and early cleavage and the generation of the characteristic small cells as described by Lönning (1977) in the reaction of plaice (*Pleuronectes platessa*) eggs to xylene and benzene are probably derived from their properties of causing membrane damage and increased membrane permeability (Roubal and Collier, 1975; Morrow *et al.*, 1975). Mechanisms located in the cell surface are important for the formation of the cleavage membrane, as shown by Rappaport (1977) in an investigation of cleavage in eggs from different invertebrates. As Roubal and Collier (1975) pointed out, aromatic hydrocarbons attack the outer surface of membranes and may thus influence the mechanism of cleavage, as apparent in the photographic evidence of Kjörsvik *et al.* (1982) in fertilized plaice (*P. platessa*) eggs. Similar effects are caused by the carcinogen benzo[a]pyrene (BAP) in flatfish embryos (sand sole, *Psettichthys melanostichus;* flathead sole, *Hippoglossoides elassodon*) (Hose *et al.*, 1982).

Strong depression of fertilization is also caused by chlorinated hydrocarbons incorporated into the egg from parental sources such as DDT and dieldrin in eggs of winterflounder (*Pseudopleuronectes americanus* (Smith and Cole, 1973) or polychlorobiphenyl (PCB) (34 ppm) in Atlantic salmon (*S. salar*) (Jensen *et al.*, 1971). In these cases, residues (4.6 ppm DDT; 1.2 ppm dieldrin) inside the egg give rise to considerable reduction of fertilization rate down to 40% and 12%, thus suggesting direct cytogenetic effects. High DDT residues in the same range may be responsible for the failure of reproduction in seatrout *Cynoscion nebulosus* (Butler *et al.*, 1972).

Alderdice *et al.* (1979a) have shown that metals such as cadmium delay the water hardening and the process of water uptake in Pacific herring eggs, while at the same time the primary bursting pressures of exposed eggs are reduced from between 700 and 1300 g to 200 and 350 g at cadmium concentrations near 1 mg/l. Brungs (1969) observed a similar effect in zinc-treated eggs of the fathead minnow, *Pimephales promelas.* Eggs in zinc concentrations higher than 0.18 mg/l remain in a flaccid condition resulting frequently in the rupturing of the egg capsules during handling. Since the hardening of the egg chorion requires calcium (Kusa, 1949; Lönning *et al.*, 1984) and the presence of an enzyme (Zotin, 1958), this process might very well be influenced by cadmium, a metal that is chemically closely related to calcium; a strong calcium/cadmium interaction could be assumed, since cadmium competes with calcium for binding sites in the egg capsule (Maljkovic and Branica, 1971; von Westernhagen *et al.*, 1975), thus interfering with the hardening process. The bound cadmium might alter the physical properties of the capsule and its jelly coat, reducing its strength (Alderdice *et al.*, 1979a) and permeability to salt and water. Although the cadmium effect on capsule strength may be explained, effects of other metals and substances are subject of problematic interpretation.

Reduced water uptake by salmonid eggs is also known to be caused by low pH. Thus Peterson and Martin-Robichaud (1982) and Eddy and Talbot (1983) as well as Rombough and Jensen (1985) report inhibited water uptake of newly fertilized eggs of Atlantic salmon and rainbow trout in water of pH 5.5 and lower. Together with the reduced water uptake goes a decrease in the ability to resist deformation when subjected to mechanical loads. Rombough and Jensen (1985) concluded that the low pH probably interferes with cortical vesicle exocytosis and affects the osmotic activity of perivitelline colloids, for instance through denaturation of the proteinaceous colloid (Rudy and Potts, 1969), so that it is no longer osmotically active. The same mechanism might apply in the action of heavy metals.

C. Early Development

Blockage of phosphorylation of ADP caused mainly by aromatic hydrocarbons and naphthalenes may lead to visible effects on early cleavage patterns. Initial irregular cleavages can be related to cytogenetic damage and can be traced through blastodisc and early gastrula formation by the appearance of opaque cell patches indicating irregular cell sizes.

In the meroblastic fish egg, the early cleavages on the surface of the yolk separate the clearly visible developing blastodisc from the yolk mass. Hence any deviation from the "typical" cleavage pattern is easily recognized and has frequently been used to describe sublethal effects on early embryogenesis. In particular, effects of temperature and salinity (Lieder, 1964; Alderdice and Forrester, 1968; von Westernhagen, 1968, 1970, 1974) have attracted the attention of scientists (see Chapter 3, this volume). In general, sublethal effects of pollutants are seldom visible at the very early cleavage stages. Experimentally only substances such as the aromatic compounds benzene and xylene provided irregular cleavages in the two- to eight-cell stages of plaice (*Pleuronectes platessa*) (Lönning, 1977) and cod (*G. morhua*) eggs (Kjörsvik *et al.*, 1982). The appearance of the early cleavage stages upon treatment with aromatic hydrocarbons clearly demonstrates the impairment of cell division. Similar disruptive early cleavage patterns in cod eggs have been reported by Dethlefsen (1977) on treatment of incubated eggs with DDT and DDE. We may find other substances that have the same effects on fish eggs, since during our current research (unpublished data) in the Baltic we have frequently found abnormal early cell stages in pelagic eggs (cod, plaice, flounder, sprat) (Fig. 1). The same observations have been made for cod eggs by Kjörs-

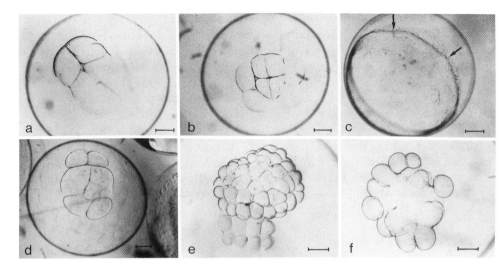

Fig. 1. Aberrant early cleavages and gastrula (arrows) in (a, b, c) cod (*Gadus morhua*) eggs, (d) sprat (*Clupea sprattus*) eggs, and (e, f) plaice (*Pleuronectes platessa*) eggs, caught with a plankton net in the Baltic in 1983. Horizontal bars 200 μm.

vik *et al.* (1984) in Norwegian waters and by Dethlefsen *et al.* (1985) for the eggs of flounder *Platichthys flesus,* dab *Limanda limanda,* and whiting *Merlangius merlangus* in the German Bight.

Sublethal effects on the developing embryo are more pronounced in the blastodisc stage and during beginning epiboly and may readily be caused experimentally. Stockard (1906) describes effects of LiCl solutions on *F. heteroclitus,* provoking unusual enlargement of the segmentation cavity under the blastodisc. Moreover, it is at the blastodisc stage that zinc-treated eggs of zebrafish (*B. rerio*), produce protoplasmic protrusions projecting abnormally from the sides of the embryo (Speranza *et al.,* 1977). When exposed to naphthalenes, methylnaphthalenes, and aromatic hydrocarbons, eggs of several marine species display retarded early cell division and differentiation into irregular blastodiscs with opaque patches, indicating different cell sizes (Stene and Lönning, 1984; Falk-Petersen *et al.,* 1982), probably originating from the initial cleavages. Only brief treatment during early cleavage frequently permits the embryo to develop normally up to midgastrulation before abnormal development becomes apparent. Typical effects during epiboly are irregular margins of the blastoderm with different size cells giving rise to malformed gastrulae, the embryo being less distinct and often surrounded by irregular cells (Falk-Petersen *et al.,* 1982). These defects become particularly evident in experiments with aromatic hydrocarbons and DDT and DDE, as reported by Dethlefsen (1977) for cod (*G. morhua*) eggs and Smith and Cole (1973) for the eggs of winter flounder (*Pseudopleuronectes americanus*). However, embryos with similar aberrations are found in the field (Fig. 1). Investigations by Longwell and Hughes (1980) provide some evidence for statistically significant associations between cytological, cytogenetic, and embryological diagnoses of the health of pelagic mackerel eggs and heavy metal and toxic hydrocarbon levels of some surface waters. Embryos from these areas (New York Bight) show increased incidence of chromosome and mitotic abnormalities, which probably led to the observed developmental aberrations. Particularly high incidences (>50%) of chromosome bridging and translocations are encountered. It may be difficult to explain the cause and significance of these aberrations of early development. There could be several reasons, but most of them are probably biochemical in origin, inhibiting metabolic processes responsible for differentiation and maintenance. Rosenthal and Alderdice (1976) suggest a blockage of phosphorylation of adenosine diphosphate (ADP), thus inhibiting the formation of adenosine triphosphate (ATP), which is a prerequisite for a multitude of metabolic processes necessary for differentiation. If the energy budget of the embryo is severely reduced either by direct blockage of the above mentioned pathway or by an overload in metabolic work required for detoxification of hydrocarbons through enzymatic degradation, no coordinated differentiation takes place and development is retarded or arrested. Similar effects can be provoked by exposure of herring embryos (*C. harengus*) to dinitrophenol (DNP) (Stelzer *et al.*, 1971), a decoupler of oxidative phosphorylation. Other causes for aberrations from the normal cleavage patterns may be the action of substances affecting cell cleavage directly, as does colchicine through inactivation of the chromosome transport mechanism of the spindle apparatus. Similar effects are exerted by cyclic hydrocarbons (Lönning, 1977), and the effects resemble those caused by DNP (Watermann, 1940). Yet short-term effects, when sublethal, might not necessarily be persistent. Particularly during the very early cleavage stages, irregularities caused by different stressors might be adjusted in the course of development. If, for instance, irregular cell patterns undergo further cleavages, the previously noticed asymmetrical patterns may disappear in the morula stage and macroscopically no traces of the initial aberration can be noticed (H. von Westernhagen *et al.*, unpublished). The only remnants of unequal initial cleavages may be chromosome aberrations. Thus, particularly in early developmental stages in situ, investigations demonstrate relatively high rates of chromosomal aberrations (anaphase aberrations) in pelagic fish eggs (Longwell and Hughes, 1980; Kjörsvik *et al.*, 1984; Dethlefsen *et al.*, 1985). The significance of these anaphase aberrations for later embryos is not clear, but Kjörsvik and co-workers suggest that this is a sign of "bad quality" eggs, yielding low rates of advanced embryos.

D. Advanced and Late Development

A multitude of observations describes the morphological reactions of advanced fish embryos to various pollutants. The most conspicious damages at this stage are abnormal development of the spinal column, abnormal head and eye development, and irregular proliferations from the main body over the yolk surface. Neither of the mentioned pollutants provokes typical, single-pollutant-specific reactions in the embryo. Morphological aberrations are not particularly pollutant-specific and may be caused also by natural stressors.

There exists relatively little experimental work on abnormal development in fish embryos before neurulation and formation of the head and optic cups. Most of the effects described are of advanced and late development, although in nature the situation is different in the sense that the rate of malformed early embryos in the sea is 4- to 10-fold that of the late stages (Kjörsvik *et al.*, 1984; Dethelfsen *et al.*, 1985). One of the reaons why most authors concentrate on later developmental stages is probably because early aberrations are very inconspicious. Besides, in one group, the salmonids, the early stages are extremely delicate to handle and most investigations study the salmonid only beginning at the eyed stage. In this context the term advanced development applies to embryos at or beyond stage II, as described by von Westernhagen (1970) for cod, flounder, and plaice, that is, with the beginning of head and eye formation, visible tail bud, and yolk two-thirds surrounded by blastoderm.

The most conspicious damages at and before this stage are irregular margins of the periblast on the yolk displaying a serrated appearance and emigrating groups of opaque cells in the blastoderm layer or in the space between blastoderm and periblast. We have found these sorts of defects in live eggs taken in the Baltic (Fig. 1c), but the period of epiboly prior to organ formation is rarely described in literature. Probably before this stage of organ differentiation a lot of early damage can be repaired and the embryo may recover if no more stress is applied. At the same time, this period (gastrulation before closure of blastopore) is considered especially sensitive, and stressed embryos die rather than compensate with aberrant development. However, at the time of organ formation, effects become particularly pronounced in head and notochord, as described for effects of copper and zinc in cod (*G. morhua*) embryos (Swedmark and Granmo, 1981) and lead at 0.05 and 0.07 mg/l in dechorionated embryos of zebrafish (*B. rerio*) (Ozoh, 1980).

These malformations are very similar to those caused by treatment of fish with petroleum hydrocarbons or derivates. Thus, Linden (1974, 1976, 1978) reported on herring eggs, which, when treated with crude oil and/or oil dispersants from 3.1 to 59 mg/l water-soluble fraction (WSF) of crude and number 1 fuel oil, display abnormal spinal columns, abnormal heads, and lack of spinal column. In comparable experiments with Ekofisk oil at concentrations of 0.1 to 1.0 mg/l, Lönning (1977) described the same effects on plaice (*Pleuroneces platessa*) embryos.

An embryo damaged in such a manner is called a typical "oil larva," with poorly differentiated head, protruding eye lenses, and a bent notochord. Inhibited pigmentation also can be taken as a sublethal effect and reaction to oil stress (Kjörsvik *et al.*, 1982). Embryos with these malformations are frequently found after oil spills in the vicinity of oil slicks, an indication of the fast action of the WSF of crude oil on fish embryos (Longwell, 1977). Figure 2 shows the different gross morphological abnormalities of cod eggs caused by treatment with Iranian crude oil.

Available literature on teratogenic effects of chlorinated hydrocarbons on advanced and late fish embryos can be summarized as follows. Cod embryos (*G. morhua*) exposed to DDT concentrations of 0.025 mg/l and more react with irregular proliferations at the yolk surface, and the embryo develops a bent or zigzag-growing spinal column (Dethlefsen, 1977). Sheepshead minnow (*Cyprinodon variegatus*) and killifish (*F. heteroclitus*) eggs, when subjected to DDT and malathion (organophosphate insecticide), carbaryl, or parathion at 10 mg/l, display developmental arrest prior to the initiation of heart beat. Blood pigmentation does not occur. *Cyprinodon variegatus* develops a malformed spine (Weis and Weis, 1974, 1976). Experiments conducted by Kaur and Toor (1977) with carp eggs (*Cyprinus carpio*) and the insecticides diazinon, fenitrotion, carbaryl, malathion, and phosphamidon produce similar effects. Upon exposure to concentrations around 0.008 (diazinon), 0.25 (fenitrothion), 1.0 (carbaryl), 2.5 (malathion), and 112 (phosphamidon) mg/l the embryos show stunted growth, curving of the tail, deformed head regions, enlargement of the pericardial sac, circulatory failure, deformed vertebral column, and

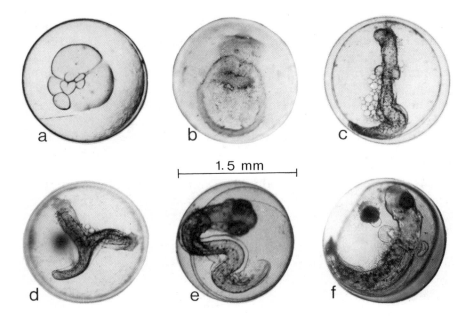

Fig. 2. Abnormalities in embryonic development of cod (*Gadus morhua*) eggs under the influence of the water-soluble fraction (WSF) of Iranian crude oil: (a) early cleavage, (b) gastrula, (c) embryo without head, (d) twinning, (e) axis deformation, and (f) microphthalmia. [From Kühnhold (1974).]

poorly developed eye pigment and chromatophores. Virtually all of these sublethal effects can also be observed when fish embryos develop under naturally stressed conditions, as described by von Westernhagen (1970) for the effects of extremes of temperature and salinity on plaice (*P. platessa*) embryos (Fig. 3).

Grossly deformed embryos may be caused by still other factors, for instance, low pH (pH 4.5) of the incubating water in fathead minnow (*Pimephales promelas*) (Mount, 1973) and Atlantic salmon (*S. salar*). Yet the typical injuries to salmon embryos by low pH, such as alterations in vascular structures, cellular dysplasia, necrosis, and sloughing of superficial ectoderm (Daye and Garside, 1980), are similar to those caused by heavy metals, detergents, halogenated organic compounds, and some petroleum fractions. Thus, it is apparent that it might be difficult if not impossible to identify any particular substances responsible for one or several sublethal morphological effects. On the one hand, it is difficult to distinguish clearly between morphological, physiological, or behavioral abnormalities, since one may result from

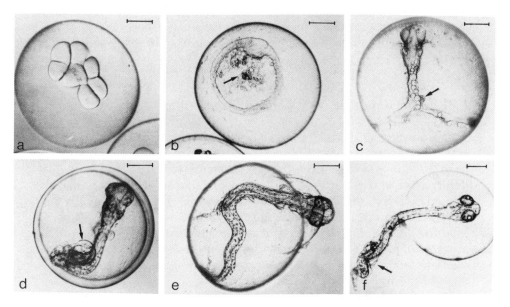

Fig. 3. Plaice (*Pleuronectes platessa*) eggs and larvae incubated under extreme temperature and salinity conditions. Arrows indicate zones of aberrant development: (a) aberrant early cleavage, 0°C, 25‰ salinity; (b) gastrula with irregular cell proliferation, 10°C, 25‰ salinity; (c) embryo not able to close blastopore, 10°C, 20‰ salinity; (d) distorted notochord, failure in pectoral fin development, 2°C, 33‰ salinity; (e, f) crippled larvae, 10°C, 15‰ salinity. Horizontal bars 200 μm.

the other, as they are frequently related. For example, fin anomalies may be related to modified (reduced) dermal respiration (Rosenthal and Alderdice, 1976). On the other hand, the detrimental action of a metal on embryogenesis may be indirect. For example, cadmium in high concentrations may alter the properties of the egg membrane and its "jelly coat," as known from herring eggs (Alderdice *et al.*, 1979c) or eggs of garpike (von Westernhagen *et al.*, 1975), ultimately impeding oxygen exchange. Thus, observed malformations in these experiments may be attributed simply to lack of oxygen, as described by Braum (1973) for herring eggs incubated experimentally under low oxygen tension, rather than to direct effects of the metal.

In fact, some of the anomalies resemble monstrosities produced during incubation at low oxygen levels as described by Alderdice *et al.* (1958) for salmon (*O. keta*) and Braum (1973) for herring eggs (*Clupea harengus*). The general retardation or arrest of development is also a phenomenon occurring at low oxygen levels (see also Hamdorf, 1961). Therefore, in general, one may say that the major morphological aberrations such as notochord distortions and head and eye malformations occurring in late embryos are not particularly pollutant-specific but are the expression of an embryo in a stressed condition.

E. Effects Other than Morphological Aberrations

As nonmorphological aberrations, effects on egg shell (chorion), embryo activity, and the hatching enzyme are prominent. Deviating from the nonspecific cause/effect relationship for gross morphological deformities and pollutants, reduction of chorion strength is mainly caused by heavy metals and at times by low pH. Low pH typically also depresses activity of the hatching enzyme, which results in low or retarded hatch. Petroleum hydrocarbons are extremely effective depressors of embryo activity measured as heart beat, pectoral fin movement, or body activity. Total embryo activity, though, is likewise reduced by other pollutants.

Aside from the occurrence of gross malformations, several functions of the chorion and the embryo are drastically impaired during and after exposure to these pollutants. As already mentioned, cadmium severely impairs hardening of the egg membrane after fertilization. Cadmium-exposed herring eggs never reach the maximum hardness attained by untreated eggs, and the egg capsules remain flaccid throughout embryogenesis (Rosenthal and Sperling, 1974; von Westernhagen et al., 1974; Alderdice et al., 1979a). In conjunction with this effect, herring eggs upon treatment with cadmium display smaller volumes than individuals incubated under uncontaminated conditions (Alderdice et al., 1979a,b), leaving a smaller perivitelline space. Also, zinc at a concentration of 6.0 mg/l causes the softening of egg membranes in C. harengus eggs (Somasundaram et al., 1984b) and brook trout (Salvelinus fontinalis) (Holcombe et al., 1979). In fathead minnows (Pimephales promelas), at a concentration of 295 or 145 μg/l and above, Zn reduces chorion strength so that eggs burst upon touching. Bursting pressure is only 15% of the normal value at zinc concentrations of 1360 μg/l (Benoit and Holcombe, 1978).

Chorion strength is also negatively influenced by low pH, as shown by the investigations of Mount (1973) with fathead minnow eggs (P. promelas) exposed to pH lower than 5.9. The same effects of low pH on capsule strength are known for eggs of rainbow trout (S. gairdneri) (Kügel, 1984), although Haya and Waiwood (1981) report hardening of Atlatnic salmon eggs in water of pH 4.5 due to a change in the physical structure of the outer mucopolysaccharide layer of the chorion. The significance of the softening of the egg shell to substrate spawning fish such as salmonids is evident since the soft eggs buried in the gravel, when subjected to movement, may easily break. In fact, the eggs of the fathead minnow at low pH become so flaccid that the cleaning action of the male on the egg clutches breaks the egg shells (Mount, 1973) and kills the embryos. Similar detrimental effects on eggs of the substrate spawning Pacific herring may be expected when high metal concentrations cause softening of the egg membranes.

When the embryo starts to develop a heartbeat, this parameter has frequently been used as a measure of pollutant effects. Typically, heartbeat frequency increases with age, and although subject to considerable variation caused by disturbances this increase is consistant with ongoing development (Rosenthal, 1967).

Metals such as cadmium reduce embryonic heart rate considerably. Thus in garpike *Belone belone* embryos, reduction in heart rate can be caused by incubation in water containing 1.0 mg/l or more of the metal, depending on the salinity of the incubating medium (Fig. 4). This is true also for the heartbeat in the Japanese medaka *Oryzias latipes* when the embryos are reared in 60 μg methylmercury/l, where heartbeat is reduced from 80–90 to 50 beats per minute (Dial, 1978). Zinc at a concentration of 2 mg/l cause a transient 2 : 1 heart block (two beats of the atrium for every beat of the ventricle) after 3–4 days of exposure in fathead minnow *P. promelas* (Pickering and Vigor, 1965).

Linden (1974, 1976, 1978) and Kühnhold (1978) demonstrated that petroleum hydrocarbons are extremely effective in the impairment of the normal functioning of the heart. Crude oil, especially in conjunction with oil dispersants, reduces heart-beat of herring embryos by 50%. Also, killifish *F. heteroclitus* embryos show sublethal responses to the water-soluble fraction of crude oil, which reduces heartbeat and overall embryo activity (Sharp *et al.*, 1979). Other hydrocarbons, such as benzene, applied at concentrations of 177 and 45 mg/l to incubating jars with Pacific herring *C. pallasi* and anchovy *Engraulis mordax* also reduce heart-beat of late embryos (Struhsaker *et al.*, 1974) or induce irregularities, as does toluene in the embryo of the Japanese medaka *O. latipes* (Stoss and Haines, 1979).

Changes in embryonic heart rate have also been caused experimentally by other pollutants, including DNP (Rosenthal and Stelzer, 1970: *C. harengus*), sulfuric acid from titanium dioxide production (Kinne and Rosenthal, 1967: *C. harengus*), or the organophosphate insecticide malathion (Weis and Weis, 1976: *Cyprinodon variegatus*).

At a stage of development where the regular heartbeat is well

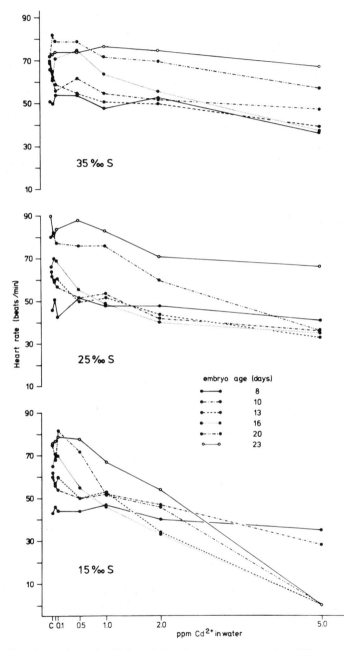

Fig. 4. Heartbeat of garpike (*Belone belone*) embryos exposed to different cadmium concentrations in the incubating water. [From von Westernhagen *et al.* (1975).]

established, the embryonic axis is developed far enough to enable the embryo to make the first wriggling movements. These movements, accomplished by slight, repetitive body flexure, are presumed to circulate the perivitelline fluid, thus improving the increasing oxygen needs of the late embryo. In some species that release very advanced larvae from their eggs (i.e., *B. belone*, *O. latipes*, *F. heteroclitus*, salmonids), the undulating movements of the body axis are supported by flapping of the pectoral fins and the opercula. As is the case for heartbeat, embryonic movements may be severely impaired either directly through the stressors or indirectly whenever egg volumes are reduced imposing mechanical blockage to movements. The influence of cadmium, for instance, on embryonic movements in herring eggs becomes apparent when the embryo has grown to encircle the yolk. The normal "wriggling" then is replaced by "trembling," a high-frequency shivering that is maintained even in more advanced embryos, which are normally performing rotations or "somersaults" within the egg shell. As Fig. 5 shows, both types of activities are influenced by cadmium in the incubating water (von Westernhagen *et al.*, 1974). Copper concentrations around 0.13 mg/l exert strong paralyzing effects on herring embryos, which, with progressing development, become more and more immobilized (von Westernhagen *et al.*, 1979). Other sublethal effects of cadmium on embryo activity are displayed by the reduced pectoral fin movements of garpike (*B. belone*) embryos (von Westernhagen *et al.*, 1975) when exposed to cadmium concentrations higher than 1.0 mg/l. Stoss and Haines (1979) as well as Leung and Bulkley (1979) report influence of toluene on the opercular movement of the late embryo of the Japanese medaka *O. latipes*, which becomes erratic, irregular, and shallow at concentrations of 80–100 μg/l (WSF). A general decrease of embryonic activity is known for salmon (*S. salar*) embryos when reared at low pH of 4.0–4.5 compared to controls in pH 6.7 (Peterson and Martin-Robichaud, 1983). A possible explanation for the reduced activity in cadmium-treated embryos may be the effects of cadmium on enzyme activity. In Pacific herring eggs, Mounib *et al.* (1976) found that exposure to 10 mg cadmium/l decreased activity of four important carbon dioxide–fixing enzymes: propionyl coenzyme A (CoA) carboxylase, nicotinamide adenine dinucleotide (NAD) and NADP malic enzymes, and phosphoenolpyruvate (PEP) carboxykinase. In control eggs, PEP carboxykinase activity increases by two orders of magnitude in the period from the early blastodisc just prior to hatching. The increase indicates the importance of the enzyme metabolism. In contrast, there is considerably less increase in PEP carboxykinase activity in cadmium-exposed eggs

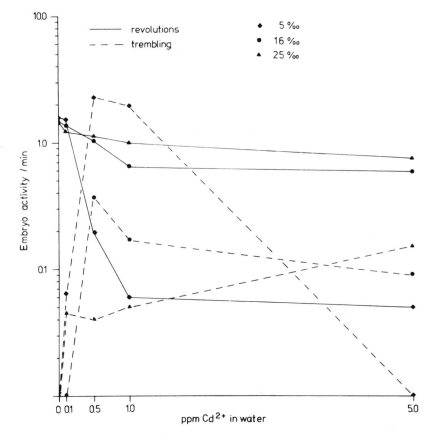

Fig. 5. Activity of herring (*Clupea harengus*) embryos influenced by different cadmium concentrations in the incubating water. [From von Westernhagen *et al.* (1974).]

up to the beginning of embryo activity, while in the controls activity increases further. Final activity of the enzyme is more than 25% less than that of the controls. Relative activity of propionyl-CoA carboxylase in the cadmium exposed eggs is about 20% of the control level prior to blastopore closure and only 32% prior to hatching. Relative activity of NAD and NADP malic enzymes remains stable up to eye pigmentation but is reduced by about 20% just prior to hatching, when the embryo has completed differentiation and growth is dominant. In view of the important role played by carbon dioxide–fixing enzymes in biosynthetic processes, the effect of cadmium in depressing enzyme activity during the developmental stages may result in lethargic embryos and small and inactive larvae.

As a consequence of the lowered activity of the embryo, the hatching process may be severely impaired—on the one hand, because the hatching enzyme is not distributed throughout the perivitelline fluid and on the other hand, after the digestion of the inner layer of the zona radiata by the hatching enzyme, the emerging larva cannot break the nondigested outer part of the egg shell (Hagenmaier, 1974a), and thus remains longer in the egg casing or never hatches.

Even though the proteolytic hatching enzyme may be produced and distributed sufficiently, heavy metals in the external and internal medium may not allow the enzyme to display its proteolytic functions fully. Hagenmaier (1974b) reports that manganese, zinc, mercury, or copper inhibits the proteolytic functions of the enzyme. From experiments with Atlantic salmon (*S. salar*), brown trout (*S. trutta*), and rainbow trout (*S. gairdneri*), Grande (1967) thinks that zinc affects the enzymatic processes that soften the egg capsule. Also, the pH of the incubating medium has a strong bearing on the functioning of the hatching enzyme, which has a maximum activity at pH 8.5 in eggs of the rainbow trout (Hagenmaier, 1974c) and from 7.5 to 8.0 in the Pacific salmon (*O. keta*) (Bell *et al.*, 1969). Thus, in experiments with perch (*Perca fluviatilis*) eggs, Runn *et al.* (1977) found strong impairment of the activity of the hatching enzyme at low pH resulting in reduced hatch. Similarly, at pH 4.5, chorionase activity of Atlantic salmon eggs reached only 49% of its activity at pH 6.5 (Haya and Waiwood, 1981).

F. Incubation Time and the Process of Hatching

Alterations in incubation time may be caused by premature or delayed hatching. Metals may either shorten or lengthen this period, and no unequivocal prediction is possible; petroleum hydrocarbons as WSF usually retard hatching and development. The effect of chlorinated hydrocarbons on hatching have not been looked into closely. While low pH retards development in general, the major effects of low pH are on inactivation of chorionase (hatching enzyme) below pH 5.5, which delays hatching or prevents it entirely.

Onset of hatching in teleost eggs begins with the secretion of proteolytic hatching enzyme from the hatching glands around the head region of the embryo/larva. The secretion and activity of the enzyme is relatively rapid, as seen in studies with *F. heteroclitus* eggs (Kaighn, 1964). Chorion treated with chorionase dissolves within 5–10 min. Thus, with the strong movements of the late embryo (Poy, 1970), hatching in *F. heteroclitus* should be completed within a few minutes, while it may take several hours in salmonids with a thicker egg shell and lower incubation temperatures (Hayes, 1942; Bell *et al.*, 1969). Incubation time, the period between fertilization and 50% hatch, is mainly dependent on temperature, higher ranges accelerating development. Low oxygen concentrations, when maintained throughout development, lengthen the total incubation period (Hamdorf, 1961; *S. gairdneri*).

Changes in time to hatch are common in fish embryos subjected to sublethal effects of pollutants. Many xenobiotic substances shorten the incubation period or cause premature hatch. Others, however, lengthen the development period or delay hatching. Due to the different effects of metals on the late embryo, incubation may be either shortened or lengthened. In most cases that have come to the attention of the reviewer, incubation time is shortened and larvae hatch prematurely. Rainbow trout (*S. gairdneri*) has frequently been used for these essays. Thus, Shabalina (1964) notes that cobaltous chloride in concentrations of up to 5 mg/l shortens the incubation period; hatching larvae are viable. Also, nickel and copper when applied in concentrations of 1 mg/l accelerate development of rainbow trout eggs by about 45% with copper and by 20% with nickel (Shaw and Brown, 1971). The same is true for vanadium (44 mg/l) (Giles and Klaverkamp, 1982). Brook trout (*Salvelinus fontinalis*) eggs incubated at 32.5 μg copper/l hatch prematurely (McKim and Benoit, 1971) as do herring at even lower concentrations (>0.1 mg/l) (Ojaveer *et al.*, 1980). Other metals such as cadmium and zinc have the same effects on fish embryos as shown by Vladimirov (1969) for carp (*Cyprinus carpio*) and by Rosenthal and Sperling (1974), von Westernhagen *et al.* (1974), and Somasundaram *et al.* (1984a) for herring (*Clupea harengus*).

However, the reverse, a prolonged incubation period, is also known to be caused by metals such as zinc, cadmium, copper, and mercury (Grande, 1967; Servizi and Martens, 1978; Swedmark and Granmo, 1981; Weis, 1984; Somasundaram *et al.*, 1984a,b). In the case of zinc, concentrations below 2.0 mg/l accelerate while higher concentrations retard development of herring eggs.

The above does not imply that embryos hatching early developed faster or that the late-hatching ones displayed delayed embryogenesis. In fact, changes in differentiation pattern, known to occur in trout (*S. gairdneri*) eggs, developing at low partial O_2 pressure (Hamdorf, 1961), are not reported. Rather, the immature embryo hatches (in the case of "acceleration") or an over-mature larva hatches with a small yolk reserve and advanced differentiation (a functional mouth in the case of cod *G. morhua*; Swedmark and Granmo, 1981).

There may be no general explanation for this phenomenon. Observed effects depend on the application of the metal, its concentration, stage of development, duration of exposure, etc. Thus the reasons for early or late hatching are frequently found in the history of the egg. On the one hand, an early hatch might be caused by a beneficial effect of low levels of, say, zinc, on the embryo (Somasundaram *et al.*, 1984a; *C. harengus*), with truly accelerated development in the sense of the "sufficient challenge" concept forwarded by Smyth (1967), or by the highly detrimental effects of copper at concentrations of 0.133 mg/l, which immoblize herring embryos totally, so that the hatching glands produce a punctiform hole in the chorion (no distribution of hatching enzyme), causing premature liberation of the embryo (von Westernhagen *et al.*, 1979) (Fig. 6). On the other hand, late hatch may be caused by a retarded development such as caused by high concentrations of zinc (>2 mg/l), with resulting inability of the embryo to break the chorion (Swedmark and Granmo, 1981), or because of malfunctioning of the hatching enzyme proper as suggested by Servizi and Martens (1978) for the delayed hatching of sockeye (*Oncorhynchus nerka*) and pink salmon (*O. gorbuscha*) exposed to copper concentrations higher than 6 μg/l. A similar explanation may apply for the other substances dealt with in this context and the sublethal effects of petroleum hydrocarbons, which may likewise lengthen the time from fertilization to hatching or may shorten it.

In contrast to the common effects of heavy metals on incubation, most authors report a delayed effect of petroleum hydrocarbons on hatching when applied as the water-soluble fraction (WSF). At low concentrations (12.5%) of the WSF of crude oil, development was accelerated, with early hatch in *F. grandis* (Ernst *et al.*, 1977); the early hatch of Japanese medaka *Oryzias latipes* is likewise considered a premature hatching resulting from stimulation of the hatching mechanism by oil components (Leung and Bulkley, 1979). However, most authors report delayed hatch of larvae after treatment of developing eggs and embryos with petroleum hydrocarbons. Thus, Linden (1978) reported delayed hatching in Baltic herring larvae *C. harengus membras* exposed to 54 mg/l WSF of light fuel oil; this is also true when herring are pulse-exposed (24, 48, 96 h) to 40–45 mg/l initial concentrations of benzene (Struhsaker *et al.*, 1974). Delay in hatching also occurs in eggs of winter flounder *Pseudopleuronectes americanus* if the parents have been exposed to 100 μg/l WSF of number 2 fuel oil during gonad maturation (Kühnhold *et al.*, 1978), and in *F. heteroclitus* exposed to 25% WSF of this oil (Sharp *et al.*, 1979). Other reports of petroleum hydrocarbons delaying hatching are given by Kühnhold

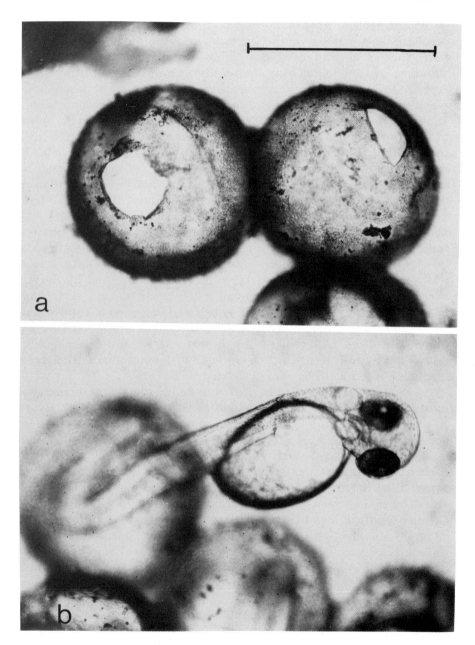

Fig. 6. *Clupea harengus.* (a) Empty egg chorions with punctiform hole after incubation in copper-contaminated water (133 μg Cu/l); (b) prematurely hatching crippled herring larva after incubation in copper-contaminated water. Horizontal bar 1 mm.

(1974) (*G. morhua*), Stoss and Haines (1979) (*O. latipes*, toluene) and Carls and Rice (1984) (*Theragra chalcogramma*). Further, the polycyclic aromatic carcinogen, benzo[a]pyrene, is known to retard development and hatching in rainbow trout (*S. gairdneri*) (Kocan and Landolt, 1984; Hannah *et al.*, 1982). The effect of high concentrations of petroleum hydrocarbons on fish embryos may sometimes be a narcotizing one that reduces metabolism, thus slowing down development (Struhsaker *et al.*, 1974; Carls and Rice, 1984) as well as exerting narcotic effects on the late ready-to-hatch embryo.

There is little information on the disturbance of the hatching process by chlorinated hydrocarbons (mainly pesticides and PCB). No direct impairment of hatching has been reported. Reports of variations of time to hatch refer to abnormally shaped embryos that hatch late due to physical failure to break the egg shell (Dethlefsen, 1977; *G. morhua*, DDT) or to premature hatch in coho salmon *Oncorhynchus kisutch* eggs, treated with Arocolor 1254 (PCB) at 4.4 and 15.0 μg/l (Halter and Johnson, 1974), where early hatching may be caused by an alteration of the chorion due to PCB treatment. Reduction in hatching time also occurs in eggs of minnow *Phoxinus phoxinus* when the parental fish have been administered PCB orally and the eggs contained high amounts (1.5–170 mg/kg fresh weight) of Clophen 50. Hatching time in eggs containing >15 mg/kg fresh weight PCB is significantly reduced compared to controls (Bengtsson, 1980).

In contrast to the effects of pesticides on the hatching process and hatching time, which are diffuse and probably relate to the general disturbed condition of the embryo, effects of pH on the hatching process are much better understood. Report of sublethal effects of pH on fish eggs and larvae are mainly restricted on the effects of low pH related to acidification of lakes in North America and northern Europe. The effect of low pH on development and hatching is very consistent in all but one of the available reports. Only Trojnar (1977b) in his experiments with brook trout (*Salvelinus fontinalis*) eggs reported faster development and hatching at pH 4.65 than at pH 8.07, where hatching took place 12 days later at the higher pH. Early hatching larvae at pH 4.65 did not appear to be premature, but were fully developed. This report is contrary to all other information on sublethal effects of low pH on early developmental stages of fish, where there is general agreement that low pH prolongs the period from fertilization to hatching. Thus, Peterson *et al.* (1980a,b) show that the eggs of Atlantic salmon, *S. salar*, exposed to water of pH 4.0–5.5, following eye pigmentation, showed delayed hatch. The same effects are reported by Swarts *et al.* (1978) for brook trout (*S. fontinalis*) eggs incubated at pH 4.75. The delay in hatching may be considerable, reaching 14 days for eyed eggs of rainbow trout (*S. gairdneri*) subjected to pH 4.0–4.5 when compared to controls incubated in pH 7.8–8.0 (Kügel, 1984). In yellow perch (*Perca fluviatilis*), incubation time increased by 29% at pH 4.0 compared to pH 6.4 (Rask, 1983). The prolongation of the period from fertilization to hatching at low pH occurs also in eggs of zebrafish (*B. rerio*) (Johansson *et al.*, 1973) and fathead minnow (*Pimephales promelas*) (Mount, 1973). Likewise, low pH retards development of Pacific herring (*C. pallasi*) embryos, as already observed by Kelley (1946), while high pH (pH 10) accelerates development slightly (Johansson *et al.*, 1973; *B. rerio*). The factors involved may be either a general retardation of development, or a delay in the process of hatching, or both. Evidence for retarded development is provided by Kelley (1946), but in most cases the prolonged incubation period is due to an impairment of the hatching process due to inhibition of enzyme (chorionase) activity at lower pH. Due to the permeability of the chorion for hydrogen ions (Peterson *et al.*, 1980a), the perivitelline fluid rapidly adjusts to the pH of the incubating medium, and the pH may be too low for maximum enzyme activity. Thus at pH 5.2, chorionase activity of rainbow trout *S. gairdneri* embryos, is reduced to 10% of the optimal rate at a pH of 8.5 (Hagenmaier, 1974a), and the process of hatching takes several days rather than hours. In salmon *S. salar,* reduction of chorionase activity at low pH is less drastic, but still, at pH 4.5 only 49% of the activity at pH 6.5 is observed (Haya and Waiwood, 1981). Alevins of Atlantic salmon (*S. salar*) delayed in hatching are thus larger than when hatched at the normal pH (Peterson *et al.*, 1980a). Other indications of incomplete lysis of the chorion at low pH are tail-hatched larvae with the chorion around the yolk or partially hatched larvae with the head still inside the egg shell (Johansson *et al.*, 1977; Brown and Lyman, 1981; Kügel, 1984).

G. Hatchability and Viable Hatch

One of the main interests scientists had in the effects of pollutants on fish eggs and larvae was the reduction in hatching success. Effects of pollutants on hatchability and viable hatch are dependent on the stage of development, and the species and type of pollutant. Exposure before closure of blastopore causes more severe reduction in hatching success than when advanced stages are exposed. "Viable hatch" is a more sensitive indicator of pollutant effects than "hatchability."

Since the total of all previously mentioned effects on eggs and embryos is expressed in the number of larvae emerging, it appears useful to present data on reduction of hatchability and viable hatch (percent) in a table (Table I). This should enable the readers to determine readily the effects of pollutants on emergence of larvae. For the user's convenience, zero hatch is included. Data in Table I show that hatchability as a parameter to assess sublethal effects is limited in its use by the differences in toxicity of the various pollutants. Thus, in saltwater species, cadmium exerts detrimental effects only at high (environmentally nonrelevant) concentrations in the range of 1000–2000 μg/l (von Westernhagen et al., 1974, 1975; Voyer et al., 1979) or more (Rosenthal and Sperling, 1974; von Westernhagen and Dethlefsen, 1975). In freshwater species reactions toward cadmium are more sensitive (Pickering and Gast, 1972; Spehar, 1976; Rombough and Garside, 1982). The situation is similar regarding the effects of zinc. Copper, in turn, even in saltwater, causes substantial effects on hatchability and viable hatch at concentrations between 30 and 90 μg/l (Blaxter, 1977; Servizi and Martens, 1978; Ojaveer et al., 1980; Cosson and Martin, 1981). These concentrations are frequently found in surface microlayers of polluted areas (Hardy et al., 1985), which have been found to cause significant reduction in hatchability of herring (*C. harengus*) eggs (von Westernhagen et al., 1987).

Sublethal effects on hatching caused by chlorinated hydrocarbons are dependent on the toxicity of the pollutant but also on the mode of application. Via a biomagnification effect, DDT reduces hatching in eggs from female fathead minnow (*Pimephales promelas*), kept in water with 2 μg DDT/l until spawning, from 89% to 74%, and the PCB Aroclor 1254 prevents hatching entirely in eggs from female fathead minnows treated with 4.6 μg/l (Nebeker et al., 1974). When Aroclor 1254 is applied to *Cypronodon variegatus* eggs at 10 μg/l, hatchability is only 57% (Schimmel et al., 1974).

In general it is evident that hatchability of eggs increases when later developmental stages are exposed to pollutants. In early-stage exposures, effects on percent hatching are more severe. This is in line with Stockard's (1921) ideas concerning effects of development arrest in fish embryos through abiotic substances. Thus, in *F. heteroclitus*, egg development may be stopped safely shortly after gastrulation is completed. Critical stages are those before closure of the blastopore, during which marked inequalities of cellular proliferations are taking place. Since body-axis formation takes place fairly early in development (beginning with neurulation and continuing throughout ontogenesis), impairment of this process has its ultimate bearing on hatch-

Table I

Sublethal Effects of Pollutants (Heavy Metals, Petroleum Hydrocarbons, Chlorinated Hydrocarbons, pH) on Hatchability and Viable Hatch of Fish Eggs[a]

Species	Pollutant	Day of exposure	Concentration (μg/l)	Hatch-ability (%)	Viable hatch (%)	°C	Salinity (‰)	Reference
			Heavy metals					
Belone belone	Cd	1	C	73	100[b]	15	25	von Westernhagen et al. (1975)
		1	100	62	100			
		1	500	68	90			
		1	1000	73	65			
		1	2000	21	0			
		1	5000	0				
Brachydanio rerio	Cu+Pb	1	C	86		26		Ozoh (1979a)
		1	36	38				
		1	72	56				
	Cu	1	72	47				
	Pb	1	72	42				
Brachydanio rerio	Zn	P	C		63			Speranza et al. (1977)
		P	5000		1			
Brachydanio rerio	Hg	P	C	41		27		Kihlström et al. (1971)
		P	C	44				
		P	1	29				
		P	0.2	30				
Cichlasoma nigro-fasciatum	Pb	P	C	95		25		Ozoh (1979b)
		Pc	50	45				
		Pc	100	30				
		Pc	300	17				
		Pc	400	27				
		Pc	500	21				
		Pc	600	44				

Species	Metal		Conc.			9–10	29–32	Reference
Clupea harengus	Cu	4	C		45			Blaxter (1977)
		4	30		44			
		4	90		0			
		4	300		0			
		4	900		0			
		1	C		25			
		1	30		22			
		1	90		12			
		1	300		0			
Clupea harengus	Cu	1	C	90	96			Ojaveer et al (1980)
		1	5	81	93			
		1	10	71				
		1	100	71				
	Cd	1	C	86	98			
		1	3	81	98			
		1	5	82	98			
		1	50	66	97			
		1	500	13	91			
Clupea harengus	Cd	1	C	94		10	16	von Westernhagen et al. (1974)
		1	100	95				
		1	500	93				
		1	1000	62				
		1	5000	0				
Clupea harengus	Cd	1	C	87	89	10	16	Rosenthal and Sperling (1974)
		1	100	83	95			
		1	1000	16	84			
		1	5000	0	75			
		1	10000	0	14			
Dicentrarchus labrax	Cu	1	C		100			Cosson and Martin (1981)
		1	5		74			
		1	10		79			
		1	50		18			
		1	100		1			

(continued)

283

Table I (*Continued*)

Species	Pollutant	Day of exposure	Concentration (µg/l)	Hatchability (%)	Viable hatch (%)	°C	Salinity (‰)	Reference
Fundulus heteroclitus	Hg	1	C	89		25	20	Sharp and Neff (1980)
		1	4	81				
		1	10	81				
		1	20	73				
		1	30	69				
		1	40	41				
		1	60	6				
		1	80	0				
Jordanella floridae	Cd, Zn	1	C, C	66, 70		25		Spehar (1976)
		1	17, 28	66, 76				
		1	4.1, 47	73, 72				
		1	8.1, 75	66, 59				
		1	16, 139	68, 73				
		1	31, 267	0, 0				
Menidia menidia	Cd	5	C	90		15–19	20	Voyer et al. (1979)
		5	170	100				
		5	390	94				
		5	750	74				
Morone saxatilis	Cu	2	C	100				O'Rear (1972)
		2	10	45				
		2	100	27				
		2	500	45				
		2	2800	0				
		2	5000	0				
Oncorhynchus nerka	Cu	2	18		98	6–9		Servizi and Martens (1978)
		2	37		96			
		2	78		50			

Species	Metal		Conc.	%			Reference
Oncorhynchus nerka	Cd	2	174	0			Servizi and Martens (1978)
		2	C	96	6–9		
		2	0.4	97			
		2	1.5	96			
		2	5.7	95			
	Hg	2	C	95	6–9		
		2	1.0	93			
		2	2.5	95			
		2	4.3	64			
		2	9.3	0			
Oncorhynchus tshawytscha	Cu	1	C	80	13–14		Hazel and Meith (1970)
		1	21	76			
		1	40	82			
		1	80	78			
Oryzias latipes	Hg	1	C	47	26	?	Heisinger and Green (1975)
		1	10	58			
		1	15	21			
		1	20	0			
		1	30	0			
Pimephales promelas	Cd	1	C	95	25		Pickering and Gast (1972)
		1	7.8	97			
		1	14	95			
		1	27	94			
		1	57	78			
Platichthys flesus	Cd	1	C	87	5	32	von Westernhagen and Dethlefsen (1975)
		1	100	83			
		1	500	84			
		1	1000	81			
		1	2000	80			
		1	3000	67			
		1	5000	22			

(continued)

Table I (*Continued*)

Species	Pollutant	Day of exposure	Concentration (μg/l)	Hatchability (%)	Viable hatch (%)	°C	Salinity (‰)	Reference
Pseudopleuronectes americanus	Cd, Ag	1	C, C		100	9	21	Voyer et al. (1982)
		1	100, 18		100			
		1	550, 99		100			
		1	1000, 180		90			
		1	1000		36			
		1	1000, 18		55			
Pseudopleuronectes americanus	Cd	2	C	93	90	10	20	Voyer et al. (1977)
		2	100	80	74			
		2	320	87	81			
		2	1000	90	19			
		2	1150	100	84			
		2	1550	98	78	5		
		2	2100	88	49			
Salmo gairdneri	Cr	1	C	98		12		Van der Putte et al. (1982), exp. at pH 6.5
		1	20	98				
		1	200	100				
		1	2000	84				
Salmo salar	Cd	1	C		76	5		Rombough and Garside (1982).
		1	2.8		82			
		1	11		70			
		1	29		80			
		1	90		76			
		1	270		47			
		1	870		0			
Salvelinus fontinalis	Zn	1	343	93		9		Holcombe et al. (1979)
		1	724	97				
		1	709	95				

Species	Toxicant	Type	C	%	Days	Reference
Salvelinus fontinalis	Pb	1	1382	90	9	Holcombe et al. (1976)
		1	1353	79		
		1	2017	67		
		1	2099	73		
		1	4336	3		
		1	4363	1.5		
Salvelinus fontinalis	mHg	P	C	96–100		McKim et al. (1976)
		P	3,4	57–95		
		P	58	79		
		P	119	85–86		
		P	235	60–73		
		P	474	28		
Salvelinus fontinalis	Cu	P	C	97	5–14	McKim and Benoit (1971)
		P	0.03	87		
		P	0.09	99		
		P	0.3	98		
		P	0.9	0,84		
		P	2.9	0		
Salvelinus fontinalis	Cu	P, 1	C	81	14	
		P, 1	3.4	99		
		P, 1	5.7	85		
		P, 1	9.5	98		
		P, 1	17.4	95		
		P, 1	32.5	26		

Petroleum hydrocarbons

Species	Toxicant	Type	C	%	Days	Reference
Clupea harengus	Oil (WSF)	1	C	32	14	Vuorinen and Axell (1980)
		1	600	50	6	
		1	1900	26		
		1	5400	28		
		1	17500	24		
		1	36000	4		

(continued)

Table I (*Continued*)

Species	Pollutant	Day of exposure	Concentration (μg/l)	Hatch-ability (%)	Viable hatch (%)	°C	Salinity (‰)	Reference
Clupea harengus membras	Oil (WSF)	1	C	100		9–14	6	Linden (1978)
		1	50	60				
		1	500	4				
		1	5000	0				
		3	50	99				
		3	500	91				
		3	500	0				
Clupea harengus pallasi	Benzene	P	C	93		11–12	22	Struhsaker (1977)
		P	800	67				
Clupea harengus pallasi	Oil (WSF)	5–6	C	53		8–9		Smith and Cameron (1979)
		5–6	680	44				8 h only exposed
		5–6	680	32				24 h only exposed
		5–6	680	27				8 h only exposed
		5–6	680	0				6 days exposed
Fundulus heteroclitus	Oil (WSF)	1	C	90		22	20	Sharp et al. (1979)
		1	10% WSF	90				
		1	20% WSF	72				
		1	25% WSF	7				
		1	25% WSF	85				4 days only exposed
		1	25% WSF	40				8 days only exposed
Fundulus heteroclitus	Oil (WSF)	1	C	100		21	20	Anderson et al. (1977)
		1	25% WSF	100				
		1	50% WSF	60				
		1	100% WSF	0				

Species	Oil/(WSF)		Concentration	%	%			Reference
Cyprinodon va-riegatus		1	C	100		21	20	
		1	25% WSF	88				
		1	50% WSF	62				
		1	100% WSF	0				
Gadus morhua	Oil	1	C	20		5–6	29–34	Kühnhold (1974)
	(WSF)	1	100	14				
		1	1000	4				
		1	10000	0				
		5	C	53				
		5	100	24				
		5	1000	19				
		5	10000	17				
Mallotus villosus	Oil	14–20	C	100		6–7	34–35	Johannessen (1976)
	(WSF)	14–20	10	100				
		14–20	25	88				
		14–20	50	75				
		14–20	100	68				
Platichthys flesus luscus	Oil	d	C	90	58	9–12		Mazmanidi and Bazhasvili (1975)
	(WSF)	d	25	89	55			
		d	50	89	44			
		d	100	86	24			
		d	200	88	15			
		d	400	89	9			
		d	1700	90	0			
			2500		0			
Pseudopleuronectes americanus	Oil	P	C		63	1–10	31	Kühnhold et al. (1978)
	(WSF)	P	10		65			
		P	100		53			
		1	10		60			
		1	100		42			

(continued)

289

Table I (*Continued*)

Species	Pollutant	Day of exposure	Concentration (µg/l)	Hatch-ability (%)	Viable hatch (%)	°C	Salinity (‰)	Reference
			Chlorinated hydrocarbons					
Cyprinus carpio	Simazine	1	20,000	79		16		Kapur and Yadav (1982)
		1	30,000	68				
		1	40,000	50				
		1	60,000	9				
		1	80,000	0				
	Gramaxone	1	40,000	86		16		
			60,000	63				
			80,000	9				
			90,000	0				
	Taficide	1	25,000	92		16		
		1	30,000	68				
			40,000	34				
			50,000	11				
			100,000	0				
Cyprinodon va-riegatus	Aroclor 1254	1	C	79		29	16–32	Schimmel et al. (1974)
		1	0.1	69				
		1	0.32	73				
		1	1.0	82				
		1	3.2	75				
		1	10.0	57				
Cyprinodon va-riegatus	Aroclor 1254	P	C	93		30	10–27	Hansen et al. (1974)
		P	0.1	88				
		P	0.32	80				
		P	1.0	98				
		P	3.2	85				
		P	10	72				

Species	Compound							Reference
Gadus morhua	DDT	1	C	80		8	35	Dethlefsen (1977)
		1	150	60				
		1	300	60				
		1	700	40				
		1	800	30				
Phoxinus phoxinus	Clophen A 50	P	C	49	12–16			Bengtsson (1980)
		P[e]	1.6	39				
		P[e]	15	43				
		P[e]	170	8				
Pimephales pro-melas	Aroclor 1254	P	C	74				Nebeker et al. (1974)
		P	0.23	55				
		P	0.52	63				
		P	1.8	79				
		P	4.6	0				
		P	15	0				
Pimephales pro-melas		P	C	89				Jarvinen et al. (1977)
		P	0.5	84				
		P	2.0	74				
Oncorhynchus kisutch	Aroclor 1254	f	0	96	12–14			Halter and John-son (1974)
		f	4.4	88				
		f	7.8	79				
		f	15.4					
		f	26.0	47				
		f	56.4	63				

pH

Catostomus com-mersoni		1	8.1	56	17–19			Trojnar (1977a)
		1	5.8	61				
		1	5.4	60				
		1	5.0	55				
		1	4.5	16				
		1	4.2	0				

(*continued*)

291

Table I (*Continued*)

Species	Pollutant	Day of exposure	Concentration (μg/l)	Hatch-ability (%)	Viable hatch (‰)	°C	Salinity (%)	Reference
Cyprinodon neva-densis	8.3	1	51					Lee and Gerking (1980)
	7.0	1	32					
	6.5	1	9					
	6.0	1	3					
	5.5	1	0					
Jordanella floridae	6.8	1	44		26			Craig and Baksi (1977)
	6.0	1	57					
	5.0	1	51					
	5.0	1	17					
	4.5	1	0					
Perca fluviatilis	7.3	1	100		14			Runn et al. (1977)
	5.5	1	96					
	5.0	1	70					
	4.5	1	8					
	5.5	14	98					
	5.0	14	44					
	4.5	14	51					
Pimephales pro-melas	7.5	P, 1	78				20–25	Mount (1973)
	6.6	P, 1	80					
	5.9	P, 1	42					
	5.2	P, 1	0					
Rutilus rutilus	7.7	1	89		16			Johansson and Milbrink (1976)
	6.1	2	82					
	5.6	2	40					
	5.2	2	30					
	4.7	2	6					
Perca fluviatilis	8.0	1	53		16			
	5.6	2	12					
	5.1	2	6					

Perca fluviatilis	4.6	2	2		
	4.0	2	0		
	6.4	2	90	15	Rask (1983)
	4.0	2	41		
	3.5	2	0		
Salmo salar	6.8	1	96		
	5.0	1	91		
	4.5	1	92	4	Daye and Garside (1979)
	4.2	1	94		
	4.0	1	89		
	3.7	1	0		
Salmo salar[g]	4.9	1	70		
	4.8	1	60	~10	Lacroix (1985)
	4.7	1	50		
	4.55	1	40		
Salmo salar	6.0–6.8	h	100		
	5.5	h	100		
	5.0	h	97	8	Peterson et al. (1980a)
	4.5	h	~55		
	4.0	h	0		
Salmo trutta[i]	8.0	1	98		
	5.5	1	98	4–6	Johansson et al. (1977)
	4.7	1	90		
	4.2	1	70		
Salvelinus fonti-nalis	8.3	35	90		
	4.75	35	91		
	4.4	35	21		
	8.3	35	100		
	4.75	35	64	10	Swarts et al. (1978) (3 different strains)
	4.4	35	12		
	8.3	35	80		
	4.75	35	75		
	4.4	35	50		

(continued)

Table I (*Continued*)

Species	Pollutant	Day of exposure	Concentration (µg/l)	Hatch-ability (%)	Viable hatch (‰)	°C	Salinity (%)	Reference
Salvelinus fonti-	7.0	P	82		9			Menendez (1976)
nalis	6.6	P	74					
	6.1	P	59					
	5.6	P	54					
	5.1	P	26					
	7.0	1	74					
	6.6	1	65					
	6.1	1	54					
	5.6	1	47					
	5.1	1	44					
	4.5	1	0					
Salvelinus fonti-	C	1	99	95[j]				Kwain and Rose
nalis	6.5	1	95	91[j]				(1985)
	6.0	1	98	96[j]				
	5.5	1	82	70[j]				
	5.0	1	65	58[j]				
	4.5	1	48	0[j]				

[a] Hatchability and viable hatch expressed as percent of successfully inseminated eggs; P, parental exposure prior to spawning; C, control.
[b] Percent of total hatch.
[c] As ppm per 100 g body weight (injected).
[d] Exposure from gastrulation.
[e] Orally dosed, concentration as mg kg^{-1} wet weight.
[f] Two weeks before hatching.
[g] Data extrapolated from graph.
[h] Exposure at eyed stage.
[i] Åvå strain.
[j] Survival to swim-up.

ability of the embryo. In the majority of the cases, whenever hatching is affected, embryos experimentally liberated from the chorion remain in a curled position unable to swim, indicating failures in development that prevented hatching. This is not the case in embryos prevented from hatching in waters of low pH. Within a time limit, embryos not hatching at low pH can be induced to hatch almost immediately after transfer to high pH (see also Rask, 1983). This indicates that only the process of hatching—via inhibition of the chorionase—is affected, while the development of the embryo has proceeded normally—an indication also for a true effect of low pH and not the secondary effect of high metal concentrations in the water due to low pH, as frequently suggested. As a consequence, though, when eggs are left throughout development in low pH, hatching success may be drastically reduced, as shown for Atlantic salmon (*S. salar*) (Daye and Garside, 1977, 1979; Lacroix, 1985), other salmonids (Johansson *et al.*, 1977), and various other species [i.e., white sucker (Trojnar, 1977a), walleye (Hulsman *et al.*, 1983), desert pupfish (Lee and Gerking, 1980), roach and yellow perch (Milbrink and Johansson, 1975); Johansson and Milbrink, 1976; Runn *et al.*, 1977; Rask, 1983)].

Even though hatchability is usually considered a measure of pollutant effects on ontogenesis, it should be remembered that this obscures the fact that within these data substantial numbers of nonviable larvae may be included. Thus reference to the rate of "viable hatch" as a means of assessing sublethal effects of pollutants would be preferred, since only the normal and viable larvae are of concern for recruitment. It should be added that species such as gar pike, *Belone belone*, *Dicentrarchus labrax*, or the tilapias (Paflitschek, 1979) even as larvae are so vigorous that they live with major damage to their skeletal system. Yet when assessing effects on the frail clupeoid and pleuronectid larvae, "viable hatch" is the more sensitive parameter, as can be seen from Table I (Rosenthal and Sperling, 1974; Mazmanidi and Bazhashvili, 1975; Voyer *et al.*, 1977; Ojaveer *et al.*, 1980). Whenever values for viable hatch and hatchability are given simultaneously, viable hatch is substantially lower. In many experiments, even the value for viable hatch overestimates the real figure for viable larvae, since, as already noted, a number of effects may not be detected with the naked eye, but require histological examination—for example, the effects of zinc on brain and muscle tissues of herring (*C. harengus*) (Somasundaram *et al.*, 1984a,b; Somasundaram, 1985). Other metabolic effects might appear after the young fish has passed the larval period. For instance, hatchability of rainbow trout (*S. gairdneri*) is not affected by cromium at concentrations at 0.2 mg/l (van der Putte *et al.*, 1982), yet the survival of alevins is affected after 32 weeks, depending on the pH of the rearing water. The inability for complete and early calcification caused by cadmium treatment is another effect that shows only later in the alevin's life in salmon (*S. salar*) (Rombough and Garside, 1984).

Similar hidden effects that become apparent only after histological examination are known from fish larvae treated with crude oil (Cameron and Smith, 1980; Hawkes and Stehr, 1982), and it is likely that a large number of hitherto undetected effects of pollutants have considerable bearing on the percentage of viable larvae.

III. SUBLETHAL EFFECTS DISPLAYED BY LARVAE HATCHED FROM TREATED EGGS

The larval stage of a fish, although very different from the egg in outer appearance, is not totally different in its physiological state. Any impairment of functions or organs afflicted in the embryo is carried over to the free-living larva. Hatching is a rather arbitrarily determined point, since it may occur at a variety of ontogenetic stages. Many larvae, after emerging from the egg shell, are still incapable of feeding, due to the fact that the mouth is not yet functioning (clupeids, gadids, pleuronectids, cyprinids, and others), or because they are not capable of swimming and remain on the bottom (salmonids) or attached to plants in the water (pike). Species such as found in the family Belonidae, Cyprinodontidae, or members of the mouth-breeding cichlidae are able to feed upon hatching. All may rely for a longer or shorter period on yolk reserves for metabolism. However, for the sake of convenience, the newly hatched larvae are treated together in this section.

A. Larval Length

Reduction of length in newly hatched larvae from eggs incubated under the influence of pollutants (notably heavy metals, petroleum hydrocarbons, and chlorinated hydrocarbons) is a commonly observed feature. Reduced length of newly hatched larvae is frequently correlated with larger yolk-sac sizes, suggesting impaired development. Reduced length in itself is not considered to lower larval fitness.

One characteristic of several abiotic factors influencing fish larvae incubation is the altered size and shape of yolk sac and the length of the newly hatched larvae. Basicly, yolk-sac size and shape as well as larval length change in relation to various abiotic factors, such as incubation salinity and temperature or oxygen. Examples for a decrease in length of newly hatched rainbow trout (S. *gairdneri*) at reduced oxygen tension are given by Hamdorf (1961). Salinity and temperature effects on length of newly hatched larvae are known to occur in the incubation of several fish species such as English sole *Parophrys vetulus* (Alderdice and Forrester, 1968), herring *C. harengus* and *C. pallasi* (von Westernhagen *et al.*, 1974; Alderdice and Velsen, 1971), and garpike *B. belone* (Fonds *et al.*, 1974). Frequently they occur in conjunction with a prolonged (larger larvae) or shortened (short larvae) incubation period until hatching.

Variations in larval size are known from rearing experiments with herring (*C. harengus, C. pallasi*) eggs in cadmium, zinc, and copper. In all effective treatments, larvae hatch early and total lengths are smaller than in controls. Effective concentrations are given by Rosenthal and Sperling (1974) to be 1.0 mg/l for cadmium and 0.1 mg/l in a pulse exposure of copper (Rice and Harrison, 1978), while Ojaveer *et al.* (1980) report that cadmium concentrations as low as 3.0 μg/l reduce larval length. Zinc increases larval total length in concentrations up to 2.0 mg/l; starting at 6.0 mg/l, the length of larvae hatched from zinc incubated eggs decreases (Somasundaram *et al.*, 1984b). Yet under chronic exposure to sublethal levels of zinc, guppies (*Poecilia reticulata*) are very sensitive to levels of only 0.88 and 1.7 μg zinc/l (Uviovo and Beatty, 1979). Offspring produced under these conditions are smaller than controls and have not absorbed the yolk completely, indicating that zinc reduces energy utilization. The authors suggest that zinc has an "uncoupling" effect in the mitochondria, similar to the effect that Hiltibran (1971) has demonstrated in the mitochondria of the bluegill *Lepomus macrochirus* liver. Effects of cadmium on length of newly hatched herring larvae are depicted in Table II, showing decreasing length with increasing cadmium at different salinities. Eaton (1974) also found that incubation of bluegill *L. macrochirus* eggs in cadmium concentrations higher than 0.08 mg/l leads to a shortening of the total length of the hatching larvae.

Reduced length of newly hatched larvae is frequently correlated with larger yolk-sac sizes, as noted for yolk sacs of herring larvae incubated at different cadmium and salinity conditions. Besides the influence of salinity on yolk-sac size (May, 1974a; Alderdice and Velsen, 1971), these findings indicate low yolk utilization under cadmium exposure (Rombough and Garside, 1982; S. *salar*). Reduced length of newly hatched larvae is also caused by the exposure of the

Table II

Clupea harengus Larvae: Total Length, Diameter of Eye, and Otic Capsule at Hatching[a]

Experimental design		Total length (mm)				Otic capsule (mm)				Eye diameter (mm)			
S (‰)	Cd. conc.	n	$\bar{x}$	s	$s_{\bar{x}}$	n	$\bar{x}$	s	$s_{\bar{x}}$	n	$\bar{x}$	s	$s_{\bar{x}}$
5	Control	80	8.27±0.34		0.04	80	0.327±0.020		0.002	55	0.288±0.018		0.002
5	0.1 ppm	91	7.77±0.54		0.06	89	0.316±0.024		0.002	64	0.280±0.011		0.001
5	0.5 ppm		Not measurable			171	0.313±0.026		0.002	125	0.251±0.016		0.001
5	1.0 ppm		Not measurable			179	0.314±0.025		0.002	137	0.252±0.018		0.002
5	5.0 ppm		Not measurable			33	0.212±0.039		0.007	39	0.220±0.016		0.003
16	Control	100	7.78±0.58		0.06	100	0.314±0.023		0.002	76	0.287±0.014		0.002
16	0.1 ppm	62	7.88±0.62		0.08	65	0.311±0.025		0.003	41	0.284±0.017		0.003
16	0.5 ppm	93	7.70±0.51		0.05	93	0.310±0.021		0.002	43	0.285±0.009		0.001
16	1.0 ppm	105	7.13±0.38		0.04	129	0.301±0.027		0.002	129	0.283±0.014		0.001
16	5.0 ppm		Not measurable			141	0.244±0.033		0.003	100	0.258±0.020		0.002
25	Control	50	7.98±0.17		0.02	49	0.311±0.015		0.002		0.288		
25	0.1 ppm	63	7.90±0.46		0.06	62	0.312±0.021		0.003	63	0.277±0.018		0.002
25	0.5 ppm	72	7.88±0.41		0.05	71	0.315±0.017		0.002	46	0.289±0.010		0.001
25	1.0 ppm	91	7.48±0.60		0.06	91	0.311±0.034		0.004	91	0.293±0.011		0.001
25	5.0 ppm		Not measurable			102	0.244±0.062		0.006	82	0.263±0.024		0.003
32	Control	50	7.05±0.28		0.04	53	0.279±0.027		0.004	25	0.274±0.008		0.002
32	0.1 ppm	54	7.00±0.34		0.05	56	0.286±0.020		0.003	50	0.265±0.020		0.001
32	0.5 ppm	61	6.94±0.42		0.05	63	0.285±0.027		0.003	38	0.268±0.014		0.002
32	1.0 ppm	74	6.82±0.41		0.05	73	0.288±0.022		0.002	74	0.272±0.018		0.002
32	5.0 ppm		Not measurable			35	0.178±0.035		0.006	45	0.234±0.019		0.003

[a] n, Number of larvae measured; $\bar{x}$, mean; s, standard deviation; $s_{\bar{x}}$, error of the mean. Larvae derived from incubation trials in 32‰ salinity originated from a second female. After von Westernhagen et al. (1974).

eggs to petroleum hydrocarbons, usually applied as the water-soluble fraction (WSF) of crude oil or its derivates. Thus Pacific herring (*C. pallasi*) eggs exposed in a 48-h pulse exposure of 12.1 mg benzene/l yield larvae with a mean standard length of 9.2 mm, compared to 10.3 mm in controls. The same is true for anchovy (*E. mordax*) eggs exposed to 40–55 ppm (Struhsaker *et al.*, 1974). Also, Baltic herring (*C. harengus membras*) eggs exposed to 5.4–5.8 mg/l total oil hydrocarbons yield significantly shorter hatching larvae than controls (Linden, 1978). The same effects are described by Carls and Rice (1984) after exposing eggs of the walleye pollock *T. chalcogramma* and embryos of the Japanese medaka *O. latipes* to the WSF of oil (Leung and Bulkley, 1979). In the walleye pollock, reduction in length amounts to 0.5 mm at a total length of only 4.5 mm. In the killifish *F. heteroclitus*, exposure of eggs to the WSF of number 2 fuel oil leads to a shortening of hatching larvae with increasing strength of the applied WSF (Sharp *et al.*, 1979; Linden *et al.*, 1980). In this species the latter authors note a simultaneous decrease in the number of vertebrae. Reduced lengths of newly hatched larvae are also known to occur after treatment of cod (*G. morhua*) eggs with DDT and DDE. Within the range of DDT applied (0, 0.0095, 0.0413, 0.09, 0.15, 0.39, 0.69 mg/l), emerging larvae are progressively smaller with increasing insecticide concentrations; the mean total length is only 4 mm at the highest DDT concentration, while control larvae measure 4.75 mm (Dethlefsen, 1977). A reduction in length of yolk-sac fry of pike *Esox lucius* incubated in concentrations as low as 0.1 ng 2,3,7,8-tetratchlorodibenzo-*p*-dioxin (TCDD)/l is also known through the experiments of Helder (1980).

The significance of the hatching size for the fitness of fish larvae is not clear, although it is generally accepted that it is of disadvantage for the larvae to hatch small. This assumption depends on an "uneasy feeling" rather than on facts. Swimming velocity, for instance, does not seem to be altered significantly in different-size herring larvae within 7.0 to 11.0 mm total length (von Westernhagen and Rosenthal, 1979), and thus prey catching behavior will not be impeded. Of course, smaller larvae have a smaller range of food availability, since organs such as eyes and otic capsules (von Westernhagen *et al.*, 1974) as well as the head and jaw apparatus are smaller, thus limiting the choice of food particles. However, the limitation to smaller food particles does not reduce survival if enough food is available, as shown by my unpublished data on larvae of *Blennius pavo*.

It seems reasonable to argue that it is not the absolute size of a larva but its size in relation to its ontogenetic stage of development and the remaining yolk volume that is important for survival. If the embryo encounters unfavorable conditions at the end of its intrachorion development and it "decides" to hatch prematurely, as reported for the effects of several abiotic factors including low oxygen, the prematurely liberated larva, unless otherwise damaged, is generally only shorter than normal, but is holding a larger yolk sac. If the causative agent of premature hatch does not have any lasting effects, development will proceed normally.

B. Yolk-Sac Size and Yolk Metabolism

A large or deformed yolk sac is taken as an indicator for metabolic or osmotic disturbances that may be caused by mitochondrial malfunction, induced by heavy metals or petroleum hydrocarbons.

It is not always true that a large yolk sac at hatching is due to premature hatching. There are indications that a large yolk sac occurs because of metabolic or osmotic disturbances in the embryo/larva that prevent proper use of the energy stored in the yolk. Lönning (1977), from observations of cod, plaice, and flounder eggs (*G. morhua, Pleuronectes platessa, Platichthys flesus*) exposed to Ekofisk oil, thinks that the use of the energy-rich substances in the yolk becomes delayed by an inhibition of the mitochondrial system. For instance, newly hatched larvae from Pacific herring *C. pallasi* eggs exposed to Prudhoe Bay crude oil for 4–144 h and then returned to uncontaminated seawater show no gross abnormalities. Yet transmission microscopy of exposed organisms reveals inter- and intracellular spaces in brain and muscle tissue that are not found in controls (Cameron and Smith, 1980). Many mitochondria (13%) in the body muscle of exposed animals are swollen, some with deteriorating cristae. Changes in mitochondrial functions would affect the total respiration and metabolism of the larvae and thus explain the previously mentioned general suppression of embryo activity and metabolism after prolonged exposure to petroleum hydrocarbons. Linden *et al.* (1980) inferred that at low hydrocarbon levels, when the homeostatic mechanisms are not overwhelmed, respiration rates reflect increased costs of homeostasis. When the stress is more severe, but still sublethal, the response would be mediated by lack of metabolic integration because of poorly functioning homeostatic mechanisms. They believe that exposure to oil predominately impedes mobilization of nutrient uptake from the yolk through the breakdown of mitochondria in the cells, leading to glycogen and lipid depletion such as demonstrated by Sabo and Stegeman (1977). This would ultimately lead to reduced tissue growth, as found in embryonic herring *C. pallasi* exposed to sublethal concentrations of benzene (Eldrige *et al.*, 1977). Yolk utilization is likewise reduced by exposure to sodium pentachlorophenate (Na PCP), as shown by Chapman and Shumway (1978), for example, in the alevins of steelhead trout *S. gairdneri*. The bioenergetic data obtained in their study are consistent with the concept that PCP disrupts energy metabolism. This is particularly deleterious to the early larva, since energy requirements increase rapidly by approximately tenfold shortly after hatching (Eldridge *et al.*, 1977; *Clupea pallasi*). Struhsaker *et al.* (1974) relate the impaired utilization of yolk at high concentrations of benzene to an increasing narcotization of the animal; this aspect will be treated later in a different context. Inhibited yolk utilization is also known under conditions of low pH. Pike *E. lucius* larvae incubated at pH 4.2 are smaller than controls but have larger yolk sacs, which persist for 14 days, indicating poor utilization of yolk reserves. Yolk frequently appears coagulated, ultimately leading to death of the affected fry (Johansson and Kihlström, 1975). Retarded yolk absorption at pH 5 is also known in brook trout (*S. fontinalis*) alevins (Menendez, 1976).

Frequently, the outer appearance of the yolk shows signs of abnormality or one notices the appearance of an empty space between yolk sac and yolk (Mazmanidi and Bazhashvili, 1975; *Platichthys flesus luscus*) or anteriorly adjacent to the pericardium as demonstrated by Linden (1978) for newly hatched herring (*C. harengus membras*) larvae. These features are similar to those observed by von Westernhagen (1970) and Alderdice and Velsen (1971) when marine fish larvae are incubated outside their optimum temperature and salinity regimes, thus indicating additional stress on the larvae, resulting in failure of osmoregulatory functions.

Cadmium and zinc are also known to interfere with osmoregulation in Pacific herring (*C. pallasi*) eggs (Alderdice *et al.*, 1979c). Cadmium exposure of eggs reduces osmolality of perivitelline fluid, probably due to the marked tendency of cadmium to form complexes (Remy, 1956), particularly with iodide, bromide, and chloride ions. The effect of the formation of these complexes is to marshal other ions into complex formation [e.g., $Cd(I_3)_2$], reducing the number of active particles in solution, thus reducing osmotic pressure. Rosenthal and Sperling (1974) report a disproportionate shortening of the yolk sacs of cadmium exposed herring *C. harengus* larvae at cadmium levels of 5.0 and 10.0 mg/l, which may be caused by reduced perivitelline fluid turgor.

Zinc has similar effects on fish larvae. Somasundaram *et al.*

(1984d), working with *C. harengus,* speculate, on the basis of their histological investigations, that the swelling of mitochondria and sarcoplasmic reticulum caused by zinc treatment suggests the creation of an osmotic imbalance through zinc. In mammals zinc causes swelling of mitochondria and appears to alter potassium permeability, uncouples oxydative phosphorylation, and inhibits the electron transport chain (Cash *et al.,* 1968; Kleiner, 1974; Bettger and O'Dell, 1981). Uncoupling of oxydative phosphorylation will ultimately lead to an energy deficit, even though the embryo may compensate with increased decomposition of carbohydrates (Stelzer *et al.,* 1971); this would have its bearing on the osmoregulatory capacities of the embryo.

C. Morphological Aberrations: Eye Deformities, Skeletal Abnormalities

Gross malformations such as eye deformation and reduction, as well as skeletal deformities, are caused by all types of pollutants and are not pollutant-specific. Typical anomalies, which may also be caused by extreme temperatures and salinities, are spirality and curvature of the notochord and abnormal development of the jaw. The severity of the effects can be generally related to the doses applied and diminishes with exposure during later stages of development. Since cadmium interferes with calcium metabolism, it is suspected to impair the calcification process directly. Petroleum hydrocarbons probably act as general stressors and do not have a specific effect on any enzyme or physiological process.

Besides aberrations of length and yolk usage, newly hatched larvae display a vast array of gross deformities, such as lack of organs, extremities, etc. and/or abnormal behavior due to the action of pollutants during their embryonic stage. These malformations and behavioral aberrations may play major or minor roles in their survival. In general, aberrations that stem from tissue injury or enzyme inhibition during earlier stages can be categorized as follows: various types of eye deformation or reduction, jaw anomalies, malformations of the vertebral column, minor morphological aberrations (i.e., fin defects, otic capsule defects, change in color pattern), impairment of swimming and prey catching behavior, and reduced growth. Some of these have obvious effects on survival. The bearing of others may be insignificant or difficult to recognize.

Eye deformations are common in fish larvae exposed to sublethal levels of stressors such as heavy metals. Microphthalmia (Ojaveer *et al.*, 1980) in Baltic herring incubated in copper and cadmium solutions of >0.01 and 0.05 mg/l have been noted, as well as cyclopia in cod (*G. morhua*) reared in sublethal (0.01–0.5 mg Cu/l; 0.5–10.0 mg Zn/l) concentrations of copper and zinc (Swedmark and Granmo, 1981). These gross abnormalities are similar to those already described for the effects of mercury on killifish (*F. heteroclitus*) embryos, by P. Weis and Weis (1977). The spectrum of eye and head defects produced in *F. heteroclitus* by exposure to inorganic mercury reflects interference with inductive processes at a relatively early stage. The severity of the response diminishes with exposure during later stages of development (J. S. Weis and Weis, 1977). *Fundulus heteroclitus* appears to have a propensity for this type of malformation; further, Stockard (1907) produced cyclopia in this species by treatment with magnesium chloride. It is interesting to note that optic abnormalities are not produced in *F. heteroclitus* by exposure to insecticides (Weis and Weis, 1974).

Other subtle deviations from the normal are displayed by herring (*C. harengus*) embryos exposed to cadmium (Rosenthal and Sperling, 1974; von Westernhagen *et al.*, 1974). Exposure to cadmium concentrations higher than 1.0 mg/l leads to a reduction in eye diameter (Table II). Also, Somasundaram *et al.* (1984a) showed a significant reduction in eye size at a concentration of 6 and 12 mg/l of zinc in herring (*C. harengus*), even considering reduced length at hatching. The same is true for the reduced otic capsule diameter. Gross eye deformations are one of the typical effects occurring after sublethal exposure to metals and other teratogenic compounds such as benzo[a]pyrene (BAP). When exposing rainbow trout (*S. gairdneri*) to the mutagen BAP, Kocan and Landolt (1984) and Hannah *et al.* (1982) always find gross physical defects in the ocular and cephalic region of larvae similar to those resulting from exposure to heavy metals. Skeletal and cephalic abnormalities of newly hatched fish, encountered most frequently upon exposure to BAP, are believed to be caused by the mutagenic action of the BAP. Implied possible mutagenic action of copper and cadmium is not supported by experimental evidence, and thus the question remains open.

Another sublethal stressor affecting the eyes of young fish is low pH. Daye and Garside (1980), when sectioning alevins from exposed eggs of Atlantic salmon (*S. salar*), found that incubation at pH 4.0 yields alevins with eye lens fibers less differentiated than that of controls. The lenses also suffer severe sloughing of epithelium, a common pathologic change due to acid environment. Anatomically, the prime sites of injury are the superficial tissues. Internal structures are affected secondarily, both in time and degree.

A wide array of skeletal malformations (i.e., jaw, head, pelvic and pectoral girdle, vertebral, and opercular anomalies) occurs commonly in freshwater and marine fish species (Wunder, 1971; Kroger and Guthrie, 1973; Dethlefsen, 1980, 1984). Accordingly, one would also expect these anomalies to occur in fish larvae, and this is the case under laboratory conditions. For example, anomalous formation of the jaw, mentioned during earlier investigations, is caused by extreme temperatures and salinities in several fish species [von Westernhagen (1970, 1974), *Platichthys flesus*, *B. belone;* Alderdice and Velsen (1971), *C. pallasi*]. These anomalies are also caused by sublethal effects of metal pollutants and are likely to be expressed as cranio–facial and mandibular malformations; particularly as a reaction toward mercury (Weis, 1984; *F. heteroclitus*) and zinc (Somasundaram *et al.*, 1984a; *C. harengus*). Different types of malformations of the head region are found in herring (*C. harengus*) larvae (Fig. 7). Due to the incomplete development of the feeding apparatus of many fish larvae at the time of hatching (mouth opening still closed), symptoms of jaw defects are not always immediately detectable, in particular when experimentally hatched larvae are not given additional time to develop their mouth apparatus before assessment. Figure 7 shows several types of jaw deformations in herring. These differ depending on the stage of development. Jaw deformations are also known to occur "spontaneously" in hatchery enterprises in North America. The open-jaw syndrome of salmonids is of particular concern in hatchery-reared salmon (Crouch *et al.*, 1973). Jaw deformities may also result from the treatment of eggs with crude oil (Tilseth *et al.*, 1984; Solberg *et al.*, 1984). At concentrations of about 150–1245 μg/l (WSF), cod (*G. morhua*) larvae suffer from deformation of the upper jaw, which may have a later bearing on feeding. Also, short-term exposure of newly spawned Pacific herring (*C. pallasi*) eggs for 24–96 h at concentrations of 4800–45,000 μg benzene/l causes severe anomalies in the head region of hatching larvae, including jaw deformations (Struhsaker *et al.*, 1974). Similar pictures of the head region of Baltic herring larvae after treatment with the WSF of crude oil (up to 59,000 μg hydrocarbons/l) are given by Linden (1978). Also, when exposing 6-day-old Pacific herring embryos to the WSF of Prudhoe Bay oil at concentrations of around 1000 μg total hydrocarbons/l for only 48 h, advanced larvae display a high incidence of gross morphological abnormalities, such as improperly formed mouth, misfit of the lower jaw into the upper, missing of the premaxillary bone, and failure of the jaw to fully

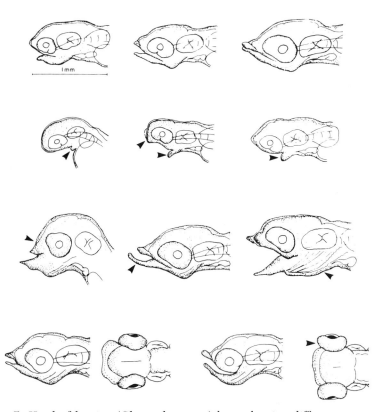

Fig. 7. Head of herring (*Clupea harengus*) larva showing different types of jaw malformations after incubation in Zn-contaminated water. Upper row, normal 1, 2, 5 days old; second row, rudimentary lower and/or upper jaw (arrow); third row, secondary pugheadedness, cross bite, and protrusion of branchial arches (arrows); fourth row, normal lateral and dorsal view, and same view of larva with exophthalmus. [From Somasundaram *et al.* (1984a).]

differentiate (Smith and Cameron, 1979). Another abnormality noticed only under the electron microscope was the absence of branchiostegal membranes, a phenomenon observed also by von Westernhagen *et al.* (1987) after treatment of herring embryos with surface microlayer hexane extracts.

Abnormal development of the jaw is seen in larvae exposed as eggs to pesticides like the moluscicides Bayluscid and Lebaycid (Paflitschek, 1979; *Tilapia leucosticta, Heterotilapia multispinosa*) or as a result of high PCB content (2.8 μg/g wet wt) in eggs of rainbow trout (*S. gairdneri*) (Hogan and Brauhn, 1975). In the natural environment these jaw anomalies are likely to interfere severely with feeding and thus reduce survival. In the most severe cases, starvation will follow the inability to feed. In my observations, I frequently noticed that swimming is greatly impaired in gaping-mouth larvae of marine fish, probably due to increased water resistance.

Mouth and jaw anomalies constitute a relatively small part of the gross abnormalities occurring after treatment of eggs with pollutants. The bulk of the symptoms observed are related to axis formation. Injury to the vertebral column or its anlage in response to pollutants is commonly seen by most investigators working with fish eggs and larvae. The range of damage is extensive: from very slight flexures to bends or spiral distortions, shortening of the body axis, or reductions of the brain. Within the range of possible damage, none of the substances seem to cause substance-specific damage, which can undoubtedly be attributed to a particular pollutant. Responses seem to be general and ubiquitous without regard to the stressor. Exposure time and substance concentration influence the severity of the symptoms. The physical appearance of affected larvae resembles that of individuals incubated under natural stress of extremes of temperature and salinity, as shown by von Westernhagen (1970) and described by several other authors.

Damage of the vertebral column expressed as curvature of the larval body axis is caused by all metals currently termed "heavy metals" when present in the incubating medium. The most common metal pollutants are cadmium, copper, mercury, lead, and zinc, employed singly or in combinations at concentrations of a few micrograms per liter in the case of the acutely toxic metals such as mercury, and up to several thousand micrograms per liter with metals such as lead or zinc.

The toxic levels of the different metals differ and depend on factors such as susceptibility of fish species, temperature and salinity, or chemical speciation of the metal. Thus cadmium causes vertebral damage in developing fish eggs at concentrations between 80 μg/l (Eaton, 1974; *Lepomis macrochirus*) and 300 μg/l (Rombough and Garside, 1982; *S. salar*) in fresh or brackish water (Voyer *et al.*, 1977; *Pseudopleuronectes americanus*), but at higher concentrations of between 1000 and 2000 μg/l in seawater (von Westernhagen *et al.*, 1974, 1975; *C. harengus, B. belone*). Since cadmium interferes with calcium metabolism—cadmium replacing calcium—the effect of cadmium on vertebrae formation might be a direct one, as suggested by the investigations of Rombough and Garside (1984) observing the impairment of the calcification process in Atlantic salmon alevins. Earlier effects of the metal, prior to ossification, must be considered general effects on metabolism. Copper is also known to be effective in producing distorted larvae at fairly low concentrations. At 30 μ copper/l, 30% of the hatching herring larvae are deformed (Blaxter, 1977). At the same level zebrafish (*B. rerio*) develops scoliosis (Ozoh, 1979a), and for *Dicentrarchus labrax* (sea bass) this level is only slightly higher (100 μg/l) (Cosson and Martin, 1981). The initial concentrations for mercury causing sublethal effects on axis formation are even lower. J. S. Weis and Weis (1977), in experiments with the killifish (*F. heteroclitus*), prove that concentrations of only 10 μg Hg/l cause lordosis and scoliosis in newly hatched larvae. Although affected larvae are still able to swim, movements are impaired. Experiments of Sharp and Neff (1980, 1982) and P. Weis and Weis (1977) with *F. heteroclitus* confirmed the low effective concentrations for mercury that cause similar effects to those described above. Still lower detrimental concentrations of mercury were reported by Servizi and Martens (1978) with sockeye (*O. nerka*) and pink salmon (*O. gorbuscha*) eggs. Apparently, mercury concentrations above 2.5 μg/l increase vertebral deformities; at 4.3 μg mercury/l, 46% of the alevins are crippled with impaired swimming.

Exposure of brook trout (*Salvelinus fontinalis*) over three generations shows that lead is also an effective teratogen at low concentrations. Alevins of the third generation from eggs exposed to 119 μg lead/l display 21% scoliosis, compared to only 2% in the controls (Holcombe *et al.*, 1976). In short exposures during embryonic development, lead causes sublethal effects on axis formation, but concentrations must be around 1000 μg/l (J.S. Weis and Weis, 1977; *F. heteroclitus*). Small aberrations from normal axis formation caused by zinc are first detectable at the micrograms per liter level. Slight bends of the tail tip of herring larvae at hatching can be observed after incubation at 50 μg/l (Somasundaram *et al.*, 1984a). Animals with this type of deformity still swim like normal larvae. With increasing zinc concentrations, damage to the vertebral column becomes more severe. Further information on damage of cod (*G. morhua*) and herring (*C. harengus*) larvae hatched from zinc-exposed eggs are given in reports by Swedmark and Granmo (1981) and Ojaveer *et al.* (1980); effective concentrations for herring are in the same range, although higher for cod. A series of photographs of various degrees of spinal malformations caused by metal is given in Fig. 8 from our own experiments with herring and flounder. In the herring larva (Figs. 8b,8c), the damage done to the vertebrae is visible.

Disturbed axis formation is also common in larvae incubated in solutions of petroleum hydrocarbons—although not a typical phenom-

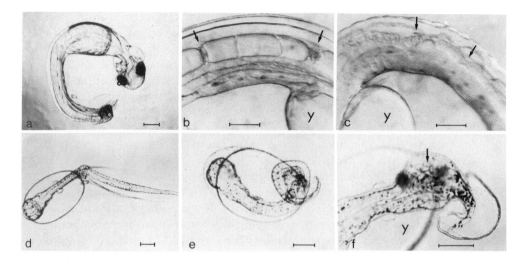

Fig. 8. Newly hatched larvae of herring (*Clupea harengus*) and flounder (*Platichythys flesus*) incubated in water contaminated with 5 mg cadmium/l. (a) Crippled herring larva with exophthalmus; (b, c) herring larvae with damaged or almost disintegrated notochord (arrows); (d–f) bent and severely crippled flounder larvae; y, yolk. Horizontal bars indicate 200 μm.

enon. According to the Sharp *et al.* (1979) interpretation of their experimental results with oil-exposed killifish (*F. heteroclitus*) embryos:

Hydrocarbon pollutants act in fish embryos as general stressors and do not have a specific effect on any single enzyme or physiological process. Thus hydrocarbon pollutants may shunt limited metabolic energy away from critical differentiation and morphogenetic processes to maintenance functions.

The depressant or retarding effects of hydrocarbons during early development may be reflected in the effects on gross morphology of emergent fry, where the degree of the effects depends on the strength of the hydrocarbon applied. A variety of species has been used in exposure studies with hydrocarbons (mostly WSF of crude oil), one of the most common being herring (*C. harengus*). Effects of 680 μg/l WSF for 48 h on developing herring (*C. harengus pallasi*) eggs leads to a significantly higher incidence of gross morphological abnormalities than in controls (Smith and Cameron, 1979). Most of the abnormalities are bent vertebral columns leading to larvae with *L, S*, or helical configurations. Affected larvae are usually unable to swim in a straight line or not able to swim at all. The same aberrations have been provoked by Linden (1976, 1978) using somewhat higher concentrations, 3100 to 11,900 μg/l, and by Struhsaker *et al.* (1974) using benzene. The eggs of several other species have been subjected to the same or similar treatment using crude oil or other petroleum hydrocarbons. Thus Mironov (1969) used anchovy *Engraulis encrasicolus ponticus;* Häkkilä and Niemi (1973) used northern pike *Esox lucius;* Kühnhold (1974) used cod *G. morhua;* Mazmanidi and Bazhashvili (1975) used flounder *Platichthys flesus luscus;* Stoss and Haines (1979) Japanese medaka *Oryzias latipes;* and Kühnhold *et al.* (1978) used the winter flounder *Pseudopleuronectes americanus.* All authors report deformation of axis to a greater or lesser extent. Effective concentrations of the WSF of petroleum hydrocarbons are in the range of 100 μg/l (Mazmanidi and Bazhashvili, 1975) to 4000 μg/l (Stoss and Haines, 1979). When combining oil with oil-spill dispersants (Wilson, 1972; Linden, 1974, 1976), deleterious effects on axis formation are usually aggravated, reaching levels known from the teratogenic effects of benzo[a]pyrene (24 μg/l; Winkler *et al.*, 1983) on notochord abnormalities in the grunion *Leuresthes tenuis.*

Body flexure occurs also in newly hatched larvae from eggs containing chlorinated hydrocarbons such as DDT [Dacre and Scott (1971); *S. gairdneri;* Smith and Cole (1973), *Pseudopleuronectes americanus*] in the range of 2.4–4.6 mg/kg wet weight. Hogan and Brauhn (1975) assume that the occurrence of 60–70% deformed rainbow trout in a trout hatchery was caused by the high PCB (Aroclor 1242) content (2.7 μg/g) in the eggs. When exposing eggs to these substances in incubating water at concentrations of about 100 μg DDT/l (Dethlefsen, 1977; *G. morhua*) or only 13 μg PCB/l (Mauk *et al.*, 1978; *Salvelinus fontinalis*), a considerable percentage of the hatching larvae displays curvature of the body in different degrees. The severity of these effects increases with the concentration of the organochlorine employed. From the data on viable hatch of cod under the influence of DDT it appears that 10 μg DDT/l or higher already reduces viable hatch. Similar effects can be caused by other pesticides, such as malathion, but at considerably higher concentrations of 10,000 μg/l for sheepshead minnow (*Cyprinodon variegatus*) eggs (Weis and Weis, 1976) or carp (*C. carpio*) eggs treated with various herbicides (Kapur and Yadav, 1982).

Occasional hatching of crippled and distorted larvae is also observed when eggs are incubated at low pH. Thus at pH 4.0 to 4.5, perch *Perca fluviatilis* hatching is delayed and the few (3%) larvae hatched show vertebral deformations (Runn *et al.*, 1977). This effect is known to occur at pH 5.0 in white sucker *Catostomus commersoni*

(Trojnar, 1977a). In Pacific herring (*Clupea pallasi*) eggs subjected to low pH during development (pH 6.7), hatching is reduced to almost zero although a few bent individuals emerge (Kelley, 1946). Runn *et al.* (1977) judged these deformities to be secondary and not caused by a disturbance of the early organogenesis but developing during the prolonged nonhatch period at low pH—possibly aggrevated by the smaller inner volume of the egg and the reduction of the diffusion of metabolites through the perivitelline fluid. The same phenomenon is known for cod (*G. morhua*) embryos that fail to hatch. When the fully developed larva (with jaws and no more yolk) is liberated by dissection, it remains curled in an embryonic posture (von Westernhagen, 1970), indicating that the growth of the larva inside the egg was responsible for the malformation.

D. Minor Morphological Aberrations

Aside from the above-mentioned obvious gross malformations, an array of minor deformities and deficiencies are known to be caused by pollutants. These cannot be dealt with in detail, although they may represent the real sublethal effects at the individual level. In other words, effects of the pollutants may be expressed as minor changes not large enough to cause immediate or ultimate death, but large enough to reduce overall fitness. Typical effects may follow different kinds of treatment. Thus, fin erosion or sloughing of epithelial tissue occurs after exposing embryos to cadmium, copper, zinc, or lead (von Westernhagen *et al.*, 1975; Ozoh, 1979a; Somasundaram, 1985), as well as after incubation in water contaminated with petroleum hydrocarbons (Kühnhold, 1972; Linden, 1975; Smith and Cameron, 1979; Vuorinen and Axell, 1980) or rearing in organochlorines or organophosphates (Paflitschek, 1979; Helder, 1980). Other symptoms are impaired blood circulation or blockage of blood vessels leading to thrombosis, as seen in larvae hatching from eggs incubated in cadmium solutions (Pickering and Gast, 1972; Eaton, 1974; Beattie and Pascoe, 1978), toluene (Stoss and Haines, 1979), or tetrachlorodibenzo-*p*-dioxin (TCDD) (Helder, 1980). The same effects are known to be caused in fish larvae by the carcinogen benzo[a]pyrene (Winkler *et al.*, 1983). Very common is the poor development of pigmentation caused by cadmium (Ozoh, 1980) or exposure to petroleum hydrocarbons (Mazmanidi and Bazhashvili, 1975; Anderson *et al.*, 1977; Johnson *et al.*, 1979; Falk-Petersen *et al.*, 1985), malathion (Weis and Weis, 1976), or low pH (Johansson and Kihlström, 1975; Nelson, 1982).

E. Metabolic Alterations

One major subcellular effect caused by metals and oil is an alteration of the internal structure of mitochondria, which leads to an impairment of the intracellular energy transfer system. Blockage or inhibition of this system may be the cause for inadequate use of yolk
reserves and retarded development.

More subtle impact of pollutants (stressors) may easily escape attention or resist interpretation. Recent studies show that there is still a
large number of phenomena that are now being looked at more closely
but are not yet fully understood. This is particularly true for events at
the subcellular levels, which are known to be affected and in turn to
affect the whole organism. Thus herring embryos, incubated in zinc-
polluted water, show, in addition to signs of epithelial necrosis,
changes in mitochondria structure, absence of the Golgi apparatus,
and reduction in smooth endoplasmic reticulum (Somasundaram,
1985). Because of reduced mitochondria internal active surface, cell
metabolism might be impaired and total energy budget of the animal
affected (Somasundaram *et al.*, 1984c); zinc is known to interfere with
oxygen uptake of mitochondria (Hiltibran, 1971) and might cause uncoupling of oxidative phosphorylation and inhibition of the electron
transport chain (Kleiner, 1974; Bettger and O'Dell, 1981), a theory
also forwarded by Uviovo and Beatty (1979).

Similar effects on mitochondria are caused by xylene on the earliest cleavage stages of cod (*G. morhua*) eggs (Kjörsvik, 1986) and the
WSF of crude oil on herring larvae hatching from oil-treated eggs
(Cameron and Smith, 1980). Enzyme activity in brook trout *S. fontinalis*, as shown by Christensen (1975), is also greatly affected by
metals such as cadmium, mercury, and lead. Activity of glutamic–
oxaloacetic transaminase (GOT), alkaline phosphatase (ALP), acetylcholine esterase (ACH), and adenosine triphosphate (ATP) have been
either significantly decreased in late embryos or increased in alevins.
Probably several malformations and developmental aberrations are
ultimately caused by a blockage of the energy-transfer system, leading
to an arrest of respiration and differentiation, or to dedifferentiation.
Inhibition of acetylcholine esterase in neuromuscular and brain tissue, for instance, as demonstrated to occur in rainbow trout (*S.
gairdneri*) exposed to organophosphate pesticides (Matton and Lattam, 1969) or in *Cyprinodon variegatus* treated with malathion (Weis
and Weis, 1976), will severely impair locomotion and/or cause death
of the organism by asphyxia.

Blockage or inhibition of intracellular energy-transfer systems or

shunting energy from differentiation metabolism to detoxification processes may also be the cause for the commonly observed retardation
in growth and the inability of the yolk sac larvae to use yolk reserves
adequately. It has already been noted that herring (*C. harengus*) larvae hatching in cadmium-contaminated water had larger yolk sacs
than those from controls (von Westernhagen *et al.*, 1974). Obviously,
resorption of yolk under the influence of metal stress is impaired. This
impairment of yolk utilization continues in the larval stage. Thus
brook trout *S. fontinalis* alevins incubated in copper-contaminated
water (32.5 μg/l) take 4 weeks longer to complete yolk resorption and
remain smaller than controls (McKim and Benoit, 1971). Impairment
of yolk utilization on incubation in cadmium-contaminated water also
occurs in Atlantic salmon alevins when reared through yolk absorption (Rombough and Garside, 1982). At 9.6°C concentrations of 0.47
μg Cd/l impaired yolk utilization to the extent that final weight of
alevins was significantly reduced when compared to controls. The
same is known for salmon fry reared at 40–55 μg copper/l (Hazel and
Meith, 1970; Servizi and Martens, 1978), and for the young of the
zebrafish *B. rerio* under the influence of lead (Ozoh, 1979a). The lasting effect of cadmium on yolk utilization is shown in rearing experiments with Atlantic salmon (*S. salar*). When reared in concentrations
of 2 μg/l, the fish display reduced growth, which continues even after
initiation of feeding (Peterson *et al.*, 1983).

Impaired yolk utilization has also been reported as an effect of
petroleum hydrocarbons on embryos of winter flounder *Pseudopleuronectes americanus* (Kühnhold *et al.*, 1978) and the killifish *F.
heteroclitus* (Sharp *et al.*, 1979). The effect of these pollutants is not
entirely a direct one; these substances express their activity through
storage in the lipid reserves of the yolk and are later mobilized during
yolk absorption. This system is particularly active with lipophilic substances such as the chlorinated hydrocarbons. Thus larvae of fathead
minnows *Pimephales promelas* hatched after exposure to 15 μg PCB/l
are severely retarded in growth compared to controls (Nebeker *et al.*,
1974), and this was also reported by Halter and Johnson (1974) working with Pacific salmon *O. nerka* eggs and Aroclor 1254. Hogan and
Brauhn (1975) observed that hatchery-reared rainbow trout *S.
gairdneri* fry display a high percentage of deformed animals with a
variety of skeletal abnormalities and impaired yolk utilization. On
chemical analysis, the eggs of parental fish showed 2.7 μg PCB/g egg
wet weight and 0.09 μg DDT/g. Also, *S. salar* incubated in DDT show
retarded alevin development, particularly in their behavior (Dill and
Saunders, 1974). Another substance, sodium pentachlorophenate

(Chapman and Shumway, 1978), when applied at 40 µg/l, decreased yolk utilization, growth, and development in steelhead trout *S. gairdneri*.

The significance of impaired yolk utilization is obvious. With a given yolk reserve, larvae must develop to a certain ontogenetic stage in a given time span. If this is not attained, the larva is likely to find itself in an environment for which it is not yet prepared (swimming speed, orientation); in the case of salmonids, larvae may emerge and fall prey to larger predators. In the limited environment of a body of fresh water, the proper timing might be crucial for survival. In the sea this factor might not be of paramount importance, but its bearing should not be underestimated.

F. Behavioral Abnormalities

Eggs incubated under the influence of metals or petroleum hydrocarbons may release larvae with reduced activity. While effects of metals are long-lasting, petroleum hydrocarbons act twofold; transiently, with subsequent recovery, and permanently, when histopathological damage took place. Also, the high chlorinated hydrocarbon contents of larval yolk may be responsible for reduced activity. Lowered larval activity is an indication of reduced fitness.

Several authors assume that an additional reason for slow yolk utilization is the reduced activity of larvae. Frequently, larvae hatching from exposed eggs lie motionless on the bottom of the experimental containers or perform only sluggish movements not equal to normal swimming activity. Hatched larvae that remain immobile often come from eggs incubated in high concentrations of copper, zinc, or cadmium (Eaton, 1974; Swedmark and Granmo, 1981); the reason for their immobilization is not quite clear. Voyer *et al.* (1982) describe winter flounder larvae *Pseudopleuronectes americanus* incubated in cadmium concentrations of up to 100 µg/l that show reduced swimming activity only in low salinity (10‰ salinity). This response suggests a potential long-term effect on larval feeding, growth, and susceptibility to predation.

The reasons for the reduced activity of metal-treated larvae are definitely different from those that occur after treatment with hydrocarbons. The effects of hydrocarbons on embryo activity can be observed also in newly hatched larvae. Crude oil fractions (WSF) are very effective in reducing activity. Hatched embryos and larvae suffer narcotic effects when swimming in water admixed with petroleum hydrocarbons (Sharp *et al.*, 1979). From several experiments by Kühnhold (1969, 1972; Kühnhold *et al.*, 1978), we know that larvae may become completely stunned when swimming into clouds of petroleum hydrocarbons or oil dispersants (Wilson, 1974) and sink to the bottom. When they reach lower WSF concentrations the process is reversed (Häkkilä and Niemi, 1973). Sometimes larvae are not fully immbolized but show only reduced swimming activity (Mazmanidi and Bazhasvili, 1975). Partly anesthetized larvae, although still swimming, may lose equilibrium (Stene and Lönning, 1984) and be unable to catch prey. When kept for prolonged periods at high concentrations of the WSF of petroleum hydrocarbons, growth is negatively affected due to nonfeeding [Struhsaker *et al.* (1974), 4000 μg/l WSF benzene; Solberg *et al.*, (1984), 30–200 μg/l WSF crude oil]. The narcotic effects of crude-oil extracts (8000 μg/l WSF) have a stronger impact on starved larvae than on individuals with fully functional yolk sacs (Davenport *et al.*, 1979).

Effects are twofold, depending on WSF concentration and duration of exposure, and may be transient with subsequent recovery or long-lasting if there is histopathological damage to the retina or forebrain, as shown for larvae of the surf smelt *Hypomesus pretiosus* by Hawkes and Stehr (1982). A larva with a damaged brain or eye is unable to survive. Acute effects may be intensified by other organic compounds accumulated from the water. Solbakken *et al.* (1984) found that cod (*G. morhua*) eggs and larvae exposed to several naphthalenes, phenanthrene, benzo[a]pyrene, and PCB for 24 h accumulated these lipophilic xenobiotics in the yolk and stored them until the yolk was used, thus causing a delayed effect on later development, resulting in reduced activity, morphological aberrations, and the like. Extensive changes in behavioral patterns, such as delay in the occurrence of certain swim positions in alevins from DDT-treated Atlantic salmon (Dill and Saunders, 1974), or reduced swimming activity, may be observed in fish larvae that contain chlorinated hydrocarbons. Experiments with the cyprinodont *Adinia xenica* confirm that eggs from DDT- and mirex-contaminated parents yield larvae that show loss in equilibrium and effects of narcotization (Koenig, 1977). Since the yolk is the main storage site for DDT and PCBs (Atchison, 1976; Guiney *et al.*, 1980), the larva, while consuming yolk, takes up more and more of the stored chlorinated hydrocarbons, producing the damaging effects.

The consequences of the behavioral abnormalities depend on their severity and duration. If swimming and prey catching behavior are impaired, increasing risk of starvation or predation would be anticipated, leading to reduced survival (see Rosenthal and Alderdice,

1976). Larvae with considerable amounts of chlorinated hydrocarbons in their yolk (see Hogan and Brauhn, 1975) are usually not viable and are characterized by a high incidence of malformations, failure to completely absorb their yolk sac, or death during the larval stage (Macek, 1968). Schimmel *et al.* (1974) observed reduced survival of sheepshead minnow *Pimephales promelas* fry after incubation in >0.32 µg Aroclor 1254/l. Reduced survival is common in larvae hatched from eggs treated with PCB or DDT during incubation (Hansen *et al.*, 1974; Freeman and Idler, 1975; Dethlefsen, 1977). Proof of the detrimental effects of DDT in yolk does not stem only from laboratory experiments but is also available in field data. Dacre and Scott (1971) and Hogan and Brauhn (1975) report substantial loss of trout fry due to high DDT in larval tissues. Effects of high levels of chlorinated hydrocarbons in eggs on percent viable hatch are also known from the investigations of von Westernhagen *et al.* (1981) with Baltic flounder *Platichthys flesus*, Hansen *et al.* (1985) with herring *C. harengus*, Westin *et al.* (1985) with striped bass *Morone saxatilis* and Cameron *et al.* (1986) with whiting, *Merlangius merlangus*. Effective chlorinated hydrocarbons are PCB at a gonad concentration of 120–180 µg/kg wet weight and DDE at 18 µg/kg.

IV. SUBLETHAL EFFECTS ON LARVAE NOT EXPOSED AS EGGS

Outstanding sublethal effects on larvae are impairment of yolk utilization and ensuing depression in growth, which may be caused either by low levels of metals, petroleum hydrocarbons, or chlorinated hydrocarbons, or by low pH. In addition, petroleum hydrocarbons are particularly effective in reducing larval activity.

In the previous section I have discussed deleterious effects of pollutants on larvae resulting from exposed eggs. All of the effects lower the individual's survival chances even if the larva is released into an uncontaminated environment after hatching.

However, pollutants may also exert their influence on normal larvae with (sublethal) reactions that result in reduced survival of the young fish. Larvae hatched in uncontamined water may come under the influence of contaminants during larval drift by encountering an oil spill or plumes of heavily polluted river water. Such experimentally mimicked posthatching encounters with pollutant stressors will now be addressed.

Larvae and alevins are generally considered more susceptible to abiotic factors than is the egg (Häkkilä and Niemi, 1973; Linden, 1974; Rice *et al.*, 1975). Effective concentrations for the production of sublethal effects should be low. For example, sublethal effects exhibited by larval salmon exposed to heavy metals are skeletal deformities after exposure to low mercury (9.3 μg/l; Servizi and Martens, 1978) or low cadmium (>0.78 μg/l; Rombough and Garside, 1984) concentrations. Other responses may be inhibition of enzymatic processes by cadmium, lead, and mercury exposure, as seen in brook trout, *S. fontinalis*; alevins (Holcombe *et al.*, 1976; McKim *et al.*, 1976); this may lead to reduced survival due to uncoupling or inhibition of unknown metabolic functions [Benoit (1975), *Lepomis macrochirus*; Spehar (1976), *Jordanella floridae*]. Thus, one sublethal reaction of larval spot *Leiostomus xanthurus* to cadmium is a lowering of its thermal maximum (Middaugh *et al.*, 1975). Most of the information on sublethal effects of metals is related to metal-induced depression of growth due to insufficient yolk utilization or decreased larval activity and reduced feeding. Blaxter (1977), using larval herring and plaice, found that in feeding plaice larvae there is a marked reduction in growth in length and development at 90 μg copper/l or above, while in herring, doses of 300 μg/l tend to inhibit activity during their dark phase of migration. The growth inhibiting effects of copper on alevins are documented also for king salmon *O. tshawytscha* (21 μg/l; Hazel and Meith, 1970) as well as brook trout (McKim and Benoit, 1971) starting at 3.4 μg/l. In the latter species the influence of copper on metabolism retards yolk absorption for about 4 weeks. Lowered rate of yolk utilization is also considered to be the reason for the slow growth of alevins of Atlantic salmon kept in water with 2 μg cadmium/l. The growth-depressing effect of cadmium continues after the initiation of feeding (Peterson *et al.*, 1983). The possible role of metal-damaged mitochondria in the impairment of energy transfer and metabolism of protein (Somasundaram *et al.*, 1984b,c) has already been mentioned and might be responsible for the inability of metal exposed larvae to make adequate use of their yolk reserves. Besides the effects on metabolism, the latter authors report considerable brain damage in herring *C. harengus* larvae resulting from zinc exposure; this will have a bearing on swimming activity and prey catching behavior.

In general, the effects of low levels of heavy metals on fish larvae are less spectacular than effects described for petroleum hydrocarbons. Due to the increasing numbers of oil spills, many reports have recently been prepared on the effects of petroleum hydrocarbons on marine life; the reaction of fish larvae to oil has aroused particular interest. Generally, the reaction of fish larvae exposed to petroleum hydrocarbons resembles that of larvae hatched from eggs incubated in oil-polluted waters. Petroleum hydrocarbons and dispersants clearly decrease larval activity, as noted by a reduction of heartbeat of yolk-sac larvae of pike *Esox lucius* (Häkkilä and Niemi, 1973). At 18 mg/l of an emulsifier (Neste A), a reduction in heartbeat from 90 beats min^{-1} to 30 beats min^{-1} is observed after 2 days, followed by a period of reduced swimming activity and narcosis. Also, under the influence of an oil dispersant, larvae of herring *C. harengus* and plaice *Pleuronectes platessa* show narcosis within 20 min of exposure to 8 mg/l; affected larvae may recover if exposure time is within the above-mentioned range (Wilson, 1972, 1974). Linden (1975) found narcotic effects caused by oil/dispersant mixtures (0.01/0.005 ml l^{-1}) using herring larvae; Rosenthal and Gunkel (1967) noted similar effects. These mixtures (>0.5 mg/l) induce impaired swimming and prey catching behavior, as do dispersants (Wilson, 1972) or other petroleum hydrocarbons alone (Kühnhold, 1969; Stene and Lönning, 1984). Concentrations of oil/dispersants of 100/50 μl^{-1} already destroy larval fin primordia (Linden, 1975) or other tissue (Kühnhold, 1972), which might later kill the individual.

Thus, one of the major effects of oil and dispersants is reducing larval activity. This may influence larval survival directly, since it has been demonstrated experimentally that anesthetized larvae are more susceptible to predation than others (see Rosenthal and Alderdice, 1976). One reason may be that under the influence of a narcotic such as the WSF of crude oil, the larva's "critical distance," that is, the greatest distance between a larva and a small object (hypothetical predator) to induce a flight reaction, is shortened (Johnson *et al.*, 1979), and flight reactions may then be initiated too late to be successful. Yet the more obvious effect on anesthetized larvae is their reduced prey catching ability due to slower movements and uncoordinated swimming and prey catching maneuvers (Rosenthal, 1969). Swimming capacity determines the volume of a water body a larva is able to search for food per unit time (Rosenthal and Hempel, 1970); thus, reduced swimming speed decreases the number of encounters with food particles. Starvation enhances detrimental effects of petroleum hydrocarbons (Davenport *et al.*, 1979). While 35-day-old starved cod *G. morhua* larvae exposed to 8 mg/l WSF (crude oil) get narcotized, larvae with functional yolk sacs remain unaffected.

An additional factor contributing to the deleterious effects of petroleum hydrocarbons is the apparent reduction in larval growth in oil-polluted waters. Since petroleum hydrocarbons are readily accumulated (Roubal *et al.*, 1977), even fairly small quantities of oil (0.075

ml/l) in the water inhibit growth in Pacific salmon *O. gorbuscha* alevins exposed for 10 days (Rice *et al.*, 1975). Length and weight were equally affected. Vuorinen and Axell (1980) speculated that the poor growth of pike *E. lucius* larvae in oil-contamined water (>0.1 mg/l) may be related to gill damage, which decreases oxygen supply and food utilization. A concentration-dependent reduction in growth is known for cod *G. morhua* larvae continuously exposed to 50–200 μg/l of a WSF of crude oil. Besides a direct impairment of yolk utilization in the presence of oil (Lönning, 1977) or benzene (Struhsaker *et al.*, 1974), the suppressed larval growth may reflect alterations in the metabolic rate with energy diverted from assimilation to detoxification (Eldridge *et al.*, 1977).

Poor yolk utilization in larvae and alevins exposed to chlorinated hydrocarbons, in particular PCBs, is the major feature observed with these pollutants. Thus, alevins of coho salmon *O. kisutch* exposed to 15 μg Aroclor 1254/l react with reduced growth and poor yolk absorption (Halter and Johnson, 1974), as do the young of the fathead minnow *P. promelas* (Nebeker *et al.*, 1974). The inhibition of the mitochondrial NADH oxidase system by PCB as shown by Pardini (1971) may very well be responsible for this phenomenon. Inhibited yolk consumption is also present in fry exposed after hatching to low pH. In pike *E. lucius*, pH$<$4.2 inhibits yolk absorption considerably (Johansson and Kihlström, 1975), and this is true also for larvae of the white sucker *Catostomus commersoni* (Trojnar, 1977a). Likewise, growth of fry is impaired and survival significantly reduced in the flagfish *J. floridae* at pH 4.5 to 5.5 (Craig and Baksi, 1977); brook trout *Salvelinus fontinalis* alevins react in a similar way (Kwain and Rose, 1985). Time for yolk absorption is increased by 21 days in brook trout alevins at pH$<$5.0 (Menendez, 1976), and larvae usually remain smaller (Swarts *et al.*, 1978). In fry of rainbow trout *Salmo gairdneri*, low pH affects the whole metabolism, as shown by the decreased cardiac rate, decreased rate of ossification, slower growth, less pigmentation, and increased mortality rate (Nelson, 1982).

This section demonstrates the outstanding effects of all types of pollutants on yolk utilization. The proper and complete utilization of the yolk reserves is prerequisite for survival and successful development. Larvae that are not able to use these reserves for differentiation to the feeding stage will perish, particularly since the period between yolk utilization and first feeding is critical in many species and requires precise "timing," as pointed out by May (1974b) in the "critical period concept."

V. DISCUSSION, PROBLEMS AND THE FUTURE

Present knowledge of sublethal responses of fish eggs and larvae to different types of pollutants is summarized in Table III. The general plan of Table III is adopted from Rosenthal and Alderdice (1976) and may be used for a quick overview of the mode of action of pollutants. Major malformations during advanced development, such as bent body axis, may be caused by varying groups of pollutants, and it is interesting to note the absence of specific effects caused by one single factor. This is surprising considering the enlarged scope for differentiated reactions in advanced stages of development, particularly since the embryonic stages from gastrulation to organogenesis have been reported to be the stages wherein most teratogenic effects occur (Wilson, 1973). Yet after completion of gastrulation the frequency of teratogenic effects declines, because organogenesis is essentially complete (Stoss and Haines, 1979). Morphologically the embryo stabilizes itself during the later stages, and the seeming lack of diversity in morphological abnormalities may be considered an artifact, due to the inability of techniques to register effects before they express themselves as gross malformations. Enzymatic and hormonal monitoring, which has received little attention in early life-stage testing (McKim, 1985) would provide a measure of the disruption of normal enzyme and/or hormone development during organogenesis in the growing embryo. Thus, gross malformations may be considered belated reactions to stress, just as elevated body temperature in mammals is an indication of infection. It does not tell us whether the cause is pneumonia, measles, an infected tooth, or appendicitis. This also explains why in Table III we find a preponderance of morphological abnormalities as indicators for sublethal responses. At present we are still far from the desirable goal of establishing clear cause–effect relationships between pollution and cellular or physiological processes. Although research has been particularly active, resulting in a large body of fundamental information, when compared to the experimental evidence available 10 years ago (Rosenthal and Alderdice, 1976) we note a lack of new approaches. The majority of the contributions still focus on a simple listing of not acutely lethal responses of eggs, embryos, or larvae to what seems to be an irrelevant enumeration of more or less toxic substances, often employed in outrageously high concentrations. Few of these contributions increase our knowledge of the nature of sublethal effects, uncovering the mechanistic cause–effect relationship between stressor and recipient. Neither are we today in a

Table III

Summary of Observed Responses to Heavy Metals, Petroleum Hydrocarbons, Chlorinated Hydrocarbons, and pH Considered as Sublethal Effects[a]

Stage of organization where pollutant stress is imposed or recognized	Observed response	Pollutant	Observed or suspected consequences of response
Ovarian eggs and egg deposition	Reduced egg number, Chromosome damage	M, PH, ClH, pH PH	Reduced recruitment Embryonic malformations
Fertilization, water uptake, and water hardening	Rate of fertilization Changes in properties of egg membranes	M, PH, ClH, pH M, ClH	Reduced recruitment Reduced capsule strength
	Water uptake during and after fertilization	M, pH	Changes in osmoregulatory capacity Changes in buoyancy of pelagic eggs (changes in transport, distribution and location in water column)
	Impaired or delayed water hardening	M, pH	Reduced capsule strength
	Changes in egg volume	M	Lowered resistance to deformation Reduced space for embryo activity
Embryonic development (early and advanced)	*Biochemical effects* Changes in ATP levels	M	Energy deficit Retarded development Necrotic tissue Dedifferentiation Organ malformation

320

Effect		
Changes in enzyme activity	M, pH	Interference with general metabolism and biosynthetic processes; Retarded development; Reduced yolk/energy conversion; Smaller larval size at hatching; Reduced hatching success; Retarded hatching
Physiological effects		
Respiration effects	PH	Changes in embryonic growth rates
Embryonic heart rate	M, PH	Retarded development
Morphological effects		
Unusual shape of blastodisc	M	Embryonic malformations (?)
Deformation of blastomeres	ClH	Embryonic malformations (?)
Irregular cleavage of blastomeres	PH, ClH	Embryonic malformations (?)
Amorphous embryonic tissue (no definite embryo formed)	M, PH, ClH, pH	No viable hatch
Necrosis	M, PH, ClH, pH	Embryonic malformations
Pigment anomalies	M, PH	Susceptible to predation (?)
Yolk-sac blood circulation not well developed	M, pH	Impaired yolk utilization (?)
Organ malformations		
Bent body axis	M, PH, ClH, pH	Respiration (?); Impaired swimming, feeding
No blood pigmentation	M, PH	Impaired respiration
Elongated heart tube	M	Impaired blood circulation (?)
Eye malformations (see also yolk sac larvae)	M, PH	Impaired vision, prey hunting, phototaxis

(continued)

321

Table III (*Continued*)

Stage of organization where pollutant stress is imposed or recognized	Observed response	Pollutant	Observed or suspected consequences of response
	Malformed otoliths and/or otic capsules	M	Impaired equilibrium, swimming, prey hunting
	Behavioral effects		
	Embryonic activity reduced	M, PH, pH	Reduced mixing of perivitelline fluid affecting respiration distribution of hatching enzyme
			Retarded development
			Abnormal hatching process
			Retarded hatching
	Pectoral fin and opercula movements reduced	M, PH	As above (embryonic activity)
	Altered hatching parameters		
Incubation time and the process of hatching	Prolonged hatching process	PH, pH	Increased susceptibility to predation
	Increased or decreased incubation time	M, PH, ClH, pH	Desynchronizing of food availability at time of first feeding
	Reduced viable hatch	M, PH, ClH, pH	Reduced survival potential at population level
Larvae (hatched from exposed eggs)	Smaller larval size at hatching	M, PH, ClH	Reduced biomass, increased susceptibility to predation reduced cruising speed
	Changes in subcellular morphology	M	Impaired metabolic functions

322

Biochemical effects		
Changes in enzyme activity	M	Impaired metabolic functions
Swimming behavior		
Impairment of swimming		
Inability to swim	M, PH, ClH	Impaired ability to maintain position in water column
Equilibrium anomalies	M, PH	Inability to capture food
Loss of avoidance reaction	M, PH	Reduced escape reaction
		Increased susceptibility to predation
Altered yolk utilization		
Reduced conversion efficiency	M, PH, ClH, pH	Reduced larval size at time of first food intake
Changed rate of yolk utilization	M, PH, pH	Shift in timing to first food intake
Malformations		
Serrated fins	M	Altered dermal respiration, swimming ability, susceptibility to disease
Eye defects (anophthalmia, cyclopia)	M, PH, pH	Reduced visual perception, phototaxis, prey hunting ability
Mouth, lower jaw, branchial apparatus	M, PH, ClH	Impaired respiration, feeding success
Vertebral column	M, PH, ClH, pH	Impaired swimming ability, escape reaction, prey hunting
Sloughing of epidermal tissue	M, PH, ClH, pH	Susceptibility to infections
Faulty pigmentation	M, PH, ClH, pH	Susceptibility to predation

(*continued*)

323

Table III (*Continued*)

Stage of organization where pollutant stress is imposed or recognized	Observed response	Pollutant	Observed or suspected consequences of response
Larvae (hatched from unexposed eggs)	*Malformations*		
	Skeletal deformations	M, PH	Impaired swimming ability, prey catching
	Skin lesions	PH	
	Morphological effects		
	Brain damage	M	Impaired swimming, feeding
	Biochemical effects		
	Metabolic depression	M	Energy deficit, impaired food conversion, increased thermal sensitivity
	Retarded ossification	pH	Reduced swimming speed
	Reduced conversion efficiency of yolk and reduced growth	M, PH, ClH, pH	Loss of competitive fitness
	Behavioral effects		
	Changed migratory behavior	M	Food deprivation, starvation, increased susceptibility to predation, reduced water volume searched for food
	Change in critical distance	PH	
	Decreased activity, reduced swimming capacity	PH	

[a] These are shown in relation to ontogenic stages of development in which they are imposed or observed and to known or possible consequences at later stages of development; M, metal; PH, petroleum hydrocarbon; ClH, chlorinated hydrocarbons, pesticides. Adapted from Rosenthal and Alderdice (1976).

better position to recognize the consequences of sublethal responses, even though there have been substantial contributions in histology, morphology, physiology, and ethology, as called for by Rosenthal and Alerdice (1976), nor are we able to bridge the gap between inferences based on observed individual responses and the implications they generate regarding the cellular level at which they were elicited on the one hand and their effect on survival potential of the population on the other hand; our ultimate concern must be for the long-term effects of contamination on the ecosystem (Bayne, 1985).

Although in the particular field of biochemistry, molecular mechanisms of the effects of trace metals on enzymatic-related metal toxicity have been identified (see Viarengo, 1985), the connection between the pollutant impact and the complete reaction of the organism is still uncertain for most of the xenobiotic substances, preventing a clear view of the more subtle events inside the cell.

With the differentiation of the embryo, its scope for responsiveness becomes progressively enlarged from cells to tissue to organs etc., while the complexity of physiological and enzymatic reactions and metabolic feedbacks is diversified. In view of the logarithmically growing numbers of conceivable responses to stressors, we are left with an unsatisfying incapability of explaining even the first of the most obvious deviations from normal development—that is, the occurrence and significance of aberrant early cell divisions. Do they have a lasting effect on differentiation or does the growing embryo compensate and repair the early disorder? We do not yet have answers for those basic questions and, with progressing development, the expanding variety of responses makes matters more difficult. This has been recognized by most workers in this field (see also McKim, 1985). In seeking a way out of this dilemma, the question arises: should we be studying the basic processes that direct ontogeny and differentiation or should we, for the time being, accept a less sophisticated approach, which may be simpler, easier to comprehend, and more workable, and accept the risk of misinterpretation? Frequently the nature of the investigation and the experimental approach do not allow a detailed explanation, since only gross morphological abnormalities are described while subtle responses at the cellular or subcellular levels are not considered. Thus, by far, more publications provide only a superficial evaluation of the causes of observed effects and refer only vaguely to such topics as metabolic inhibitors, for example.

Much has been said in this review about energy pathways, their inhibition, and blockage leading to malformations, yet with much too little experimental background. Too little is known about how substances or their metabolites interfere with energy transfer through oxidative phosphorylation, the citric acid cycle, or protein metabolism. No information is available as to why certain pathways are not passable for the embryo or the larva after the application of certain stressors. Observations of reduced heartbeat, dedifferentiation of tissues, or incomplete yolk absorption suggest impaired energy metabolism. Although this may be correct and the reason for almost identical effects from different types of pollutants, this answer does not tell us anything about the mechanisms involved or the true nature of the processes impaired by lack of cellular energy.

Experiments designed to study metabolic processes at the subcellular level are rare and the situation is seldom so simple that we can understand the action of a pollutant in terms of a cause–effect relationship, as appears to be the case in the chorionase inhibition at low pH. There are several possible ways for heavy metals to interfere with metabolic processes, although their significance for individual ontogenetic processes can only be postulated. Thus, zinc interferes with lysosomal membrane stability, while mercury and cadmium are able to disrupt ionic balance and interfere with osmoregulation. Other metals such as copper have been shown to be very effective inhibitors of behavioral patterns at pollutant levels present in the today's environment. At concentrations of a few micrograms per liter in the water they impair feeding in yolk-sac larvae, thus reducing survival.

Water-soluble fractions of petroleum hydrocarbons are effective narcotics, anesthetizing fish embryos and larvae at concentrations found after accidental oil spills. However, in the environment effective concentrations are not usually maintained over long periods, since oil-degrading bacteria quickly reduce toxicity; the probability of petroleum hydrocarbons significantly affecting fish larvae is minimal.

Experimentally chlorinated hydrocarbons also reduce percent viable hatch at low concentrations in the lipids of parental gonads. Our investigations with marine fish and those of others with salmonids have demonstrated drastic reduction of viable hatch due to chlorinated hydrocarbons accumulated in gonads from the natural environment.

It appears from this review that most sublethal effects are biochemical in origin and that they are expressed as histological, morphological, physiological, or behavioral respones. In conjunction with a better understanding of physiological and biochemical events at the subcellular level goes the necessity to concentrate more on cytopathological and ultrastructural effects on cells, tissues, and organs. From recent investigations we learn that even without an obvious external damage, pollutants such as metals (zinc) and petroleum hydrocarbons may severely influence cellular and tissue organization, disrupting cell structure in brain and eyes or causing damage to ultrastructural cell bodies such as mitochondria or endoplasmic reticulum; possible effects on orientation or locomotion give clues to the mechanism of the pollutants. As yet there are too few investigations of detailed scrutiny in this field.

The significance of the gross malformations is generally better understood than the alterations at the subcellular level. This is particularly true for the individual and represents a substantial part of Table III, where the majority of the observable responses to pollutants ultimately cause the death of the individual. There are few responses, such as changes in color pattern or slight body flexure due to metal, that may not be crucial for survival or may be repaired as development proceeds.

Particularly hardy species such as cyprinodontids, belonids, or some salmonids when reared in captivity may even survive major damages to fins, jaw, or vertebrae. In the field, however, the percentage survivors with gross morphological abnormalities is negligible, since fish larvae are relatively fragile and natural mortality is high even in healthy individuals.

At this point an important question arises. What is the significance to natural populations of the pollution-induced sublethal responses? The only one of the four factors reviewed in this chapter that at present shows clear effects on fish populations is the low pH of freshwater lakes and rivers in the northern hemisphere. From the available data on the susceptibility of fish eggs to low pH (pH 4.0–5.0), it is evident that the prevailing pH of many lakes and rivers in northern Europe and America (Beamish, 1974, 1976; Jensen and Snekvik, 1972; Schofield, 1976; Wright and Snekvik, 1978; Harvey, 1980; Rosseland *et al.*, 1980; Overrein *et al.*, 1980; Sevaldrud *et al.*, 1980; Haines, 1981; Gunn, 1982) is already too low to guarantee normal ontogenetic processes in fish eggs. In these lakes, although adult populations of fish may still survive for a few years, the impairment of the hatching enzyme through acid waters is too great to allow more than a small percent of larval hatch (Beamish and Harvey, 1972). From field and laboratory experiments, Muniz and Leivestad (1980) conclude that the most common mechanism of population extinction in acidified lakes is postembryonic mortality and subsequent lack of recruitment. As we know (Harvey and Lee, 1982), other lakes in particularly sensitive areas such as the La Cloche area in Canada or southern Norway are already seriously depleted of the original fish populations or are devoid of fish altogether (Beamish, 1974; Sevaldrud *et al.*, 1980; Gunn, 1982). Since Swarts *et al.* (1978) showed that no short-term adaptation of eggs or alevins to low pH can be expected, it seems probable that more and more lakes with low buffering capacity will be deprived of their fish populations. The effect of low pH is usually exacerbated by elevated metal contents. Several of these metals, such as copper (McKim and Benoit, 1971; Horning and Neiheisel, 1979) or mercury (Huckabee and Griffith, 1974), occur at concentrations (3.0–18.0 $\mu g/l$) toxic to embryos and larvae of freshwater fish (Haines, 1981). The same may be true for aluminum (Grahn, 1980), although in the case of aluminum we are dealing probably with asphyxia due to hydroxide precipitation.

This is a social rather than a biological problem (see also Haines, 1981; LaZerte and Dillon, 1984). As long as society and industry are not prepared to take measures against the increasing acidification of lakes, promoting a drastic reduction of sulfuric (SO_2) and nitric (NO_x) exhaust-gas emission, lake acidification will progress rapidly; for many fish populations the sublethal effects of decreased hatching will turn into a lethal effect when no viable hatch is possible. Aside from an ecological catastrophe, some areas must face substantial economic losses due to decreasing revenues from recreational fishing (see Tuomi, 1981). For the marine environment, pH has no bearing, since the effective low pH is far below the level that can be observed even at extreme conditions in the sea (Kelley, 1946; von Westernhagen and Dethlefsen, 1983).

Petroleum hydrocarbons are likely to produce short-term effects on fish populations. These substances manifest their effects in two ways: (1) as acute effects resulting from oil spills, blow-outs, or the like, these effects may be locally confined and, even if severe (Longwell, 1978), will not have far reaching consequences because only a limited number of eggs or larvae is affected in time and space; (2) long-term sublethal effects may be expected for coastal and estuarine substrate spawning populations, since petroleum hydrocarbons and their metabolites concentrate in and on the substrate (sand) that is used by demersal spawners (Chapman *et al.*, 1982; Landolt and Kocan, 1984). Some substances or their metabolites are highly mutagenic (i.e., benzo[a]pyrene) and in the long run produce chronic effects by increasing genetic load in a population (Hose *et al.*, 1982; Kocan and Landolt, 1984). This applies particularly for harbors and heavily industrialized areas of both fresh and saltwater bodies where small indigenous fish populations exist. Although the mechanisms of the process are comprehensible, nothing is known about its dynamics and the time scale involved. It may be that a fish population employing the principle of "low expenditure per progeny" in its reproductive strategy requires a certain—relatively high—percentage of nonviable eggs and larvae before recruitment becomes detectibly impaired resulting in collapse of the population. Fluctuations within natural populations of fish are known to surpass the factor 4. Thus in Arctic cod the level of recruitment was the same in 1953, 1955, and 1960 as that in 1944–1946, even though the size of the spawning stock of the years 1953, 1955, and 1960 was only 25% of that in 1944–1946. If, on the other hand, the population concerned is more of the "high expenditure per progeny" type, as is the case in salmonids with relatively few eggs per female, then a 20% reduction in reproductive success may have a bearing on the population. This has been shown in a seatrout (*Cynoscion nebulosus*) population, which declined drastically due to impaired reproductive success because of high DDT residues in female gonads (Butler *et al.*, 1972). Also, Willford *et al.* (1981) concluded from the reproductive failure of lake trout *Salvelinus namaycush* in Lake Michigan that increased mortality of fry due to high DDE and PCB levels impedes restoration of the lake trout population to self-sustainability.

Although the concentrations of these substances in the water are not high enough to cause acute effects (Brügmann and Luckas, 1978) and are constantly declining (DDT, PCB; Olsson and Reutergardh, 1986), we believe that chlorinated hydrocarbons are causative agents triggering such a process. The involved mechanism acts by biomagnification: the accumulated substances become effective on the embryo or larva through the parental gonads and the yolk of the eggs. Investigations of von Westernhagen *et al.* (1981), Hansen *et al.* (1985), Westin *et al.* (1985), and Cameron *et al.* (1986), conducted with several fish species, have shown that chlorinated hydrocarbons [PCB, DDT, DDE, hexachlorobiphenyl (HCB), chlordane] in gonads reduce viable hatch down to 2% and less in flounder *Platichthys flesus*, herring *C. harengus*, striped bass *M. saxatilis*, and whiting (*Merlangius merlangus*). However, the overall reproductive capacity of the population does not seem to be endangered—the mean viable hatch of flounder as determined from data of von Westernhagen *et al.* (1981) is still around 55%. Similarly high values are found for herring (59%) (Hansen *et al.*, 1985). Yet results from the North Sea indicate that viable hatch of whiting *Merlangius merlangus* (Cameron *et al.*, 1986) experimentally incubated, is usually low (11%). These data fall in line with the high percentage of aberrations found in whiting eggs in the North Sea and low hatching success of "wild" eggs (Dethlefsen *et al.*, 1985).

Thus, although at present there appears to be no direct impact on recruitment in these species, the investigations of Hogan and Brauhn (1975) and Butler *et al.* (1972) show that other species have already yielded to high gonad burden of chlorinated hydrocarbons displaying failures in recruitment.

A great variety of heavy metals has been investigated, but very few relevant concentrations have been employed. In most cases inordinately high metal levels have been tested. Even the lower concentrations, such as the low levels of copper that affect feeding of herring and plaice larvae (Blaxter, 1977), are still high when compared to actual concentrations in river water or seawater (Anonymous, 1983; Kremling and Petersen, 1978; Duinker and Nolting, 1978). Only rarely do we find concentrations in nature that can be related to information provided in this review and that might cause sublethal effects on eggs and larvae (Smith *et al.*, 1981). Thus, for the time being we must conclude that heavy metals in the sea or natural fresh waters (except for mine tailings, etc.) do not influence recruitment. Yet there are indications that concentrations in nature are only one order of magnitude or less away from metal levels known to cause sublethal effects, and sediment concentrations of metals if mobilized are in most cases high enough to unleash a series of sublethal and lethal effects (Förstner and Müller, 1973; Hershelman *et al.*, 1981).

In brief, the detrimental effects of low pH on fish reproduction in fresh water are clearly discernable, while those of petroleum hydrocarbons and chlorinated hydrocarbons are about to show in fresh water and in the sea. For the time being, metals at concentrations existing in the field do not yet exert sublethal effects on fish eggs and larvae that have a bearing on recruitment.

REFERENCES

Akiyama, A. (1970). Acute toxicity of two organic mercury compounds to the teleost, *Oryzias latipes*, in different stages of development. *Bull. Jpn. Soc. Sci. Fish.* **36**, 563–570.

Alderdice, D. F., and Forrester, C. R. (1968). Some effects of salinity and temperature on early development and survival of the English sole (*Parophrys vetulus*). *J. Fish. Res. Board Can.* **25**, 495–521.

Alderdice, D. F., and Velsen, F. P. J. (1971). Some effects of salinity and temperature on early development of Pacific herring (*Clupea pallasi*). *J. Fish. Res. Board Can.* **28**, 1545–1562.

Alderdice, D. F., Wickett, W. P., and Brett, R. (1958). Some effects of temporary exposure to low dissolved oxygen levels on Pacific salmon eggs. *J. Fish. Res. Board Can.* **15**, 229–249.

Alderdice, D. F., Rosenthal, H., and Velsen, F. P. J. (1979a). Influence of salinity and cadmium on capsule strength of Pacific herring eggs. *Helgol. Wiss. Meeresunters.* **32**, 149–162.

Alderdice, D. F., Rosenthal, H., and Velsen, F. P. J. (1979b). Influence of salinity and cadmium on the volume of Pacific herring eggs. *Helgol. Wiss. Meeresunters.* **32**, 163–178.

Alderdice, D. F., Rao, T. R., and Rosenthal, H. (1979c). Osmotic responses of eggs and larvae of the Pacific herring to salinity and cadmium. *Helgol. Wiss. Meeresunters.* **32**, 508–538.

Alink, G. M., Frederix-Wolters, E. M. H., van der Gaag, M. A., van de Kerkhoff, J. F. J., and Poels, C. L. M. (1980). Induction of sister-chromatid exchanges in fish exposed to Rhine water. *Mutat. Res.* **78**, 369–374.

Anderson, J. W., Dixit, D. B., Ward, G. S., and Foster, R. S. (1977). Effects of petroleum hydrocarbons on the rate of heart beat and hatching success of estuarine fish embryos. *In* "Physiological Responses of Marine Biota to Pollution" (F. J. Vernberg, A. Calabrese, F. P. Thurberg, and W. B. Vernberg, eds.), pp. 241–258. Academic Press, New York.

Anonymous (1972a). "Ausgewählte Methoden der Wasseruntersuchung D. Guppy Test," Vol. 2. Fischer, Jena.

Anonymous (1972b). "Water Quality Criteria 1972." U.S. Environmental Protection Agency, Washington, D.C.

Anonymous (1976). "Quality Criteria for Water." U.S. Environmental Protection Agency, Washington, D.C.

Anonymous (1983). "Gewässerüberwachung. 2. Halbjahr 1982." Landesamt für Wasserhaushalt und Küsten Schleswig-Holstein, Kiel.

Atchison, G. J. (1976). The dynamics of lipids and DDT in developing brook trout eggs and fry. *J. Great Lakes Res.* **2**, 13–19.

Bayne, B. L. (1985). Cellular and physiological measures of pollution effect. *Mar. Pollut. Bull.* **16**, 127–128.

Beamish, R. J. (1974). Loss of fish populations from unexploited remote lakes in Ontario, Canada, as a consequence of atmospheric fallout of acid. *Water Res.* **8**, 85–95.

Beamish, R. J. (1976). Acidification of lakes in Canada by acid precipitation and resulting effects on fishes. *Water, Air, Soil Pollut.* **6**, 501–514.

Beamish, R. J., and Harvey, H. H. (1972). Acidification of the La Cloche mountain lakes, Ontario, and resulting fish mortalities. *J. Fish Res. Board Can.* **29**, 1131–1143.

Beamish, R. J., Lockart, W. L., van Loon, J. C., and Harvey, H. H. (1975). Long term acidification of a lake and resulting effects on fishes. *Ambio* **4**, 98–102.

Beattie, J. H., and Pascoe, D. P. (1978). Cadmium uptake by rainbow trout, *Salmo gairdneri* Richardson, eggs and alevins. *J. Fish Biol.* **13**, 631–637.

Bell, G. R., Hoskins, G. E., and Bagshaw, J. W. (1969). On the structure and enzymatic degradation of the external membrane of the salmon egg. *Can. J. Zool.* **47**, 146–148.

Bengtsson, B. E. (1974). The effects of zinc on the mortality and reproduction of the minnow, *Phoxinus phoxinus.* *Arch. Environ. Contam. Toxicol.* **2**, 342–355.

Bengtsson, B. E. (1980). Long term effects of PCB (Clophen A 50) on growth, reproduction and swimming performance in the minnow, *Phoxinus phoxinus.* *Water Res.* **14**, 681–687.

Benoit, D. A. (1975). Chronic effects of copper on survival, growth and reproduction of the blue gill (*Lepomis macrochirus*). *Trans. Am. Fish. Soc.* **104**, 353–358.

Benoit, D. A., and Holcombe, G. W. (1978). Toxic effects of zinc on fathead minnows (*Pimephales promelas*) in soft water. *J. Fish Biol.* **13**, 701–708.

Bettger, W. J., and O'Dell, B. L. (1981). A critical physiological role of zinc in the structure and function of biomembranes. *Life Sci.* **28**, 1425–1438.

Blaxter, J. H. S. (1977). The effect of copper on the eggs and larvae of plaice and herring. *J. Mar. Biol. Assoc. U.K.* **57**, 849–858.

Braum, E. (1973). Einflüsse chronischen exogenen Sauerstoffmangels auf die Embryogenese des Herings (*Clupea harengus*). *Neth. J. Sea Res.* **7**, 363–375.

Brown, D. J. A., and Lyman, S. (1981). The effect of sodium and calcium concentrations on the hatching of eggs and the survival of yolk sac fry of brown trout, *Salmo trutta* L. at low pH. *J. Fish Biol.* **19**, 205–211.

Brügmann, L., and Luckas, B. (1978). Zum Vorkommen von polychlorierten Biphenylen und DDT-Metaboliten im Plankton und Wasser der Ostsee. *Fischereiforsch. Wiss. Schriftenr.* **16**, 31–37.

Brungs, W. A. (1969). Chronic toxicity of zinc to the fathead minnow, *Pimephales promelas* Rafinesque. *Trans. Am. Fish. Soc.* **98**, 272–279.

Butler, P. A., Childress, R., and Wilson, A. J. (1972). The association of DDT residues with losses in marine productivity. *In* "Marine Pollution and Sea Life" (M. Ruivo, ed.), pp. 262–266. Fishing News (Books), London.

Cameron, J. A., and Smith, R. L. (1980). Ultrastructural effects of crude oil on early life stages of Pacific herring. *Trans. Am. Fish. Soc.* **109**, 224–228.

Cameron, P., von Westernhagen, H., Dethlefsen, V., and Janssen, D. (1986). Chlorinated hydrocarbons in North Sea whiting (*Merlangius merlangus*) and effects on reproduction. *Int. Counc. Explor. Sea,* Counc. Meet. **E 25**, 1–10.

Carls, M. G., and Rice, S. D. (1984). Comparative stage sensitivities of walleye pollock, *Theragra chalcogramma*, to external hydrocarbon stressors. *NOAA Tech. Mem. NMFS F/NWC* **NMFS F/NWC-67**, 69.

Carlson, A. R. (1971). Effects of long-term exposure of carbaryl (sevin) on survival, growth, and reproduction of the fathead minnow (*Pimephales promelas*). *J. Fish Res. Board Can.* **29**, 583–587.

Cash, W. D., Aanning, H. L., Carlson, H. E., Cox, S. W., and Ekong, E. A. (1968). Role of zinc(II) in the mitochondrial swelling action of insulin. *Arch. Biochem. Biophys.* **128**, 456–459.

Chapman, G. A., and Shumway, D. L. (1978). Effects of sodium pentachlorophenate on survival and energy metabolism of embryonic and larval steelhead trout. *In* "PCP—Chemistry, Pharmacology and Environmental Toxicology" (K. Rang-Rao, ed.), pp. 285–299. Plenum, New York.

Chapman, P. M., Vigers, G. A., Farrell, M. A., Dexter, R. N., Quinlan, E. A., Kocan, R. M., and Landolt, M. (1982). Survey of biological effects of toxicants upon Puget Sound biota I. Broad-scale toxicity survey. *NOAA Tech. Mem. OMPA* **OMPA-25**, 1–98.

Christensen, G. M. (1975). Biochemical effects of methylmercuric chloride, cadmium chloride, and lead nitrate on embryos and alevins of the brook trout, *Salvelinus fontinalis*. *Toxicol. Appl. Pharmacol.* **32**, 191–197.

Cosson, R. P., and Martin, J.-L. M. (1981). The effects of copper in the embryonic development, larvae, alevins and juveniles of *Dicentrarchus labrax* (L.). *Rapp. P.-V. Reun., Cons. Int. Explor. Mer* **178**, 71–75.

Craig, G. R., and Baksi, W. F. (1977). The effects of depressed pH on flagfish reproduction, growth and survival. *Water Res.* **11**, 621–626.

Crouch, D. E., Yasutake, W. T., and Rucker, R. R. (1973). Open-jaw syndrom in chinook salmon (*Oncorhynchus tshawytscha*) at a hatchery. *J. Fish. Res. Board Can.* **30**, 1890–1892.

Dacre, J. C., and Scott, D. (1971). Possible DDT mortality in young rainbow trout. *N. Z. J. Mar. Freshwater Res.* **5**, 58–65.

Davenport, J., Lönning, S., and Saethre, L. J. (1979). The effects of Ekofisk oil extract upon oxygen uptake in eggs and larvae of the cod *Gadus morhua* L. *Astarte* **12**, 31–34.

Daye, P. G., and Garside, E. T. (1977). Lower lethal levels of pH for embryos and alevins of Atlantic salmon, *Salmo salar* L. *Can. J. Zool.* **55**, 1504–1508.

Daye, P. G., and Garside, E. T. (1979). Development and survival of embryos and alevins of the Atlantic salmon, *Salmo salar* L., continuously exposed to acidic levels of pH from fertilization. *Can. J. Zool.* **57**, 1713–1718.

Daye, P. G., and Garside, E. T. (1980). Structural alterations in embryos and alevins of the Atlantic salmon, *Salmo salar* L., induced by continuous or short-term exposure to acidic levels of pH. *Can. J. Zool.* **58**, 27–43.

Dethlefsen, V. (1977). The influence of DDT and DDE on the embryogenesis and the mortality of larvae of cod (*Gadus morhua* L.). *Ber. Dtsch. Wiss. Komm. Meeresforsch.* **25**, 115–148.

Dethlefsen, V. (1980). Observations on fish diseases in the German Bight and their possible relation to pollution. *Rapp. P.-V. Reun., Cons. Int. Explor. Mer* **179**, 110–117.

Dethlefsen, V. (1984). Diseases of North Sea fishes. *Helgol. Meeresunters.* **37**, 353–374.

Dethlefsen, V., Cameron, P., and von Westernhagen, H. (1985). Untersuchungen über die Häufigkeit von Missbildungen in Fischembryonen der südlichen Nordsee. *Inf. Fischwirtsch.* **32**, 22–27.

Dial, N. A. (1978). Methylmercury: Some effects on embryogenesis in the Japanese medaka, *Oryzias latipes. Teratology* **17**, 83–92.

Dill, P. A., and Saunders, R. C. (1974). Retarded behavioral development and impaired balance in Atlantic salmon (*Salmo salar*) alevins hatched from gastrulae exposed to DDT. *J. Fish. Res. Board Can.* **31**, 1936–1938.

Duinker, J. C., and Nolting, R. F. (1978). Mixing, removal and mobilization of trace metals in the Rhine estuary. *Neth. J. Sea Res.* **12**, 205–223.

Eaton, J. G. (1973). Chronic toxicity of a copper, cadmium and zinc mixture to the flathead minnow (*Pimephales promelas* Rafinesque). *Water Res.* **7**, 1723–1736.

Eaton, J. G. (1974). Chronic cadmium toxicity to the bluegill (*Lepomis macrochirus* Rafinesque). *Trans. Am. Fish. Soc.* **103**, 729–735.

Eddy, F. B., and Talbot, C. (1983). Formation of the perivitelline fluid in Atlantic salmon eggs (*Salmo salar*) in fresh water and in solutions of metal ions. *Comp. Biochem. Physiol. C* **75C**, 1–4.

Eldridge, M. B., Echeverria, T., and Whipple, J. A. (1977). Energetics of Pacific herring (*Clupea harengus pallasi*) embryos and larvae exposed to low concentrations of benzene, a monoaromatic component of crude oil. *Trans. Am. Fish. Soc.* **106**, 452–461.

Ernst, V. V., Neff, J. M., and Anderson, J. W. (1977). The effect of water-soluble fractions of No. 2 fuel oil on the early development of the estuarine fish, *Fundulus grandis* Baird and Girard. *Environ. Pollut.* **14**, 25–36.

Falk-Petersen, I.-B., Saethre, L. J., and Lönning, S. (1982). Toxic effects of naphthalenes and methylnaphthalenes on marine plankton organisms. *Sarsia* **67**, 171–178.

Falk-Petersen, I.-B., Kjörsvik, E., Lönning, S., Möller Naley, A., and Sydnes, L. K. (1985). Toxic effects of hydroxylated aromatic hydrocarbons on marine embryos. *Sarsia* **70**, 11–16.

Fonds, M., Rosenthal, H., and Alderdice, D. F. (1974). Influence of temperature and salinity on embryonic development, larval growth and vertebral number in garfish (*Belone belone*). *In* "The Early Life History of Fish" (J. H. S. Blaxter, ed.), pp. 509–525. Springer-Verlag, Berlin and New York.

Förstner, U., and Müller, G. (1973). Heavy metal accumulation in river sediments: A response to environmental pollution. *Geoforum* **14**, 53–61.

Freeman, H. C., and Idler, D. R. (1975). The effect of polychlorinated biphenyl on steroidogenesis and reproduction in the brook trout (*Salvelinus fontinalis*). *Can. J. Biochem.* **53**, 666–670.

Giles, M. A., and Klaverkamp, J. F. (1982). The acute toxicity of vanadium and copper to eyed eggs of rainbow trout (*Salmo gairdneri*). *Water Res.* **16**, 885–889.

Grahn, O. (1980). Fish kills in two moderately acid lakes due to high aluminum concentrations. *Ecol. Impact Acid Precip. Proc. Int. Conf., 1980*, pp. 310–311.

Grande, M. (1967). Effect of copper and zinc on salmonid fishes. *Adv. Water Pollut. Res., Proc. Int. Conf., 3rd, 1966*, Vol. 1, pp. 97–111.

Guiney, P. D., Lech, J. J., and Peterson, R. E. (1980). Distribution and elimination of a polychlorinated biphenyl during early life stages of rainbow trout (*Salmo gairdneri*). *Toxicol. Appl. Pharmacol.* **53**, 521–529.

Gunn, J. M. (1982). Acidification of lake trout (*Salvelinus namaycush*) lakes near Sudbury, Ontario. *Acid Rain/Fish., Proc. Int. Symp., 1981*, p. 351.

Hagenmaier, H. E. (1974a). The hatching process in fish embryos. IV. The enzymological properties of a highly purified enzyme (chorionase from the hatching fluid of the rainbow trout. *Salmo gairdneri* Rich.). *Comp. Biochem. Physiol. B* **49B**, 313–324.

Hagenmaier, H. E. (1974b). The hatching process in fish embryos. V. Characterization of the hatching protease from the perivitelline fluid of the rainbow trout, *Salmo gairdneri* Rich., as a metalloenzyme. *Wilhelm Roux' Arch. Entwicklungsmech. Org.* **175**, 157–162.

Hagenmaier, H. E. (1974c). Zum Schlupfprozess bei Fischen. VI. Entwicklung, Struktur und Funktion der Schlupfdrüsenzellen bei der Regenbogenforelle, *Salmo gairdneri* Rich. *Z. Morphol. Tiere* **79**, 233–244.

Haines, T. A. (1981). Acidic precipitation and its consequences for aquatic ecosystems: A review. *Trans. Am. Fish. Soc.* **110**, 669–707.

Häkkilä, K., and Niemi, A. (1973). Effects of oil and emulsifiers on eggs and larvae of northern pike (*Esox lucius*) in brackish water. *Aqua Fenn.*, pp. 44–59.

Halter, M. T., and Johnson, H. E. (1974). Acute toxicities of a polychlorinated biphenyl (PCB) and DDT, alone and in combination, to early life stages of coho salmon (*Oncorhynchus kisutch*). *J. Fish. Res. Board Can.* **31**, 1543–1547.

Hamdorf, K. (1961). Die Beeinflussung der Embryonal- und Larvalentwicklung der Regenbogenforelle durch die Umweltfaktoren, O_2-Partialdruck und Temperatur. *Z. Vergl. Physiol.* **44**, 523–549.

Hannah, J. B., Hose, J. E., Landolt, M. L., Miller, B. S., Felton, S. P., and Iwaoka, W. T. (1982). Benzo(a)pyrene-induced morphologic and developmental abnormalities in rainbow trout. *Arch. Environ. Contam. Toxicol.* **11**, 727–734.

Hansen, D. J., Schimmel, S. C., and Forester, J. (1974). Aroclor 1254 in eggs of sheepshead minnows: Effect on fertilization success and survival of embryos and fry. *Proc. Annu. Conf. Southeast. Assoc. Game Fish Comm.* **27**, 420–426.

Hansen, P.-D., von Westernhagen, H., and Rosenthal, H. (1985). Chlorinated hydrocarbons and hatching success in spring spawners of Baltic herring. *Mar. Environ. Res.* **15**, 59–76.

Hardy, J. T., Apts, C. W., Crecilius, E. A., and Bloom, N. S. (1985). Sea-surface microlayer metals enrichment in an urban and rural bay. *Estuarine Coastal Shelf Sci.* **20**, 299–312.

Harvey, H. H. (1980). Widespread and diverse changes in the biota of North American Lakes and rivers coincident with acidification. *Ecol. Impact Acid Precip., Proc. Int. Conf., 1980*, pp. 92–98.

Harvey, H. H., and Lee, C. (1982). Historical fisheries changes related to surface water pH changes in Canada. *Acid Rain/Fish., Proc. Int. Symp., 1981*, pp. 45–55.

Hawkes, J. W., and Stehr, C. M. (1982). Cytopathology of the brain and retina of embryonic surf smelt (*Hypomesus pretiosus*) exposed to crude oil. *Environ. Res.* **27**, 164–178.

Haya, K., and Waiwood, B. A. (1981). Acid, pH and chorionase activity of Atlantic salmon (*Salmo salar*) eggs. *Bull Environ. Contam. Toxicol.* **27**, 7–12.

Hayes, F. R. (1942). The hatching mechanism of salmon eggs. *J. Exp. Zool.* **89**, 357–373.

Hazel, C. R., and Meith, S. J. (1970). Bioassay of king salmon eggs and sac fry in copper solutions. *Calif. Fish Game* **56**, 121–124.

Heisinger, J. F., and Green, W. (1975). Mercuric chloride uptake by eggs of the ricefish and resulting teratogenic effects. *Bull. Environ. Contam. Toxicol.* **14**, 665–673.

Helder, T. (1980). Effects of 2,3,7,8-Tetrachlorodibenzo-*para*-dioxin (TCDD) on early life stages of the pike (*Esox lucius* L.). *Sci. Total Environ.* **14**, 255–264.

Hershelman, G. P., Schafer, H. A., Jan, T.-K., and Young, D. R. (1981). Metals in marine sediments near a large Califonia municipal outfall. *Mar. Pollut. Bull.* **12**, 131–134.

Hiltibran, R. C. (1971). Effects of cadmium, zinc, manganese and calcium on oxygen and phosphate metabolism of bluegill liver mitochondria. *J. Water Pollut. Control Fed.* **43**, 818–823.

Hogan, J. W., and Brauhn, J. L. (1975). Abnormal rainbow trout fry from eggs containing high residues of a PCB (Aroclor 1242). *Prog. Fish-Cult.* **37**, 229–230.

Holcombe, G. W., Benoit, D. A., Leonard, E. N., and McKim, J. M. (1976). Long-term effects of lead exposure on three generations of brook trout (*Salvelinus fontinalis*). *J. Fish. Res. Board Can.* **33**, 1731–1741.

Holcombe, G. W., Benoit, D. A., and Leonard, E. N. (1979). Long-term effects of zinc exposure in brook trout (*Salvelinus fontinalis*). *Trans. Am. Fish. Soc.* **108**, 76–87.

Horning, W., and Neiheisel, T. (1979). Chronic effect of copper on the bluntnose minnow, *Pimephales notatus* (Rafinesque). *Arch. Environ. Contam. Toxicol.* **8**, 545–552.

Hose, J. E., Hannah, J. B., DiJulio, D., Landolt, M. L., Miller, B. S., Iwaoka, W. T., and Felton, S. P. (1982). Effects of benzo(a)pyrene on early development of flatfish. *Arch. Environ. Contam. Toxicol.* **11**, 167–171.

Huckabee, J. W., and Griffith, N. A. (1974). Toxicity of mercury and selenium to the eggs of carp (*Cyprinus carpio*). *Trans. Am. Fish. Soc.* **103**, 822–825.

Hulsman, P. F., Powles, P. M., and Gunn, J. M. (1983). Mortality of walleye eggs and rainbow trout yolk sac larvae in low pH waters of the LaCloche mountain area, Ontario. *Trans. Am. Fish. Soc.* **112**, 680–688.

Inaba, D., Nomuara, M., and Suyama, M. (1958). Studies on the improvement of artifical propagation in trout culture. II. On the pH values of eggs, milt, coelomic fluid and others. *Bull. Jpn. Soc. Sci. Fish.* **23**, 762–765.

Jarvinen, A. W., Hoffman, M. J., and Thorslund, T. W. (1977). Long-term toxic effects of DDT food and water exposure on fathead minnows (*Pimephales promelas*). *J. Fish. Res. Board Can.* **34**, 2089–2103.

Jensen, K., and Snekvik, E. (1972). Low pH levels wipe out salmon and trout populations in southernmost Norway. *Ambio* **1**, 223–225.

Jensen, S., Johansson, N., and Alsson, M. (1971). PCB—Indications of effects on salmon. *Proc. PCB Conf., 1970*, pp. 1–9.

Johannessen, K. J. (1976). Effects of seawater extract of Ekofisk oil on hatching success of Barents Sea capelin. *Int. Counc. Explor. Sea, Counc. Meet.* E:29, 1–12.

Johansson, N., and Kihlström, J. (1975). Pikes (*Esox lucius* L.) shown to be affected by low pH values during first weeks after hatching. *Environ. Res.* **9**, 12–17.

Johansson, N., and Milbrink, G. (1976). Some effects of acidified water on the early development of roach (*Rutilus rutilus* L.) and perch (*Perca fluviatilis*L.). *Water Res. Bull.* **12**, 39–48.

Johansson, N., Kihlström, J. E., and Wahlberg, A. (1973). Low pH values shown to affect developing fish eggs (*Brachydanio rerio* Ham.-Buch.). *Ambio* **2**, 42–43.

Johansson, N., Runn, P., and Milbrink, G. (1977). Early development of three salmonid species in acidified water. *Zoon* **5**, 127–132.

Johnson, A. G., Williams, T. D., Messinger, J. F., and Arnold, C. R. (1979). Larval spotted seatrout (*Cynoscion nebulosus*) a bioassay subject for marine subtropics. *Contrib. Mar. Sci.* **22**, 57–62.

Kaighn, M. E. (1964). A biochemical study of the hatching process in *Fundulus heteroclitus. Dev. Biol.* **9**, 58–80.

Kapur, K., and Yadav, N. K. (1982). The effects of some herbicides on the hatching of eggs in common carp, *Cyprinus carpio* var. *communis. Acta Hydrobiol.* **24**, 87–92.

Kaur, K., and Toor, H. S. (1977). Toxicity of pesticides to embryonic stages of *Cyprinus carpio communis* Linn. *Indian J. Exp. Biol.* **15**, 193–196.

Kelley, A. M. (1946). Effect of abnormal carbondioxide tension on development of herring eggs. *J. Fish. Res. Board Can.* **6**, 435–440.

Kihlström, J. E., Lundberg, C., and Hulth, L. (1971). Number of eggs and young produced by zebra fishes (*Brachydanio rerio,* Hamilton–Buchanan) spawning in water containing small amounts of phenyl mercuric acetate. *Environ. Res.* **4**, 355–359.

Kinne, O., and Rosenthal, H. (1967). Effects of sulfuric water pollutants on fertilization, embryonic development and larvae of the herring, *Clupea harengus. Mar. Biol. (Berlin)* **1**, 65–83.

Kjörsvik, E. (1986). Morphological and ultrastructural effects of xylenes upon the embryonic development of the cod (*Gadus morhua* L.). *Sarsia* **71**, 65–71.

Kjörsvik, E., and Lönning, S. (1983). Effects of egg quality on normal fertilization and early development of the cod, *Gadus morhua* L. *J. Fish Biol.* **23**, 1–12.

Kjörsvik, E., Saethre, L. J., and Lönning, S. (1982). Effects of short-term exposure to xylenes on the early cleavage stages of cod eggs (*Gadus morhua* L.). *Sarsia* **67**, 299–308.

Kjörsvik, E., Stene, A., and Lönning, S. (1984). Morphological, physiological and genetical studies of egg quality in cod (*Gadus morhua* L.). *Floedevigen Rapp.* **1**, 67–86.

Kleiner, D. (1974). The effect of Zn^{2+} ions on mitochondrial electron transport. *Arch. Biochem. Biophys.* **165**, 121–125.

Kocan, R. M., and Landolt, M. L. (1984). Alterations in patterns of excretion and other metabolic functions in developing fish embryos exposed to benzo(a)pyrene. *Helgol. Meeresunters.* **37**, 493–504.

Koenig, C. C. (1977). The effects of DDT and Mirex alone and in combination on the reproduction of a salt marsh cyprinodont fish (*Adinia xenica*). *In* "Physiological Responses of Marine Biota to Pollutants" (F. J. Vernberg, A. Calabrese, F. P. Thurberg, and W. B. Vernberg, eds.), pp. 357–376. Academic Press, New York.

Kremling, K., and Petersen, H. (1978). The distribution of Mn, Fe, Zn, Cd and Cu in Baltic seawater; A study on the basis of one anchor station. *Mar. Chem.* **6**, 155–170.

Kroger, R. L., and Guthrie, J. F. (1973). Additional anomalous menhaden and other fish. *Chesapeake Sci.* **14**, 112–116.

Kügel, B. (1984). Einfluss von Schmelzwasser auf Eientwicklung und Schlüpfvorgang bei Regenbogenforellen (*Salmo gairdneri* R.). Diploma Thesis, pp. 1–73. Fakultät für Biologie, Albert-Ludwig-Universität, Freiburg i. Br., Germany.

Kühnhold, W. W. (1969). The influence of water-soluble compounds of crude oils and their fraction on the ontogenetic development of herring fry (*Clupea harengus* L.). *Ber. Dtsch. Wiss. Komm. Meeresforsch.* **20**, 165–171.

Kühnhold, W. W. (1972). The influence of crude oils on fish fry. *In* "Marine Pollution and Sea Life" (M. Ruivo, ed.), 315–318. Fishing News (Books), London.

Kühnhold, W. (1974). Investigation on the toxicity of seawater-extracts of three different crude oils in eggs of cod (*Gadus morhua* L.). *Ber. Dtsch. Wiss. Komm. Meeresforsch.* **23**, 165–180.

Kühnhold, W. (1978). Effects of water soluble fraction of a Venezuelan heavy fuel oil (No. 6) on cod eggs and larvae. *Proc. Symp. Wake Argo Merchant, 1978,* pp. 126–130.

Kühnhold, W. W., Everich, D., Stegeman, J. J., Lake, J., and Wolke, R. E. (1978). Effects of low levels of hydrocarbons on embryonic, larval and adult winter flounder (*Pseudopleuronectes americanus*). *Proc. Conf. Assess. Ecol. Impact Oil Spills, 1978,* pp. 677–711.

Kusa, M. (1949). Hardening of the chorion of salmon egg. *Cytologia* **15**, 131–137.

Kwain, W., and Rose, G. A. (1985). Growth of brook trout *Salvelinus fontinalis* subject to sudden reduction of pH during their early life history. *Trans. Am. Fish. Soc.* **114**, 564–570.

Lacroix, G. L. (1985). Survival of eggs and alevins of Atlantic salmon (*Salmo salar*) in relation to the chemistry of interstitial water in redds in some acidic streams of Atlantic Canada. *Can. J. Fish. Aquat. Sci.* **42**, 292–299.

Landolt, M. L., and Kocan, R. M. (1984). Lethal and sublethal effects of marine sediment extracts on fish cells and chromosomes. *Helgol. Meeresunters.* **37**, 479–491.

LaZerte, B. O., and Dillon, P. J. (1984). Relative importance of anthropogenic versus natural sources of acidity in lakes and streams of central Ontario. *Can J. Fish. Aquat. Sci.* **41**, 1664–1677.

Lee, R. M., and Gerking, S. D. (1980). Sensitivity of fish eggs to acid stress. *Water Res.* **14**, 1679–1681.

Leung, T. S., and Bulkley, R. V. (1979). Effects of petroleum hydrocarbons on length of incubation and hatching success in the Japanese medaka. *Bull. Environ. Contam. Toxicol.* **23**, 236–243.

Lieder, U. (1964). Polyploidisierungsversuche bei Fischen mittels Temperaturschock und Colchizinbehandlung. *Z. Fisch. Deren Hilfswiss.* [N. S.] **12**, 247–257.

Linden, O. (1974). Effects of oil spill dispersants on the early development of Baltic herring. *Ann. Zool. Fenn.* **11**, 141–148.

Linden, O. (1975). Acute effects of oil and oil/dispersant mixture on larvae of Baltic herring. *Ambio* **4**, 130–133.

Linden, O. (1976). The influence of crude oil and mixtures of crude oil/dispersants on the ontogenetic development of the Baltic herring, *Clupea harengus membras* L. *Ambio* **5**, 136–140.

Linden, O. (1978). Biological effects of oil on early development of the Baltic herring, *Clupea harengus membras*. *Mar. Biol. (Berlin)* **45**, 273–283.

Linden, O., Laughlin, R. B., Sharp, J. R., and Neff, J. M. (1980). The combined effect of salinity, temperature and oil on the growth pattern of embryos of the killifish *Fundulus heteroclitus* Walbaum. *Mar. Environ. Res.* **3**, 129–144.

Longwell, A. C. (1977). A genetic look at fish eggs and oil. *Oceanus* **20**, 46–58.

Longwell, A. C. (1978). Field and laboratory measurements of stress responses at the chromosome and cell levels in planktonic fish eggs and the oil problem. *Proc. Symp. Wake Argo Merchant, 1978*, pp. 116–125.

Longwell, A. C., and Hughes, J. B. (1980). Cytologic, cytogenetic, and developmental state of Atlantic mackerel eggs from sea surface waters of the New York Bight, and prospects for biological effects monitoring with ichthyoplankton. *Rapp. P.-V. Reun., Cons. Int. Explor. Mer* **179**, 275–291.

Lönning, S. (1977). The effects of crude Ekofisk oil and oil products on marine fish larvae. *Astarte* **10**, 37–47.

Lönning, S., Kjörsvik, E., and Davenport, J. (1984). The hardening process of the egg chorion of the cod, *Gadus morhua*, and the lump-sucker, *Cyclopterus lumpus* L. *J. Fish Biol.* **24**, 505–522.

Macek, K. J. (1968). Reproduction of brook trout (*Salvelinus fontinalis*) fed sublethal concentrations of DDT. *J. Fish Res. Board Can.* **25**, 1787–1796.

McIntyre, J. D. (1973). Toxicity of methylmercury for steelhead trout sperm. *Bull. Environ. Contam. Toxicol.* **9**, 98–99.

McKim, J. M. (1985). Early life stage toxicity tests. *In* "Fundamentals of Aquatic Toxicology" (G. M. Rand and S. R. Petrocelli, eds.), pp. 48–95. Hemisphere, New York.

McKim, J. M., and Benoit, D. A. (1971). Effects of long-term exposures to coppper on survival, growth, and reproduction of brook trout (*Salvelinus fontinalis*). *J. Fish Res. Board Can.* **28**, 655–662.

McKim, J. M., Arthur, J. W., and Thorslund, T. W. (1975). Toxicity of the linear alkylate sulfonate detergent to larvae of four species of freshwater fish. *Bull. Environ. Contam. Toxicol.* **14**, 1–7.

McKim, J. M., Olson, G. F., Holcombe, G. W., and Hunt, E. P. (1976). Long-term effects of methylmercuric chloride on three generations of brook trout (*Salvelinus fontinalis*): Toxicity, accumulation, distribution, and elimination. *J. Fish. Res. Board Can.* **33**, 2726–2739.

McKim, J. M., Eaten, J. G., and Holcombe, G. W. (1978). Metal toxicity to embryos and larvae of eight species of freshwater fish. II. Copper. *Bull. Environ. Contam. Toxicol.* **19**, 608–616.

Maljkovic, D., and Branica, M. (1971). Polarography of seawater. II. Complex formation of cadmium with EDTA. *Limnol. Oceanogr.* **16**, 779–785.

Matton, P,. and Lattam, Q. N. (1969). Effect of the organophosphate Dylox on rainbow trout larvae. *J. Fish. Res. Board Can.* **26**, 2193–2200.

Mauk, W. L., Mehrle, P. M., and Mayer, F. L. (1978). Effects of polychlorinated biphenyl Aroclor 1254 on growth, survival and bone development in brook trout (*Salvelinus fontinalis*). *J. Fish Res. Board Can.* **35**, 1084–1088.

May, R. C. (1974a). Effects of temperature and salinity on yolk utilization in *Bairdiella icistia* (Jordan and Gilbert) (Pisces, Sciaenidae). *J. Exp. Mar. Biol. Ecol.* **16**, 213–225.

May, R. C. (1974b). Larval mortality in marine fishes and the critical period concept. *In* "The Early Life History of Fish" (J. H. S. Blaxter, ed.), pp. 3–19. Springer-Verlag, Berlin and New York.

Mazmanidi, N. D., and Bazhasvili, T. R. (1975). Effects of dissolved petroleum products on the embryonic development of the Black Sea flounder. *Hydrobiol. J.* **11**, 39–43.

Menendez, R. (1976). Chronic effects of reduced pH on brook trout. *J. Fish Res. Board Can.* **33**, 118–123.

Middaugh, D. P., Davis, W. R., and Yoakum, R. L. (1975). The response of larval fish, *Leiostomus xanthurus*, to environmental stress following sublethal cadmium exposure. *Contrib. Mar. Sci.* **19**, 13–19.

Milbrink, G., and Johansson, N. (1975). Some effects of acidification on roe of roach *Rutilus rutilus* L. and perch, *Perca fluviatilis* L.—with special reference to the Avaa lake system in eastern Sweden. *Fish. Board Swed., Inst. Freshwater Res., Drottningholm, Rep.* **54**, 52–62.

Mironov, O. G. (1969). The development of some Black Sea fishes in seawater polluted by petroleum products. *Vopr. Ikhtiol.* **19**, 1136–1139.

Morrow, J. E., Gynitz, R. L., and Kirton, M. P. (1975) Effects of some components of crude oil in young coho salmon. *Copeia* **2**, 326–331.

Mounib, M. S., Rosenthal, H., and Eisan, J. S. (1976). Effect of cadmium on developing eggs of the Pacific herring with particular reference to carbon dioxide fixing enzymes. *Biol. Reprod.* **15**, 423–428.

Mount, D. I. (1968). Chronic toxicity of copper to fathead minnows (*Pimephales promelas*, Rafinesque). *Water Res.* **2**, 215–233.

Mount, D. I. (1973). Chronic effect of low pH on fathead minnow survival, growth and reproduction. *Water Res.* **7**, 987–993.

Mount, D. I., and Stephan, C. E. (1969). Chronic toxicity of copper to fathead minnow (*Pimephales promelas*) in soft water. *J. Fish. Res. Board Can.* **26**, 2449–2457.

Muniz, J. P., and Leivestad, H. (1980). Acidification—Effects on freshwater fish. *Ecol. Impact Acid Precip., Proc. Int. Conf., 1980*, pp. 84–92.

Nebeker, A. V., Puglisi, F. A., and DeFoe, D. L. (1974). Effect of polychlorinated biphenyl compounds on survival and reproduction of the fathead minnow and flagfish. *Trans. Am. Fish. Soc.* **103**, 562–568.

Nelson, J. A. (1982). Physiological observations on developing rainbow trout, *Salmo gairdneri* (Richardson), exposed to low pH and varied calcium ion concentrations. *J. Fish Biol.* **20**, 359–372.

Niblett, P. D., and McKeown, B. A. (1980). Effect of the lamprey-larvicide TFM (3-trifluoromethyl-4-nitrophenol) on embryonic development of the rainbow trout (*Salmo gairdneri* Richardson). *Water Res.* **14**, 515–519.

Nimmo, D. R. (1985). Pesticides. *In* "Fundamentals of Aquatic Toxicology" (G. M. Rand and S. R. Petrocelli, eds.), pp. 335–373. Hemisphere, New York.

Ojaveer, E., Annist, J., Jankowski, H., Palm, T., and Raid, T. (1980). On effects of copper, cadmium and zinc on the embryonic development of Baltic spring spawning herring. *Finn. Mar. Res.* **247**, 135–140.

Olsson, M., and Reutergardh, L. (1986). DDT and PCB pollution trends in the Swedish aquatic environment. *Ambio* **15**, 103–109.

O'Rear, C. W. (1972). The toxicity of zinc and copper to striped bass eggs and fry with methods for providing confidence limits. *Proc. Annu. Conf. Southeast Assoc. Game Fish Comm.* **26**, 484–489.

Overrein, L., Seip, H., and Tollan, A. (1980). Acid precipitation—effects on forest and fish. *Ecol. Impact Acid Precip., Proc. Int. Conf., 1980*, pp. 19–24.

Ozoh, P. T. E. (1979a). Malformations and inhibitory tendencies induced to *Brachydanio rerio* (Hamilton–Buchanan) eggs and larvae due to exposures in low concentrations of lead and copper ions. *Bull. Environ. Contam. Toxicol.* **21**, 668–675.

Ozoh, P. T. E. (1979b). Studies on intraperitoneal toxicity of lead to *Cichlasoma nigrofasciatum* (Guenther) development. *Bull. Environ. Contam. Toxicol.* **21**, 676–682.

Ozoh, P. T. E. (1980). Effects of reversible incubations of zebrafish eggs in copper and lead ions with or without shell membranes. *Bull. Environ. Contam. Toxicol.* **24**, 270–275.

Paflitschek, R. (1979). Untersuchungen über die toxische Wirkung von Bayluscid und Lebaycid auf Ei-, Jugend- und Adultstadien der Buntbarsche *Tilapia leucosticta* (Trewawas, 1933) und *Heterotilapia multispinosa* (Guenther, 1898). *Z. Angew. Zool.* **66**, 143–172.

Pardini, R. S. (1971). Polychlorinated biphenyls (PCB): Effect on mitochondrial enzyme systems. *Bull. Environ. Contam. Toxicol.* **6**, 539–545.

Peterson, R. H., and Martin-Robichaud, D. J. (1982). Water uptake by Atlantic salmon ova as affected by low pH. *Trans. Am. Fish. Soc.* **111**, 772–774.

Peterson, R. H., and Martin-Robichaud, D. J. (1983). Embryo movements of Atlantic salmon (*Salmo salar*) as influenced by pH, temperature and state of development. *Can. J. Fish. Aquat. Sci.* **40**, 777–782.

Peterson, R. H., Daye, P. G., and Metcalfe, J. L. (1980a). Inhibition of Atlantic salmon (*Salmo salar*) hatching at low pH. *Can. J. Fish. Aquat. Sci.* **37**, 770–774.

Peterson, R. H., Daye, P. G., and Metcalfe, J. L. (1980b). The effects of low pH on hatching of Atlantic salmon eggs. *Ecol. Impact Acid Precip., Proc. Int. Conf., 1980*, p. 328.

Peterson, R. H., Metcalfe, J. L., and Ray, S. (1983). Effects of cadmium on yolk utilization, growth and survival of Atlantic salmon alevins and newly feeding fry. *Arch. Environ. Contam. Toxicol.* **12**, 37–44.

Petit, J., Jalabert, B., Chevassus, B., and Billard, R. (1973). L'insemination artificielle de la truite (*Salmo gairdneri* Richardson) I. Effects du taux de dilution, du pH et de la pression osmotique du dileur sur la fécondation. *Ann. Hydrobiol.* **4**, 201–210.

Pickering, Q. H., and Gast, M. (1972). Acute and chronic toxicity of cadmium to the fathead minnow (*Pimephales promelas*). *J. Fish. Res. Board Can.* **29**, 1099, 1106.

Pickering, Q. H., and Vigor, W. N. (1965). The acute toxicity of zinc to eggs and fry of the fathead minnow. *Prog. Fish-Cult.* **27**, 153–157.

Potts, W. T. W., and Eddy, F. B. (1973). The permeability to water of the eggs of certain marine teleosts. *J. Comp. Physiol.* **82**, 305–315.

Poy, A. (1970). Über das Verhalten der Larven von Knochenfischen beim Ausschlüpfen aus dem Ei. *Ber. Dtsch. Wiss. Komm. Meeresforsch.* **21**, 377–392.

Prein, A. E., Thie, G. M., Alink, G. M., Koeman, J. H., and Poels, C. L. M. (1978). Cytogenetic changes in fish exposed to water of the river Rhine. *Sci. Total Environ.* **9**, 287–291.

Rand, G. M., and Petrocelli, S. R. (1985). Introduction. In "Fundamentals of Aquatic Toxicology" (G. M. Rand and S. R. Petrocelli, eds.), pp. 1–28. Hemisphere, New York.

Rappaport, R. (1977). Experiments concerning the cleavage furrow in invertebrate eggs. *J. Exp. Zool.* **161**, 1–8.

Rask, M. (1983). The effect of low pH on perch, *Perca fluviatilis* L. I. Effects of low pH on the development of eggs of perch. *Ann. Zool. Fenn.* **20**, 73–76.

Remy, H. (1956). "Treatise on Inorganic Chemistry," Vol. 2. Elsevier, Amsterdam.

Rice, D. W., and Harrison, F. L. (1978). Copper sensitivity of Pacific herring, *Clupea harengus pallasi*, during its early life history. *Fish. Bull.* **76**, 347–357.

Rice, S. D., Moles, D. A., and Short, J. (1975). The effect of Prudhoe Bay crude oil on survival and growth of eggs, alevins and fry of pink salmon (*Oncorhynchus gorbuscha*). *Proc. Conf. Prev. Control Oil Pollut., 1975*, pp. 503–507.

Rombough, P. J., and Garside, E. T. (1982). Cadmium toxicity and accumulation in eggs and alevins of Atlantic salmon *Salmo salar*. *Can. J. Zool.* **60**, 2006–2014.

Rombough, P. J., and Garside, E. T. (1984). Disturbed ion balance in alevins of Atlantic salmon *Salmo salar* chronically exposed to sublethal concentrations of cadmium. *Can. J. Zool.* **62**, 1443–1450.

Rombough, P. J., and Jensen, J. O. T. (1985). Reduced water uptake and resistance to deformation in acid-exposed eggs of steelhead *Salmo gairdneri*. *Trans. Am. Fish. Soc.* **114**, 571–576.

Rosenthal, H. (1967). Die Herztätigkeit von Heringsembryonen bei verschiedenen Temperaturen. *Helgol. Wiss. Meeresunters.* **16**, 112–118.

Rosenthal, H. (1969). Untersuchungen über das Beutefangverhalten bei Larven des Herings *Clupea harengus*. *Mar. Biol. (Berlin)* **3**, 208–221.

Rosenthal, H., and Alderdice, D. F. (1976). Sublethal effects of environmental stressors, natural and pollutional, on marine fish eggs and larvae. *J. Fish. Res. Board Can.* **33**, 2047–2065.

Rosenthal, H., and Gunkel, W. (1967). Wirkungen von Rohöl-Emulgatorengemischen auf marine Fischbrut und deren Nährtiere. *Helgol. Wiss. Meeresunters.* **16**, 315–320.

Rosenthal, H., and Hempel, G. (1970). Experimental studies in feeding and food requirements of herring larvae (*Clupea harengus*). *In* "Marine Food Chains" (J. H. Steele, ed.), pp. 344–364. Oliver & Boyd, London.

Rosenthal, H., and Sperling, K.-R. (1974). Effects of cadmium on development and survival of herring eggs. *In* "The Early Life History of Fish" (J. H. S. Blaxter, ed.), pp. 383–396. Springer-Verlag, Berlin and New York.

Rosenthal, H., and Stelzer, R. (1970). Wirkung von 2,4-und 2,5-Dinitrophenol auf die Embryonalentwicklung des Herings *Clupea harengus*. *Mar. Biol. (Berlin)* **5**, 325–336.

Rosseland, B., Sevaldrud, I., Svalastog, D., and Muniz, I. (1980). Studies on freshwater fish populations—Effects of acidification on reproduction, population structure, growth and food selection. *Ecol. Impact Acid Precip., Proc. Int. Conf., 1980*, pp. 336–337.

Roubal, W. T., and Collier, T. K. (1975). Spin-labelling techniques for studying mode of action of petroleum hydrocarbons on marine organisms. *Fish. Bull.* **73**, 299–305.

Roubal, W. T., Collier, T. K., and Malins, D. C. (1977). Accumulation and metabolism of carbon-14 labeled benzene, naphthalene and athracene by young coho salmon (*Oncorhynchus kisutch*). *Arch. Environ. Contam. Toxicol.* **5**, 513–529.

Ruby, S. M., Azel, J., and Craig, G. R. (1977). The effects of depressed pH on oogenesis in flagfish, *Jordanella floridae*. *Water Res.* **11**, 575–762.

Rudy, P. P., and Potts, W. T. W. (1969). Sodium balance in the eggs of Atlantic salmon, *Salmo salar. J. Exp. Biol.* **50**, 239–246.

Runn, P., Johansson, N., and Milbrink, G. (1977). Some effects of low pH on the hatchability of eggs of perch, *Perca fluviatilis* L. *Zoon* **5**, 115–125.

Russo, R. C. (1985). Ammonia, nitrite and nitrate. In "Fundamentals of Aquatic Toxicology" (G. M. Rand and S. R. Petrocelli, eds.), pp. 455–471. Hemisphere, New York.

Sabo, D. J., and Stegeman, J. J. (1977). Some metabolic effects of petroleum hydrocarbons in marine fish. In "Physiological Responses of Marine Biota to Pollutants" (F. J. Vernberg, A. Calabrese, F. P. Thurberg, and W. B. Vernberg, eds.), pp. 279–287. Academic Press, New York.

Schimmel, S. C., Hansen, D. J., and Forester, J. (1974). Effects of Aroclor 1254 on laboratory-reared embryos and fry of sheepshead minnows (Cyprinodon variegatus). Trans. Am. Fish. Soc. 103, 582–586.

Schofield, C. L. (1976). Acid precipitation: Effects on fish. Ambio 5, 228–230.

Servizi, J. A., and Martens, D. W. (1978). Effects of selected heavy metals on early life of sockeye and pink salmon. Int. Pac. Salmon Fish. Comm., Prog. Rep. 39, 1–26.

Sevaldrud, J., Muniz, J., and Kalvenez, S. (1980). Loss of fish populations in southern Norway. Dynamics and magnitude of the problem. Ecol. Impact Acid Precip., Proc. Int. Conf., 1980, pp. 350–351.

Shabalina, A. A. (1964). Effect of cobalt chloride on growth and development in the rainbow trout (Salmo irideus). Izv. Gos. Nauchno-Issted. Inst. Ozern. Rechn. Rybn. Khoz. 58, 139–149.

Sharp, J. R., and Neff, J. M. (1980). Effects of duration of exposure to mercuric chloride on the embryogenesis of the estuarine teleost, Fundulus heteroclitus. Mar. Environ. Res. 3, 195–213.

Sharp, J. R., and Neff, J. M. (1982). The toxicity of mercuric chloride and methylmercuric chloride to Fundulus heteroclitus embryos in relation to exposure conditions. Environ. Biol. Fishes 7, 277–284.

Sharp, J. R., Fucik, K. W., and Neff, J. M. (1979). Physiological basis of differential sensitivity of fish embryonic stages to oil pollution. In "Marine Pollution: Functional Responses" (W. B. Vernberg, A. Calabrese, F. P. Thurberg, and F. J. Vernberg, eds.), pp. 85–108. Academic Press, New York.

Shaw, T. L., and Brown, V. M. (1971). Heavy metals and the fertilization of rainbow trout eggs. Nature (London) 230, 251.

Smith, J. D., Butler, E. C. V., Grant, B. R., Little, G. W., Millis, N., and Milne, P. J. (1981). Distribution and significance of copper, lead, zinc and cadmium in the Corio Bay ecosystem. Aust. J. Mar. Freshwater Res. 32, 151–164.

Smith, R. L., and Cameron, J. A. (1979). Effect of water soluble fractions of Prudhoe Bay crude oil on embryonic development of Pacific herring. Trans. Am. Fish. Soc. 108, 70–75.

Smith, R. M., and Cole, C. F. (1973). Effects of egg concentrations of DDT and dieldrin on development in winter flounder (Pseudopleuronectes americanus). J. Fish. Res. Board Can. 30, 1894–1898.

Smyth, H. F. (1967). Sufficient challenge. Food Cosmet. Toxicol. 5, 51–58.

Solbakken, J. E., Tilseth, S., and Palmork, K. H. (1984). Uptake and elimination of aromatic hydrocarbons and a chlorinated biphenyl in eggs and larvae of cod (Gadus morhua L.). Mar. Ecol.: Prog. Ser. 16, 297–301.

Solberg, T. S., Tilseth, S., Westrheim, K., and Klungsöyr, J. (1984). Effects of different fractions of Ekofisk crude oil on eggs and yolk-sac larvae of cod (Gadus morhua L.). Pap., Early Life Hist. Symp. 1984, pp. 1–22.

Somasundaram, B. (1985). Effects of zinc on epidermal ultrastructure in the larvae of Clupea harengus. Mar. Biol. (Berlin) 85, 199–207.

Somasundaram, B., King, P. E., and Shackley, S. E. (1984a). Some morphological effects of zinc upon the yolk sac larvae of Clupea harengus L. J. Fish Biol. 25, 333–343.

Somasundaram, B., King, P. E., and Shackley, S. E. (1984b). The effect of zinc on post-fertilization development in eggs of *Clupea harengus* L. *Aquat. Toxicol.* **5**, 167–178.

Somasundaram, B., King, P. E., and Shackley, S. E. (1984c). The effect of zinc on the ultrastructure of the brain cell of the larva of *Clupea harengus* L. *Aquat. Toxicol.* **5**, 323–330.

Somasundaram, B., King, P. E., and Shackley, S. E. (1984d). The effect of zinc on the ultrastructure of the trunk muscle of the larva of *Clupea harengus* L. *Comp. Biochem. Physiol. C.* **79C**, 311–315.

Spehar, R. L. (1976). Cadmium and zinc toxicity to flagfish, *Jordanella floridae*. *J. Fish Res. Board Can.* **33**, 1939–1945.

Spehar, R. L., Leonard, E. N., and DeFoe, D. L. (1978). Chronic effects of cadmium and zinc mixtures on flagfish (*Jordanella floridae*). *Trans. Am. Fish. Soc.* **107**, 354–360.

Speranza, A. W., Seeley, R. J., Seeley, V. A., and Perlmutter, A. (1977). The effect of sublethal concentrations of zinc on reproduction in the zebra fish, *Brachydanio rerio* Hamilton–Buchanan. *Environ. Pollut.* **12**, 217–222.

Sprague, J. B. (1969). Measurement of pollutant toxicity to fish. I. Bioassay methods for acute toxicity. *Water Res.* **3**, 793–821.

Stelzer, R., Rosenthal, H., and Siebers, D. (1971). Einfluss von 2,4-Dinitrophenol auf die Atmung und die Konzentration einiger Metabolite bei Embryonen des Herings *Clupea harengus*. *Mar. Biol. (Berlin)* **11**, 369–378.

Stene, A., and Lönning, S. (1984). Effects of 2-methylnaphthalene on eggs and larvae of six marine fish species. *Sarsia* **69**, 199–203.

Stephan, C. E., and Mount, D. J. (1973). Use of toxicity tests with fish in water pollution control. *In* "Biological Methods for the Assessment of Water Quality" (J. Cairns and K. L. Dickson, eds.), pp. 164–177. Am. Soc. Test. Mater., Philadelphia, Pennsylvania.

Stockard, C. R. (1906). The development of *Fundulus heteroclitus* in solutions of lithium chloride with appendix on its development in fresh water. *J. Exp. Zool.* **3**, 99–120.

Stockard, C. R. (1907). The artificial production of a single median cyclopean eye in the fish embryo by means of sea water solutions of magnesium chloride. *Arch. Entwicklungs Mech. Org.* **23**, 249–258.

Stockard, C. R. (1921). Development rate and structural expression; an experimental study of twins "double monsters" and single deformities, and the interaction among embryonic organs during their origin and development. *Am. J. Anat.* **28**, 115–277.

Stoss, F. W., and Haines, T. A. (1979). The effects of toluene on embryos and fry of the Japanese medaka (*Oryzias latipes*) with a proposal for rapid determination of maximum acceptable toxicant concentration. *Environ. Pollut.* **20**, 139–148.

Struhsaker, J. W. (1977). Effects of benzene (a toxic component of crude oil) on spawning Pacific herring, *Clupea harengus pallasi*. *Fish. Bull.* **75**, 43–49.

Struhsaker, J. W., Eldridge, M. B., and Echeverria, T. (1974). Effect of benzene (a water soluble component of crude oil) in eggs and larvae of Pacific herring and northern anchovy. *In* "Pollution and Physiology of Marine Organisms" (F. J. Vernberg and W. B. Vernberg, eds.), pp. 253–284. Academic Press, New York.

Swarts, F. A., Dunson, W. A., and Wright, J. E. (1978). Genetic and environmental factors involved in increased resistance of brook trout to sulphuric acid solutions and mine acid polluted waters. *Trans. Am. Fish. Soc.* **107**, 651–677.

Swedmark, M., and Granmo, A. (1981). Effects of mixtures of heavy metals and a surfactant on the development of cod (*Gadus morhua* L.). *Rapp. P.-V. Reun., Cons. Int. Explor. Mer* **178**, 95–103.

Tam, W. H., and Payson, P. D. (1986). Effects of chronic exposure to sublethal pH on growth, egg production and ovulation in brook trout, *Salvelinus fontinalis*. *Can. J. Fish. Aquat. Sci.* **43**, 275–280.

Tilseth, S., Solberg, T. S., and Westrheim, K. (1984). Sublethal effects of the water-soluble fraction of Ekofisk crude oil on the early larval stages of cod (*Gadus morhua* L.). *Mar. Environ. Res.* **11**, 1–16.

Trojnar, J. (1977a). Egg and larval survival of white sucker (*Catostomus commersoni*) at low pH. *J. Fish. Res. Board Can.* **34**, 262–266.

Trojnar, J. (1977b). Egg hatchability and tolerance of brook trout (*Salvelinus fontinalis*) fry at low pH. *J. Fish. Res. Board Can.* **34**, 574–579.

Tuomi, A. (1981). Socio-economic impacts of acid rain on Canada's Atlantic salmon. *IASF Spec. Publ. Ser.* **10**, 47–55.

Urist, M. R., and Schjeide, A. O. (1961). The partition of calcium and protein in the blood of oviparous vertebrates during estrus. *J. Gen. Physiol.* **44**, 743–756.

Uviovo, E. J., and Beatty, D. D. (1979). Effects of chronic exposure to zinc on reproduction in the guppy (*Poecilia reticulata*). *Bull. Environ. Contam. Toxicol.* **23**, 650–657.

van der Putte, I., Galieen, W., and Strik, J. J. T. W. A. (1982). Effects of hexavalent chromium in rainbow trout (*Salmo gairdneri*) after prolonged exposure at two different pH levels. *Ecotoxicol. Environ. Saf.* **6**, 246–257.

Viarengo, A. (1985). Biochemical effects of trace metals. *Mar. Pollut. Bull.* **16**, 153–158.

Vladimirov, V. J. (1969). Dependence of the embryonic development and viability of the carp on the trace element zinc. *Vopr. Ikhtiol.* **9**, 687–696.

von Westernhagen, H. (1968). Versuche zur Erbrütung der Eier des Schellfisches (*Melanogrammus aeglefinus* L.) unter kombinierten Salzgehalts- und Temperaturbedingungen. *Ber. Dtsch. Wiss. Komm. Meeresforsch.* **19**, 270–287.

von Westernhagen, H. (1970). Erbrütung der Eier vom Dorsch (*Gadus morhua*), Flunder (*Pleuronectes flesus*) und Scholle (*Pleuronectes platessa*) unter kombinierten Temperatur- und Salzgehaltsbedingungen. *Helgol. Wiss. Meeresunters.* **21**, 21–102.

von Westernhagen, H. (1974). Incubation of garpike eggs (*Belone belone* Linne) under controlled temperature and salinity conditions. *J. Mar. Biol. Assoc. U.K.* **54**, 625–634.

von Westernhagen, H. and Dethlefsen, V. (1975). Combined effects of cadmium and salinity on development and survival of flounder eggs. *J. Mar. Biol. Assoc. U.K.* **55**, 945–957.

von Westernhagen, H., and Dethlefsen, V. (1983). North Sea oxygen deficiency 1982 and its effects on the bottom fauna. *Ambio* **12**, 264–266.

von Westernhagen, H., and Rosenthal, H. (1979). Laboratory and in situ studies on larval development and swimming performance of Pacific herring *Clupea harengus pallasi*. *Helgol. Wiss. Meeresunters.* **32**, 539–549.

von Westernhagen, H., Rosenthal, H., and Sperling, K.-R. (1974). Combined effects of cadmium and salinity on development and survival of herring eggs. *Helgol. Wiss. Meeresunters.* **26**, 416–433.

von Westernhagen, H., Dethlefsen, V., and Rosenthal, H. (1975). Combined effects of cadmium and salinity on development and survival of garpike eggs. *Helgol. Wiss. Meeresunters.* **27**, 268–282.

von Westernhagen, H., Dethlefsen, V., and Rosenthal H. (1979). Combined effects of cadmium, copper and lead on developing herring eggs and larvae. *Helgol. Wiss. Meeresunters.* **32**, 257–278.

von Westernhagen, H., Rosenthal, H., Dethlefsen, V., Ernst, W., Harms, U., and Hansen, P.-D. (1981). Bioaccumulating substances and reproductive success in Baltic flounder *Platichthys flesus. Aquat. Toxicol.* **1**, 85–99.

von Westernhagen, H., Kocan, R., Landolt, M., Fürstenberg, G., Janssen, D., and Kremling, K. (1987). Toxicity of sea surface microlayer: Effects on herring and turbot embryos. *Mar. Environ. Res.* (in press).

Voyer, R. A., Wentworth, C. E., Berry, E., and Hennekey, R. J. (1977). Viability of embryos of the winter flounder, *Pseudopleuronectes americanus,* exposed to combinations of cadmium and salinity at selected temperatures. *Mar. Biol. (Berlin)* **44**, 117–124.

Voyer, R. A., Heltsche, J. F., and Kraus, R. A. (1979). Hatching success and larval mortality in an estuarine teleost, *Menidia menidia* (Linnaeus), exposed to cadmium in constant and fluctuating salinity regimes. *Bull. Environ. Contam. Toxicol.* **23**, 475–481.

Voyer, R. A., Cardin, J. A., Heltsche, J. F., and Hoffman, G. L. (1982). Viability of embryos of the winter flounder *Pseudopleuronectes americanus* exposed to mixtures of cadmium and silver in combination with selected fixed salinities *Aquat. Toxicol.* **2**, 223–233.

Vuorinen, P., and Axell, M.-B. (1980). Effects of the water soluble fraction of crude oil on herring eggs and pike fry. *Int. Counc. Explor. Sea, Counc. Meet.* **E:30**, 1–10.

Watermann, A. J. (1940). Effects of colchicine on the development of the fish embryo, *Oryzias latipes. Biol. Bull. (Woods Hole, Mass.)* **78**, 29–34.

Weis, J. S., and Weis, P. (1977). Effects of heavy metals on development of the killifish, *Fundulus heteroclitus. J. Fish Biol.* **11**, 49–54.

Weis, P. (1984). Metallothionein and mercury tolerance in the killifish, *Fundulus heteroclitus. Mar. Environ. Res.* **14**, 153–166.

Weis, P., and Weis, J. S. (1974). Cardiac malformations and other effects due to insecticides in embryos of the killifish, *Fundulus heteroclitus. Teratology* **10**, 263–268.

Weis, P., and Weis, J. S. (1976). Abnormal locomotion associated with skeletal malformation in the sheepshead minnow *Cyprinodon variegatus,* exposed to malathion. *Environ. Res.* **12**, 196–200.

Weis, P., and Weis, J. S. (1977). Methylmercury teratogenesis in the killifish, *Fundulus heteroclitus. Teratology* **16**, 317–326.

Westin, D. T., Olney, C. E., and Rogers, B. A. (1985). Effects of parental and dietary organochlorines on survival and body burdens of striped bass larvae. *Trans. Am. Fish. Soc.* **114**, 125–136.

Willford, W. A., Bergstedt, R. A., Berlin, W. H., Foster, N. R., Hesselberg, R. J., Mac, M. J., Passino, D. R. M., Reinert, R. E., and Rottiers, D. V. (1981). Introduction and summary—Chlorinated hydrocarbons as a factor in the reproduction and survival of lake trout (*Salvelinus namaycush*) in Lake Michigan. *Tech. Pap. U. S. Fish Wildl. Ser.* **105**, 1–4.

Wilson, J. G. (1973). "Environment and Birth Defects." Academic Press, New York.

Wilson, K. W. (1972). Toxicity of oil-spill dispersants to embryos and larvae of some marine fish. *In* "Marine Pollution and Sea Life" (M. Ruivo, ed.), pp. 318–322. Fishing News (Books), London.

Wilson, K. W. (1974). The ability of herring and plaice larvae to avoid concentrations of oil dispersants. *In* "The Early Life History of Fish" (J. H. S. Blaxter, ed.), pp. 589–602. Springer-Verlag, Berlin and New York.

Wilson, K. W. (1976). Effects of oil dispersants on the developing embryos of marine fish. *Mar. Biol. (Berlin)* **36**, 259–268.

Winkler, D. L., Duncan, K. L., Hose, J. E., and Puffer, H. W. (1983). Effects of benzo(a)pyrene on the early development of California grunion, *Leuresthes tenuis* (Pisces, Atherinidae). *Fish. Bull.* **81**, 473–481.

Wright, R., and Snekvik, E. (1978). Acid precipitation: Chemistry and fish populations in 700 lakes in southernmost Norway. *Verh.—Int. Ver. Theor. Angew. Limnol.* **20**, 765–775.

Wunder, W. (1971). Missbildungen beim Kabeljau (*Gadus morhua*) verursacht durch Wirbelverkürzung. *Helgol. Wiss. Meeresunters.* **20**, 201–212.

Zotin, A. J. (1958). The mechanism of hardening of the salmonid egg membrane after fertilization or spontaneous activation. *J. Exp. Morphol.* **6**, 549–568.

5

VITELLOGENESIS AND OOCYTE ASSEMBLY

THOMAS P. MOMMSEN

Department of Zoology
University of British Columbia
Vancouver, British Columbia, Canada V6T 2A9

PATRICK J. WALSH

Rosenstiel School of Marine and Atmospheric Science
University of Miami
Miami, Florida 33149

FISH PHYSIOLOGY, VOL. XIA

I. INTRODUCTION

At certain stages of their life histories, the females of egg-laying vertebrates, including most species of fishes, enter a phase of maturation of their oocytes in preparation for ovulation and spawning. Under the mutifaceted influence of hormonal centers such as the hypothalamus and the pituitary gland (reviewed by Peter, 1983), the growing follicles synthesize and excrete into the systemic circulation steriod hormones that govern a variety of different metabolic processes. One of the primary target organs for these steriods, particularly 17β-estradiol, is the liver. This organ, which possesses highly specific binding proteins for 17β-estradiol, in turn responds to such hormonal stimulus with the synthesis and export of vitellogenin. First named by Pan *et al.* (1969), vitellogenin constitutes the carrier molecule for various classes of compounds accumulated by the developing oocyte. While the backbone of the vitellogenin molecule is a protein chain of substantial size (molecular weight 250,000–600,000), it also carries copious amounts of lipid material, carbohydrate components, phosphate groups, and mineral salts. Following highly selective uptake into the oocyte, the transport molecule vitellogenin is broken up and accumulated as egg-specific yolk constituents, such as phosvitin and lipovitellin.

In addition to these well-known egg components, growing fish oocytes accumulate a variety of other compounds, sometimes in substantial amounts, which play an integral part in the proper development of the embryonic and larval fish. Some compounds serve as a reservoir for energy-demanding processes. In some cases, however, the physiological importance of these compounds is not yet known or can only be inferred from circumstantial evidence. Substances that fall into these categories include glycogen, carotenoids, lectins, sialoglycoproteins, wax esters, and sterol esters.

The early developmental stages of many species of fish entail long periods, sometimes weeks, of starvation prior to their first exogenous feeding. Therefore, the maternal production of vitellogenin and the deposition of adequate supplies of yolk, as well as the proper assembly of the oocyte, are essential to subsequent embryonic and larval survival. This chapter covers several aspects of vitellogenesis and oocyte assembly, such as hormonal induction of vitellogenin synthesis in the liver, the transport of vitellogenin and its later fate in the developing oocyte, and the origin and nature of other egg components. While these various topics for fishes have been addressed in the primary literature only during the last two decades, this chapter will focus attention on recent developments and touch on comparative aspects that deserve future attention. Since this chapter was completed, a review appeared in press covering aspects of vitellogenesis and oocyte developments, albeit from a widely differing angle (Wallace, 1985).

At this point it should be emphasized that research focusing on piscine systems has been a relatively recent addition to a field that has long been established for amphibians and birds as experimental animals. Readers interested in more comparative facets or specific molecular principles are therefore referred to recent reviews of those areas (Soreq, 1985; Browder, 1985).

II. VITELLOGENESIS

A general framework for the hormonal control of exogenous vitellogenesis was first proposed by Bailey (1957), and over the years has been tested experimentally and modified accordingly. Environmental cues such as photoperiod, temperature, feeding, and social factors all regulate the production of gonadotropic hormones (GtH) in the pituitary (Peter, 1983; Liley and Stacey, 1983). In females, the ovaries respond to increased levels of GtH by enhancing estrogen (17β-estradiol and possibly estrone) production and release into the bloodstream (Fostier et al., 1983). Estrogens are transported to the possible target tissues bound to specific blood proteins or attached to the ubiquitous blood albumins. They are thought to enter the tissues by facilitated diffusion and in the livers specifically induce the synthesis of vitellogenin. Much of the experimental evidence in support of this scheme has been reviewed recently (Ng and Idler, 1983). These studies, as well as a considerable amount of continued effort, fall into several categories:

1. In vitro studies make it possible to identify and characterize highly specific estrogen binding proteins in fish hepatic tissue (Le Menn et al., 1980; Turner, et al., 1981; Lazier et al., 1985; Maitre et al., 1985b).
2. The "natural" dynamics of hormone titers and vitellogenesis have been studied and indicate that estradiol levels and vitellogenesis are positively correlated and increase in parallel (de Vlaming et al., 1984; Ueda et al., 1984; van Bohemen and Lambert, 1981), following photoperiod-related increases in gonadotropins (Bromage et al., 1982; Breton et al., 1983).

3. The effects of experimental treatment of fish with various hormones, hormonal metabolites, and analogs have been well documented. Note that in these and many of the studies discussed later (Section II,H), the observed effects are inducible in male and/or immature fish. Injection with pituitary extracts or certain purification fractions [concanavalin A (Con A)–Sepharose/carbohydrate–rich fraction] induces estrogen synthesis and subsequent vitellogenesis (Idler and Campbell, 1980). Injection with estrogen alone (preferentially estradiol) or pharmacological doses of androgen induces the appearance of vitellogenin in the plasma (Plack *et al.*, 1971; Korsgaard and Petersen, 1979; Korsgaard *et al.*, 1983; Hori *et al.*, 1979; van Bohemen *et al.*, 1982a,b) as well as follicular fluid (Korsgaard, 1983).

4. In vitro studies of ovaries have confirmed the cells of the follicular epithelium as responsible for the production of estrogen (van Bohemen and Lambert, 1981; Sundararaj *et al.*, 1982b; Nagahama, 1983). However, certain parts of the fish brain have also been shown to be capable of synthesizing estrogen (van Bohemen and Lambert, 1981; Callard *et al.*, 1981), whereas the quantitative contribution of brain-derived estrogens to circulating estrogens and hence vitellogenesis has not been resolved.

The major components of fish oocytes are derived from the blood-borne high-molecular-weight compound vitellogenin, which is synthesized in the liver of oviparous vertebrates. This supply of oocyte components—especially yolk—from extraovarian sources has been termed exogenous vitellogenesis. The classification of vitellogenin as a phospholipoglycoprotein already indicates the crucial functional groups that are carried on the protein backbone of the molecule, namely, lipids, some carbohydrates, and phosphate groups. In addition, vitellogenin also has strong ion-binding properties and thus may serve as a major supply of minerals to the oocyte.

A. Hormonal Induction

In the course of the normal events accompanying early vitellogenesis, the follicle cells surrounding the developing oocyte are stimulated to synthesize estradiol. Androgens constitute the prevailing precursors of estradiol, and a slow increase in estradiol during the annual cycle indicates early exogenous vitellogenesis. After the hormone has entered the target organs, in this case the liver, it binds to highly specific estrogen receptors. In a subsequent step, the receptor/estradiol complex will bind to high affinity sites on the chromatin. While the actual biochemistry of the hormone–receptor complex and the mechanism of its binding to nuclear structures has not yet been entirely elucidated (Miesfeld *et al.*, 1984), it has been well documented that the interaction of the hormone–receptor complex with the DNA leads to modulation of the expression of specific genes. In the case of estradiol administration to immature female or male fish, specific activation is directed toward the vitellogenin gene, which is located somewhat downstream on the DNA of the actual binding site for the receptor/hormone complex. Thus, the estrogen receptor can be regarded as a gene regulatory protein. A diagrammatic representation of the feedback system between the ovary and the liver involving estradiol and vitellogenin is presented in Fig. 1.

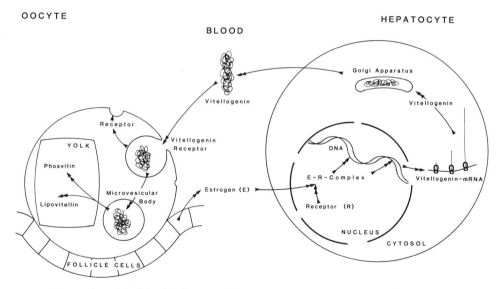

Fig. 1. Simplified feedback system between ovary and liver during exogenous vitellogenesis. Under the influence of pituitary hormones, the follicle cells release estrogen into the bloodstream. The estrogen enters the target cells (hepatocyte) by facilitated diffusion. After the events depicted in Figs. 2 and 3, vitellogenin is synthesized on the rough endoplasmatic reticulum, modified, packed into the Golgi apparatus, and excreted into the bloodstream. Blood-borne vitellogenin binds to specific receptor proteins on the oocyte membrane, is taken up by micropinocytosis, and is moved in microvesicular bodies. Before final deposition in the yolk, vitellogenin is broken up into lipovitellin and phosvitin components. Estrogen is not involved in the uptake and processing of vitellogenin by the growing oocyte.

The conversion of androgens into estrogens is catalyzed by the enzyme aromatase and occurs not only in the ovary but also in the brain of all vertebrates; aromatase activity is exceptionally high in teleosts (Callard *et al.*, 1981; Lambert and van Oordt, 1982). However, the estrogen thus produced in the fish brain apparently is not released into the circulation, unless chemically altered, which results in its ultimate removal from the circulation by gill tissue. Exertion of the multifaceted biological effects of this hormone will consequently be restricted to the local brain level (Callard, 1982) and therefore cannot be implicated in the initiation of exogenous vitellogenesis.

B. Estrogen Receptors in Target Cells

The prevailing model for the cellular distribution and action of steriod hormone receptors has been the so-called "two-step model," which was first proposed by Gorski, Jensen, and co-workers (Gorski *et*

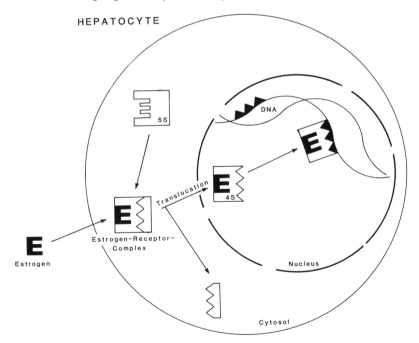

Fig. 2. Two-step model of estrogen–receptor mechanism. Estrogen binds to cytosolic 5-S receptor proteins in the target cell (liver). During translocation from the cytosol to the nuclear matrix, the estrogen-occupied receptor (estrogen–receptor complex) changes from 5 S to 4 S and acquires high affinity for the chromatin. The estrogen–receptor complex binds to specific sites on the DNA and among other things (cf. Table II) leads to activation of the gene coding for vitellogenin.

al., 1968; Jensen *et al.*, 1968). According to this model, which is summarized in Fig. 2, estradiol—or any other steriod hormone for that matter—enters the target cell by diffusion and binds to a specific cytosolic form of the receptor protein. Subsequently, the estrogen–receptor complex undergoes a transformation from a non-DNA binding form to a species that does and is translocated to the nucleus. Receptor transformation is also reflected in a decrease in receptor size from a sedimentation coefficient of around 5 S to about 4 S. The model implies that the cytosolic form of the receptor can be present with or without bound steriod, while the nuclear form only exists in the hormone-occupied state.

Ever since the inception of the two-step model, experimental evidence has been accumulating, which is difficult to reconcile with the model. Recently, a different model of steriod receptor localization has been proposed, disbanding the existence of the cytosolic receptor. Using powerful immunocytochemical techniques, the proponents of this "nuclear model" (Fig. 3) localized all receptor molecules, unoccupied or occupied, in the nuclear region (King and Greene, 1984;

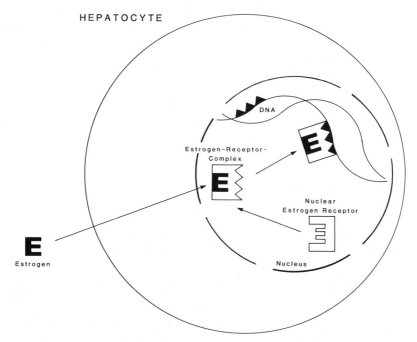

Fig. 3. Nuclear model of estrogen–receptor mechanism. Estrogen enters the target cell and binds to highly specific nuclear receptor proteins. Estrogen–receptor complex displays high affinity for chromatin, binds to specific sites on the DNA, and, among other things (cf. Table II), leads to activation of the vitellogenin gene.

Welshons *et al.*, 1984). Since these authors found the cytosol to be devoid of appreciable amounts of steriod binding sites, they came to the conclusion that the larger, cytosolic receptor may be an experimental artifact. In this model, no nuclear translocation of receptor/hormone complex occurs and the receptor attains its increase in affinity for nuclear structures by binding hormone directly within the nuclear compartment. Highlighting the merits of the comparative approach, the work of Callard and Mak (1985) on the estrogen receptors of the testes in an elasmobranch fish (*Squalus acanthias*) support this novel working model; both occupied and unoccupied receptors are exclusively associated with the nuclear compartment.

In recent years, evidence has accumulated that many steriod receptors fall into an interesting group of proteins that are regulated by reversible covalent modifications through phosphorylation/dephosphorylation. Experimental data suggest that phosphorylation is a prerequisite for the hormone binding activity of the receptor protein and that dephosphorylation leads to an inactivation of the receptor molecule. While most proteins, including the chicken progesterone receptor (Dougherty *et al.*, 1982) and the glucocorticoid receptor of fibroblasts (Housley and Pratt, 1983), are phosphorylated at serine residues within the polypeptide chain, the calf uterine *estrogen* receptor protein is phosphorylated on tyrosine (Migliaccio *et al.*, 1984), a relatively unusual phosphorylation site for covalent modification (Krebs, 1985).

In light of the present controversy over the actual cellular localization of unoccupied estrogen receptors, where the evidence points to the cytosolic receptor as a possible experimental artifact (King, 1984), it is somewhat confusing to reconcile the new interpretation with the fact that phosphorylation of the calf estrogen receptor is brought about by a specific, calcium- and calmodulin-dependent cytosolic receptor kinase. Not surprisingly, however, the dephosphorylation-dependent inactivation of the receptor is caused by a phosphatase (Migliaccio *et al.*, 1984) with exclusive localization within the nuclear compartment.

While the validity of the existing models has not been tested for hepatic tissue in piscine systems, the presence of highly specific estrogen receptors has been verified for a number of different species of fish, namely, *Gobius niger* (Le Menn *et al.*, 1980), Pacific hagfish (*Eptatretus stouti;* Turner *et al.*, 1981), Atlantic salmon (*Salmo salar;* Lazier *et al.*, 1985; Mommsen and Lazier, 1986), sea-raven (*Hemitripterus americanus*), winter flounder (*Pseudopleuronectes americanus*), sculpin (*Myoxocephalus octodecimspinosus;* Mann *et al.*, 1988), rainbow trout (*S. gairdneri;* Maitre *et al.*, 1985b; Mann *et al.*, 1988) and most recently in the brown trout (*S. trutta;* Pottinger, 1986).

The livers of nonvitellogenic or male Atlantic salmon (*S. salar*) contain specific high-affinity estrogen binding proteins in the cytosolic fraction, while the liver nuclei reveal low concentrations of high-affinity estradiol-binding components. Injection of pharmacological doses of estradiol in the fish leads to a transient depletion of the cytosol binder and the appearance of exceptionally high concentrations of estradiol binding sites in nuclear salt extracts (Fig. 4). In naive fish, the concentration of binding sites in the cytosol was 640 fmol/g liver, while their concentration in liver nuclei was 150 fmol/g liver. After the administration of a single dose of estradiol, the concentration of nuclear binding sites increases almost 80-fold to above 11 pmol/g liver (Lazier *et al.*, 1985). The induction is apparently highly specific for estradiol and, after a single injection of estradiol into an experimental fish, reaches a maximum of receptor response after some 24 h. Such response can just as easily be elicited in cultured hepatocytes, using physiological, rather than the more usual pharmacological, doses of hormone (Fig. 4A). The binding characteristics are identical for the in vivo or the in vitro response (Fig. 4B).

Put into the proper perspective of the cellular environment, the maximum concentration of accumulated estrogen receptor is dwarfed in comparison with the products of some commonly expressed genes. Most enzymes involved in liver cell metabolism, for instance, occur in concentrations ranging from 10 to 100 nmol per gram of liver, that is, at least two orders of magnitude higher than the maximum amount of estrogen receptor. One gram of liver contains about 0.9 μg estrogen receptor, assuming a molecular mass of about 80 kDa, compared with almost 500 μg pyruvate kinase alone.

A similar increase in the total level of cellular estrogen-binding sites, where the increase in nuclear binding sites more than offsets the decrease in cytosolic sites, is observed when cultured salmon liver cells are exposed to physiological doses of estradiol (Mommsen and Lazier, 1986; cf. Fig. 4). The time course and the peak concentration reached differ somewhat between the in vivo and the in vitro systems, possibly due to the amounts of estradiol administered and enhanced metabolism in the isolated liver cells. While this particular subject was not the focus of any of the studies on fish, it is interesting to note that the metabolism of the steriod hormone itself and not so much the kinetics of the hormone–receptor decay influences the effectiveness of the hormone administration. In tissue cultures of *Xenopus* hepatocytes, for instance, estradiol turnover is rather rapid, especially in cells derived from male toads. Here, the half-life of the molecule is in the range of only 40 min (Tenniswood *et al.*, 1983) at 20°C. It is obvious that if similar conditions are found in isolated fish cells, very

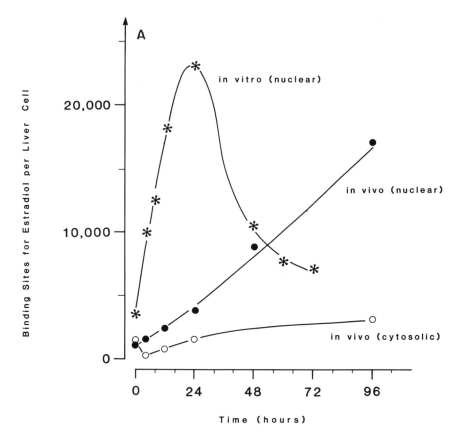

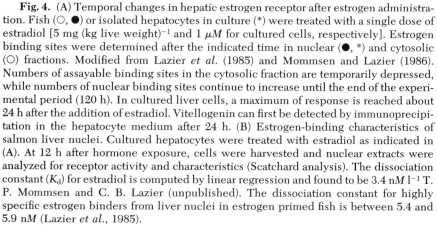

Fig. 4. (A) Temporal changes in hepatic estrogen receptor after estrogen administration. Fish (○, ●) or isolated hepatocytes in culture (*) were treated with a single dose of estradiol [5 mg (kg live weight)$^{-1}$ and 1 μM for cultured cells, respectively]. Estrogen binding sites were determined after the indicated time in nuclear (●, *) and cytosolic (○) fractions. Modified from Lazier *et al.* (1985) and Mommsen and Lazier (1986). Numbers of assayable binding sites in the cytosolic fraction are temporarily depressed, while numbers of nuclear binding sites continue to increase until the end of the experimental period (120 h). In cultured liver cells, a maximum of response is reached about 24 h after the addition of estradiol. Vitellogenin can first be detected by immunoprecipitation in the hepatocyte medium after 24 h. (B) Estrogen-binding characteristics of salmon liver nuclei. Cultured hepatocytes were treated with estradiol as indicated in (A). At 12 h after hormone exposure, cells were harvested and nuclear extracts were analyzed for receptor activity and characteristics (Scatchard analysis). The dissociation constant (K_d) for estradiol is computed by linear regression and found to be 3.4 nM l^{-1} T. P. Mommsen and C. B. Lazier (unpublished). The dissociation constant for highly specific estrogen binders from liver nuclei in estrogen primed fish is between 5.4 and 5.9 nM (Lazier *et al.*, 1985).

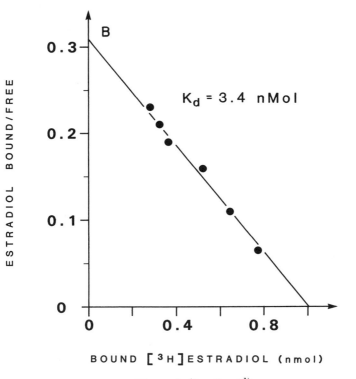

Figure 4. (*Continued*)

little exposure is needed to initiate the transcription of estradiol-dependent genes, in this case the estrogen-receptor gene, thus setting up an interesting positive feedback system.

Compared with other vitellogenic vertebrates, the teleost liver appears to be the richest source of highly specific estrogen receptors, making the fish liver an ideal model system to study the induction of receptor and analyze in detail the mechanism of hormone–receptor and receptor–chromatin interactions in lower vertebrates. The values listed in Table I compare the magnitude of the receptor induction in *S. salar* with other oviparous vertebrates utilized in the analysis of the biochemistry of the estrogen receptor. Recently, it was found that the salmon is not unique in this respect, nor is such largesse of response restricted to salmonid fish. Liver nuclei isolated from the sea-raven (*H. americanus*), the longhorn sculpin (*M. octodecimspinosus*), winter flounder (*P. americanus*), and the rainbow trout (*S. gairdneri*) all accumulate similarly high concentrations of estrogen binding proteins as

Table I
Magnitude of Estrogen-Receptor Response in
Vitellogenic Vertebrates[a]

	Naive animals	After induction with estradiol
Salmo salar	0.15	> 12
Xenopus laevis	0.2	2.5
Gallus domesticus	0.1	0.4

[a] Values are given as 10^{-12} moles of nuclear binding sites for estradiol per gram of liver. Concentration of cytosolic binding sites in the untreated salmon was about 0.6×10^{-12} mol (g liver)$^{-1}$. For time courses of receptor abundance see Fig. 4. The maximum number of nuclear binding sites in vivo and in vitro computes to around 25,000 nuclear estrogen receptor molecules per liver cell. Sources: Hayward *et al.* (1980), Lazier (1975), and Lazier *et al.* (1985).

the Atlantic salmon (Mann *et al.*, 1988). In a goby (*Gobius niger*), an elasmobranch (*Potamotrygon, ssp.*), and a hagfish (*Eptatretus stouti*), on the other hand, numbers of estrogen-binding sites in the liver are more than an order of magnitude lower than in the salmon (Le Menn *et al.*, 1980; Callard and Mak, 1985; Turner *et al.*, 1981).

The magnitude of the receptor response, which resembles that in some mammalian systems (Walter *et al.*, 1985), and the reported stability of the nuclear binding protein in many teleost fishes (Lazier *et al.*, 1985) should make it possible to attain receptor preparations of highest purity. To date, this goal has been hampered by the poor response in all other oviparous vertebrates (see Table I). The anticipated availability of receptor preparation of extreme purity may help to shed light on the ongoing controversy over the cellular distribution of hepatic receptor (King, 1984; Szego and Pietras, 1985) and the complex fate of occupied estrogen receptors in the cell nucleus (Shapiro, 1982). Furthermore, comparisons with estrogen receptors from mammalian tissues will furnish insights into the evolutionary trends of the receptor gene (Greene *et al.*, 1986).

While the actual source of these comparatively large amounts of the gene-regulatory estrogen receptor in the piscine system has not been elucidated yet, some indirect evidence in other vitellogenic vertebrates indicates that de novo synthesis of receptor protein is involved (Lazier, 1975; Perlman *et al.*, 1984). Again, the highly sensitive piscine system seems ideally suited to supply mechanistic insight into the source of "induced" receptor molecules by molecular biology techniques, similar to the ones used to assess the transcriptional activity and mRNA longevity of the albumin gene in *Xenopus* (Wolffe *et al.*, 1985), for instance.

In all parameters analyzed, the fish receptor proteins resemble the receptors from other vertebrates. The salmon receptors are characterized by a high specificity for extradiol and they do not bind progesterone, hydrocortisone, or dihydrotestosterone. In agreement with studies of receptors from many other vertebrate sources, the fish receptors display high affinity for the nonsteroidal estrogen diethylstilbestrol as well as for the nonsteroidal antiestrogen 4-hydroxytamoxifen. The estrogen binding proteins in the liver of the hagfish *E. stouti*, on the other hand, display unique features, different from the global vertebrate picture. This species possesses nuclear estrogen receptors with a lower affinity for extradiol than other vertebrate counterparts (dissociation constant $K_d=38$ nM versus $K_d=$ 3–6 nM in the salmon; see Fig. 4B). However, the hagfish system is unusual in that estrone or estriol displaced estradiol from the nuclear binding components as efficiently as estradiol or diethylstilbestrol (Turner *et al.*, 1981). In other vertebrates, binding affinities for estriol or estrone are usually more than an order of magnitude lower than for estradiol. In the rainbow trout, estrone administration leads to the induction of vitellogenin synthesis in the liver and its release into the bloodstream, but estrone displays only 5% to 12% of the potency of estradiol (van Bohemen *et al.*, 1982a,b). It seems that one of the functions of estrone in vivo may be to prime hepatic tissue for subsequent exposure to estradiol and thus to potentiate the vitellogenic response of the hepatocytes to estradiol. The interesting characteristics of the hagfish receptor, together with the positioning of the cyclostomes within the vertebrate line, might in future shed some light on the evolution of steroid sex hormones, their interactions, and receptor specificity.

In the annual cycle of the rainbow trout, the liver is exposed to differing concentrations and ratios of estradiol and estrone. While blood concentrations of both estrogens increase during early vitellogenesis, the first phase of vitellogenesis is dominated by estrone, which increases by a factor of 10 altogether (van Bohemen and Lambert, 1981). During the later stages of exogenous vitellogenesis, estradiol reaches blood concentrations of 60 ng/ml, reflecting an increase of 60-fold. It would be interesting to analyze whether similar changes are reflected in the abundance or preference of specific estrogen receptors in the nuclei of the liver, the main target organ for ovarian estrogens.

In addition to the estrogens, androgens are able to elicit a vitellogenic response in teleost fish, albeit only when administered in pharmacological doses, (Le Menn, 1979; Hori et al., 1979). Interestingly, at
least in G. niger this response appears to be mediated by androgen
binding to the estrogen receptor rather than through the nuclear androgen receptor itself (Le Menn et al., 1980). Similarly, high doses of
androgen fed to juvenile salmon may lead to a pronounced feminiza-
tion of some fish (Solar et al., 1984), although the molecular mechanisms for these phenomena remain to be analyzed.

The sedimentation coefficient of the salmon receptor protein of 3.6
S indicates that it may be a little smaller than the nuclear estrogen
receptors of birds or mammals, but it falls into the same range as the
receptor isolated from Xenopus laevis (Lazier, 1978; Wright et al.,
1983).

C. Plasma Binding Proteins

The steroid hormones produced and released by the ovarian cells
are transported to their target tissue in the systemic circulation. Although probably not a target issue in itself, fish plasma displays a
certain degree of steroid-binding capacity. For a variety of fishes as
well as for other vertebrate groups, such "sex-steroid binding proteins" have been characterized numerous times (Wingfield, 1980).
Their specificities and properties clearly distinguish plasma binders
from the cellular steroid receptors, while their exact physiological
function, over and above the suggested role in steroid transport, is still
under debate. Since steroid hormones exert their biological functions
only in the free and not in the bound form, such plasma steroid-
binding proteins may serve to buffer free steroid concentrations in
conditions of high steroid turnover, thus obviating time consuming de
novo synthesis.

In the plasma of S. salar, two differing estradiol binding components are abundant, one with high affinity and one with low affinity for
estradiol (Lazier et al., 1985). In contrast to the highly estradiol-spe-
cific nuclear receptors inducible in the salmon liver, neither of the
plasma-binding components are competed for by the nonsteroidal estrogen diethylstilbestrol. Furthermore, again differing from the situation in the liver, the antiestrogen 4-hydroxytamoxifen does not compete with estradiol for binding to the high-affinity ($K_d=13$ nM)
estrogen binder in plasma. Experiments also indicate that the androgen dihydrotestosterone as well as progesterone and estrone reveal

considerable affinity for the plasma binder and are likely to compete with extradiol in vivo as they do in vitro (Lazier *et al.*, 1985).

D. Hepatic Events

With a time delay of a few hours following the binding of the estrogen/receptor complex to the nuclear DNA, a variety of changes in liver cells are initiated that are consistent with a substantial increase in the capacity for protein synthesis and export—plasma concentration of vitellogenin may reach 50 mg/ml (Ng and Idler, 1983). Indeed, naturally vitellogenic fish reveal much higher rates of hepatic protein synthesis than nonvitellogenic fish (e.g., Haschemeyer and Mathews, 1983; Yu *et al.*, 1980; Emmersen and Korsgaard, 1983), a phenomenon that can be provoked by estrogen administration in vivo as well as in vitro. Several ultrastructural differences are observed between liver cells from immature and vitellogenic fish. In the red grouper (*Epinephelus akaara*), vitellogenic livers are characterized by expanded nuclear envelope cisternae, swollen mitochondria, and much enhanced rough endoplasmic reticulum, Golgi apparatus and secretory vesicles (Ng *et al.*, 1984). Hepatocytes of naive fish treated with estradiol showed similar, but not entirely identical ultrastructural changes (Ng *et al.*, 1984). Several studies indicate an increase in hepatosomatic index (van Bohemen *et al.*, 1982a,b; Dasmahapatra *et al.*, 1981) and, at least in the red grouper, this appears to be due to a rise in cell lipid and water content rather than proliferation of cell numbers (Ng *et al.*, 1984). In the Atlantic salmon, as in the flounder, estradiol administration leads to increases in liver protein, total RNA, and total nuclear count (Korsgaard *et al.*, 1986; Korsgaard and Petersen, 1976). Since at the same time the liver volume and weight increase, calculated on a unit weight basis, only the amount of cellular RNA is augmented significantly. It can be concluded that in these two species of fish, differing from the grouper, hyperplasia rather than hypertrophy appears to be responsible for the enhanced liver weight (Korsgaard *et al.*, 1986). The more than 30% increase in cellular RNA content (on a unit weight basis) is yet another indication of the increased biosynthetic activity of the liver (Korsgaard *et al.*, 1986), where the de novo synthesis of messenger RNA for vitellogenin may account for some of the observed increase in total RNA.

A larger proportion of the newly synthesized RNA in hepatic tissue following the exposure to estradiol is due to apparent increases in the amounts of ribosomal RNA, which can be explained by the massive increases in rough endoplasmatic reticulum observable in liver micro-graphs of fish and other oviparous vertebrates (Bast *et al.*, 1977; Selman and Wallace, 1983a). Since estrogen administration is respon-sible for the proliferation of cell structures, such as endoplasmic retic-ulum (ER), Golgi vesicles (and turnover), and mitochondria, genes coding for any of these structures must have been activated or estro-gen administration must have at least led to increased translational activity involving existing mRNAs. Obviously estradiol is able to or-chestrate cell metabolism and biosynthetic activities at a number of different levels.

An ancillary question concerns the actual localization of hepatocy-tes active in the synthesis and export of vitellogenin. Contrary to the widespread belief that all hepatocytes are metabolically identical, it has been shown rather conclusively that rat hepatocytes in the perive-nous and periportal regions of the liver possess differiong metabolic functions, with anabolic pathways such as gluconeogenesis, fat syn-thesis, and proteins synthesis being favored in the better oxygenated periportal cells (Jungermann and Katz, 1982). It would be interesting to know whether a similar hepatic zonation exists in the lower verte-brates in general and extends to estrogen receptors and the vitello-genic response.

Estrogen treatment of fish also appears to result in a general gear-ing up of metabolism to provide the large amounts of energy and reducing power (NADPH) necessary for protein and lipid synthesis (Ng *et al.*, 1984; Petersen and Korsgaard, 1977). Ng *et al.* (1984) fur-ther report significant and substantial increases in transaminases and enzymes of the Krebs cycle and glycolysis. On the other hand, natu-rally vitellogenic female sockeye salmon (*Oncorhynchus nerka*) on their spawning migration do not increase any specific metabolic ma-chinery in liver (on a weight basis, Mommsen *et al.*, 1980), apart from the general augmentation due to an—probably estradiol-dependent—increase in liver weight (Idler and Clemens, 1959).

In vivo treatment of male flounder (*Platichthys flesus*) results in an increase of protein synthetic activity when assessed in an in vitro system (Korsgaard *et al.*, 1983). That such stimulation may be a direct effect of estrogen on liver cells was recently demonstrated in hepato-cytes isolated from juvenile coho salmon (*O. kisutch*; Bhattacharya *et al.*, 1985). Hepatocytes treated with 17β-estradiol and exposed to either [^{14}C]serine or [^{14}C]glycine exhibit an increase in radioactivity precipitable by trichloroacetic acid (TCA), a decrease in TCA-soluble radioactivity, and enhanced release of TCA-precipitable radioactivity into the medium compared with untreated controls (Bhattacharya *et*

Table II
Cellular Events Associated with the Estrogen-Dependent Induction
of Vitellogenesis in Teleost Hepatic Tissue

Transient decrease in cytosolic estrogen receptor protein
Induction of nuclear estrogen receptor protein
Increase in hepatosomatic index due to hyperplasia or hypertrophy
Proliferation of Golgi apparatus
Increase in cisternae of the nuclear envelope
Synthesis of ribosomes
Polysome assembly
Increase in rough endoplasmatic reticulum
Swelling of mitochondria
Appearance of a new species of mRNA (vitellogenin)
Increase in protein synthetic activity
Synthesis of vitellogenin
Increase in cellular RNA
Increase in lipid metabolism (?)[a]
Augmented output of very low density lipoproteins (VLDL)
Decrease in glycogen content per cell
Increase in metabolic enzymes
Higher amount of hepatic DNA
Increase in hepatic water content

[a] For the fishes, to date only circumstantial evidence suggests this particular alteration. See text for relevant references.

al., 1985). Originally it was observed that liver slices from cod (*G. morhua*) treated with estradiol incorporated labeled [^{14}C]leucine into "egg proteins" (Plack and Fraser, 1971). All this evidence confirmed that in the teleost fishes, just as in other oviparous vertebrates, estradiol leads to the rapid and specific hepatic synthesis of the egg-yolk precursor vitellogenin. The most apparent ultrastructural and biochemical changes that hepatocytes experience during exogenous vitellogenesis are summarized in Table II.

E. Vitellogenin

In the last 10 years vitellogenin molecules from a number of different fishes have been isolated and partially characterized biochemically. Interestingly, the fishes display a much higher variability in the different parameters, such as molecular weight, degree of phosphorylation, degree of lipidation, or subunit composition than their amphibian or avian counterparts (Table III). As the example of the tetrameric vitellogenin in the Japanese eel shows (Hara *et al.*, 1980), not all fish vitellogenins are dimers—for instance, and in the case of the brown

Table III

Molecular Weights of Native Vitellogenin and Subunits from Fish Other Than Rainbow Trout

Species	Native molecular mass (kDa)	Subunit molecular mass (kDa)	Method[a]	Reference
Carassius auratus	326	140–156	Native/SDS–PAGE	Hori et al. (1979)
(goldfish)	380	140–147	Native/SDS–PAGE	de Vlaming et al. (1980)
Gadus morhua (cod)	400	—	Gel filtration	Plack et al. (1971)
Anguilla japonica (Japanese eel)	350	85	Gel filtration and SDS–PAGE	Hara et al. (1980)
Fundulus heteroclitus (killifish)	—	200	SDS–PAGE	Selman and Wallace (1983a)
Ameiurus nebulosus (brown bullhead)	—	145	[b]	Roach and Davies (1980)
Platichthys flesus (flounder)	550	—	Gel filtration	Korsgaard and Petersen (1976)
Oncorhynchus kisutch (coho salmon)	390[c]	—	Gel filtration	Markert and Van-stone (1971)
Salmo salar (Atlantic salmon)	495 and 520	—	Gel filtration	So et al. (1985)
Salmo trutta (brown trout)	440	—	Gel filtration	Norberg and Haux (1985)
Heteropneustes fossilis (catfish)	550	—	Gel filtration	Nath and Sundararaj (1981)

[a] SDS–PAGE: sodium dodecyl sulfate polyacrylamide gel electrophoresis.

[b] Translation product of the prevailing mRNA induced by estradiol treatment of bullhead catfish.

[c] Lipovitellin from coho salmon eggs.

bullhead, the messenger RNA for vigellogenin is substantially smaller than that for any other vertebrate (Table III). Even within the same species, in this case the rainbow trout (S. gairdneri), a large variation in the apparent molecular weight of the vitellogenin is noticed, which may be due to different methodologies used, different degrees of proteolytic breakdown, or dephosphorylation occurring during the isolation (Table IV). Some degree of heterogeneity in vitellogenin may be due to the fact that in the fishes, as in other vertebrates, vitellogenin is not encoded by a single gene, but rather by a family of slightly different genes.

As long as the translation of one isolated vitellogenin messenger RNA (mRNA) in this salmonid fish leads to different molecular-weight estimates for the vitellogenin monomer, more attention will have to

Table IV
Molecular Weight of Native Vitellogenin and Subunits from
Rainbow Trout *S. gairdneri*

Native molecular mass (kDa)	Subunit molecular mass (kDa)	Method	Reference
342	—	Ultracentrifugation	Campbell and Idler (1980)
440	—	Gel filtration	
440	—	Gel filtration	Norberg and Haux (1985)
470	—	Gradient PAGE	Campbell and Idler (1980)
—	170[a]	Gradient PAGE	Chen (1983)
—	200	SDS–PAGE of poly(A)+ translation product	Valotaire et al. (1984)
500	—	Gel filtration/SDS–PAGE	Sumpter (1981)
500	250	Not stated	Y. Valotaire (1984)[b]
535	16–103[c]	SDS and gradient PAGE	Maitre et al. (1985a)
600	220	Gel filtration/SDS–PAGE	Hara and Hirai (1978)

[a] Polypeptide plus posttranslational modifications, polypeptide alone = 160 kDa.

[b] Unpublished observation, cited in Valotaire et al. (1984).

[c] Eleven polypeptides that represent breakdown products from handling or precursors.

be devoted to multiple vitellogenin genes, to strain differences within one species, or to possible partial degradation of this large molecule or its mRNA—which do not critically affect the immunological reactivity—before a definitive answer with respect to molecular weight and phosphorylation sites can be given.

Estradiol treatment leads to the appearance of a specific high-molecular-weight species of messenger RNA (6300 or 7200 nucleotides) in the rainbow trout (Chen, 1983; Valotaire et al., 1984; see Table IV). Using cytoplasmic polyadenylated RNA isolated from these estradiol-exposed trout in a cell-free translation system, Chen (1983) was able to synthesize a 160,000-Da polypeptide that was chemically, immunologically, and electrophoretically identical to the authentic vitellogenin monomer. Similar results were obtained for the same species by Valotaire and co-workers (1984), although their larger mRNA (7200 nucleotides) upon translation yielded a considerably larger (200,000 Da) polypeptide, which was immunoprecipitable with antibodies against trout serum vitellogenin. The same authors also synthesized DNA complimentary to the estrogen-stimulated mRNA and back-hybridized with liver RNA to determine the increase in RNA due to estrogen treatment; the treatment increased it by 9%.

F. Posttranslational Modifications

The biochemical information concerning vitellogenin clearly indicates that a great deal of posttranslational modification must occur in the liver cell to reach the finished product seen in the serum. First the protein backbone of the vitellogenin is synthesized on membrane-bound ribosomes, a feature that it shares with other proteins destined to be secreted from the hepatocyte (Lewis *et al.*, 1976). In subsequent steps, the molecule must be lipidated, glycosylated, and phosphorylated. It has been suggested that all these processes occur on the membranes of the endoplasmatic reticulum and that they are already initiated while the polypeptide chain is being translated (Tata and Smith, 1979), although this view has been debated by Gottlieb and Wallace (1982). Finally, existing "pro" sequences or signal peptides have to be removed before vitellogenin is packaged into Golgi vesicles and secreted into the bloodstream.

While some information exists concerning the nature and extent of modifications on the vitellogenin molecule, rather limited information is available for fish with respect to the mechanism, sequential events, or locale of these transformations. Therefore, the following discussion has to be confined to a description of nonprotein components found on the circulating vitellogenin molecule.

Just as do the vitellogenins from other oviparous vertebrates, fish vitellogenins carry a certain number of phosphate groups, some of it as protein phosphorus, in a region that in the mature oocytes becomes deposited as phosvitin. Generally, the molecule is phosphorylated on serine moieties, and since the degree of phosphorylation of delipidated piscine vitellogenins ranges around 0.6–0.7% (by weight) (i.e., only about 50% of the protein phosphate content in other vertebrates), the serine content must be comparatively lower. Experimentally, this alkaline-labile protein phosphorus, which is specific to naturally or induced vitellogenic animals, has been utilized repeatedly for the determination of the degree of the vitellogenic response in fish. Vitellogenic female fish contain between 20 and 100 μg of protein phosphorus per milliliter of plasma, while untreated males contain less than 5μg/ml (Craik and Harvey, 1984). In spite of the large amounts of protein phosphate moved through the plasma compartment during vitellogenesis, in the unfertilized egg inorganic phosphate, and phospholipid, and not protein-bound phosphate, make up the bulk of the nuclear magnetic resonance (NMR) visible [^{31}P]phosphate (Grasdalen and Jørgensen, 1985). This observation indicates that additional maternal sources must supply phosphate and phospholipids to the oocyte. Also substantial dephosphorylation of vitellogenin-derived phosphoproteins during transmit through the oocyte or following deposition in the yolk may explain the low protein phosphate content of mature oocytes (Craik, 1982). The highly charged phosphate component also gives the vitellogenin molecule its high ion-binding capacity. Teleost vitellogenins are known to bind ions such as calcium, magnesium, or iron efficiently (Hara, 1976; Hara and Hirai, 1978; Hara *et al.*, 1980) and thus may designate an important vehicle for mineral supply to the growing oocyte. In fact, the competition of vitellogenin with chelating substances has been used successfully to isolate fish vitellogenins from other plasma proteins (Ng and Idler, 1983).

In contrast to the phosphate content of fish vitellogenins, which is lower than that of other oviparous vertebrates, the amounts of lipid material carried on the vitellogenin molecule are generally about twice as high as for other vertebrate groups. The lipid content of vitellogenin ranges around 20% by weight in fishes as different in lifestyle and feeding preferences as the goldfish (21%; Hori *et al.*, 1979), rainbow trout (21%; Wiegand and Idler, 1982; Frémont *et al.*, 1984; 18%, Norberg and Haux, 1985), sea-trout (19%; Norberg and Haux, 1985), or the elasmobranch dogfish (18%; Craik, 1978a). The bulk of this lipid material, which later forms the lipovitellin moiety of the yolk, can be classified as polar lipid (Hori *et al.*, 1979). In rainbow trout vitellogenin, for instance, polar lipids make up some 82% of the total (Wiegand and Idler, 1982). Generally the mature oocyte, however, contains much larger percentages of triglyceride, and it is therefore reasonable to assume that sources other than vitellogenin must supply the oocyte with nonpolar lipids, such as triglycerides, steryl esters, sterols, and wax esters. In this context it is interesting to note that dietary manipulation of free fatty acids in trout is reflected in altered fatty acid composition of serum lipids, but not of the lipoproteins, which are most important during vitellogenesis (Frémont *et al.*, 1984).

While fish vitellogenin is known to contain carbohydrate groups, little concrete information on the amount, nature, and linkages of the carbohydrates is available. However, it is known that for many proteins, successful glycosylation is a prerequisite for excretion of the export protein. In other cases, such as the chicken ovalbumin, which usually occurs in glycosylated form, no glycosylation is required for excretion. Experiments utilizing tunicamycin, a specific inhibitor of N-glycosylation, revealed that the absence of the oligosaccharide side chains from the ovalbumin molecule had no effect on its secretion (Colman *et al.*, 1981). For comparative purposes and from an evolutionary perspective, it would be a rewarding task to determine the group of glycoproteins to which the fish vitellogenins belong and whether successful glycosylation is a prerequisite for excretion from the hepatocyte. Obviously, there is a large information gap between the process of glycosylation of the vitellogenin molecule in the liver and the presence of large amounts of sialoglycoproteins in fish eggs (Inoue and Iwasaki, 1980a,b).

Vitellogenin could be detected in the blood but not in the livers of estradiol-treated rainbow trout, a result that was first interpreted to indicate that vitellogenin is rapidly secreted following synthesis (van Bohemen *et al.*, 1982b). However, Nunomora *et al.* (1983), using the peroxidase–antiperoxidase complex method (immunologically specific for vitellogenin), were able to localize significant amounts of vitellogenin in livers of estradiol treated rainbow trout (*Salmo gairdneri*), chum salmon (*O. keta*), or charr (*Salvelinus leucomaenis*). Similarly, So and co-workers (1985) detected cross-reactivity of antibodies against salmon (*Salmo salar*) vitellogenin with liver extracts of vitellogenic fish. These results can be reconciled by the fact that van Bohemen *et al.* (1982b) assayed for vitellogenin by molecular weight determination on sodium dodecyl sulfate (SDS) polyacrylamide gels and thus screened for mature vitellogenin rather than immunoreactive components (see below).

More information regarding posttranslational modification is available from other vertebrate systems. Recently, rooster hepatocytes were used to determine the probable sequence of events in hepatic vitellogenesis (Wang and Williams, 1982). Precursors (pVTG I and pVTG II) for each of the two types of avian vitellogenin (VTG I and VTG II) were found in hepatocytes of roosters treated with estrogen by pulse-labeling with [^{3}H]serine and pulse-chase experiments. The molecular weights of the precursors were lower than those of the mature vitellogenins as determined by SDS gel electrophoresis. However, further analysis of these polypeptides by immunological methods, peptide mapping, and molecular-weight determinations by gel chromatography revealed that the precursors are similar to mature vitellogenin in size and degree of glycosylation, but are not phosphorylated. Wang and Williams (1982) could also show that highly phosphorylated proteins, such as mature avian vitellogenin, will yield erroneously high molecular weights on SDS gels. The very small quantities of phosphorylated vitellogenin inside of these hepatocytes led Wang and Williams (1982) to suggest that phosphorylation is rapidly followed by secretion. Their determination of vitellogenin molecular weight by gel chromatography caused the same authors (Wang and Williams, 1982) to revise the "accepted" molecular weight for the avian vitellogenin monomer from above 235,000 to 180,000. In light of the obvious controversies about the actual molecular weights of piscine vitellogenins, even within the same species (see Table III), this type of multifaceted approach represents a fertile area for research on vitellogenesis in fish. In this field, the isolated hepatocyte systems would appear to be an excellent, as yet underutilized, experimental tool (Moon et al., 1985). Recently, several laboratories have been able to prove that fish hepatocytes in suspension or in primary culture are highly responsive to estrogen (Bhattacharya et al., 1985; Mommsen and Lazier, 1986), and that hepatocytes isolated from primed fish will synthesize and excrete large amounts of vitellogenin in vitro (Haschemeyer and Mathews, 1983).

From the reviewed studies on fishes and other egg-laying vertebrates, a preliminary picture of the sequence of events implicated in exogenous vitellogenesis can be synthesized (Table V). Unfortunately, especially for the fishes, many parts of the scheme require a major concerted research effort from biochemists, physiologists, and molecular biologists alike to replace speculation and add information on actual mechanisms. The major task would be to successfully utilize the vast potential of piscine system to elucidate and understand the estrogen receptor mechanism, its interaction with the nuclear DNA, and not least the subsequent gene activation. Further challenging topics include the diverse posttranslational modifications of the vitellogenin molecule occurring in the liver cell. Furthermore, the particular intracellular structures where the individual steps occur have to date eluded identification.

Recent work on the *number* of similar vitellogenin molecules of *Xenopus* has revealed that the situation is not quite as clear-cut or simple as it first appeared. Rather than being just one protein, coded for by one type of mRNA, a whole family of vitellogenin genes is in existence (Wahli et al., 1981), all of which give rise to slightly different vitellogenin molecules. These in turn supply the growing oocyte with the different building blocks for at least five different types of yolk polypeptides, namely lipovitellins 1 and 2, phosvitin, and phosvettes 1 and 2 [nomenclature of Wiley and Wallace (1981)], which in themselves are somewhat heterogeneous. Lipovitellins 1 and 2 can each be resolved into three differing polypeptide components, while dephosphorylated phosvitin yields two polypeptide bands of different molecular weights on SDS electrophoresis. Phosvettes are relatively small phosphorylated components with single polypeptide chains. From their study on the vitellogenin of *Xenopus*, Wiley and Wallace

Table V
Suggested Sequence of Events during Hepatic Synthesis of Vitellogenin[a]

Nuclear compartment
  Activation of vitellogenin gene through binding of receptor–hormone complex to
    specific regions of the nuclear DNA
  Transcription and presence of primary transcript in the nuclear compartment
  Processing of primary transcript
  Translocation to cytoplasm
Rough endoplasmatic reticulum
  Polysome assembly
  Translation of vitellogenin mRNA
  Processing of previtellogenin subunits
  Phosphorylation at serine residues[b]
  Lipidation[c]
  Translocation to smooth endoplasmatic reticulum
Smooth endoplasmatic reticulum
  Further phosphorylation at serine residues[b]
  Translocation to Golgi apparatus
Golgi apparatus
  Glycosylation
    Mannose
    N-Acetylglucosamine
    N-Acetylneuraminic acid, etc.
  Lipidation[c]
  Removal of existing signal peptides
  Dimerization
  Phosphorylation at serine residues[d]
  Excretion into systemic circulation

[a] Adapted from Tata and Smith (1979), Wang and Williams (1982), and Gottlieb and Wallace (1982).
[b] In *Xenopus*, phosphorylation occurs in the rough endoplasmatic reticulum and the smooth endoplasmatic reticulum only.
[c] The exact cellular site of the noncovalent attachment of lipid to vitellogenin is still under debate.
[d] In the chicken, vitellogenin is phosphorylated during its time in the Golgi apparatus, followed by rapid excretion from the hepatocyte.

(1981) came to the conclusion that the whole gamut of yolk proteins is derived from multiple vitellogenin molecules and also that the phosvettes are alternate cleavage products from homologous regions of different parent vitellogenins. Using complimentary DNA copies of *Xenopus* vitellogenin mRNA, Wahli *et al.* (1981) deduced that vitellogenin is encoded in a family of at least four expressed genes.

A similar situation appears to exist in the chicken, where three different genes of the vitellogenin family are expressed, producing three polypeptide chains with molecular weights ranging from

Table VI
Selected Estradiol-Dependent Genes in Lower Vertebrates[a]

Proteins	Organ	Organism(s)	Effect
Ovalbumin	Oviduct	Birds	Induced
Lysozyme			Induced
Conalbumin			Induced
Ovomucoid			Induced
Avidin			Induced
Vitellogenin	Liver	Oviparous vertebrates	Induced
Albumin		*Xenopus, Oncorhy-*	
		nchus nerka (?)	Depressed
ApoB, ApoII (VLDL)		Chicken, teleosts (?)	Induced
Vitamin-binding proteins		Birds	
Biotin			Induced
Thiamin			Induced
Cobalamin			Induced
Riboflavin			Induced
Transferrin			Induced
Estrogen receptor		Teleosts, chicken	Induced

[a] For structural and further biochemical changes initiated in the hepatocyte under the influence of estradiol, see Table II. References: Muniyappa and Adiga (1980), Leger *et al.* (1981), Lazier *et al.* (1985), White (1985), Wolffe *et al.* (1985).

170,000 to 190,000. Moreover, the three subunits possess different degrees of phosphorylation, and subsequently make up the native vitellogenin dimer (Wang *et al.*, 1983). By analogy, a similar multi-gene family can be expected to code for vitellogenin in the fishes. The observed multitude of differing egg phosphoproteins further suggests widespread heterogeneity within the vitellogenin molecule.

G. Other Actions of Estradiol

The specific action of estradiol at the nuclear level in hepatic tissue is by no means restricted to the activation of the vitellogenin gene, although, at least in the fishes, vitellogenin constitutes the single most important de novo synthesis of protein. In the chicken, which has attracted most attention in this respect, a number of other genes are activated concomitantly, including those coding for apoVLDLII—the major VLDL (very-low-density lipoproteins) in laying chickens (Deeley, *et al.*, 1985)—as well as various vitamin-binding proteins. A preliminary (i.e., constantly growing) list of genes that are directly affected by estradiol in various vertebrates is given in Table VI. In a mechanistically unknown fashion, estradiol also induces the multitude of ultrastructural changes occuring in the liver of all vertebrates actively undergoing vitellogenesis (cf. Table II). As pointed out above, the teleost fishes add another estrogen-induced gene to this growing list, namely, the gene for the nuclear estrogen receptor protein.

Estradiol also exerts negative effects on the synthesis of other export proteins, among them the ubiquitous serum albumins. In many vertebrates, estradiol administration leads to a pronounced reduction in the concentration of albumin circulating in blood, an effect that is especially apparent in chronically estradiol-exposed male *Xenopus*. Experiments conducted by Tata and co-workers (Wolffe *et al.*, 1985) led to the conclusion that in this amphibian, two levels of estrogen action on albumin synthesis can clearly be distinguished. First, estradiol administration leads to a deinduction of transcription of the two genes coding for the larger (74 kDa) and more abundant albumin. Second, it also causes a substantial destabilization of the messenger RNA for albumin, which is reflected in a decrease in the actual half-life of the messenger RNA by two-thirds (Wolffe *et al.*, 1985). The same deinduction of albumin synthesis in the presence of estradiol can also be observed *in vitro* using isolated hepatocytes (Wangh, 1982).

In fishes, a similar reduction in the amount of circulating albumin is apparent in naturally vitellogenic sockeye salmon (*O. nerka*) during their spawning migration (T. Mommsen, S. Mookerjea, and C. French, unpublished results). Conversely, in the rooster, estradiol *withdrawal* results in the destabilization of vitellogenin and apoVLDLII mRNAs, while the stability of the serum albumin mRNA is not affected (Wiskocil *et al.*, 1980). In addition to inducing the de novo synthesis of vitellogenin mRNA, estradiol has been shown to accelerate the rate of transcription of other genes, while not necessarily altering the *amounts* of mRNAs coding for different genes.

In all lower vertebrates, estradiol exerts a pronounced lipogenic action on peripheral tissues, while in *Xenopus* it also enhances the activities of enzymes involved in the hepatic synthesis of lipids (Phillips and Shapiro, 1981). It can be speculated that the de novo synthesized lipid will be partly destined for the lipidation of vitellogenin and partly for inclusion in the increased output of VLDL by the liver. To date, only one study has addressed this topic in fishes: female capelin (*Mallotus villosus*) displayed considerably higher total activities of fatty acid–catabolizing enzymes than did their male counterparts (Henderson *et al.*, 1984). However, as long as only one part of fatty acid metabolism (either anabolic or catabolic direction) is analyzed, no conclusive statement can be made about the lipid turnover in the respective tissues. Ultimately, the ratio of fluxes in the two directions will influence the actual *net* flux, and hence determine net import or export. The increased potential in vitellogenic grouper to generate cytosolic NADPH furnishes circumstantial evidence supporting the notion of increased hepatic fat synthesis during this period (Ng *et al.*, 1984). In light of the general observation of increased lipid content in the blood of vitellogenic fishes (Plack and Pritchard, 1968; Petersen and Korsgaard, 1976; Sand *et al.*, 1980), it can be speculated that while lipid turnover is stepped up, net flux is increased in the direction of lipid export from hepatic tissue. Strong lipogenic action of estradiol has been reported for *S. gairdneri* and *Heteropneustes fossilis* (Haux and Norberg, 1985; Dasmahapatra and Medda, 1982). Micrographs of vitellogenic livers of *Fundulus heteroclitus* contain less lipid depositions than livers from male fish (Selman and Wallace, 1983b), while the livers of two other teleosts (*Notemigonus crysoleucas* and *Brachydanio rerio*) increased the amounts of lipid under the influence of estradiol in a dose-dependent fashion (de Vlaming *et al.*, 1977; Peute *et al.*, 1978).

In the blenny (*Zoarces viviparus*), lipid is accumulated in the liver before vitellogenesis is hormonally induced, and subsequently, the lipid is mobilized and can be found in the bloodstream during vitellogenesis. Also, estradiol treatment during the course of pregnancy—a nonvitellogenic period in the blenny—leads to a dose-dependent accumulation in vitellogenin and a concomitant increase of lipids in the blood (Korsgaard and Petersen, 1979).

In conclusion, two different strategies can be envisaged with respect to lipid mobilization and estradiol action in different species of fish. The simpler situation exists in fishes that accumulate lipids within the liver, such as the cod or the blenny. Here, estradiol is likely to first cause a mobilization of intrahepatic lipid stores and later increase the output of VLDL from the liver. In fishes that use extrahepatic sites for lipid deposition, such as salmonids or a sculpin (*Leptococcus armatus*; de Vlaming *et al.*, 1984), estradiol first induces the mobilization of extrahepatic lipids, and perhaps subsequently paces their uptake into the liver leading to increased hepatic output of VLDLs.

The treatment of goldfish with salmon gonadotropin leads to an augmentation of plasma triglycerides and cholesterol (Wiegand and Peter, 1980) in goldfish with undeveloped ovaries, a phenomenon that is most likely mediated by gonadotropin-dependent estradiol production by the ovary. In animals undergoing the final stages of ovarian development, the same treatment decreases plasma lipid concentration, which is possibly due to a gonadotropin-enhanced (progesterone-dependent?) lipid uptake into the ovary.

Varied results are reported for the changes in intracellular glycogen content following estradiol treatment, although the generally observed trend seems to support the notion that hepatic glycogen is decreased in vitellogenic females. However, variable results for the contents of hepatic glycogen can be expected, since of all storage materials they are the most likely to be dependent on the preexperimental state of the experimental organism with respect to variables such as diet, photoperiod, and temperature.

Vitellogenic females of the killifish *F. heteroclitus* or estrogen-injected males contained less glycogen in their livers than uninjected male fish (Selman and Wallace, 1983a). A similar picture can be found in many other teleost fishes [*H. fossilis*, Dasmahapatra and Medda (1982); *Z. viviparus*, Korsgaard and Petersen (1979); *S. gairdneri*, Haux and Norberg (1985); *Anguilla anguilla*, Olivereau and Olivereau (1979)], where estradiol-primed vitellogenic fish generally contain less glycogen in their livers than do vehicle-injected controls. In the grouper, in contrast, induction of exogenous vitellogenesis leads to a marked increase in hepatic glycogen (Ng *et al.*, 1984). Sockeye salmon (*O nerka*) build up maximum liver glycogen levels at the end of the spawning migration, when exogenous vitellogenesis is approaching completion, and the fish subsequently call upon liver glycogen to fuel the exhausting spawning process (French *et al.*, 1983).

An integral part of estradiol action is the observed hypercalcemia in vitellogenic fish, which can largely be ascribed to the calcium-binding properties of phosphorylated, and hence highly charged, components of the native vitellogenin molecule. Furthermore, this hypercalcemia has been employed to confirm the vitellogenic state of experimental animals. Fish scales have been singled out as the suggested source of the bound calcium (Mugiya and Watabe, 1977), while, for once, estradiol does not seem to be implicated in the uptake of environmental calcium, neither through the gills nor through the intestine (Mugiya and Ichii, 1981). As in the case of carotenoid binding, the actual site for the attachment of calcium to the vitellogenin molecule has not been identified, although liver seems the most likely candidate. If in the future it can be confirmed that other metals, such as copper or cadmium, travel from their hepatic deposition site to the ovary bound to vitellogenin (Shackley *et al.*, 1981), it will be appreciated how easily heavy metals will be able to impair the fine-tuned ion balance of the growing oocyte.

It is interesting to note that cortisol, a steroid hormone, which is known to exert direct metabolic effects by way of enzyme induction and permissive effects on peptide hormones such as glucagon, also possesses a pronounced enhancing influence on the estrogen-induced synthesis of vitellogenin in the catfish *H. fossilis* (Sundararaj *et al.*, 1982a). This situation is somewhat reminiscent of the estrone-dependent priming of vitellogenesis through estradiol. Glucocorticoid administration to cultured hepatocytes curtails the vitellogenic response to estradiol, while at the same time enhancing the production of albumin, a protein whose synthesis may be suppressed in the presence of estrogen (cf. Table VI). Furthermore, recent experiments also suggest an important role for thyroxine, which is tightly bound by isolated fish liver nuclei (Bres and Eales, 1986), as an accelerating factor in exogenous vitellogenesis in the guppy (*Poecilia reticulata*; Lam and Loy, 1985). Evidently, a number of other hormones interact in an as yet undetermined manner with estradiol during exogenous vitellogenesis (cf., Leatherland, 1985); these interactions should provide a multitude of challenging topics of study for endocrinologists and molecular biologists.

H. Male Fish

It is an interesting facet of the induction of vitellogenesis that the estrogenic response can also be elicited in males of oviparous vertebrates, including fish (Emmersen *et al.*, 1979; Korsgaard *et al.*, 1983; Maitre *et al.*, 1985a). It clearly indicates that the administration of estradiol can activate normally silent genes. The complete absence of products of these unexpressed genes has made male animals a prime model for the analysis of gene regulation and activation. Basically, the male liver can be "reprogrammed" to synthesize and export large amounts of vitellogenin and other proteins, a process that appears to occur without involving DNA replication. Since an appropriate deposition site is lacking in the male, the fate of the vitellogenin in the bloodstream differs: it builds up to rather high concentrations and eventually is taken up by the liver and degraded along with other blood proteins.

The actual process of vitellogenesis is accompanied by identical patterns of hepatocyte differentiation in both sexes, including the proliferation in Golgi vesicles, rough endoplasmatic reticulum, and RNA mentioned (cf. Table II). In the male Atlantic salmon (*S. salar*), the estrogenic response also includes an increase in the amount of assayable nuclear estrogen receptor to levels characteristic of induced female fish, which is probably due to de novo synthesis of the receptor protein (Lazier et al., 1985). The identical situation in hepatocytes from male *Xenopus* has made it possible to unequivocally identify receptor synthesis as the rate-limiting step in vitellogenin gene transcription (Perlman et al., 1984).

In male fish, vitellogenin synthesis cannot be stimulated by the administration of pituitary extracts, indicating two specific properties of the vitellogenic response in fishes: (1) with regard to exogenous vitellogenesis, the liver is not a direct target organ for pituitary hormones, and (2) in males, vitellogenesis is specifically dependent on estrogen administration, because of the inability of the gonad to produce estrogen (Idler and Campbell, 1980).

I. Elasmobranch Fishes

In general, vitellogenesis and its hormonal control in the elasmobranch fishes have received much less attention than in teleost fishes. The few studies on elasmobranchs suggest that, even in species from temperate zones, vitellogenesis and oviposition appear to occur throughout the year, with a maximum during winter (Sumpter and Dodd, 1979). As a consequence, vitellogenin is detectable in dogfish (*Scyliorhinus canicula*; Craik, 1978b) and skate (*Raja erinacea*; T. P. Mommsen, unpublished) blood throughout the year, albeit in a low concentration compared with vitellogenic teleosts. The biochemical properties of the elasmobranch vitellogenins and their relationship to vitellogenins from other vertebrates remain to be analyzed.

Injection of estradiol results in a much smaller vitellogenic response than in teleosts (Craik, 1978a). While the synthesis of vitellogenin in the female dogfish is a slow process compared with teleosts, its uptake into the ovary is fine-tuned to the rate of its synthesis. This results in an unusually long half-life for vitellogenin (9 days; Craik, 1978b) in dogfish plasma, and a similar result can be expected for other elasmobranchs that are vitellogenic throughout the year. The only other systems where such long half-lives for vitellogenin represent the rule rather than the exception are the estrogen-primed males of other vertebrates that possess no tissue that would recognize vitellogenin for uptake. In male, estrogen-injected *Xenopus*, for instance, vitellogenin is removed from the bloodstream at a rate of less than 1% per day—which resembles plasma protein turnover—compared to more than 12% per day in the vitellogenic female (Wallace and Jared, 1968).

III. OOCYTE ASSEMBLY

A. Transport of Vitellogenin

After the Golgi vesicles of the hepatocyte have unloaded the mature vitellogenin into the plasma, the circulatory system delivers it to the ovary. It appears that vitellogenin is dissolved freely in the plasma, since no special carrier molecule for it could be identified in teleosts or amphibians. A different situation is found in the blood of birds. In these vertebrates, vitellogenin is carried from its hepatic site of synthesis to the gonad as part of the high-density lipoproteins (HDL), which are regularly synthesized and excreted by the avian liver.

B. Uptake of Vitellogenin

In the females of oviparous vertebrates, with the possible exception of elasmobranch fishes, circulating vitellogenin is rapidly and specifically cleared from the bloodstream by the growing oocyte. In vitellogenic *Xenopus*, some 12% of the vitellogenin circulating in the blood is taken up by the gonad per day (Wallace and Jared, 1968). In the absence of a specific target tissue in estrogen-primed males, the vitellogenin continues to exist in the circulatory system until it is finally removed by the liver and degraded along with other plasma proteins.

The mechanism of vitellogenin recognition and the selectivity of its uptake into the oocyte remain open questions, especially for the fishes. Here only a single study on the rainbow trout has critically looked into these mechanisms (Campbell and Jalabert, 1979), with conclusions that do not support the picture that has emerged from a multitude of studies on *Xenopus* and the chicken.

In the latter two experimental animals, it appears that vitellogenin is bound on the oocyte membrane by specific, high-molecular-weight receptors (molecular weight ~500,000), which are taken up into the oocyte and turn over independent of vitellogenin binding. The receptor proteins display low nonspecific binding, are saturable, and apparently specifically recognize and bind the phosvitin region of the vitellogenin molecule; again, phosphorylation is crucial to the process of receptor recognition and vitellogenin uptake (Opresko *et al.*, 1981; Yusko *et al.*, 1981). Other studies, in addition, have implicated the

importance of N-glycosylation of the vitellogenin molecule on its uptake by the oocyte (Lane *et al.*, 1983). Similar receptors are presumed
to exist for VLDL in avian oocytes, and it has been pointed out elsewhere that the vitamin-binding proteins are only recognized and
taken up into the oocyte if adequately phosphorylated (Miller *et al.*,
1982). After its binding to the oocyte surface receptor, the vitellogenin
molecule, possibly in conjunction with its receptor, is taken up into
the oocyte by micropinocytosis. In the fishes, vitellogenin contains
only about half the protein phosphorus of other vertebrates, and the
phosvitins comprise a more heterogeneous group altogether. Therefore, a critical analysis of the involvement of phosphate groups in
these lower vertebrates may lead to interesting insights into receptor
recognition and receptor mechanism in general.

In subsequent steps, the vitellogenin is directed toward different
yolk sites within the oocyte, depending on the stage during vitellogenesis. While the receptor molecule appears to be recycled, the vitellogenin molecule is cleaved proteolytically into the main yolk components in the course of its translocation from the oocyte surface to the
yolk deposition sites. From an enzymatic point of view, the system
responsible for the cleavage of the vitellogenin molecule is poorly
characterized, but the lysosomal system seems to be implicated (see
below). Finally, the components such as the lipovitellins, phosvitins,
and phosvettes are deposited within membrane-bound spherical yolk
bodies, in many marine teleosts constituting fluid yolk globules rather
than the well-known insoluble platelets. Such yolk bodies form the
so-called "extravesicular yolk," which may fuse at some point during
oocyte development (Wallace and Selman, 1981).

The "intravesicular yolks" that have been described for growing
teleost oocytes are supposedly precursors of the cortical alveoli,
which shed their endogenously synthesized "yolk" at fertilization (te
Heesen, 1977; Wallace and Selman, 1981). The discussion in this
chapter will be restricted to the egg components derived from exogenous vitellogenesis and will therefore not be concerned with the auto-
synthetic intravesicular yolk as defined above.

Considering the rapidity and specificity of vitellogenin deposition
in teleost oocytes in the course of exogenous vitellogenesis, Campbell
and Jalabert (1979) obtained surprising results: developing trout oocytes in vitro did not take up vitellogenin selectively over serum albumin and at a rate that amounted to less than 10% of that observed in
Xenopus under comparable experimental conditions (Campbell and
Jalabert, 1979). Obviously, more research is needed before any general statements about diversity or conservation in the mechanism of

vitellogenin uptake in the evolution of the vertebrate line can be made. Interestingly, *Xenopus* oocytes are selective for vitellogenin over albumin or ferritin, while the vertebrate source of the vitellogenin—which included a teleost—possessed little influence on the rate of uptake (Wallace *et al.*, 1980). Similarly, microinjected vitellogenin mRNAs from different species gave rise to mature vitellogenins and led to subsequent export from the *Xenopus* oocyte. Amphibian vitellogenin was later taken back up and deposited in the yolk platelets. Locust vitellogenin, on the other hand, was synthesized and exported but not sequestered from the medium (Lane *et al.*, 1983).

With respect to the hormonal regulation of vitellogenin uptake by the oocyte, a rather scant body of information is available, apart from the fact that estrogen does not seem to be involved. Instead, uptake may be dependent on the presence of progesterone, with its exact mode of action on the surface of the oocyte and not on the transcriptional level still being under debate. This steroid may possess some general maturation function or act specifically to induce micropinocytosis in vitellogenin (Tata and Smith, 1979). Studies on *Xenopus* have indicated that once the oocytes have entered into the vitellogenic state, the rate of vitellogenin sequestering is regulated by the follicle cells and not by the oocyte itself (Wallace, 1983). There is an ongoing debate on the number of gonadotropins present in fishes, but independent of the outcome of this perceived controversy, two statements can be made with respect to exogenous vitellogenesis. One pituitary hormone, rich in carbohydrate, leads estrogen production in the female gonad and thus initiates the events outlined in Fig. 1. Another pituitary hormone, which is characteristically low in carbohydrate content, specifically enhances the uptake of vitellogenin from the bloodstream into the growing oocyte while at the same time being devoid of vitellogenic action per se (Burzawa-Gerard, 1982).

Only if the oocyte has taken up the vitellogenin by micropinocytosis will the molecule be processed correctly, cleaved at predetermined sites, and directed toward specific sites in the yolk. On the other hand, if microinjected into the oocyte, the vitellogenin molecule is rapidly degraded in its entirety and degradation products never reach the yolk platelet (in *Xenopus;* Wallace and Hollinger, 1979). These findings reconcile the observations made when messenger RNA for vitellogenin is microinjected into growing oocytes during translational or modification studies. In this case, after the mRNA has been translated and the molecule has undergone the required posttranslational modifications, the mature vitellogenin is excreted from the oocyte and subsequently sequestered from the medium by micropinocytosis and only then directed toward the yolk, where it is stored as phosvitin and lipovitellin (Lane *et al.*, 1983). Campbell and Idler (1976) found that some degree of dephosphorylation of piscine vitellogenin may occur during incorporation into ovarian yolk. At the later stage of meiotic maturation, many fish eggs take up substantial amounts of water, and this hydration may be accompanied by a marked drop in protein phosphorus assayable in the oocyte (Craik, 1982).

During the previtellogenic part of oocyte development in the trout, microvesicular bodies (MVB) accumulate and later occupy the larger part of the cell. These bodies contain acid hydrolase activity and can be classified as a lysosomal-like compartment. In the course of exogenous vitellogenesis, large yolk vesicles form, which contain yolk as well as the remnants of the microvesicular bodies. At the completion of vitellogenesis, the microvesicular bodies have disappeared (Busson-Mabillot, 1984) and as a general observation, acid phosphatases are absent from fully developed oocytes (Korfsmeier, 1980), while cathepsin and α-glucosidase activities are present in unfertilized eggs (Vernier and Sire, 1977). Although lysosomal activities occur associated with yolk platelets in most lower vertebrates, this rule is not without exception. For example, the oocytes of two species of marine fishes with polylecithal egg cells (herring and plaice) are altogether devoid of acid hydrolases (Korfsmeier, 1980).

The exact role of the lysosomes in the proteolytic cleavage of vitellogenin that has been sequestered from the bloodstream by micropinocytosis remains an enigma to date. From the information that has been gathered from other vertebrate systems, it is not obvious what type of enzymes are responsible for the breakdown of vitellogenin and where they originate. The observation of efficient, nonspecific breakdown and subsequent removal of microinjected vitellogenin in the *Xenopus* oocyte suggests that micropinocytosed vitellogenin is not available for full lysosomal attack and may be only exposed to enzymes that will specifically cleave it into lipovitellins, phosvitins, and phosvettes. Obviously, the vitellogenin molecule itself is not resistant to other types of proteolytic attack. On the other hand, it seems that the microvesicular bodies transporting vitellogenin and its products to the yolk are somewhat related to the lysosomal system, since they clearly display some enzyme activities with characteristic acidic maxima. The microvesicular bodies, however, do not display the full complement of lysosomal enzymes, which would most likely lead to the degradation of vitellogenin and its receptor.

C. Phosvitin and Lipovitellin

The general picture of cleavage products of the vitellogenin molecule in piscine systems is not quite as clear-cut as in the other nonmammalian vertebrates. Although it has long been known that fish eggs contain lipovitellins and phosvitins with by and large similar properties to those from other vertebrates, a much more pronounced interspecific variation on the theme is apparent. Such variability is reflected in high numbers of different components as well as sometimes unusual chromatographic behavior. The most extreme example to date is found in the eggs of an Antarctic fish (*Chaenocephalus aceratus*) that possesses no less than nine different phosphorylated proteins (Shigeura and Haschemeyer, 1984). Fish lipovitellins are much more heterogeneous than those from other vertebrate eggs, are smaller, contain higher concentrations of lipids, and possess little or no protein phosphorus. Numerous low-molecular-weight phosvitins have been found in teleost eggs, characterized by widely varying, albeit generally low, amounts of alkali-labile protein phosphorus (Mano and Yoshida, 1969; Markert and Vanstone, 1971; Inoue *et al.*, 1971; de Vlaming *et al.*, 1980; Craik, 1982). In the killifish *Fundulus heteroclitus*, the native vitellogenin molecule (200 kDa; cf. Table III) cannot be localized within the oocyte, indicative of a rapid and efficient degradation into smaller components (Wallace and Selman, 1985). In fact, five major protein bands (122, 103, 45, 26, and 20 kDa) are abundant in growing oocytes, all of which are allegedly derived from proteolytic breakdown of vitellogenin. During final maturation, this pattern is further changed in that the 122- and 45-kDa proteins are degraded to yield a number of lower-molecular-weight proteins. It has been implied that these new proteins are involved in the hydration process during final maturation (Wallace and Selman, 1985), but their physiological function(s) and their relationship to phosvitin or the nature of the proteolytic machinery responsible await identification.

D. Oocyte Lipids

In the course of exogenous vitellogenesis, teleost oocytes accumulate large amounts of lipids in addition to the polar lipids delivered as part of the vitellogenin molecule. In spawned eggs, which contain between 8 and 32% lipid (based on dry weight), several classes of lipids are represented, where the emphasis varies strongly between different species of fish. Depending on the preferred type of lipid accumulated in the eggs, three strategies can be distinguished: the first group, which includes rainbow trout, sole (*Solea vulgaris*), and a whitefish (*Coregonus albula*), is characterized by equally high levels of polar lipids and triglycerides (Kaitaranta and Ackman, 1981; Devauchelle *et al.*, 1982). Baltic herring, roach, and turbot (*Scophthalmus maximus*) belong to a second group, which accumulates mainly polar lipids (75–90%) (Kaitaranta and Ackman, 1981; Devauchelle *et al.*, 1982). A third group, encompassing a wide variety of species such as the gourami (*Trichogaster cosby*; Sand *et al.*, 1971), sea bass (*Dicentrarchus labrax*; Devauchelle *et al.*, 1982), striped bass (*Morone saxatilis*; Eldridge *et al.*, 1983), perch (*Perca fluviatilis*), burbot (*Lota lota*; Kaitaranta and Ackman, 1981), and many others, accumulates large amounts (>80%) of wax and steryl esters in the so-called egg oil globules. In fact, all fish eggs harboring oil globules, which are distinct from the yolk or yolk platelets, have been shown to contain substantial amounts of wax and sterol esters (Kaitaranta and Ackman, 1981). In species such as *M. saxatilis*, the oil globule consists almost entirely of steryl esters and wax esters (90%) as well as some triglycerides (10%), whereas the small bulk of the yolk lipids is dominated by phospholipids (79%, Eldridge *et al.*, 1983). With wax esters generally belonging in the domain of the marine environment, the above list shows that the occurrence of these compounds in fish eggs is by no means restricted to marine species. The physiological advantages of accumulating large amounts of wax esters in eggs (71% of the total caloric value of the egg in *M. saxatilis*; Eldridge *et al.*, 1983) have not been analyzed yet, although it can be hypothesized that in addition to serving as an energy supply, they will play an important role in buoyancy control for the embryo and developing larva.

Unfortunately, a large gap exists in our knowledge with respect to the maternal source of these wax and steryl esters. It appears that the lipid components of vitellogenins from species accumulating oil globules in their eggs have not been given any attention to date. Vitellogenin from other fishes is known to transport some 20% (by weight) of lipid, the bulk of which consists of phospholipids (Wiegand, 1982; Norberg and Haux, 1985). If this is verified for vitellogenins of fishes that synthesize oil globules in their eggs, vitellogenin can be ruled out as the transport form for their unique lipid complement. Alternatively, the wax esters may be synthesized endogenously in the oocyte from fatty acids delivered as part of lipoproteins or bound to serum albumins. In the gourami *T. cosby*, where wax esters constitute the major lipids of the egg, the ovarian fatty acyl alcohols can be synthesized de novo from dietary acetate or longer dietary carbon chains, but the site of biosynthesis could not be identified (Sand *et al.*, 1971). Since Wiegand and Idler (1982) reported that the ovary of the rainbow trout possesses the metabolic machinery to reduce exogenously administered fatty acids to the corresponding alcohols, the endogenous synthesis of wax esters cannot be excluded. Another interesting facet of fish accumulating wax esters in their eggs is the fact that none of the adults of these species use wax esters as their lipid storage forms, but generally rely on triglycerides instead.

Comparable to the situation in avian systems, some experimental evidence for rainbow trout suggests that circulating lipoproteins may enter the ovary and serve as the major source of polyunsaturated free fatty acids, the bulk of which is transported in lipoproteins and not on the vitellogenin molecule (Frémont *et al.*, 1984). The experimentally induced or naturally occurring vitellogenesis in fishes is accompanied by large increases in liver biosynthesis and export of VLDL (cf. Tables II and VI). Just as in the hepatic synthesis of vitellogenin, the synthesis of lipoproteins may be initiated in vivo through the influence of circulating estradiol. In the annual cycle of fishes, increases in blood VLDL are positively correlated with vitellogenesis.

Comparative data on avian vitellogenesis and VLDL metabolism indicate that the ovary is capable of the uptake of lipoproteins directly from the bloodstream and that this process may be independent of the sequestration of vitellogenin through micropinocytosis. While this general scheme may not be applicable to all serum lipoproteins, it has been proven to hold for VLDL in the hen (Holdsworth *et al.*, 1974), where the basal lamina appears to be permeable to circulating VLDL (Evans *et al.*, 1979).

While Wiegand and Idler (1982) determined some capacity for endogenous triglyceride biosynthesis from acetate in the ovary in the rainbow trout, the results of Leger and co-workers (1981) on the same species suggest serum lipoproteins such as VLDL or LDL as the more likely sources for the triglycerides accumulated in the egg.

Lipid material, composed to a large extent of phospholipids, is first accumulated in the perinuclear cytoplasm of the oocyte. However, these early lipid bodies, the source of which still awaits identification, do not constitute true yolk since they are known to disappear before or during exogenous vitellogenesis. The study of Wiegand and Idler (1982), which showed for an in vitro system that labeled acetate was incorporated into ovarian polar lipids, remained inconclusive with regard to the cell fraction with which polar lipid was associated.

E. Carotenoids

Fish eggs are known to contain other secondary products such as carotenoids, which sometimes contribute to the colorful appearance of the eggs. In the chum salmon (*O. keta*), almost 1% of the fresh weight of the spawned egg consists of carotenoids, mainly astaxanthin (Kitahara, 1984). In this species, as in many other salmonids, the egg carotenoids are presumably derived from previous depositions in the muscle (Crozier, 1970). In view of the absence of a specific transporting vehicle, it can be hypothesized that the carotenoids are passively transported out of the tissue together with storage lipids according to their lipid solubility. In the course of the spawning migration, the lipid deposits within the salmon's body are mobilized in their entirety, partly for energy production during migration and in the female also as part of the estrogenic response. Depending on the composition of the individual lipids that the carotenoids are associated with, differing routes can be envisaged for their transport into the gonad of the vitellogenic female: the carotenoids may form part of the vitellogenin molecule itself or, alternatively, they may travel with the abundant lipoproteins, especially VLDL, in salmon blood (cf., Skinner and Rogie, 1978; Sire and Vernier, 1983). The light pink hue of highly purified sockeye salmon vitellogenin suggests that at least some of the carotenoids travel to the ovary bound to the lipid component of vitellogenin (T. P. Mommsen and C. J. French, unpublished). In fact, it has been reported that a crustacean lipovitellin moiety possesses a covalent binding site for carotenoids (Zagalsky *et al.*, 1983). Obviously, more research on posttranslational modifications of the vitellogenin polypeptide and on possible association of carotenoids with other lipophilic components of the fish blood is required before even a preliminary picture will emerge.

Another interesting facet of the carotenoid deposition in the oocyte is the fact that not all of the carotenoids are localized in the yolk, but some 20% is associated with other structures in the oocyte (Kitahara, 1984) leaving the question of the physiological function of such a heterologous group as carotenoids in embryo nutrition and survival wide open.

F. Glycoproteins

In addition to their ubiquitous glycogen stores, fish eggs are known to contain protein-bound carbohydrate moieties, but their exact localization and their biochemical nature have only been given rather cursory attention. Even more surprisingly, the physiological function of these carbohydrate-containing proteins, which have lately been charaterized as sialoglycoproteins, is entirely unknown, despite the fact that these compounds may surpass the egg phosvitins in abundance by almost an order of magnitude (Inoue and Iwasaki, 1980a,b). Sialoglycoproteins are apparently rather common constituents of fish eggs, since they have been described and partially characterized for a number of species, namely Pacific herring (*Clupea pallasii*), Alaskan pollack (*Theragra chalcogramma*), Japanese common charr (*Salvelinus leucophaenus*), rainbow trout (*Salmo gairdneri*), and three species of Pacific salmon (*Oncorhynchus keta, O. masou*, and *O. nerka*) (Inoue and Iwasaki 1978, 1980a,b; Iwasaki and Inoue, 1985).

The sialoglycoproteins are associated with the soluble fraction of the egg, namely, the cortical vesicles, and thus do not form part of the membrane superstructure of the egg. The purified molecules are devoid of phosphorus and derive their acidity from the abundant sialic, glutamic, and aspartic acids. In fact, these three strongly acidic components make up more than 20% (by weight) of the sialoglycoproteins in the herring egg (Inoue and Iwasaki, 1980a). In this species, the molecular weights of the three main sialoglycoproteins range from 8800 to 13,000. Their protein backbone, comprising some 40–50% of the molecules, is unique in its amino acid composition and displays very little variability. The carbohydrate composition, in contrast, is variable, especially in the content of *N*-acetylglucosamine, constituting between 12.5 and 6.2% of the total weight of the sialoglycoproteins. Other abundant carbohydrates, in addition to *N*-acetylneuraminic acid (sialic acid) and *N*-acetylglucosamine, are *N*-acetylgalactosamine, fucose, galactose, and mannose.

More recent work by Inoue and co-workers (Iwasaki and Inoue, 1985; Inoue *et al.*, 1985) on polysialoglycoproteins isolated from unfertilized egg from different species of salmonid fishes can be summarized as follows:

1. The glycoproteins are characterized by high molecular weights, ranging from 150 to 300 kDa.
2. More than 50% of their weight is comprised of sialic acid, and total carbohydrate content may reach 85%.
3. They contain poly(oligo)sialyl groups linked to *O*-glycosidic carbohydrate units.
4. The polypeptide backbone is made up of seven acidic or neutral amino acids only, namely, alanine, aspartatic acid, glutamic acid, glycine, proline, threonine, and serine.

5. Amino acids are arranged in two kinds of repeated polypeptide sequences (13 amino acid residues).
6. All of the threonine and half of the serine residues are glycosylated.

It is interesting to note that the accumulated polysialoglycoproteins in the rainbow-trout egg undergo a drastic reduction in size upon fertilization, at which time they decrease from 260 to 9 kDa (Inoue *et al.*, 1985). A similar reduction in size can be expected for the large glycoproteins prevalent in other salmonid fishes, and the situation is somewhat reminiscent of the further breakup of vitellogenin breakdown products into even smaller units during final oocyte maturation in *F. heteroclitus* (Wallace and Selman, 1985). In both cases, the drastic decreases in size are due to highly specific proteolytic attack, and the carbohydrate moieties on the polysialoglycoproteins remain unaltered. In the case of the glycoproteins, the drastic reduction in size occurs simultaneously with cortical vesicle breakdown and exocytosis (Inoue *et al.*, 1985).

Unfortunately, despite the fact that the sialoglycoproteins compounds are prevalent in fish eggs and that the timing of their breakdown implies some involvement upon fertilization (block to polyspermy?), no data exist on such important aspects as their physiological function, their source, transport form or mechanism, and timing of uptake into the developing oocyte. If these multitudes of carbohydrates are synthesized in the maternal liver as part of the posttranslational modification of the vitellogenin molecule, the codes for the small, but unique, polypeptide chains should be identifiable with relative ease within the recently purified vitellogenin messenger RNA (Chen, 1983; Valotaire *et al.*, 1984).

Biochemical and histochemical studies have identified lectins as an integral part of the soluble fraction of fish oocytes (Nosek *et al.*, 1983). However, just as in the case of the sialoglycoproteins found in mature eggs, sources or physiological function is a matter of speculation (Nosek *et al.*, 1983).

G. Vitamin-Binding Proteins

As mentioned before for the chicken, estradiol induces the hepatic synthesis of a number of vitamin-binding proteins destined for uptake into the growing oocyte (Table IV). One of these vitamin-binding proteins is the well-characterized riboflavin-binding protein, which is glycosylated as well as phosphorylated and is responsible for the transport of riboflavin to the oocyte. While in many instances carbohydrate side chains are important in the recognition of glycoproteins by their target cells, it was recently demonstrated that, in the case of the riboflavin-binding protein, correct phosphorylation and not glycosylation is crucial for the uptake of the molecule into the oocyte, as was shown for the uptake of experimentally administered phosvitin (Miller *et al.*, 1982). For the fishes, however, information on occurrence of such vitamin-binding proteins is limited to the observation that the specific riboflavin-binding protein is absent from salmon (*O. nerka*) oocytes. Riboflavin, on the other hand, occurs in salmon oocytes in similar concentrations as in the chicken egg (H. B. White and M. A. Letavic, unpublished), leaving the question of its source and its possible transport open to speculation.

Vitamin-binding proteins can be expected to play an integral role in the survival of embryos or larvae, supplying them with vitamins at critical stages of their development. Such proteins may further possess antimicrobial action by rendering vitamins stored in the egg unavailable to infesting bacteria.

H. Hormones

While it has been known for some time that fish larvae respond to exogenously administered hormones, the physiological relevance of such observations remained unclear, especially at a time when the *intraembryonic* existence and availability of such hormones had not been established. In the context of the hormonal status of fish oocytes, an avian concept may deserve attention by researchers interested in embryonic fish metabolism and morphogenesis. In addition to known nutrients and secondary compounds, the chicken egg contains significant amounts of thyroxin, and the embryonic chick liver already displays highly specific steriod receptor activities for hormones such as 17β-estradiol (Lazier, 1978) and 3,5,3'-triiodothyronine (T_3) (Bellabarba and Lehoux, 1981).

Recent analyses by Brown *et al.* (1987) and Kobuke *et al.* (1987) unequivocally established the presence of substantial amounts of thyroxin and T_3 in unfertilized ova and embryos of salmonids (*Oncorhynchus sp.*) and striped bass (*M. saxatilis*). These hormones, which are localized preferentially in the embryonic yolk, are apparently of maternal origin. The suggested route of transfer from the maternal circulatory system into the growing oocyte is through vitellogenin, since this compound displays appreciable binding capacity for thyroid hormones in the plasma collected from vitellogenic coho salmon (*O. kisutch;* A. Hara, W. W. Dickhoff, and C. V. Sullivan, unpublished results). The absolute amount of hormone transferred into the ova, however, is small (in the range of 5 ng/oocyte, Brown *et al.*, 1987; Kobuke *et al.*, 1987), and it is therefore unlikely that such transfer will be reflected in concentration changes of hormone in the maternal circulatory system.

While the presence of a hormone does not necessarily imply its physical availability or functionality, the data of Brown *et al.* (1987) and Kobuke *et al.* (1987) already indicate that yolk thyroxin and T_3 undergo turnover during early development. Thus both hormones can be assumed to be available to the embryo and to influence physiological functions. Considering how many other lipophilic compounds from the maternal system reach the growing oocyte, the deposition of thyroxin is not surprising, and the same principle is likely to be applicable to other steriod hormones. However, as the present discussion reveals (cf., III,F, G, and I), similar arguments can also be made for the potential transfer of peptide hormones (insulin, glucagon, etc.) from the maternal system into the oocyte. Two important conclusions can be drawn from these novel findings: 1. The fact that thyroxin and T_3 are present in the growing oocyte and undergo changes during oocyte development long before a functional hypothalamo-adenohypophysial-thyroid axis is established, implies that these hormones— yet unidentified—exert physiological roles during early larval morphogenesis and 2. If, as it seems possible, hormone stores supplied by the maternal system are a common feature in fish eggs, an entirely new window on the endocrinology and physiology of developing fish has been opened.

I. Yolk–DNA

The main, if not exclusive, route for the uptake of vitellogenin from the blood into the growing oocyte is micropinocytosis (Brummett and Dumont, 1977). As pointed out elsewhere in this review, the subsequent fate of the vitellogenin molecule inside of the oocyte is cleavage into different components which are later stored in the yolk. However, it can be hypothesized that the uptake of the large molecule vitellogenin by micropinocytosis may not entirely exclude smaller, blood-borne molecules, such as sugars, lipids, plasma proteins, or even DNA. To exemplify this phenomenon, a short comparative excursion to amphibian systems is required, since to date no data on similar phenomena have been analyzed for fishes.

In addition to the DNA of the oocyte chromatin, the yolk platelets of the amphibian *Xenopus laevis* contain yolk-bound DNA. This yolk–DNA is found to be double-stranded, and characterized by a high molecular weight (Hanocq *et al.*, 1972), but its actual concentration is small at about 20 ng per oocyte, compared with some 280 μg vitellogenin derived protein contained in the fully grown oocyte. When experimentally exposed to *Xenopus*, bovine, or bacterial DNA, isolated vitellogenic oocytes of *Xenopus* sequestered it from the incubation medium, and the DNA was later found to be associated with the yolk platelets (Opresko *et al.*, 1979). However, there was no discrimination in uptake rates for DNA from the different sources, and furthermore yolk–DNA was determined to undergo relatively rapid turnover. This circumstantial evidence suggests that in this particular amphibian system, the DNA associated with the yolk is not involved in information transfer during the embryonic development. On the other hand, the indisciminate uptake of DNA from the maternal bloodstream, which has been shown to contain small (25 μg ml^{-1}) amounts of DNA (Opresko *et al.*, 1979), presents a good example of an adventitious uptake of maternal blood components into the growing oocyte, which is solely a byproduct of the mode of vitellogenin uptake by micropinocytosis. It may also help to explain the otherwise surprising presence of other components of maternal plasma or their derivatives in growing oocytes.

J. Metabolism

In addition to the uptake and processing of vitellogenin and other blood-borne proteins such as VLDL, the growing fish oocyte synthesizes and accumulates a number of high-molecular-weight components. First, the oocyte displays a whole complement of RNA (some 10^4 more than in somatic cells), mainly rRNA (95%), mRNA (2–3%), (values for *Xenopus*) and tRNA, including an oocyte-specific 5-S RNA (Denis and le Maire, 1983), which are likely to be of importance in early embryonic development. Second, the oocyte can perform protein biosynthesis as well as a multitude of posttranslational modifications, specifically glycosylation, phosphorylation, and lipidation. In the course of their development, *Xenopus* oocytes increase their biosynthetic activity by more than 100-fold, from 4.3 ng protein per day in stage 1 oocytes to over 0.5 μg per day in stage 6 oocytes (Taylor and Smith, 1985). Each of the mentioned activities requires specific subsets of enzymes. This high biosynthetic potential made the *Xenopus* oocyte the system of choice to study translation and posttranslational modification of microinjected messenger RNAs from a variety of animal sources, including insect mRNAs (Lane *et al.*, 1983; Soreq, 1985). As the example with microinjected vitellogenin shows (see above), the oocyte is also capable of totally degrading "foreign" proteins. On account of these results and some histochemical studies, the oocyte usually gets credited with a limited complement of lysosomal enzymes that are supposedly also involved in the breakdown of vitellogenin into phosvitin, etc. Furthermore, adventitiously sequestered yolk–DNA has been shown to undergo turnover (Opresko *et al.*, 1979), again a metabolic activity that demands a specific set of enzymes.

Other metabolic activities of the growing teleost oocyte apparently include the synthesis of urea, which is absent in most adult teleosts, and results in oocyte urea concentrations surpassing those of the maternal system by two- to fivefold (Dépêche *et al.*, 1979).

All of these actions, as well as acid–base regulation and the active vesicle transport through the cell, require energy in the form of ATP. The ATP must somehow be generated inside of the oocyte, since it is unlikely that it is furnished by the follicle cells. Unfortunately, the questions concerning the energy supply and preferred substrates for the growing oocyte have yet to be investigated, particualarly for the fishes. This is a deplorable situation, especially since the answers to these questions may have particular relevance to the early survival of the fish embryos and larvae.

Even before exogenous vitellogenesis has been initiated, the oocytes of *Misgurnus fossilis* augment their contents of metabolic enzymes, specifically those involved in glycolysis, the pentose shunt, and gluconeogenesis. During the entire course of oocyte development, glucose sequestered from the maternal circulation serves as an important energy source and also supplies the building blocks for accumulating glycogen. In fact, the activity of one of the key enzymes in this pathway, glycogen synthetase, increases 100-fold during vitellogenesis (Yurowitzky and Milman, 1975). Following maturation, the *Misgurnus* oocyte completely loses hexokinase activity and with it the ability to use exogenously administered glucose. At the same time, the switch from exogenous to endogenous energy use, at least as far as carbohydrate metabolism is concerned, is reflected in alterations of the enzymes regulating glycogen synthesis and its degradation. The moment that hexokinase is lost from the oocyte, glycogen synthetase activity decreases by half, while glycogen phosphorylase activity increases by an order of magnitude (Yurowitzky and Milman, 1972). It can be concluded from the presence and high activities of enzymes involved in glucose catabolism that during oocyte development maternal glucose may constitute one of the major energy sources for the different ATP-demanding reactions. It would also be interesting to confirm whether vitellogenesis might be correlated with increased glucose output from the maternal liver, as the decrease in hepatic glycogen suggests. The exact extent of its importance is not clear yet, mainly because data for enzymes involved in other pathways, such as fatty acid utilization, are lacking. It should be recalled that during exogenous vitellogenesis, the availability of lipid material through increased VLDL output by the liver is stepped up, as is lipid turnover in general. Once the oocyte has been matabolically "cut off" from the maternal continuum of energy supplies, as the disappearance of hexokinase from the oocyte suggests, it exists as a metabolically distinct, and closed, unit, which from this point on has to rely on accumulated substances for survival. It can be speculated that glycogen will serve as the first supplier of metabolic energy, because of the relative ease with which it can be mobilized. Considering their overall bulk and their caloric contents, yolk lipids will be of overwhelming importance during ensueing parts of embryonic and larval development, while the accumulated amounts of protein and amino acids are most likely to be funneled into anabolic and not ATP-delivering pathways.

IV. EPILOGUE

The processes of hepatic vitellogenin synthesis and yolk-component deposition in the oocyte in the fishes represent a wide-open field for researchers in a variety of fields. From comparative and evolutionary viewpoints, studies on piscine systems are likely to supply valuable insights into hepatic steriod receptor mechanisms, estrogen interactions with a multitude of genes, and mechanisms of posttranslational modifications, as well as into the nature of hormone interactions on the receptor and gene level. On the level of the oocyte, central topics will include the vitellogenin receptor mechanism, the regulation and control of the enzymatic machinery involved in the limited breakdown of vitellogenin, and the sources—maternal or internal—of such differing compounds as wax esters, lectins, sialoglycoproteins, or hormones, or vitamins. In each case, the fishes seem to present the experimenter with a variety of species ideally suited for the individual topic, not least because of the ease with which massive vitellogenesis can be induced by the administration of 17β-estradiol. The apparent variability among different species of fish with respect to each theme will help to prevent the emergence of generalizing statements from the study of just one species. This approach is unfortunately prevalent in the literature on other vertebrates, where one toad (*Xenopus laevis*) represents all amphibians, or the chicken portrays the birds.

Finally, it is imperative to put described biochemical events into the context of the life history of fishes. In contrast to most other egg-laying vertebrates, the fishes are known to invest large amounts of their body reserves into the production of gonadal products. The most extreme examples of the striking metabolic effort exerted by fishes during the time leading up to the spawning period are some anadromous fishes such as the Pacific salmons (*Oncorhynchus* spp.) or the American shad (*Alosa sapidissima*) (Idler and Clemens, 1959; Glebe and Leggett, 1981).

It can be appreciated that only an unperturbed sequence of the outlined events in the maternal system will lead to mature oocytes with an optimized starting point for embryonic development. The fine-tuning of the orchestrated maternal events leading to mature oocytes makes it necessary to consider potential points of environmental interference. While potentially interfering influences range from acid–base disturbances (Tam *et al.*, 1987) and thermal pollution to anything that will invoke stress reactions in fish, the following will focus on two specific cases with potentially detrimental effects on the survival of the young of the ensuing generation, namely, lipophilic toxicants and heavy metals.

Although carotenoids are possibly rather ancillary compounds in the egg per se, the case of the accumulated carotenoids shall serve to emphasize the point of the potential importance that the maternal history and events may bear to the formation of egg components. Carotenoids are taken up by fish in their food and usually deposited due to their chemical properties together with functional lipids—in the case of salmonid fishes, usually in the white muscle tissue. As part of the general lipogenic action of estradiol and thus during the course of vitellogenesis, extrahepatic lipid stores are mobilized and transported to the liver; due to their hydrophobicity, carotenoids are translocated to the liver together with mobilized lipids. At this point it should be recalled that during exogenous vitellogenesis, hepatic tissue constitutes the central organ with respect to lipid metabolism, in that it takes up triglycerides and phospholipids to utilize them for different metabolic tasks: (1) fatty acids serve as major oxidative substrates to fuel metabolic processes; (2) as part of the posttranslational modifications performed by the liver, lipids are attached to that particular part of the vitellogenin molecule that later forms the lipovitellin of the oocyte; and (3) the rate of hepatic lipoprotein synthesis and export is sharply increased during vitellogenesis. Carotenoids may associate passively with any of the lipid utilized in these processes, or it may actively be bound to a covalent binding site on the vitellogenin molecule. The first route will lead to carotenoid deposition in the liver. The second option will result in carotenoid-colored vitellogenin, as the example of the pink hue of sockeye-salmon vitellogenin shows. The third alternative will also deliver carotenoids from the liver to the gonad, which during vitellogenesis displays possibly the highest rates of uptake for lipoproteins, especially VLDL, from the bloodstream. All lipophilic substances accumulated in the maternal system are likely to behave like the carotenoids.

It is known that chlorinated hydrocarbons and many other lipophilic pesticides are transported in the bloodstream by lipoproteins—DDT, for instance, has been found associated with serum lipoproteins in exposed rainbow trout (*Salmo gairdneri;* Plack *et al.*, 1979). A similar behavior can be anticipated for other lipophilic environmental toxicants, such as aliphatic or polycyclic hydrocarbons and many of their derivatives. As a consequence, it is reasonable to assume that such lipophilic compounds that have found their way into adult fish will eventually be translocated—just as the carotenoids are—to the gonad during exogenous vitellogenesis. Considering the facts that under the influence of estradiol, hepatic lipoprotein synthesis is increased (cf. Table II) and that vitellogenin itself contains a highly lipophilic region, it does not come as a surprise that DDT and other hydrophobic pesticides are accumulated in fish eggs (Plack *et al.*, 1979). Subsequently they will severely impair egg survival and hatchability (Johnson and Pecor, 1969). The massive oil globules, composed of wax esters and steryl esters, prevalent in the eggs of a large number of fish, designate a potentially detrimental sink for pesticides, petrochemicals, or other lipophilic environmental toxicants. Furthermore, at the level of the gonad, exposure of vitellogenic fish to sublethal concentration of pesticides led to a significant decrease in the uptake of [^{32}P]phosphate by the growing oocytes, thus probably compromising their normal composition (Singh and Singh, 1981).

To compound the problems posed by halogenated hydrocarbons, it has been reported that such compounds not only bind to the vitellogenin molecule, but also decrease the estradiol-dependent vitellogenic response in the rainbow-trout liver (Chen and Sonstegard, 1984). Inducers of the hepatic mixed-function oxidase system, such as beta-naphthoflavone, exert an inhibitory influence on the production of vitellogenin mRNA in the rainbow trout (Chen and Sonstegard, 1984). It should be recalled that during exogenous vitellogenesis, the matabolic demand put on the liver is enormous (cf. Table II). Consequently, it can be expected that any additional metabolic requirements placed on the liver, such as the synthesis of elements involved in detoxification, are likely to reduce the effort expended on vitellogenesis and thus may imbalance the maturing of the oocytes. A further example for the costly metabolic expenditure incurred is the occurrence of a novel vitellogenin-like protein in the blood of pesticide-exposed fish (Denison et al., 1981). Also on the level of the liver, vitellogenesis may be impaired or its timing imbalanced by the known estrogenic action of some insecticides. Examples in mammals and birds show that the chlorinated insecticide chlordecone interacts directly and rather persistently with the uterine estrogen receptor (Hammond et al., 1979). As pointed out, mammalian and piscine estrogen receptors reveal numerous similarities, making the exertion of biological effects highly likely in fish systems.

A similar line of reasoning applies to the exposure of fish to environmental heavy metals. In *Blennius pholis,* cadmium and copper are known to accumulate in hepatic tissue, and in the course of the final oocyte maturation and massive yolk deposition, these heavy metals are transferred from the liver to the gonad and accumulate in the egg (Shackley et al., 1981). Whereas this designates one passive way for the female fish to decrease its own hepatic concentration of these trace metals, it may develop into an important, potentially lethal, strategy for the oocyte. It is not too far-fetched to suggest that in situations where the environmental load of these or other heavy metals to the adult is increased from trace amounts to sublethal levels, transfer to the gonad in the course of oocyte maturation may result in the accumulation of highly toxic levels in the oocyte. While such flow of potential toxicants may presently not affect marine fish, it is already frightfully relevant for freshwater and brackish-water fishes in many parts of the world. The vitellogenin molecule itself may be implicated in the transport of hepatic heavy metals to the gonad due to its protein phosphorus–dependent charge and ion-binding capacity (Hara et al., 1980; Hara and Hirai, 1978; Lange, 1981). An additional problem may be introduced through the potential competition of hepatically accumulated heavy metals for those metal ions that are transported to the gonad during undisturbed vitellogenesis, namely magnesium, calcium, and iron.

Although adult fish are able to bind and detoxify heavy metals quite efficiently through the specific hepatic synthesis of metallothionein (Roch and McCarter, 1984), the process does not rid the parent body of the heavy metal load rapidly and thus sets the stage for the potential poisoning of the oocyte. Also, since metallothionein is induced in the liver, its synthesis effectively competes with vitellogenin (cf. Seguin *et al.*, 1984) and therefore can be expected to impair the balanced flow of vitellogenin to the gonad.

The green tinge of *Xenopus* and some other amphibian eggs is a colorful reflection of the maternal biliverdin deposited adventitiously. It also presents an additional example of how the maternal system may dispose of an excretory product via the eggs. However, as the above examples show, not all compounds deposited in the maternal liver and eventually accumulating in the eggs are as inocuous as biliverdin in *Xenopus*.

ACKNOWLEDGMENTS

We would like to thank Dr. Catherine B. Lazier (Dalhousie University) and Dr. Harold B. White III (University of Delaware) for helpful discussions. We are grateful to Dr. Bodil Korsgaard (Odense University) for critically reading the manuscript.

REFERENCES

Bailey, R. E. (1957). The effect of estradiol on serum calcium, phosphorus and protein of goldfish. *J. Exp. Zool.* **136**, 455–469.

Bast, R. E., Garfield, S. A., Gehrke, L., and Ilan, J. (1977). Coordination of ribosome content and polysome formation during estradiol stimulation of vitellogenin synthesis in immature male chick liver. *Proc. Natl. Acad. Sci. U.S.A.* **74**, 3133–3137.

Bellabarba, D., and Lehoux, J.-G. (1981). Triiodothyronine nuclear receptor in chick embryo: Nature and properties of hepatic receptor. *Endocrinology* **109**, 1017–1025.

Bhattacharya, S., Plisetskaya, E., Dickhoff, W. W., and Gorbman, A. (1985). The effects of estradiol and triiodothyronine on protein synthesis by hepatocytes of juvenile coho salmon (*Oncorhynchus kisutch*). *Gen. Comp. Endocrinol.* **57**, 103–109.

Bres, O., and Eales, J. G. (1986). Thyroid hormone binding to isolated trout (*Salmo gairdneri*) liver nuclei *in vitro*: Binding affinity, capacity, and chemical specificity. *Gen. Comp. Endocrinol.* **61**, 29–39.

Breton, B., Fostier, A., Zohar, Y., Le Bail, P. Y., and Billard, R. (1983). Gonadotropine glycoprotéique maturante et oestradiol-17β pendant le cycle reproducteur chez la truite Fario (*Salmo trutta*) femelle. *Gen. Comp. Endocrinol.* **49**, 220–231.

Bromage, N. R., Whitehead, C., and Breton, B. (1982). Relationships between serum levels of gonadotropin, oestradiol-17β, and vitellogenin in the control of ovarian development in the rainbow trout. II. Effects of alterations in environmental photoperiod. *Gen. Comp. Endocrinol.* **47**, 366–376.

Browder, L., ed. (1985). "Developmental Biology," Vol. 1. Pergamon, New York.

Brown, C. L., Sullivan, C. V., Bern, H. A., and Dickhoff, W. W. (1987). Occurrence of thyroid hormones in early developmental stages of teleost fish. *Trans. Am. Fish. Soc.* (in press).

Brummett, A. R., and Dumont, J. N. (1977). Intracellular transport of vitellogenin in *Xenopus* oocytes: An autoradiographic study. *Dev. Biol.* **60**, 482–486.

Burzawa-Gerard, E. (1982). Existe-t'il plusieurs gonadotropines (GTH) chez les poissons? Données biochimiques et vitellogenèse exogène. *In* "Reproductive Physiology of Fish" (C. J. J. Richter and H. J. T. Goos, compilers), pp. 19–22. Pudoc, Wageningen.

Busson-Mabillot, S. (1984). Endosomes transfer yolk proteins to lysosomes in the vitellogenic oocyte of the trout. *Biol. Cell.* **51**, 53–66.

Callard, G. V. (1982). Aromatase in the teleost brain and pituitary: Role in hormone action. *In* "Reproductive Physiology of Fish" (C. J. J. Richter and H. J. T. Goos, compilers), pp. 40–43. Pudoc, Wageningen.

Callard, G. V., and Mak, P. (1985). Exclusive nuclear localization of estrogen receptors in *Squalus* testis. *Proc. Natl. Acad. Sci. U.S.A.* **82**, 1336–1340.

Callard, G. V., Petro, Z., Ryan, K. J., and Claiborne, J. B. (1981). Estrogen synthesis *in vitro* and *in vivo* in the brain of a marine teleost (*Myoxocephalus*). *Gen. Comp. Endocrinol.* **43**, 243–255.

Campbell, C. M., and Idler, D. R. (1976). Hormonal control of vitellogenesis in hypophysectomized winter flounder (*Pseudopleuronectes americanus* Walbaum). *Gen. Comp. Endocrinol.* **28**, 143–150.

Campbell, C. M., and Idler, D. R. (1980). Characterization of an estradiol-induced protein from rainbow trout serum as vitellogenin by the composition and radioimmunological cross reactivity to ovarian yolk fractions. *Biol. Reprod.* **22**, 605–617.

Campbell, C. M., and Jalabert, B. (1979). Selective protein incorporation by vitellogenic *Salmo gairdneri* oocytes *in vitro*. *Ann. Biol. Anim., Biochim., Biophys.* **19**, 429–437.

Chen, T. T. (1983). Identification and characterization of estrogen-responsive gene products in the liver of rainbow trout. *Can. J. Biochem. Cell Biol.* **61**, 802–810.

Chen, T. T., and Sonstegard, R. A. (1984). Development of a rapid, sensitive and quantitative test for the assessment of the effects of xenobiotics on reproduction in fish. *Mar. Environ. Res.* **14**, 429–430.

Colman, A., Lane, C., Craig, R., Boulton, A., Mohun, T., and Morser, J. (1981). The influence of topology and glycosylation on the fate of heterologous secretory proteins made in *Xenopus* oocytes. *Eur. J. Biochem.* **113**, 339–348.

Craik, J. C. A. (1978a). The effects of oestrogen treatment on certain plasma constituents associated with vitellogenesis in the elasmobranch *Scyliorhinus canicula* L. *Gen. Comp. Endocrinol.* **35**, 455–464.

Craik, J. C. A. (1978b). Kinetic studies of vitellogenin metabolism in the elasmobranch *Scyliorhinus canicula* L. *Comp. Biochem. Physiol. A* **61A**, 355–361.

Craik, J. C. A. (1982). Levels of phosphoprotein in the eggs and ovaries of some fish species. *Comp. Biochem. Physiol. B* **72B**, 507–510.

Craik, J. C. A., and Harvey, S. M. (1984). The magnitudes of three phosphorus-containing fractions in the blood plasma and mature eggs of fishes. *Comp. Biochem. Physiol. B* **78B**, 539–543.

Crozier, G. F. (1970). Tissue carotenoids in prespawning and spawning sockeye salmon (*Oncorhynchus nerka*). *J. Fish. Res. Board Can.* **27**, 973–975.

Dasmahapatra, A. K., and Medda, A. K. (1982). Effect of estradiol dipropionate on the glycogen, lipid and water contents of liver, muscle, and gonad of male and female (vitellogenic and nonvitellogenic) Singi fish (*Heteropneustes fossilis* Bloch). *Gen. Comp. Endocrinol.* **48**, 476–484.

Dasmahapatra, A. K., Ray, A. K., and Medda, A. K. (1981). Temperature dependent action of estrogen in Singi fish (*Heteropneustes fossilis* Bloch). *Endokrinologie* **78**, 107–110.

Deeley, R. G., Tam, S.-P., and Archer, T. K. (1985). The effects of estrogen on apoprotein synthesis. *Can. J. Biochem. Cell Biol.* **63**, 882–889.

Denis, H., and le Maire, M. (1983). Thesaurisomes, a novel kind of nucleoprotein particle. *Subcellular Biochem.* **9**, 263–297.

Denison, M. S., Chambers, J. E., and Yarbrough, J. D. (1981). Persistent vitellogenin-like protein and binding of DDT in the serum of insecticide-resistant mosquitofish (*Gambusia affinis*). *Comp. Biochem. Physiol. C* **69C**, 109–112.

Dépêche, J., Gilles, R., Daufresne, S., and Chiapello, H. (1979). Urea content and urea production via the ornithine-urea cycle pathway during the ontogenic development of two teleost fishes. *Comp. Biochem. Physiol. A* **63A**, 51–56.

Devauchelle, N., Brichon, G., Lamour, F., and Stephan, G. (1982). Biochemical composition of ovules and fecund eggs of sea bass (*Dicentrarchus labrax*), sole (*Solea vulgaris*) and turbot (*Scophthalmus maximus*). In "Reproductive Physiology of Fish" (C. J. J. Richter and H. J. T. Goos, compilers), pp. 155–157. Pudoc, Wageningen.

de Vlaming, D. L., Shing, J., Paquette, G., and Vuchs, R. (1977). *In vivo* and *in vitro* effects of oestradiol-17-β on lipid metabolism in *Notemigonus crysoleucas*. *J. Fish Biol.* **10**, 273–285.

de Vlaming, V., FitzGerald, R., Delahunty, G., Cech, J. J., Selman, K., and Barkley, M. (1984). Dynamics of oocyte development and related changes in serum estradiol-17β, yolk precursor, and lipid levels in the teleostean fish, *Leptocottus armatus*. *Comp. Biochem. Physiol. A* **77A**, 599–610.

de Vlaming, V. L., Wiley, H. S., Delahunty, G., and Wallace, R. A. (1980). Goldfish (*Carassius auratus*) vitellogenin: Induction, isolation, properties and relationship to yolk proteins. *Comp. Biochem. Physiol. B* **67B**, 613–623.

Dougherty, J. J., Puri, R. K., and Toft, D. (1982). Phosphorylation *in vivo* of chicken oviduct progesterone receptor. *J. Biol. Chem.* **257**, 14226–14230.

Eldridge, M. B., Joseph, J. D., Taberski, K. M., and Seaborn, G. T. (1983). Lipid and fatty acid composition of the endogenous energy sources of striped bass (*Morone saxatilis*) eggs. *Lipids* **18**, 510–513.

Emmersen, J., and Korsgaard, B. (1983). Measurements of amino acid incorporation capacity of liver cell-free systems through the breeding season and following estradiol treatment of flounder (*Platichthys flesus* L.). *Mol. Physiol.* **2**, 181–192.

Emmersen, J., Korsgaard, B., and Petersen, I. M. (1979). Dose response kinetics of serum vitellogenin, liver DNA, RNA, protein and lipid after induction by estradiol-17β in male flounders (*Platichthys flesus* L.). *Comp. Biochem. Physiol. B* **63B**, 1–6.

Evans, A. J., Perry, M. N., and Gilbert, A. B. (1979). Demonstration of very low density lipoprotein in the basal lamina of the granulosa layer in the hen's ovarian follicle. *Biochim. Biophys. Acta* **573**, 184–195.

Fostier, A., Jalabert, B., Billard, R., Breton, B., and Zohar, Y. (1983). The gonadal steroids. In "Fish Physiology" (W. S. Hoar, D. J. Randall, and E. Donaldson, eds.), Vol. 9, pp. 277–372. Academic Press, New York.

Frémont, L., Leger, C., Petridou, B., and Gozzelino, M. T. (1984). Effects of a (n-3) polyunsaturated fatty acid deficient diet on profiles of serum vitellogenin and lipoprotein in vitellogenic trout (*Salmo gairdneri*). *Lipids* **19**, 522–528.

French, C. J., Hochachka, P. W., and Mommsen, T. P. (1983). Metabolic organization of liver during spawning migration of sockeye salmon. *Am. J. Physiol.* **245**, R827–R830.

Glebe, B. D., and Leggett, W. C. (1981). Latitudinal differences in energy allocation and use during the freshwater migrations of American shad (*Alosa sapidissima*) and their life history consequences. *Can. J. Fish. Aquat. Sci.* **38**, 806–820.

Gorski, J., Toft, D., Shyamala, G., Smith, D., and Notides, A. (1968). Hormone receptors: Studies on the interaction of estrogen with the uterus. *Recent Prog. Horm. Res.* **24**, 45–80.

Gottlieb, T. A., and Wallace, R. A. (1982). Intracellular glycosylation of vitellogenin in the liver of estrogen-stimulated *Xenopus laevis*. *J. Biol. Chem.* **257**, 95–103.

Grasdalen, H., and Jørgensen, L. (1985). ^{31}P-NMR studies on developing eggs and larvae of plaice. *Comp. Biochem. Physiol. B* **81B**, 291–294.

Greene, G. L., Gilna, P., Waterfield, M., Baker, A., Hort, Y., and Shine, J. (1986). Sequence and expression of human estrogen receptor complementary DNA. *Science* **231**, 1150–1154.

Hammond, B., Katzenellenbogen, B. S., Krauthammer, N., and McConnell, J. (1979). Estrogenic activity of the insecticide chlordecone (Kepone) and interaction with uterine estrogen receptors. *Proc. Natl. Acad. Sci. U.S.A.* **76**, 6641–6645.

Hanocq, F. M., Kirsch-Voiders, J., Hanocq-Quertier, J., Baltus, E., and Steinert, G. (1972). Characterization of yolk-DNA from *Xenopus laevis* oocytes ovulated *in vitro*. *Proc. Natl. Acad. Sci. U.S.A.* **69**, 1322–1326.

Hara, A. (1976). Iron-binding activity of female-specific serum proteins of rainbow trout (*Salmo gairdneri*) and chum salmon (*Oncorhynchus keta*). *Biochim. Biophys. Acta* **437**, 549–557.

Hara, A., and Hirai, H. (1978). Comparative studies on immunochemical properties of female specific serum proteins in rainbow trout (*Salmo gairdneri*). *Comp. Biochem. Physiol. B* **59B**, 339–343.

Hara, A., Yamauchi, K., and Hirai, H. (1980). Studies on female-specific serum protein (vitellogenin) and egg yolk protein in Japanese eel (*Anguilla japonica*). *Comp. Biochem. Physiol. B* **65B**, 315–320.

Haschemeyer, A. E. V., and Mathews, R. W. (1983). Temperature dependency of protein synthesis in isolated hepatocytes of Antarctic fish. *Physiol. Zool.* **56**, 78–87.

Haux, C., and Norberg, B. (1985). The influence of estradiol-17β on the liver content of protein, lipids, glycogen and nucleic acids in juvenile rainbow trout, *Salmo gairdnerii*. *Comp. Biochem. Physiol. B* **81B**, 275–279.

Hayward, M. A., Mitchell, T. A., and Shapiro, D. J. (1980). Induction of estrogen receptor and reversal of the nuclear/cytoplasmic receptor ratio during vitellogenin synthesis and withdrawal in *Xenopus laevis*. *J. Biol. Chem.* **255**, 11308–11312.

Henderson, R. J., Sargent, J. R., and Pirie, B. J. S. (1984). Fatty acid catabolism in the capelin, *Mallotus villosus* (Muller), during sexual maturation. *Mar. Biol. Lett.* **5**, 115–126.

Holdsworth, G., Michell, R. H., and Finean, J. B. (1974). Transfer of very low density lipoprotein from hen plasma into egg yolk. *FEBS Lett.* **39**, 275–277.

Hori, S. H., Kodama, T., and Tanahashi, K. (1979). Induction of vitellogenin synthesis in goldfish by massive doses of androgen. *Gen. Comp. Endocrinol.* **37**, 306–320.

Housley, P. R., and Pratt, W. B. (1983). Direct demonstration of glucocorticoid receptor phosphorylation by intact L-cells. *J. Biol. Chem.* **258**, 4630–4635.

Idler, D. R., and Campbell, C. M. (1980). Gonadotropin stimulation of estrogen and yolk precursor synthesis in juvenile rainbow trout. *Gen. Comp. Endocrinol.* **41**, 384–391.

Idler, D. R., and Clemens, W. A. (1959). The energy expenditures of Fraser River sockeye salmon during spawning migration to Chilko and Stuart Lakes. *Int. Pac. Salmon Fish. Comm., Prog. Rep.* **6**, 1–80.

Inoue, S., and Iwasaki, M. (1978). Isolation of a novel glycoprotein from the eggs of rainbow trout—Occurrence of disialosyl groups on all carbohydrate chains. *Biochem. Biophys. Res. Commun.* **83**, 1018–1023.

Inoue, S., and Iwasaki, M. (1980a). Sialoglycoproteins from the eggs of Pacific herring. Isolation and characterization. *Eur. J. Biochem.* **111**, 131–135.

Inoue, S., and Iwasaki, M. (1980b). Characterization of a new type of glycoprotein saccharide containing polysialosyl sequence. *Biochem. Biophys. Res. Commun.* **93**, 162–165.

Inoue, S., Kaneda-Hayashi, T., Sugiyama, H., and Ando, T. (1971). Studies on phosphoproteins from fish eggs. I. Isolation and characterization of a phosphoprotein from the eggs of Pacific herring. *J. Biochem. (Tokyo)* **69**, 1003–1011.

Inoue, S., Iwasaki, M., Kitajima, K., Inoue, Y., and Kudo, S. (1985). Localization and activation-induced proteolysis of polysialoglycoprotein in rainbow trout eggs. *Proc. Int. Symp. Glycoconjugates, 8th, 1985*.

Iwasaki, M., and Inoue, S. (1985). Structures of the carbohydrate units of polysialoglycoproteins isolated from the eggs of four species of salmonid fishes. *Glycoconjugate J.* **2**, 209–228.

Jensen, E. V., Suzuki, T., Kawashima, T., Stumpf, W. E., Jungblut, P. W., and De Sombre, E. R. (1968). A two-step mechanism for interaction of estradiol with rat uterus. *Proc. Natl. Acad. Sci. U.S.A.* **59**, 632–638.

Johnson, H., and Pecor, C. (1969). DDT and Great Lakes cohos. In "Sport Fishing Institute Bulletin," No. 203, pp. 4–5. Sport Fish. Inst., Washington, D.C.

Jungermann, K., and Katz, N. (1982). Functional hepatocellular heterogeneity. *Hepatology* **2**, 385–395.

Kaitaranta, J. K., and Ackman, R. G. (1981). Total lipids and lipid classes of fish roe. *Comp. Biochem. Physiol.* **69B**, 725–729.

King, R. J. B. (1984). Enlightenment and confusion over steroid hormone receptors. *Nature (London)* **312**, 701–702.

King, W. J., and Greene, G. L. (1984). Monoclonal antibodies localize oestrogen receptor of the nuclei of target cells. *Nature (London)* **307**, 745–747.

Kitahara, T. (1984). Behaviour of carotenoids in the chum salmon *Oncorhynchus keta* during development. *Bull. Jpn. Soc. Sci. Fish.* **50**, 531–536.

Kobuke, L., Specker, J. L., and Bern, H. A. (1987). Thyroxine content of eggs and larvae of coho salmon, *Oncorhynchus kisutch. J. Exp. Zool.* **242**, 89–94.

Korfsmeier, K.-H. (1980). Lysosomal enzymes and yolk platelets. *Eur. J. Cell Biol.* **22**, 208.

Korsgaard, B. (1983). The chemical composition of follicular and ovarian fluids of the pregnant blenny (*Zoarces viviparus* (L.)). *Can. J. Zool.* **61**, 1101–1108.

Korsgaard, B., and Petersen, I. M. (1976). Natural occurrence and experimental induction by estradiol-17β of a lipophosphoprotein (vitellogenin) in flounder (*Platichthys flesus* (L). *Comp. Biochem. Physiol. B* **54B**, 443–446.

Korsgaard, B., and Petersen, I. M. (1979). Vitellogenin, lipid and carbohydrate metabolism during vitellogenesis and pregnancy, and after hormonal induction in the blenny *Zoarces viviparus* (L.). *Comp. Biochem. Physiol. B* **63B**, 245–251.

Korsgaard, B., Emmersen, J., and Petersen, I. M. (1983). Estradiol-induced hepatic protein synthesis and transaminase activity in the male flounder, *Platichthys flesus* (L). *Gen. Comp. Endocrinol.* **50**, 11–17.

Korsgaard, B., Mommsen, T. P., and Saunders, R. L. (1986). The effect of temperature on the vitellogenic response in Atlantic salmon post-smolts (*Salmo salar*). *Gen. Comp. Endocrinol.* **62**, 193–201.

Krebs, E. G. (1985). The phosphorylation of proteins: A major mechanism for biological regulation. *Biochem. Soc. Trans.* **13**, 813–820.

Lam, T. J., and Loy, G. L. (1985). Effect of L-thyroxine on ovarian development and gestation in the viviparous guppy, *Poecilia reticulata. Gen. Comp. Endocrinol.* **60**, 324–330.

Lambert, J. G. D., and van Oordt, P. G. W. J. (1982). Catecholestrogens in the brain of the female rainbow trout, *Salmo gairdneri. Gen. Comp. Endocrinol.* **46**, 401.

Lane, C. D., Champion, J., Colman, A., James, T. C., and Applebaum, S. W. (1983). The fate of *Xenopus* and locust vitellogenins made in *Xenopus* oocytes. An export-import processing model. *Eur. J. Biochem.* **130**, 529–535.

Lange, R. H. (1981). Are yolk phosvitins carriers for specific cations? Comparative microanalysis in vertebrate yolk platelets. *Z. Naturforsch., C: Biosci.* **36C**, 686–687.

Lazier, C. B. (1975). [^{3}H]-Estradiol binding by chicken liver nuclear extracts—Mechanism of increase in binding following estradiol injection. *Steroids* **26**, 281–298.

Lazier, C. B. (1978). Ontogeny of the vitellogenic response to oestradiol and of the soluble nuclear oestrogen receptor in embryonic chick liver. *Biochem. J.* **174**, 143–152.

Lazier, C. B., Lonergan, K., and Mommsen, T. P. (1985). Hepatic estrogen receptors and plasma estrogen-binding activity in the Atlantic salmon. *Gen. Comp. Endocrinol.* **57**, 234–245.

Leatherland, J. F. (1985). Effects of 17β-estradiol and methyl testosterone on the activity of the thyroid gland in rainbow trout, *Salmo gairdneri* Richardson. *Gen. Comp. Endocrinol.* **60**, 343–352.

Leger, C., Frémont, L., Marion, D., Nassour, I., and Desfarges, M.-F. (1981). Essential fatty acids in trout serum. Lipoproteins, vitellogenin and egg lipids. *Lipids* **16**, 593–600.

Le Menn, F. (1979). Induction de vitellogénine par l'oestradiol et par des androgènes chez un téléostéen marin: *Gobius niger* L. *C. R. Hebd. Seances Acad. Sci.* **289**, 413–416.

Le Menn, F., Rochefort, H., and Garcia, M. (1980). Effect of androgen mediated by estrogen receptor of fish liver: Vitellogenin accumulation. *Steroids* **35**, 315–328.

Lewis, J. A., Clemens, M. J., and Tata, J. R. (1976). Morphological and biochemical changes in the hepatic endoplasmatic and golgi apparatus of male *Xenopus laevis* after induction of egg-yolk protein synthesis by estradiol-17β. *Mol. Cell. Endocrinol.* **4**, 311–329.

Liley, N. R., and Stacey, N. E. (1983). Hormones, pheromones, and reproductive behavior in fish. *In* "Fish Physiology" (W. S. Hoar, D. J. Randall, and E. M. Donaldson, eds.), Vol. 9B, pp. 1–63. Academic Press, New York.

Maitre, J. L., Le Guellec, C., Derrien, S., Tenniswood, M., and Valotaire, Y. (1985a). Measurement of vitellogenin from rainbow trout by rocket immunoelectrophoresis: Application to the kinetic analysis of estrogen stimulation in the male. *Can. J. Biochem. Cell Biol.* **63**, 982–987.

Maitre, J. L., Mercier, L., Dolo, L., and Valotaire, Y. (1985b). Caractérisation de récepteur spécifique à l'oestradiol, induction de la vitellogénine et de son mRNA dans la foie de truite arc-en-ciel (*Salmo gairdneri*). *Biochimie* **67**, 215–225.

Mann, M., Lazier, C., and Mommsen, T. (1988). In preparation.

Mano, Y., and Yoshida, M. (1969). A novel composition of phosvitins from salmon and trout roe. *J. Biochem. (Tokyo)* **66**, 105–108.

Markert, J. R., and Vanstone, W. E. (1971). Eggproteins of coho salmon (*Oncorhynchus kisutch*): Chromatographic separation and molecular weights of the major proteins in the high density fraction and their presence in salmon plasma. *J. Fish. Res. Board Can.* **28**, 1853–1856.

Miesfeld, R., Okret, S., Wikström, A.-C., Wrange, O., Gustafsson, J.-A., and Yamamoto, K. R. (1984). Characterization of a steroid hormone receptor gene and mRNA in wild-type and mutant cells. *Nature (London)* **312**, 779–781.

Migliaccio, A., Rotondi, A., and Auricchio, F. (1984). Calmodulin-stimulated phosphorylation of 17β-estradiol receptor on tyrosine. *Proc. Natl. Acad. Sci. U.S.A.* **81**, 5921–5925.

Miller, M. S., Benore-Parsons, M., and White, H. B. (1982). Dephosphorylation of chicken riboflavin-binding protein and phosvitin decreases their uptake by oocytes. *J. Biol. Chem.* **257**, 6818–6824.

Mommsen, T. P., and Lazier, C. B. (1986). Stimulation of estrogen receptor accumulation by estradiol in primary cultures of salmon hepatocytes. *FEBS Lett.* **195**, 269–271.

Mommsen, T. P., French, C. J., and Hochachka, P. W. (1980). Sites and pattern of protein and amino acid utilization during the spawning migration of salmon. *Can. J. Zool.* **58**, 1785–1799.

Moon, T. W., Walsh, P. J., and Mommsen, T. P. (1985). Fish hepatocytes: A model metabolic system. *Can. J. Fish. Aquat. Sci.* **42**, 1772–1782.

Mugiya, Y., and Ichii, T. (1981). Effects of estradiol-17β on branchial and intestinal calcium uptake in the rainbow trout, *Salmo gairdneri. Comp. Biochem. Physiol. A* **70A**, 97–101.

Mugiya, Y., and Watabe, N. (1977). Studies on fish scale formation and resorption. II. Effect of estradiol on calcium homeostasis and skeletal tissue resorption in the goldfish, *Carassius auratus* and the killifish, *Fundulus heteroclitus. Comp. Biochem. Physiol. A* **57A**, 197–202.

Muniyappa, K., and Adiga, P. R. (1980). Estrogen-induced synthesis of thiamin-binding protein in immature chicks—Kinetics of induction, hormonal specificity and modulation. *Biochem. J.* **186**, 201–210.

Nagahama, Y. (1983). The functional morphology of teleost gonads. *In* "Fish Physiology" (W. S. Hoar, D. J. Randall, and E. M. Donaldson, eds.), Vol. 9A, pp. 223–275. Academic Press, New York.

Nath, P., and Sundararaj, B. I. (1981). Isolation and identification of female-specific serum phospholipoproteins (vitellogenin) in the catfish *Heteropneustes fossilis. Gen. Comp. Endocrinol.* **43**, 184–190.

Ng, T. B., and Idler, D. R. (1983). Yolk formation and differentiation in teleost fishes. *In* "Fish Physiology" (W. S. Hoar, D. J. Randall, and E. M. Donaldson, eds.). Vol. 9, pp. 373–404. Academic Press, New York.

Ng, T. B., Woo, N. Y. A., Tam, P. P. L., and Au, C. Y. W. (1984). Changes in metabolism and hepatic ultrastructure induced by estradiol and testosterone in immature female *Epinepholus akaara* (Teleostei, Serranidae). *Cell Tissue Res.* **236,** 651–659.

Norberg, B., and Haux, C. (1985). Induction, isolation and a characterization of the lipid content of plasma vitellogenin from two *Salmo* species: Rainbow trout (*Salmo gairdneri*) and sea trout (*Salmo trutta*). *Comp. Biochem. Physiol. B* **81B,** 869–876.

Nosek, J., Krajhanzl, A., and Kocourek, J. (1983). Studies on lectins. LVII. Immuno-fluorescence localization of lectins present in fish ovaries. *Histochemistry* **79,** 131–139.

Nunomura, W., Hara, A., Takano, K., and Hirai, H. (1983). Immunohistochemical localization of vitellogenin in hepatic cells of some salmonid fishes. *Bull. Fac. Fish., Hokkaido Univ.* **34,** 79–87.

Olivereau, M., and Olivereau, J. (1979). Effect of estradiol-17β on the cytology of the liver, gonad and pituitary, and on the plasma electrolytes in the female freshwater eel. *Cell Tissue Res.* **199,** 431–454.

Opresko, L., Wiley, H. S., and Wallace, R. A. (1979). The origin of yolk-DNA in *Xenopus laevis. J. Exp. Zool.* **209,** 367–376.

Opresko, L., Wiley, H. S., and Wallace, R. A. (1981). Receptor-mediated binding and internalization of vitellogenin by *Xenopus* oocytes. *J. Cell Biol.* **91,** 218a.

Pan, M. L., Bell, W. J., and Telfer, W. H. (1969). Vitellogenic blood protein synthesis by insect fat body. *Science* **165,** 393–394.

Perlman, A. J., Wolffe, A. P., Champion, J., and Tata, J. R. (1984). Regulation of estrogen receptor of vitellogenin gene transcription in *Xenopus* hepatocyte cultures. *Mol. Cell. Endocrinol.* **38,** 151–161.

Peter, R. E. (1983). The brain and neurohormones in teleost reproduction. *In* "Fish Physiology" (W. S. Hoar, D. J. Randall, and E. M. Donaldson, eds.), Vol. 9, pp. 97–137. Academic Press, New York.

Petersen, I. M., and Korsgaard, B. (1977). Changes in serum glucose and lipids in liver glycogen and phosphorylase during vitellogenesis in nature in the flounder (*Platichthys flesus*, L.). *Comp. Biochem. Physiol. B* **58B,** 167–171.

Peute, J., van der Gaag, M. A., and Lambert, J. G. D. (1978). Ultrastructure and lipid content of the liver of the zebrafish, *Brachydanio rerio*, related to vitellogenin synthesis. *Cell Tissue Res.* **186,** 297–308.

Phillips, B. W., and Shapiro, D. J. (1981). Estrogen regulation of hepatic 3-hydroxy-3-methylglutaryl coenzyme A reductase and acetyl-CoA carboxylase in *Xenopus laevis. J. Biol. Chem.* **256,** 2922–2927.

Plack, P. A., and Fraser, N. W. (1971). Incorporation of L-(^{14}C)-leucine into egg proteins by liver slices from cod. *Biochem. J.* **121,** 857–862.

Plack, P. A., and Pritchard, D. J. (1968). Effects of oestradiol 3-benzoate on concentrations of retinal and lipids in cod plasma. *Biochem. J.* **106,** 257–262.

Plack, P. A., Pritchard, D. J., and Fraser, N. W. (1971). Egg proteins in cod serum. Natural occurrence and induction by injections of oestradiol-3-benzoate. *Biochem. J.* **121,** 847–856.

Plack, P. A., Skinner, E. R., Rogie, A., and Mitchell, A. I. (1979). Distribution of DDT between the lipoproteins of trout serum. *Comp. Biochem. Physiol. C* **62C,** 119–125.

Pottinger, T. G. (1986). Estrogen-binding sites on the liver of sexually mature male and female brown trout, *Salmo trutta* L. *Gen. Comp. Endocrinol.* **61,** 120–126.

Roach, A. H., and Davies, P. L. (1980). Catfish vitellogenin and its messenger RNA are smaller than their chicken and *Xenopus* counterparts. *Biochim. Biophys. Acta* **610**, 400–412.

Roch, M., and McCarter, J. A. (1984). Hepatic metallothionein production and resistance to heavy metals by rainbow trout (*Salmo gairdneri*). I. Exposed to an artifical mixture of zinc, copper and cadmium. *Comp. Biochem. Physiol. C* **77C**, 71–75.

Sand, D. M., Hehl, J. H., and Schlenk, H. (1971). Biosynthesis of wax esters in fish. Metabolism of dietary alcohols. *Biochemistry* **10**, 2536–2541.

Sand, O., Petersen, I. M., and Korsgaard, B. (1980). Changes in some carbohydrate metabolizing enzymes and glycogen in liver, glucose and lipid in serum during vitellogenesis and after induction by estradiol-17-β in the flounder (*Platichthys flesus* L.). *Comp. Biochem. Physiol. B.* **65B**, 327–332.

Seguin, C., Felber, B. K., Carter, A. D., and Hamer, D. H. (1984). Competition for cellular factors that activate metallothionein gene transcription. *Nature (London)*, **312**, 781–785.

Selman, K., and Wallace, R. A. (1983a). Oogenesis in *Fundulus heteroclitus*. II. The transition from vitellogenesis into maturation. *Gen. Comp. Endocrinol.* **42**, 345–354.

Selman, K., and Wallace, R. A. (1983b). Oogenesis in *Fundulus heteroclitus*. III. Vitellogenesis. *J. Exp. Zool.* **226**, 441–457.

Shackley, S. E., King, P. E., and Gordon, S. M. (1981). Vitellogenesis and trace metals in a marine teleost. *J. Fish Biol.* **18**, 349–352.

Shapiro, D. (1982). Steroid hormone regulation of vitellogenin gene expression. *CRC Crit. Rev. Biochem.* **8**, 187–203.

Shigeura, H. T., and Haschemeyer, A. E. V. (1984). Purification of egg-yolk proteins from the Anarctic fish, *Chaenocephalus aceratus*. *Comp. Biochem. Physiol. B* **80B**, 935–939.

Singh, H., and Singh, T. P. (1981). Effect of parathion and aldrin on survival, ovarian 32P uptake and gonadotrophic potency in a freshwater catfish, *Heteropneustes fossilis* (Bloch). *Endokrinologie* **77**, 173–178.

Sire, M.-F., and Vernier, J.-M. (1983). Enterocytes et synthèse de lipoprotéines chez les poissons. Chylomicrons ou VLDL. Étude ultrastructurale et biochimique. *Biol. Cell.* **49**, A24.

Skinner, E. R., and Rogie, A. (1978). The isolation and partial characterization of the serum lipoproteins and apolipoproteins of the rainbow trout. *Biochem. J.* **173**, 507–520.

So, Y. P., Idler, D. R., and Hwang, S. J. (1985). Plasma vitellogenin in landlocked Atlantic salmon (*Salmo salar* Ouananiche): Isolation, homologous radioimmunoassay and immunological cross-reactivity with vitellogenin from other teleosts. *Comp. Biochem. Physiol. B* **81B**, 63–71.

Solar, I. G., Donaldson, E. M., and Hunter, G. A. (1984). Optimization of treatment regimes for controlled sex differentiation and sterilization in wild rainbow trout (*Salmo gairdneri* Richardson) by oral administration of 17α-methyltestosterone. *Aquaculture* **42**, 129–139.

Soreq, H. (1985). The biosynthesis of biologically active proteins in mRNA-microinjected *Xenopus* oocytes. *CRC Crit. Rev. Biochem.* **18**, 199–238.

Sumpter, J. P. (1981). The purification and radioimmunoassay of vitellogenin from the rainbow trout, *Salmo gairdneri*. *Abstr., Int. Symp. Comp. Endocrinol., 9th, 1981*, p. 243.

Sumpter, J. P., and Dodd, J. M. (1979). The annual reproductive cycle of the female lesser spotted dogfish, *Scyliorhinus canicula* L., and its endocrine control. *J. Fish Biol.* **15**, 687–695.

Sundararaj, B. I., Nath, P., and Burzawa-Gerard, E. (1982a). Synthesis of vitellogenin and its uptake by the ovary in the catfish, *Heteropneustes fossilis* (Bloch) in response to carp gonadotropin and its subunits. *Gen. Comp. Endocrinol.* **46**, 93–98.

Sundararaj, B. I., Goswami, S. V., and Lamba, V. J. (1982b). Role of testosterone, estradiol-17β, and cortisol during vitellogenesis in the catfish, *Heteropneustes fossilis* (Bloch). *Gen. Comp. Endocrinol.* **48**, 390–397.

Szego, C. M., and Pietras, R. J. (1985). Subcellular distribution of oestrogen receptors. *Nature (London)* **317**, 88–89.

Tam, W. H., Birkett, L., Makaran, R., Payson, P. D., Whitney, D. K., and Yu, C. K.-C. (1987). Modification of carbohydrate metabolism and liver vitellogenic function in brook trout (*Salvelinus fontinalis*) by exposure to low pH. *Can. J. Fish. Aquat. Sci.* **44**, 630–635.

Tata, J. R., and Smith, D. F. (1979). Vitellogenesis: A versatile model for hormonal regulation of gene expression. *Recent Prog. Horm. Res.* **35**, 47–95.

Taylor, M. A., and Smith, L. D. (1985). Quantitative changes in protein synthesis during oogenesis in *Xenopus laevis*. *Dev. Biol.* **110**, 230–237.

te Heesen, D. (1977). Immunologische Untersuchungen an exo- und endogenen Dotterproteinen von *Brachydanio rerio* (Teleostei, Cyprinidae) und verwandten Arten. *Zool. Jahrb., Abt Anat. Ontog. Tiere,* **97**, 566–582.

Tenniswood, M. P. R., Searle, P. F., Wolffe, A. P., and Tata, J. R. (1983). Rapid estrogen metabolism and vitellogenin gene expression in *Xenopus* hepatocyte cultures. *Mol. Cell Endocrinol.* **30**, 329–345.

Turner, R. T., Dickhoff, W. W., and Gorbman, A. (1981). Estrogen binding to hepatic nuclei of pacific hagfish *Eptatretus stouti*. *Gen. Comp. Endocrinol.* **45**, 26–29.

Ueda, H., Hiroi, O., Hara, A., Yamauchi, K., and Nagahama, Y. (1984). Changes in serum concentrations of steroid hormones, thyroxine, and vitellogenin during spawning migration of the chum salmon, *Oncorhynchus keta*. *Gen. Comp. Endocrinol.* **53**, 203–211.

Valotaire, Y., Tenniswood, M., Le Guellec, C., and Tata, J. R. (1984). The preparation and characterization of vitellogenin messenger RNA from rainbow trout (*Salmo gairdneri*). *Biochem. J.* **217**, 73–77.

van Bohemen, C. G., and Lambert, J. G. D. (1981). Estrogen synthesis in relation to estrone, estradiol and vitellogenin plasma levels during the reproductive cycle of the female rainbow trout, *Salmo gairdneri*. *Gen. Comp. Endocrinol.* **45**, 105–114.

van Bohemen, C. G., Lambert, J. G. D., Goos, H. J. T., and van Oordt, P. G. W. J. (1982a). Estrone and estradiol participation during exogenous vitellogenesis in the female rainbow trout, *Salmo gairdneri*. *Gen. Comp. Endocrinol.* **46**, 81–92.

van Bohemen, C. G., Lambert, J. G. D., and van Oordt, P. G. W. J. (1982b). Vitellogenin induction by estradiol in estrone-primed rainbow trout, *Salmo gairdneri*. *Gen. Comp. Endocrinol.* **46**, 136–139.

Vernier, J.-M., and Sire, M.-F. (1977). Plaquettes vitellines et activité hydrolasique acide au cours du développement embryonnaire de le truite arc-en-ciel. Etude ultrastructurale et biochimique. *Biol. Cell.* **29**, 99–112.

Wahli, W., Dawid, I. B., Ryffel, G. U., and Weber, R. (1981). Vitellogenesis and the vitellogenin gene family. *Science* **212**, 298–304.

Wallace, R. A. (1983). Interactions between somatic cells and the growing oocyte of *Xenopus laevis*. *In* "Current Problems in Germ Cell Differentiation" (A. McLaren and C. L. Wylie, eds.), pp. 285–306. Cambridge Univ. Press, London and New York.

Wallace, R. A. (1985). Vitellogenesis and oocyte growth in non-mammalian vertebrates. *In* "Developmental Biology" (L. Browder, ed.), Vol. 1, pp. 127–177. Pergamon, New York.

Wallace, R. A., and Hollinger, T. G. (1979). Turnover of endogenous, microinjected, and sequestered protein in *Xenopus* oocytes. *Exp. Cell Res.* **119**, 277–287.

Wallace, R. A., and Jared, D. W. (1968). Studies on amphibian yolk. VII. Serum-phosphoprotein synthesis by vitellogenic females and estrogen-treated males of *Xenopus laevis*. *Can. J. Biochem.* **46**, 953–959.

Wallace, R. A., and Selman, K. (1981). Cellular and dynamic aspects of oocyte growth in teleosts. *Am. Zool.* **21**, 325–343.

Wallace, R. A., and Selman, K. (1985). Major changes during vitellogenesis and maturation of *Fundulus* oocytes. *Dev. Biol.* **110**, 492–498.

Wallace, R. A., Deufel, R. A., and Misulovin, Z. (1980). Protein incorporation by isolated amphibian oocytes. VI. Comparison of autologous and xenogeneic vitellogenins. *Comp. Biochem. Physiol. B* **65B**, 151–155.

Walter, P., Green, S., Greene, G., Krust, A., Bornert, J.-M., Jeltsch, J.-M., Staub, A., Jensen, E., Scrace, G., Waterfield, M., and Chambon, P. (1985). Cloning of the human estrogen receptor cDNA. *Proc. Natl. Acad. Sci. U.S.A.* **82**, 7889–7893.

Wang, S.-Y., and Williams, D. L. (1982). Biosynthesis of the vitellogenins. Identification and characterization of nonphosphorylated precursors to avian vitellogenin I and vitellogenin II. *J. Biol. Chem.* **257**, 3837–3846.

Wang, S.-Y., Smith, D. E., and Williams, D. L. (1983). Purification of avian vitellogenin III: comparisons with vitellogenins I and II. *Biochemistry* **22**, 6206–6212.

Wangh, L. J. (1982). Glucocorticoids act together with estrogens and thyroid hormones in regulating the synthesis and secretion of *Xenopus* vitellogenin, serum albumin and fibrinogen. *Dev. Biol.* **89**, 294–298.

Welshons, W. V., Lieberman, M. E., and Gorski, J. (1984). Nuclear localization of unoccupied oestrogen receptors. *Nature (London)* **307**, 747–749.

White, H. B. (1985). Biotin-binding proteins and biotin transport to oocytes. *Ann. N.Y. Acad. Sci.* **447**, 202–211.

Wiegand, M. D. (1982). Vitellogenesis in fishes. *In* "Reproductive Physiology of Fish" (C. J. J. Richter and H. J. T. Goos, compilers), pp. 136–146. Pudoc, Wageningen.

Wiegand, M. D., and Idler, D. R. (1982). Synthesis of lipids by the rainbow trout (*Salmo gairdneri*) ovary in vitro. *Can. J. Zool.* **60**, 2683–2693.

Wiegand, M. D., and Peter, R. E. (1980). Effects of the salmon gonadotropin (SG-G100) on plasma lipids in the goldfish, *Carassius auratus*. *Can. J. Zool.* **58**, 967–972.

Wiley, H. S., and Wallace, R. A. (1981). The structure of vitellogenin. Multiple vitellogenins in *Xenopus laevis* give rise to multiple forms of the yolk proteins. *J. Biol. Chem.* **256**, 8626–8634.

Wingfield, J. C. (1980). Sex-steroid binding proteins in vertebrate blood. *In* Hormones: Adaptation and Evolution" (S. Ishii, T. Hirono, and M. Wada, eds.), pp. 135–144. Jpn. Sci. Soc. Press, Tokyo.

Wiskocil, R., Bensky, P., Dower, W., Goldberger, R. F., Gordon, J. I., and Deeley, R. G. (1980). Coordinate regulation of two estrogen-dependent genes in avian liver. *Proc. Natl. Acad. Sci. U.S.A.* **77**, 4474–4478.

Wolffe, A. P., Glover, J. F., Martin, S. C., Tenniswood, M. P. R., Williams, J. L., and Tata, J. R. (1985). Deinduction of transcription of *Xenopus* 74-kDa albumin genes and destabilization of mRNA by estrogen *in vivo* and in hepatocyte cultures. *Eur. J. Biochem.* **146,** 489–496.

Wright, C. V. E., Wright, S. C., and Knowland, J. (1983). Partial purification of estradiol receptor from *Xenopus laevis* liver and levels of receptor in relation to estradiol concentration. *EMBO J,* **2,** 973–977.

Yu, J. Y.-L., Dickhoff, W. W., and Gorbman, A. (1980). Sexual patterns of protein metabolism in liver and plasma of hagfish, *Eptatretus stouti,* with special reference to vitellogenesis. *Comp. Biochem. Physiol. B* **65B,** 111–117.

Yurowitzky, Y. G., and Milman, L. S. (1972). Changes in enzyme activity of glycogen and hexose metabolism during oocyte maturation in a teleost, *Misqurnus fossilis* L. *Wilhelm Roux' Arch. Entwicklungsmech. Org.* **171,** 48–54.

Yurowitzky, Y. G., and Milman, L. S. (1975). Changes in activity of enzymes of glycogen metabolism in loach oocytes and embryos. *Biochemistry (Engl. Transl.)* **40,** 821–825.

Yusko, S., Roth, T. F., and Smith, T. (1981). Receptor-mediated vitellogenin binding to chicken oocyte. *Biochem. J.* **200,** 43–50.

Zagalsky, P. F., Gilchrist, B. M., Clark, R. J. H., and Fairclough, D. P. (1983). The canthaxanthin-lipovitellin of *Branchipus stagnalis* (L). (Crustacea: Anostraca): A resonance Raman and circular dichroism study. *Comp. Biochem. Physiol. B* **73B,** 163–167.

6

YOLK ABSORPTION IN EMBRYONIC
AND LARVAL FISHES

THOMAS A. HEMING

Pulmonary Division
Department of Internal Medicine
University of Texas Medical Branch
Galveston, Texas 77550-2780

RANDAL K. BUDDINGTON

Department of Physiology
University of California
Los Angeles, California 90024

407

I. INTRODUCTION

A major portion of our knowledge regarding yolk absorption is based on species possessing large, demersal eggs, which are adapted for colder waters and long incubation periods. This is evident in previous reviews of yolk utilization by Hayes (1949), Smith (1957, 1958), Williams (1967), Blaxter (1969), Terner (1979), and Boulekbache (1981). Development of culture techniques for other species and increasing ecological concerns, however, have elicited research of additional groups, particularly marine fishes. This information has been incorporated and contrasted in the present review.

A fish egg can be considered a semiclosed system. Once the egg membrane(s) has been hardened by exposure to water, the membrane(s) permits gas exchange but is relatively impervious to most solutes. As a consequence, the majority of fish embryos are dependent on endogenous yolk reserves to supply the substrates for energy production and growth. Viviparous fishes are an exception (see Section VI,C). Both the rate of yolk absorption and the efficiency of yolk utilization are important determinants of early development, growth, and survival. Larval survival is ultimately dependent on the availability of food in sufficient quantity and of adequate quality after yolk reserves are exhausted. It follows that there are strong selective pressures synchronizing completion of yolk absorption, development of the capability of feeding, and the availability of suitable food (Bams, 1969; Rosenthal and Alderdice, 1976). As well, since large size confers certain advantages on larvae, there are strong selective pressures to maximize the efficiency with which yolk is converted into tissues. Larger larvae of a given species can be expected to be stronger swimmers (Hunter, 1972), less affected by competition (Hulata *et al.*, 1976), more resistant to starvation (Blaxter and Hempel, 1963), less susceptible to predation (Ware, 1975), able to commence feeding earlier (Wallace and Aasjord, 1984a), and able to have increased success at first feeding (Braum, 1967; Ellertsen *et al.*, 1980).

The rate and efficiency of yolk absorption are influenced by a number of environmental factors, including temperature, light, oxygen concentration, and salinity. Fish eggs are not motile, however, and thus developing embryos are unable to actively exploit the most favorable environments available, at least until after hatching. Only species that utilize reproductive strategies such as viviparity or mouth brooding may, through parental behaviors, be able to manipulate egg incubation conditions. It is selectively advantageous, therefore, for a species to produce eggs that can develop successfully within a range of "expected" incubation conditions. The scope of these "expected" conditions will depend on those conditions experienced during evolution of the species. For some fishes, the fluctuation in environmental parameters may be relatively slight (e.g., abyssal marine habitats), while for others it may be large (e.g., some temperature freshwater habitats). From an applied standpoint, there has been an interest in determining the influence of environmental factors on yolk absorption—particularly the effects of temperature, since it is generally the most variable parameter and the most easily controlled in culture settings.

A review of present knowledge regarding yolk absorption in fish is hindered somewhat by the use of many different and often imprecisely defined terminologies for developmental staging. For the purpose of our review, we have adopted the generalized terms defined below.

1. Embryo—the developing fish prior to hatch.
2. Eleutheroembryo—the developing fish after hatch until the initiation of feeding or, in the case of viviparous fish, parturition (Balon, 1975). For our purposes, feeding refers to the ingestion of exogenous matter into the stomach or the capability to do so, rather than behavioral responses to potential food items. Defined in this way, feeding is independent of complete yolk absorption and may be independent of the capability to digest and utilize ingested material.
3. Larva—the developing fish after initiation of feeding or parturition until a juvenile, characterized by a full complement of minute adult features, is attained.
4. Yolk—the nutritional reserves provided in the ovum, including those associated with the yolk platelets and oil globules. A number of authors categorize the yolk sac contents into "yolk" and "oil" on the basis of visual appearance. For simplicity, we have regarded these categories as equivalent to yolk platelets and oil globules, respectively.
5. Tissues—the body of the developing fish including the yolk sac but without the yolk.

Throughout this review, we use the common and scientific names of fishes listed by Robins *et al.* (1980).

II. STRUCTURAL ASPECTS OF YOLK ABSORPTION

A. Yolk Morphology

The structural components of fish yolk include yolk platelets and oil globules. The majority of yolk platelets are round or oval in shape, flattened in one plane, and 4–15 μm in length. Larger platelets appear to be characteristic of species possessing larger eggs (Grodziński, 1973). Platelet size also varies within each egg, with the deeper, more centrally located platelets tending to be larger and more homogenous than the superficial peripheral ones (Vernier and Sire, 1977; Hamlett and Wourms, 1984). Each platelet consists of an outer sheath and a central core (Fig. 1). The sheath forms a semipermeable bilayer around the core (Grodziński, 1973) and contains mucopolysaccharides (Ohno *et al.*, 1964). The core is composed of lipovitellin and phosvitin, or analogous lipoproteins and phosphoproteins (Fujii, 1960; Wallace *et al.*, 1966; Jared and Wallace, 1968). These core proteins may or may not be arranged in a crystalline lattice (Lange, 1981, 1982;

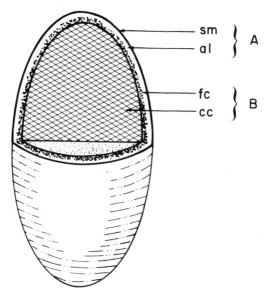

Fig. 1. Model of a yolk platelet based mainly on the structure of *Amia* platelets. The cuts reveal its interior: (A) outer sheath; (B) main body; sm, superficial membrane; al, amorphous superficial layer; fc, fibrillar cortex; cc, crystalline core. [From Grodziński (1973).]

Lange *et al.*, 1982). Moreover, the crystalline structure may be lost as the ova mature (Balinsky, 1970).

Oil globules are located among the yolk platelets. Globule number and size vary greatly among species, from innumerable small globules in the micrometer diameter range to singular large globules in the millimeter diameter range. The globules contain primarily triglycerides, although proteins (Grodziński, 1973), wax esters (Vetter *et al.*, 1983), and carotenoid pigments (Nakagawa and Tsuchiya, 1971) are also present in some species.

B. Meroblastic Fishes

Meroblastic cleavage, typical of most fishes (elasmobranchs and teleosts), results in the formation of an extraembryonic yolk sac. A characteristic feature of this extraembryonic sac is the yolk syncytium, a specialized tissue responsible for absorption of yolk. The presumptive yolk syncytium, the periblast, is recognizable in the fertilized teleost egg at the one-cell stage (Yamamoto, 1982). As cleavage proceeds, numerous free nuclei appear in the periblast, thus transforming the layer into a true syncytium.

In teleost eggs, the yolk syncytium together with overlaying mesoderm and ectoderm spreads to enclose the entire yolk mass. Endoderm does not follow the movement of the teleost blastodisc rim and, consequently, the yolk is not enclosed by an endodermal layer. Absorption of yolk nutrients in teleosts, therefore, occurs without any involvement of endodermal cells or the gut (Bachop and Schwartz, 1974). A system of blood vessels, the vitelline circulation, develops within the walls of the yolk sac. In some areas, the endothelial wall of vitelline capillaries is incomplete and embryonic blood is in direct contact with the syncytium (Shimizu and Yamada, 1980). Absorption of yolk involves endocytosis by the syncytium, intrasyncytial digestion and synthesis, and finally the release of yolk metabolites to the vitelline circulation. When yolk reserves are exhausted, the syncytium is resorbed; it does not take part in formation of the permanent fish body (Yamada, 1959; Yamamoto, 1982).

Two regions of the yolk syncytium can be distinguished on the basis of their fine structure (Shimizu and Yamada, 1980). One region, characterized by smooth endoplasmic reticulum, numerous mitochondria, and glycogen granules, is proposed to be responsible for carbohydrate and/or lipid metabolism. This region extends throughout the syncytium. The second region is characterized by rough endoplasmic reticulum and Golgi complexes, and extends in portions across the syncytium forming a stratified structure. This latter region is thought to be involved in the synthesis and transport of proteinaceous substances. Yolk protein must be dephosphorylated to become soluble. Amirante (1972) suggested that fish yolk proteins are solubilized by the action of calcium and phosphoprotein phophatase. Syncytial Golgi complexes probably supply acid hydrolases for the degradation of yolk platelets (Vernier and Sire, 1977; Hamlett, *et al.*, 1987).

In addition to the syncytial layer, the yolk itself contains enzymes (Hamor and Garside, 1973) that probably facilitate the breakdown of yolk into its constituent nutrients. Vernier and Sire (1977) described two types of yolk platelets with different enzyme contents. One form, the embryonic platelet type, has an enzyme load that allows nutrients to be released prior to establishment of the syncytium. The second or

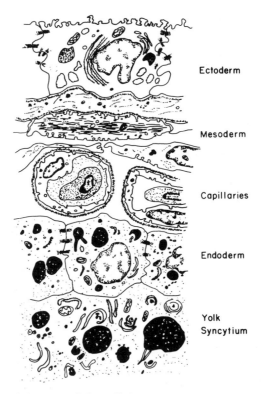

Fig. 2. Idealized diagram of the cellular organization in a preimplantation shark yolk sac. The teleost yolk sac is similar in structure except that it lacks endoderm. [From Hamlett and Wourms (1984).]

usual platelet type lacks this enzyme load and is digested by syncytial enzymes.

While the extraembryonic yolk sac with its yolk syncytium is the sole site of yolk absorption in teleosts, this is not the case in chondrichthyean fishes (sharks, skates, rays and ratfish). In holocephalians, for instance, only a small portion of the yolk mass is enclosed by the yolk sac (Dean, 1906). The remainder breaks up into a viscous fluid, which is first absorbed via the external gills of the embryo and later ingested through the mouth. This ingestion of yolk nutrients is comparable, in a general sense, to that exhibited by oophagous sharks, whose viviparous embryos ingest ova present in the same uteri (Fujita, 1981; Gilmore et al., 1983).

In elasmobranchs, the formation of an archenteron at the posterior edge of the blastodisc during gastrulation results in a yolk sac that possesses an endodermal layer (Fig. 2). This endodermal layer mediates the transfer of yolk metabolites from the syncytium to the vitelline circulation (Hamlett et al., 1987). Moreover, the elasmobranch yolk sac is continuous with the alimentary tract via a yolk stalk, and thus the majority of yolk is digested within the intestine. Yolk platelets are moved by ciliary action from the yolk sac through the yolk stalk and into the spiral intestine (Te Winkel, 1943; Baranes and Wendling, 1981). An internal yolk storage organ may or may not be present. Enzymatic activity in the gut is established relatively early in development, when the embryo is approximately one-quarter its size at parturition (Te Winkel, 1943).

C. Holoblastic Fishes

A few fish species develop holoblastically (e.g., lampreys and chondrosteans). In these species, the endodermal and lateral plates fuse along the midventral line forming an intraembryonic yolk sac. As a result, all three germ layers enclose the yolk mass (Ballard and Ginzburg, 1980). The resultant intraembryonic yolk sac directly participates in formation of the alimentary canal. During posthatch development of chondrosteans, the yolk sac is separated into two major regions, each of which develops separate blood drainages (Ballard and Needham, 1964). The distal region comprises the intestine and spiral valve, and the corresponding blood supply proceeds to the liver. Yolk within this region is the first to be utilized. The second region includes the stomach and esophagus and develops a blood supply that proceeds directly to the sinus venosus. This region is the last portion of the alimentary canal to differentiate, and yolk is retained there longer (Buddington and Christofferson, 1985).

Although hydrolytic enzymes are present within the developing alimentary canal, their activities are low (Korzhuev and Sharkova, 1967; Buddington and Doroshov, 1986). The existence of yolk material within endodermal cells lining the yolk sac implies that endocytosis and intracellular digestion may be the primary mechanisms by which yolk nutrients are made available (Krayushkina, 1957; Buddington and Christofferson, 1985). Thus, the intraembryonic yolk sac of holoblastic fishes and the extraembryonic yolk sac of meroblastic fishes exhibit similar mechanisms for mobilization of yolk nutrients.

III. YOLK COMPOSITION DURING DEVELOPMENT

Selective pressures have resulted in fish exploiting a wide diversity of reproductive strategies. As a consequence, egg size and fecundity vary among species, in oviparous fishes from about 0.7 mm egg diameter (e.g., convict surgeon fish *Acanthurus triostegus*) to greater than 10 mm diameter (e.g., chinook salmon *Oncorhynchus tshawytscha*), with spawns varying from less than 100 eggs per female (e.g., mouth-brooding cichlid *Labeotropheus fuelleborni*) to more than 9,000,000 (e.g., Atlantic cod *Gadus morhua*). Viviparous fish tend to produce fewer but proportionally larger eggs. These differences in egg size and number imply maternal investment per egg differs widely among species. The deposition of nutrients into the egg during oogenesis has been reviewed in this volume by Mommsen and Walsh (this volume, Chapter 5).

The nutrient composition of fish eggs is species-specific (Table I). Within a given species, as well, egg quality varies as a function of maternal age, weight, and diet (Kamler, 1976; Kuznetsov and Khalitov, 1979). Despite these differences, the dynamics of yolk absorption are similar among groups. Following fertilization, the developing embryo begins to utilize yolk nutrients. This is accompanied by increasing consumption of oxygen, particularly after the blastula stage is reached. As development proceeds, the absolute and relative composition of the yolk changes (Nakagawa and Tsuchiya, 1972, 1974, 1976).

Various approaches (proximal analysis, respiratory quotients, radiolabeled substrates) have been used to investigate the sequence with which yolk nutrients are catabolized for energy production. Generally, carbohydrate, lipid, and protein are consumed prior to hatch-

Table I
Chemical Composition of Fish Eggs

Species	Dry weight		Percentage of dry weight				Sources
	mg	%	Protein	Lipid	Carbohydrate	Ash	
Acipenser transmontanus (white sturgeon)	6.25[a]	23.8	67	30	—	3	Wang et al. (1987)
Coregonus albula (vendace)	16.27[b]	—	64.4	25.8	—	8.5	Dabrowski and Luczynski (1984)
Coregonus lavaretus (whitefish)	15.6[b]	—	60.3	27.7	—	9.8	Dabrowski and Luczynski (1984)
Cyprinus carpio (carp)	—	30.4[a]	64.3	5.9	3.7	—	Moroz and Luzhin (1976)
	—	10.2[b]	58.3–59.2	5.4–29.3	1.5–6.2	6.3	Moroz and Luzhin (1976); Kamler (1976)
Eleginus navaga (navaga)	0.283[a]	22.1	66.4	20.5	—	2.1	Lapin and Matsuk (1979)
	0.298[b]	10.4	56.7	16.8	—	8.4	Lapin and Matsuk (1979)
Morone saxatilis (striped bass)	0.232[a]	46.3	—	52.0	—	3.0	Eldridge et al. (1981a)
Pseudopleuronectes americanus (winter flounder)	0.051AF[a]	—	79.3	15.4	5.3	—	Cetta and Capuzzo (1982)
	0.049AF[b]	—	77.4	19.4	3.2	7.2	Cetta and Capuzzo (1982)
Salmo gairdneri (rainbow trout)	42.1[a]	41.3	56.2	—	—	—	Zeitoun et al. (1977)
	—	33.8[b]	59.8–71.3	11.4	0.6	3.8–3.9	Smith (1957); Satia et al. (1974)
Salmo salar (Atlantic salmon)	49.7[b]	36.0	52.2	36.1	1.0	2.8	Hamor and Garside (1977a)
Sardinops caerulea (Pacific sardine)	—	29.3[a]	71.6	13.0	<1	7	Lasker (1962)
Sciaenops ocellata (red drum)	—	7.0[b]	28.1	33.7	0.4	—	Vetter et al. (1983)

[a] Unfertilized.
[b] Fertilized; AF, ash-free.

ing, while lipid and protein catabolism predominate after hatching. However, the precise sequence of nutrient consumption varies both qualitatively and quantitatively. This is not surprising, since it is unlikely that any one energetic scheme is adequate to describe the sequential utilization of energy substrates in all fishes, in other than general terms. More likely, the precise scheme varies among fishes in relation to absolute egg composition, which in turn has probably evolved within the "expected" constraints of a given rearing habitat and/or reproductive strategy.

A. Dry Matter and Water Content

The absolute amount of dry matter in a fish egg exhibits both inter- and intraspecific variation. The percentage dry weight per egg, and hence egg water content, is also variable (Table I). Following fertilization, egg water content comes under osmoregulatory control. The specific gravity of fish eggs is inversely related to water content. Thus, the buoyant nature of pelagic eggs is associated with a higher water content relative to that of dense, less hydrated demersal eggs. The specific gravity of eggs and eleutheroembryos can decrease with time [e.g., Atlantic herring *Clupea harengus harengus*, Blaxter and Ehrlich (1974); Atlantic salmon *Salmo salar*, Peterson and Metcalfe (1977)] or increase with time [e.g., plaice *Pleuronectes platessa*, Blaxter and Ehrlich (1974); northern anchovy *Engraulis mordax*, Hunter and Sanchez (1976)]. These changes are a function of osmoregulatory adjustments and changes in the quantity and composition of yolk and tissues.

The percent water content of yolk and tissue cannot be assumed to remain constant throughout development. The percent water content of salmonid yolk, for example, has been reported to decrease during development (Hayes and Armstrong, 1942; Harvey, 1966), to increase (Escaffre and Bergot, 1984), or to remain unchanged (Gray, 1926; Peterson and Metcalfe, 1977). Similarly, the percent water content of salmonid tissues has been found to increase during development (Harvey, 1966; Escaffre and Bergot, 1984) or to remain unchanged (Peterson and Metcalfe, 1977). The documented changes are large in some cases [e.g., 20% increase in the relative water content of rainbow trout (*S. gairdneri*) yolk from hatch to complete yolk absorption; (Escaffre and Bergot,1984)].

Figure 3 illustrates typical changes in the weight of embryonic components (yolk and tissues) with time. Generally, as high-density,

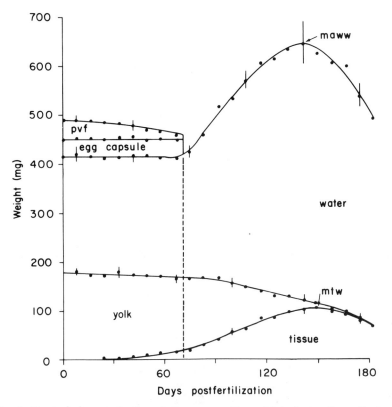

Fig. 3. Typical changes in the relative composition of fish during the yolk absorption period. Data are from chinook salmon (*Oncorhynchus tshawytscha*) at 8°C. The broken line represents 50% hatching. Variability about some means is indicated by 95% confidence limits; pvf, perivitelline fluid; maww, maximum eleutheroembryo wet weight; mtw, maximum tissue weight. [From Heming (1982).]

low-moisture yolk is converted into low-density, high-moisture tissues, there is a decrease in yolk dry weight, an increase in tissue dry weight, and an increase in bulk water (yolk plus tissues). In some marine eleutheroembryos, however, yolk absorption is accompanied by a decrease in bulk water content [e.g., plaice *P. platessa*, Blaxter and Ehrlich (1974)], suggesting that the yolk of these species has a higher moisture content than the developing tissues.

Due to the catabolism of some yolk materials for energy production, the conversion of yolk to tissues is less than 100% efficient (see Section V). As embryonic tissues and their associated maintenance costs grow, the amount of catabolic loss increases. This results in an increasing loss of bulk dry matter (yolk plus tissues) (Fig. 3). Tissue dry weight continues to increase, however, as yolk is absorbed and converted into tissues. As yolk reserves near exhaustion, the metabolic demands of maintenance and activity exceed the supply of yolk nutrients, and tissues begin to be catabolized for energy production. The resultant reduction in body weight causes a maximum tissue weight to be reached before yolk absorption is completed. Wallace and Aasjord (1984b) found that Arctic char (*Salvelinus alpinus*) at 3°C reached maximum tissue weight at complete yolk absorption, whereas at 12°C maximum tissue weight was reached with 1.5 mg of dry yolk remaining (12% of yolk reserves at hatching). The timing of maximum tissue weight, and hence the end of growth utilizing yolk alone, would appear to be influenced by temperature, occurring earlier in development at higher temperatures. This temperature dependency may reflect changes in maintenance costs (see Section V) or shifts in the relative rates of protein and lipid mobilization from yolk (see Section IV). The temperature effect may also explain why a maximum tissue weight has not been evident prior to complete yolk absorption in some studies.

The bulk wet weight of the embryo/eleutheroembryo increases during development despite the concurrent loss in bulk dry weight, because of uptake of water. Gray (1928) modeled these early weight changes and predicted that, dependent on metabolic costs, the bulk wet weight would reach a maximum before the tissue growth cycle was completed. Thus, toward the end of the endogenous nutrition period, eleutheroembryo wet weight can be expected to decrease despite a continued increase in tissue weight; the resultant maximum eleutheroembryo wet weight will be reached before the maximum tissue weight (Fig. 3). Maximum eleutheroembryo wet weight is reached earlier in development when relative metabolic costs are elevated, such as at higher rearing temperatures (Heming, 1982) or in smaller eggs of a given species (Smith, 1958; Escaffre and Bergot, 1984; Rombough, 1985).

B. Protein

Protein is the most abundant dry constitutent of many fish eggs (Table I). Assuming the data in Table II reflect general trends, the majority of egg protein resides in the yolk; the remainder is associated primarily with the perivitelline fluid. Yolk protein serves two primary functions: it provides amino acids for tissue growth and supplies energy via catabolic processes. As a result, there is a continual loss of

Table II
Relative Distribution of the Constituents of Freshly Fertilized and Water-Hardened
Eggs of Atlantic Salmon (*Salmo salar*)[a]

Constituent	Total weight (mg)	Percentage of total weight		
		Egg membranes	Perivitelline fluid	Yolk
Water	88.3	5.2	23.3	71.8
Dry matter	49.7	0.8	16.9	82.1
Protein	26.0	0.3	22.7	76.9
Lipid	18.0	0.1	7.7	92.2
Carbohydrate	0.5	48.0	46.0	2.0
Ash	1.4	7.1	7.1	85.7

[a] From Hamor and Garside (1977a).

protein from the yolk mass as it is transferred to the developing tissues. In addition, the catabolism of protein for energy production results in a decline of bulk protein quantities (tissue plus yolk). The importance of yolk protein is exemplified by the significant positive correlation between egg protein content and early survival (Satia *et al.*, 1974).

During early development, when the embryo is small and total metabolic activity is low, little if any protein is utilized for energy production; bulk protein quantities remain relatively constant. As growth proceeds and the metabolic rate increases, a larger portion of yolk protein is shunted into energy production and bulk protein quantities decline. This is especially evident after hatch, in accordance with the higher levels of activity and energy demand. This trend has been reported for salmonids (Zeitoun *et al.*, 1977; Dabrowski and Luczynski, 1984), cyprinids (Kamler, 1976), and gadids (Lapin and Matsuk, 1979; Buckley, 1981), among others. Studies of oxygen consumption and nitrogen metabolism (Kaushik *et al.*, 1982; Cetta and Capuzzo, 1982; Dabrowski *et al.*, 1984) also support an increased use of protein for energy production after hatching. It is difficult, however, to directly determine protein catabolism before hatching because of the low permeability of egg membranes to nitrogenous metabolites (Yarzhombek and Maslennikova, 1971; Rice and Stokes, 1974).

There are three major classes of yolk proteins. In order of abundance, these are lipoproteins, glycoproteins, and phosphoproteins (Nakagawa and Tsuchiya, 1969; Nakagawa, 1970). During development, there appears to be little change in the relative proportion of these yolk proteins (Nakagawa and Tsuchiya, 1972). This suggests the three protein types are utilized at the same relative rate. However, the physical and chemical properties of yolk lipoprotein change during development (Nakagawa and Tsuchiya, 1974); molecular weight increases, the prosthetic lipid groups decrease, and specific amino acids are released. Fish embryos develop at the expense of yolk protein degradation and the liberation of amino acids (Monroy et al., 1961). Changes in amino acid concentrations during development appear to parallel changes in the concentrations of DNA and RNA (Zeitoun et al., 1977). This suggests that the changes are associated with periods of varying tissue growth.

It is interesting to speculate about possible preferential retention of specific, particularly essential, amino acids. Preferential retention might be expected to ensure that sufficient amounts of essential amino acids are available for protein synthesis. Presently, we do not have an adequate understanding of how specific amino acids might be retained. It is possible that processes associated with their catabolism are suppressed or not developed until later in development, thereby preventing early losses.

C. Lipid

Following protein, lipids are the next most abundant dry constituent of most fish eggs (Tables I and II). There is considerable interspecific variation, however; Balon (1977) lists lipid quantities ranging from 0.1% of egg weight in the plaice *P. platessa* to 45% in the mouthbrooding cichlid *Labeotropheus*. This variation has been considered an adaptation to different reproductive strategies (Balon, 1977) or to the duration of the endogenous nutrition period (Kaitaranta and Ackman, 1981). Lipid content and the presence of large oil globules are poorly correlated with the pelagic or demersal nature of eggs, and thus lipids probably do not function primarily in buoyancy (Smith, 1957; Peterson and Metcalfe, 1977). During development, lipids are converted into structural components such as cell membranes or channeled into energy production. The amount of catabolic loss increases as the fish body grows, but is relatively insignificant until hatch and thereafter. This is probably related to increased activity and the resultant higher energy requirements following hatch. On a caloric basis, lipids, especially triglycerides and wax esters (neutral lipids), are the most important energy reserve of developing fish.

Nakagawa and Tsuchiya (1971, 1972) describe two major lipid

types: (1) free lipids associated with the oil globule, and (2) bound lipids associated with the high-density fraction (HDF) and therefore primarily within the yolk platelets. The oil globules are composed primarily of triglycerides; phospholipids are not detectable in salmonid globules (Nakagawa and Tsuchiya, 1971) but may be present in the oil globule of thermophilic fishes (Grodziński, 1973). In contrast, lipids of the HDF (yolk platelets) are dominated by phospholipids with triglycerides being secondary in abundance. It is interesting to note that wax esters are a major lipid component of the yolk of some species (Nevenzel, 1970; Rahn *et al.*, 1977; Kaitaranta and Ackman, 1981; Vetter *et al.*, 1983). In red drum eggs (*Sciaenops ocellata*), for example, wax esters account for 29% of the total lipid pool and provide 53% of the total calories consumed between fertilization and hatching (Vetter *et al.*, 1983). Vetter *et al.* (1983) noted that species incorporating large quantities of wax esters into their ova produce buoyant eggs that commonly encounter reduced salinities. They hypothesized that, since wax esters have a lower specific gravity than triglycerides, these esters play an role in egg buoyancy.

The yolk of many species contains lipid-soluble carotenoid pigments (Balon, 1977; Kitahara, 1984). These pigments are concentrated in the oil globules but are also present to a lesser extent in the HDF (yolk platelets) (Nakagawa and Tsuchiya, 1976). Yolk carotenoids may represent a nutrient and/or may function for protection from sunlight (Eisler, 1957) and for respiration [see Balon (1977) for discussion and references].

Similar to the protein data, the majority of studies concerning lipids have examined bulk changes within the whole embryonic system and have not distinguished between yolk and tissues. It is generally difficult, therefore, to determine if there is any selection, retention, or utilization of specific yolk lipids (triglyceride, phopholipid, wax esters), and whether interconversion between yolk lipids is possible. Data from bulk studies reveal that there is little change in relative lipid composition during embryonic development (Takama *et al.*, 1969; Lapin and Matsuk, 1979; Vetter *et al.*, 1983). This indicates that prior to hatch the major lipid classes are utilized at the same relative rate with little or no preference for specific lipids. Following hatch, the decline in bulk lipid quantities is mainly due to catabolism of triglycerides, while phospholipids, which would be incorporated into structural components (membranes) of the developing fish, are conserved (Terner *et al.*, 1968; Atchison, 1975; Rahn *et al.*, 1977). Analyses of yolk separate from tissues and egg membranes, however, indicate that after hatch the lipids of the yolk platelets (phospholipids) are preferentially consumed over those of the oil globule (triglycerides) (Nakagawa and Tsuchiya, 1972, 1976).

Shifts in fatty acid composition of the lipid classes also occur following hatch. This demonstrates the existence of preferential retention and utilization of certain fatty acids and the possibility of some fatty acid synthesis by eleutheroembryos. These trends are evident in bulk studies, but are made more lucid when yolk is analyzed separately from the tissues (Hayes et al., 1973; Atchison, 1975; Nakagawa and Tsuchiya, 1976; Rahn et al., 1977). For example, essential fatty acids of the linolenic series (e.g., 22:6) and other polyunsaturated fatty acids that cannot be metabolically synthesized by fish are concentrated within the tisuses to levels higher than those in the yolk (Hayes et al., 1973; Atchison, 1975; Rahn et al., 1977). In contrast, the concentrations of some fatty acids (e.g., 18:1 and 18:2) are higher in the yolk than in the tissues. As a result, fatty acid composition of the tissues is not directly related to amounts within the ova or within the yolk at later developmental stages.

Some of the changes in fatty acid composition, particularly at hatch, reflect the establishment of fatty acid synthetase systems (Hayes et al., 1973). This late onset of synthetic activity may result in embryos having a more extensive fatty acid requirement than juvenile or adult fish. Gourami embryos, for example, apparently require both linolenic and linoleic acids, whereas only linolenic is essential to older fish (Rahn et al., 1977). De novo synthesis of fatty acids by the embryo, however, may not be energetically efficient (Hayes et al., 1973). Additional work is needed to clarify the dynamic changes in fatty acid composition.

D. Carbohydrates

Fish eggs contain relatively little carbohydrate when compared with the amounts of protein and lipid (Table I). Moreover, a large proportion of total egg carbohydrate is associated with the egg membranes (Table II) and, therefore, is probably unavailable for use by the developing fish, at least until hatching (see Section VI,B). Yolk carbohydrates are present in both a free state and complexed with yolk proteins. In rainbow trout (*Salmo gairdneri*) yolk, glycoproteins represent more than 10% of the total protein pool (Nakagawa and Tsuchiya, 1969; Nakagawa, 1970). Glycogen is the primary carbohydrate in all fish eggs studied to date (Terner, 1979).

Carbohydrate tends to be the most utilized yolk nutrient. In red drum (*Sciaenops ocellata*), for example, the glycogen pool decreased 63% between fertilization and hatching, yet because of the small size of the pool accounted for only 1.7% of the total calories consumed (Vetter *et al.*, 1983). Intense catabolism of carbohydrate commences at fertilization (Moroz and Luzhin, 1976), indicating that carbohydrate plays an important nutritive role during initial cleavage. Its importance to later developmental stages has probably been underestimated by past bulk studies because of the early establishment of gluconeogensis (Terner, 1979; Boulekbache, 1981). Glycogen, for example, is accumulated in embryonic liver cells in significant amounts prior to hatching (Takahashi *et al.*, 1978). Such carbohydrate reserves might be expected to be important energy sources during periods of tissue hypoxia when aerobic lipid metabolism is not possible.

E. Caloric Content

Figure 4 summarizes the caloric content of the eggs or ripe ovaries of 54 species. The overall mean content is 5.697 cal mg^{-1} dry egg weight (SE = 0.096, n = 70 observations). This bulk value includes calories of the egg membranes (3.61 cal mg^{-1} dry weight), yolk platelets (5.398–5.720 cal mg^{-1} dry weight), and oil globules (9.459–9.866 cal mg^{-1} dry weight) (Rogers and Westin, 1981; Eldridge *et al.*, 1981a,b, 1982; Kamler and Kato, 1983).

The chemical composition of yolk changes during the course of development (Nakagawa and Tsuchiya, 1972, 1974, 1976) and, hence, its caloric content can be expected to change. Kamler and Kato (1983) recorded a decrease in the energy content (calories per milligram dry weight) of rainbow trout (*Salmo gairdneri*) yolk from 6.675 at fertilization to 6.344 at hatching. This suggests a decrease in the relative proportion of yolk lipid to protein prior to hatch. On a dry-weight basis, oil globules contain approximately 1.7 times the energy of yolk platelets, in accordance with the predominance of lipid in the former and protein in the latter. Yolk platelets are mobilized more rapidly than the oil globule from the yolk mass, especially after hatching (see Section IV). Consequently, after hatching, the relative caloric content of the yolk mass (platelets plus globules) can be expected to increase. This has yet to be directly ascertained. Nonetheless, the available data do not support Lasker's (1962) conclusion that the chemical composition and, hence, the relative energy content of yolk remain constant during development. This argues against the use of a single caloric

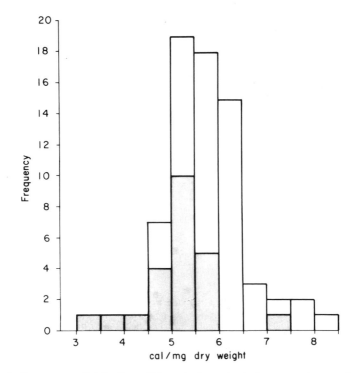

Fig. 4. Frequency distribution of the caloric content of eggs or ripe ovaries of fish, based on 70 observations from 54 species. Shaded areas designate marine fish eggs. (From numerous sources.)

content value (calories per milligram yolk) in energetic calculations of the rate and efficiency of yolk absorption.

IV. RATE OF YOLK ABSORPTION

The rate of yolk absorption can be determined by following the change with time in yolk calories, dry weight, wet weight, volume, or planar area. Each of these methods is valid within certain limiting conditions. Volume and area determinations are generally made from only two measured dimensions, which introduces unknown error. These techniques also require that the entire yolk mass be visible and, thus, have limited application for heavily pigmented eleutheroembryos. As well, yolk volume, area, and wet weight can be influenced by environmental factors (e.g., salinity; Alderdice *et al.*, 1979) and specimen preservation (Heming and Preston, 1981). Yolk wet weight will be influenced by the relative water content of yolk, which may change during development. From an energetic point of view, measurement of total yolk calories is the best approach. However, since yolk composition and caloric content per unit weight change during development (see Section III), this approach requires separate determinations of yolk caloric content at each sampling time.

The rate at which yolk reserves are depleted must be a function of the surface area of the absorptive layer (e.g., yolk syncytium) and the metabolic activity of that layer. The absorptive surface area changes during development, being minute at fertilization and then expanding to enclose the entire yolk mass in most fishes. In teleosts, the absorptive surface area is approximately equal to the area of the yolk sac. Hence, as yolk reserves are depleted and the yolk sac decreases in size, the absorptive surface area must also diminish. The reduction in surface area can be temporarily ameliorated by concurrent changes in yolk mass shape. Thus, at any given time, the surface area available for yolk absorption is dependent on the size and, to a lesser degree, shape of the yolk mass. The following mathematical formulas are useful in calculating the absorptive surface area (A) and volume (V) of yolk masses of various shapes:

1. Spherical mass:

$$A = \pi D^2 \qquad V = 0.1667\pi D^3$$

where D is yolk diameter.

2. Pyiform and conical masses:

$$A = 0.250\pi[H_1^2 + H_2^2 + (H_1 + H_2)\ [(H_2 - H_1)^2 + 4L^2]^{0.5}]$$
$$V = 0.0833\pi L(H_1^2 + H_2^2 + H_1 H_2)$$

where H_1 is the height of the smaller end ($H_1 = 0$ in a conical-shaped mass), H_2 is the height of the larger end, and L is the length of the yolk mass.

3. Cylindrical mass:

$$A = 0.50\pi(H^2 + 2HL)$$
$$V = 0.250\pi H^2 L$$

where H is the height and L is the length of the mass.

4. Ellipsoidal mass:

$$A = 0.50\pi H^2 + \pi HL^2(L^2 - H^2)^{-0.5}\sinh^{-1}\ [(L^2 - H^2)/L]$$
$$V = 0.1667\pi LH^2$$

where H is the height and L is the length of the mass.

The surface area and volume of asymmetric yolk masses, such as that of the white sucker (*Catostomus commersoni*), which has a bulbous anterior segment and an elongated posterior segment, can be estimated by addition of the appropriate formulas, in the case of white sucker the formulas for a sphere and a cylinder. In nonteleost fishes that utilize the alimentary tract for yolk absorption, the surface area available for yolk absorption is independent of yolk sac size or shape.

The relationship between yolk volume and absorptive surface area explains, in part, why the absorption rate of a given species is more rapid in larger eggs than in smaller eggs (Fig. 5). In salmonids, the rate of yolk absorption appears to vary in a 1:1 relationship with egg size. Thus, all other factors being equal, eggs of a given salmonid species reach complete yolk absorption within a span of several days, despite large differences in egg size. This may be a unique adaptation related to the reproductive strategy of salmonids. In cod and herring, the relationship between relative absorption rate and relative egg size is closer to 1:2 (Fig. 5). Thus, a doubling of egg size in these latter

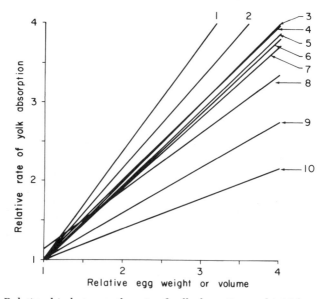

Fig. 5. Relationship between the rate of yolk absorption and initial egg size of fish: (1) *Oncorhynchus keta* (Beacham and Murray, 1985); (2) *O. keta* (Beacham *et al.*, 1985); (3) *O. kisutch* (Beacham *et al.*, 1985); (4) *Salvelinus alpinus* (Wallace and Aasjord, 1984a); (5) *O. gorbuscha* (Yastrebkov, 1966); (6) *Salmo salar* (Kazakov, 1981); (7) *S. gairdneri* (Escaffre and Bergot, 1984); (8) *O. tshawytscha* (Rombough, 1985); (9) *Clupea harengus harengus* (Blaxter and Hempel, 1963); (10) *Gadus morhua* (Knutsen and Tilseth, 1985).

species prolongs the period of endogenous nutrition (fertilization to complete yolk absorption) by about 1.3 times. In these latter species, the rate of yolk absorption per unit area of syncytium must decrease as egg size increases.

In terms of the rate of consumption, many teleosts exhibit three distinct phases of yolk absorption (Fig. 6). The first or prehatch phase is characterized by slow but steadily increasing rates of yolk absorption. Yolk platelets and oil globules are consumed at approximately the same relative rate during this phase (Nakagawa and Tsuchiya, 1972). Shortly before and at hatching, the rate of yolk absorption increases rapidly, probably in response to both an increase in absorptive surface area due to changes in yolk sac shape and an increase in the metabolic activity of the yolk syncytium. This marks the beginning of the second or posthatch phase of absorption, which is characterized by a relatively high and constant rate of absorption. During the posthatch phase, yolk platelets are preferentially consumed over the oil globule

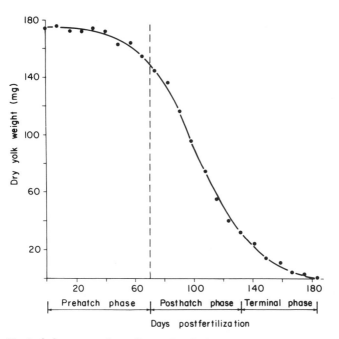

Fig. 6. Typical changes in dry yolk weight of teleost fish. Data from chinook salmon (*Oncorhynchus tshawytscha*) at 8°C (T. A. Heming, unpublished data). The period of endogenous nutrition (fertilization to complete yolk absorption) has been divided into three phases based on trends in the rate of yolk absorption. The broken line represents 50% hatching.

(May, 1974; Eldridge *et al.*, 1982; Li and Mathias, 1982; Quantz, 1985). As the reserve of yolk platelets nears exhaustion, the rate of yolk absorption slows, probably in response to both a decrease in absorptive surface area as the yolk sac shrinks and the changing composition of yolk. This marks the beginning of the terminal phase of absorption, during which the remaining yolk, predominantly oil globules, is consumed.

Factors that increase or decrease the metabolic activity of the yolk syncytium can be expected to increase or decrease, respectively, the rate of yolk absorption. The rate of yolk absorption is reduced, for example, by low dissolved oxygen concentrations (Brannon, 1965; Hamor and Garside, 1977b), sub- and supraoptimal salinities (May, 1974; Santerre, 1976), high ammonia concentrations (Fedorov and Smirnova, 1978), and sublethal concentrations of toxic xenobiotics (Crawford and Guarino, 1985). Some xenobiotics induce deformities in the yolk sac (e.g., crude oil fractions; Kühnhold, 1972). The structure of yolk itself may be sensitive to some chemicals; fuel-oil fractions can cause coalescence of the oil globules in fish yolk (Ernst *et*

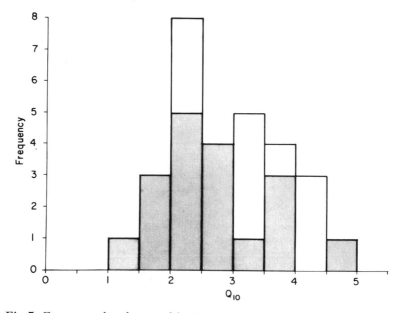

Fig. 7. Frequency distribution of the Q_{10} values for yolk absorption in fish, based on 29 observations from 23 species. Shaded areas designate marine fish eggs. (From numerous sources.)

al., 1977). The extent to which yolk absorption is influenced by such structural abnormalities is unclear.

The rate of absorption increases with temperature throughout most of the range of thermal tolerance. Figure 7 summarizes the Q_{10} values of 23 species of fish, at temperatures spanning the overall range of 1–30°C. The overall mean value is 2.916 (SE = 0.166, n = 29 observations). As the upper limit of thermal tolerance is approached, the rate of yolk absorption and hence the Q_{10} value decrease (Fig. 8), probably due to a breakdown of normal metabolic processes.

Temperature has a differential effect on the absorption of yolk platelets and oil globules. Oil absorption appears to be affected more than platelet consumption by increases in temperature (Kuo *et al.*, 1973; May, 1974; Ehrlich and Muszynski, 1982). Thus, the Q_{10} value for oil absorption is greater than that for platelet absorption (Fig. 8). Near the lower limit of the tolerated thermal range, early life stages

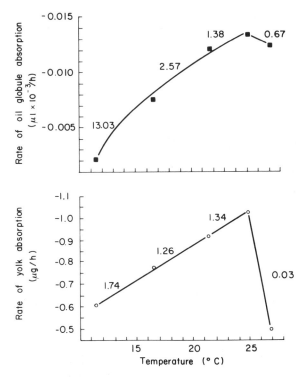

Fig. 8. Temperature-specific yolk (= yolk platelets) and oil globule absorption rates in California grunion (*Leuresthes tenuis*). The Q_{10} values for each incremental increase in temperature are shown in the figure. [From Ehrlich and Muszynski (1982).]

may encounter problems with oil absorption and metabolism, and platelet consumption may dominate (Ehrlich and Muszynski, 1982).

In some species, the rate of yolk absorption is also sensitive to mixed feeding (exogenous plus endogenous). In striped bass (*Morone saxatilis*), for example, mixed feeding results in increased consumption of the oil globule (Rogers and Westin, 1981; Eldridge *et al.*, 1981a,b, 1982); starved larvae conserve their oil globules, yet still experience tissue resorption and eventually die with oil remaining (see Section V). The effect of mixed feeding on oil consumption probably reflects an increased catabolism of oil reserves to meet the energetic demands associated with feeding activity. On the other hand, feeding activity has no effect on the yolk absorption rate of chinook salmon (*Oncorhynchus tshawytscha*) (Heming *et al.*, 1982). In still other studies, mixed feeding has been shown to slow the rate of yolk absorption [e.g., walleye pollock *Theragra chalcogramma* (Hamai *et al.*, 1974), Arctic cod *Boreogadus saida* (Aronovich *et al.*, 1975)]. In this latter group of fishes, utilization of exogenous nutrients not only satisfies the metabolic costs associated with feeding activity but would also appear to influence the utilization of endogenous nutrients.

Tsukamoto and Kajihara (1984) found that swimming activity dramatically accelerated the rate of yolk absorption in ayu eleutheroembryos (*Plecoglossus altivelis*). After swimming at a cruising speed of 0.3 body lengths s^{-1} for 60 min, the mean yolk volume of active ayu eleutheroembryos was 36% smaller than that of unexercised control fish. This is not a general effect, however, because in other species the rate of yolk absorption is independent of the energetic demands of activity. In *Oncorhynchus* spp. particularly, swimming activity has no effect on yolk absorption rate (Brannon, 1965). Moreover, Hansen and Møller (1985) found the opposite effect in Atlantic salmon (*S. salar*); active eleutheroembryos absorbed their yolk reserves more slowly than inactive eleutherembryos.

V. EFFICIENCY OF YOLK UTILIZATION

The efficiency of yolk utilization is measured in terms of the growth sustained by yolk absorption. Efficiency is commonly calculated as the ratio between the change in tissue dry weight or calories and the concurrent change in yolk weight or calories (see Blaxter, 1969, or Kamler and Kato, 1983, for equations). In addition to growth, absorbed yolk supports differentiation, maintenance, and activity. Yolk utilization is less than 100% efficient, therefore, due primarily to the metabolic costs of maintenance and activity. The costs of differentiation are probably constant among individuals of a given species and relatively small, and hence can be ignored.

The rate and pattern of embryonic growth are functions of the following: yolk composition; yolk digestion by the syncytium or analogous tissue; the uptake and transport of yolk nutrients from the yolk mass to the developing tissues; activity of the somatic synthetic machinery; and the metabolic demands of maintenance and activity. Factors acting at the level of the yolk sac can be expected to manifest themselves as changes in yolk absorption rate but need not influence utilization efficiency. In other words, yolk absorption is slower, but the fish size ultimately attained is unchanged. On the other hand, factors acting at the level of the somatic tissues can be expected to manifest themselves as changes in utilization efficiency but need not influence absorption rate, that is, the timing of yolk exhaustion is unchanged, but the ultimate fish size is reduced. Generally, early life stages utilize their yolk reserves more efficiently than later life stages utilize exogenous food (Klekowski and Duncan, 1975). The available data indicate that, under optimal conditions, yolk utilization efficiencies can be as high as 60–90% for both dry-matter conversion and caloric conversion. In some studies, caloric conversion efficiencies are reported to be greater than dry-matter efficiencies (Ehrlich and Muszynski, 1982), while in other studies caloric efficiencies are lower (From and Rasmussen, 1984).

A. Biotic Factors

The metabolic demands of maintenance and activity vary during development. Maintenance costs increase as growth proceeds. The costs associated with activity are less predictable, varying among individuals depending on the level of spontaneous activity. It follows that the efficiency of yolk utilization is not constant throughout development. Efficiency, in fact, reaches zero and then becomes negative as the maximum tissue weight is reached and tissues are then resorbed during the terminal phase of yolk absorption. For this reason, comparison of gross efficiencies calculated from differing segments of development (e.g., fertilization to hatching versus hatching to complete yolk absorption) are of questionable validity (Marr, 1966). There is little agreement, however, as to exactly what constitutes an equivalent segment of development. Hatching is not a developmental event per se and therefore should be used with caution when calculating efficiencies. Effects of temperature on the developmental timing of maximum tissue weight and maximum eleutheroembryo wet weight complicate matters still further. If one assumes that larval size at first feeding is an important determinant of subsequent growth and survival, perhaps the most relevant determination is the gross efficiency between fertilization and the time of 50% feeding, independent of the developmental stage at which 50% feeding occurs.

Yolk utilization is influenced by egg quality and yolk composition. Rogers and Westin (1981) found, for example, that unfed striped bass (*Morone saxatilis*) conserved their oil reserves, yet still experienced a metabolic deficit during the terminal phase of yolk absorption. The data suggest that tissue resorption during the terminal phase of yolk absorption was due to preferential depletion of yolk protein nitrogen rather than the onset of a caloric deficit. Egg size is also an important factor for yolk utilization since maintenance costs are directly related to tissue weight. Thus, because fish produced from larger eggs are themselves larger and have correspondingly greater maintenance costs, they use their yolk less efficiently (Yastrebkov, 1966; Kamler and Kato, 1983).

The influence of intraspecific genetic differences remains largely unexplored, yet it is probably a significant factor in yolk utilization. In a study with coho salmon (*O. kisutch*), Childs and Law (1972) compared the embryonic development of progeny of normal males (maximum lifespan 36 months) and normal females with progeny of precocious males (maximum lifespan 24 months) and normal females. They found that offspring of precocious males developed and grew more rapidly and utilized their yolk reserves more efficiently.

B. Abiotic Factors

Interpretation of environmental effects on yolk utilization is complex. Growth during the endogenous nutrition period has been found to be reduced by extremes in pH (Nelson, 1982), sub- and supraoptimal temperatures (Blaxter and Hempel, 1966; Laurence, 1973), adverse salinities (May, 1974; Santerre, 1976), low dissolved oxygen concentrations (Brannon, 1965; Hamor and Garside, 1977b), exposure to light (Eisler, 1957; Hamor and Garside, 1975), and exposure to sublethal concentrations of toxic xenobiotics (Henderson *et al.*, 1983; Tilseth *et al.*, 1984). Within the tolerated range of each environmental parameter, decreases in utilization efficiency probably reflect increased costs of homeostasis and maintenance. As the upper and lower limits of each tolerated range are approached, however, deactivation of somatic synthetic systems can be expected (Hamor and Garside, 1977b; Nelson, 1982).

This interpretation is complicated by concurrent changes in activity. Some environmental factors, particularly some xenobiotics, suppress activity and reduce the associated energetic costs, effectively freeing more yolk nutrients for growth and increasing utilization efficiency. Leduc (1978), for example, found the efficiency of yolk utilization in Atlantic salmon (*Salmo salar*) eleutheroembryos exposed to hydrogen cyanide (HCN) to increase with increasing cyanide concentrations in the range of 0 to 0.1 mg l^{-1}. He attributed this effect to a reduction in activity at higher cyanide concentrations. Conversely, since embryonic activity circulates the perivitelline fluid, aiding gas exchange and ensuring distribution of the hatching enzyme, environmental factors that reduce activity might in some instances reduce yolk utilization efficiency and decrease hatching success (Rosenthal and Alderdice, 1976). It is possible, therefore, for a particular environmental factor to either reduce or enhance the rate and efficiency of yolk absorption, depending on the exposure regime.

Temperature is probably the most variable environmental parameter affecting yolk utilization efficiency and as such has received much of the research attention, most recently in studies by Howell (1980), Johns and Howell (1980), Johns *et al.* (1981), Ehrlich and Muszynski (1982), Heming (1982), Kamler and Kato (1983), From and Rasmussen (1984), and Luczynski *et al.* (1984), among others. Overall, the available data indicate that utilization efficiency reaches a maximum within the range of thermal tolerance of a given species; efficiency decreases toward both upper or lower limits of the tolerated thermal range. The exact shape of the curve describing the effect of temperature on utilization efficiency varies among species, probably in relation to differences in reproductive strategy and rearing habitat. In this regard, the work of Ehrlich and Muszynski (1982) is innovative. By investigating both the behavioural and physiological responses to temperature, these authors were able to map the relationship between yolk utilization and temperature selection in California grunion (*Leuresthes tenuis*). More work of a comparative nature is required before definitive statements about the relationship between yolk utilization and reproductive strategy can be made.

VI. NONYOLK NUTRIENT SOURCES DURING EARLY DEVELOPMENT

A. Pütter's Theory

Pütter's theory, the direct utilization of dissolved organic constituents of water by the early life stages of fish, has been largely discounted (see Morris, 1955, for discussion and references) but deserves some comment. Although fish eggs are generally regarded as cleidoic, some embryos possess a limited ability to assimilate dissolved organic matter from the water. Embryos of a number of species [rainbow trout S. gairdneri (Terner, 1968); Atlantic salmon S. salar (Mounib and Eisan, 1969); Atlantic herring C. harengus harengus (Siebers and Rosenthal, 1977)] have been shown to take up and metabolize external substrates such as ^{14}C-labeled pyruvate, acetate, glyoxylate, and glycine. Uptake of exogenous substrates increases during embryonic development, reaching a maximum rate just prior to hatch. It is unlikely that exogenous substrates make a significant contribution to the general needs of developing oviparous fish, however, because of the paucity of dissolved organic matter in natural waters and the relatively slow transfer rates. Siebers and Rosenthal (1977) calculated that uptake of dissolved amino acids from a 2-μM solution provided only 1.1% of the energy requirements of developing Atlantic herring (C. harengus harengus) embryos.

Since juvenile fish are capable of absorbing dissolved glucose (Lin and Arnold, 1982) and albumin (Amend and Fender, 1976) via the gills and lateral line system, assimilation of dissolved organic matter from the water can be expected after hatch. This is comparable to the nutrient absorptive role served by the external gill filaments of viviparous shark embryos (Hamlett et al., 1985). Assimilation of dissolved organic matter may in fact be important for the survival of species such as Pacific sardine (Sardinops caerulea) (Lasker, 1962) that encounter a metabolic deficit prior to acquiring the capability of exogenous feeding. Wiggins et al. (1985) proposed assimilation of dissolved organic matter as a possible reason for the low incidence of food ingestion in first-feeding larvae of American shad (Alosa sapidissima). Imada (1984) found the growth of larval thalli (Porphyra tenera) was improved by addition of sugars and salts of some organic acids, especially arabinose, to the culture water.

B. Egg Membranes and Perivitelline Fluid

The nutritive role of egg membranes and perivitelline fluid is uncertain (Laale, 1980). Analyses of salmonid perivitelline fluid (Eddy, 1974; Hamor and Garside, 1977a) have demonstrated the presence of protein (25% of fluid wet weight), lipid (5–12%), and carbohydrate (1–2%). Assimilation of these nutrients may be of some importance, especially prior to the formation of the yolk syncytium (Hamor and Garside, 1977a).

The external membranes of fish eggs contain (in descending order of abundance) carbohydrate, protein, and lipid (Hamor and Garside, 1977a) (Table II). Smith (1958) proposed that embryos assimilate nutrients released from the egg membranes by the action of the hatching enzyme during the hatching process. Cetta and Capuzzo (1982) found energetic evidence suggesting winter flounder (*Pseudopleuronectes americanus*) embryos utilize nutrients of the egg membranes. While a nutritive function of perivitelline fluid and egg membranes may be of some significance during certain segments of development (i.e., prior to formation of the yolk syncytium and at hatch), these materials do not represent substantial nutrient reserves when compared with yolk (Table II). This is attested to by the normal development of embryos in the absence of perivitelline fluid and egg membranes following mechanical or enzymatic dechorionation.

C. Viviparity

Wourms (1981) defined viviparity as "a process in which eggs are fertilized internally and are retained within the maternal reproductive system for a significant period of time, during which they develop to an advanced stage and then are released." This definition does not distinguish between ovoviviparity and viviparity. As such, viviparity in fish can be seen to present an almost continuous progression, from a primitive pattern in which the egg contains sufficient yolk for complete embryonic development and the female provides only protection, to an advanced pattern in which the egg has little yolk and the embryo develops connections to maternal tissues at an early stage in order to satisfy its nutritional, respiratory, and excretory requirements (volume XIB). Nonetheless, all fishes with the possible exception of surfperches (embiotocids), whose eggs may lack yolk reserves (de

Vlaming *et al.*, 1983), rely on yolk nutrients for energy and growth during at least the initial portion of their early development.

D. Mixed Feeding

Most fishes studied under laboratory conditions are capable of mixed feeding (exogenous plus endogenous) before incurring a metabolic deficit during the terminal phase of yolk absorption. The notable exception is Lasker's (1962) work with Pacific sardines (*Sardinops caerulea*) at 14°C. He found that Pacific sardine eleutheroembryos experienced a metabolic deficit before complete yolk absorption and prior to functional development of the jaws and eyes. The relationship between yolk absorption and structural development is sensitive to temperature, however. Santerre (1976) found that at 22°C the development of functional eyes and jaws in the jack *Caranx mate* coincided with complete yolk absorption, whereas at 30°C the eyes and jaws became functional 20 h before complete yolk absorption. It is possible, therefore, that Pacific sardine reared at temperatures other than 14°C may be capable of mixed feeding and so may be able to offset any potential metabolic deficit prior to complete yolk absorption.

An understanding of mixed feeding is important in examination of the critical period concept and in fish culture. For these reasons, mixed feeding and/or delayed first feeding has been the subject of a large number of studies, most recently by Rogers and Westin (1981), Eldridge *et al.* (1981b, 1982), Heming *et al.* (1982), McGurk (1984), Wallace and Aasjord (1984b), Powell and Chester (1985), and Wiggins *et al.* (1985), among others. The available data demonstrate that mixed feeding offsets any potential metabolic deficit prior to complete yolk absorption and enhances growth and survival, especially during the terminal phase of yolk absorption. Early contact with food may also influence initial feeding behavior, resulting in increased food consumption and consequently greater larval growth (Hurley and Brannon, 1969; Wallace and Aasjord, 1984b). Grigorosh (cited in Yastrebkov, 1966) reported that, during mixed feeding, fish larvae with larger yolk reserves exhibited a diurnal feeding pattern while larvae with smaller yolk reserves fed continuously. He considered continuous feeding to be a disadvantage since it made larvae more prone to predation.

Mixed feeding has been difficult to corroborate in field surveys, perhaps due to a rapid rate of digestion, a low requirement for exogenous nutrients, diurnal feeding patterns, or defecation and regurgitation of ingested material upon capture. It is possible that to a certain extent the evidence of mixed feeding in laboratory studies represents an abnormal behavioral response to abnormal types and amounts of food under abnormal circumstances. Salmonids, in particular, exhibit a phase of "precocious" feeding when offered exogenous food from shortly after hatching (Harvey, 1966; Hurley and Brannon, 1969; Heming et al., 1982). Ingestion of food during this phase does not benefit growth or survival when compared to unfed controls. Precocious feeding may in actuality be disadvantageous, resulting in increased mortalities (Hurley and Brannon, 1969). Ochiai et al. (1977) observed that premature feeding by ayu (*Plecoglossus altivelis*) resulted in some fish swallowing food into the pneumatic duct of the swim bladder. Death ultimately ensued, apparently caused by bacterial and fungal infection of the swim bladder and adjacent viscera. Further examination of the physiology and ethology of mixed feeding is required to understand the importance of the timing of initial feeding.

VII. NUTRITION OF EMBRYOS AND LARVAE

The nutritional requirements of fish embryos and eleutheroembryos are virtually unknown. It is possible, however, that optimal feed formulations for first-feeding fish might be similar to yolk composition and reflect to some extent the nutrient requirements and metabolic capacities of prefeeding fish.

The digestive and metabolic processes of first-feeding vertebrates are often undeveloped relative to those of juveniles or adults (Henning, 1981). It is known that the digestive physiology of eleutheroembryos is different from that of juvenile and adult fish (Buddington and Christofferson, 1985). It is highly likely, therefore, that the nutritional requirements of early life stages are distinct from those of older fish. Moreover, since the liver and its complement of synthetase systems do not develop until some time after the yolk syncytium has been formed (Takahashi et al., 1978), prefeeding fish probably have a broader set of nutritional requirements than later life stages. Normally, this would not present a problem since the required nutrients would be provided by the yolk. Under certain circumstances (e.g., inadequate maternal diet), however, yolk reserves may be deficient in some essential component. The relationships among maternal diet, egg quality, and embryo survival warrant further research.

A major barrier to defining the nutrient requirements of prefeeding fish has been the inability to alter the composition of the food resource, that is, the endogenous yolk reserve. In this regard, the use of defined media to rear embryos that have been separated surgically from their yolk reserves deserves further consideration. It may also be possible to remove the yolk, or a portion thereof, and replace it with a defined media. This would maintain the integrity of the yolk syncytium and minimize physical trauma to the embryo. Direct incorporation of radiolabeled substrates into the yolk sac using a replacement technique would eliminate the need for epidermal uptake, as used by Terner (1968), and would prevent maternal metabolism of the labelled substrates, as can occur when labels are incorporated into the yolk during oogenesis. Another potential method for defining the nutritional requirements of early life stages could be based on viviparous teleost embryos (see Section VI,C). By rearing viviparous embryos and eleutheroembryos on defined media, it may be possible to determine their nutrient requirements. To our knowledge, this approach has not been exploited.

REFERENCES

Alderdice, D. F., Rosenthal, H., and Velsen, F. P. J. (1979). Influence of salinity and cadmium on the volume of Pacific herring eggs. *Helgol. Wiss. Meeresunters.* **32,** 163–178.

Amend, D. F., and Fender, D. C. (1976). Uptake of bovine serum albumin by rainbow trout from hyperosmotic solutions: A model for vaccinating fish. *Science* **192,** 793–794.

Amirante, G. A. (1972). Immunochemical studies on rainbow trout lipovitellin. *Acta Embryol. Exp.*, Suppl., pp. 373–383.

Aronovich, T. M., Doroshev, S. I., Spectorova, L. V., and Makhotin, V. M. (1975). Egg incubation and larval rearing of navaga (*Eleginus navaga* Pall.), polar cod (*Boreogadus saida* Lepechin) and Arctic flounder (*Liopsetta glacialis* Pall.) in the laboratory. *Aquaculture* **6,** 233–242.

Atchison, G. J. (1975). Fatty acid levels in developing brook trout (*Salvelinus fontinalis*) eggs and fry. *J. Fish. Res. Board Can.* **32,** 2513–2515.

Bachop, W. E., and Schwartz, F. J. (1974). Quantitative nucleic acid histochemistry of the yolk sac syncytium of oviparous teleosts: Implications for hypotheses of yolk utilization. *In* "The Early Life History of Fish" (J. H. S. Blaxter, ed.), pp. 345–353. Springer-Verlag, Berlin and New York.

Balinsky, B. I. (1970). "An Introduction to Embryology." Saunders, Philadelphia, Pennsylvania.

Ballard, W. W., and Ginzburg, A. S. (1980). Morphogenetic movements in acipenserid embryos. *J. Exp. Zool.* **213,** 69–103.

Ballard, W. W., and Needham, R. G. (1964). Normal embryonic stages of *Polyodon spatula* (Walbaum). *J. Morphol.* **114**, 465–478.

Balon, E. K. (1975). Terminology of intervals in fish development. *J. Fish. Res. Board Can.* **32**, 1663–1670.

Balon, E. K. (1977). Early ontogeny of *Labeotropheus* Ahl, 1927 (Mbuna, Cichlidae, Lake Malawi), with a discussion on advanced protective styles in fish reproduction and development. *Environ. Biol. Fishes* **2**, 147–176.

Bams, R. A. (1969). Adaptations of sockeye salmon associated with incubation in stream gravels. *In* "Symposium on Salmon and Trout in Streams" (T. G. Northcote, ed.), pp. 71–87. Univ. of British Columbia, Vancouver.

Baranes, A., and Wendling, J. (1981). The early stages of development in *Carcharhinus plumbeus*. *J. Fish Biol.* **18**, 159–175.

Beacham, T. D., and Murray, C. B. (1985). Effect of female size, egg size, and water temperature on developmental biology of chum salmon (*Oncorhynchus keta*) from the Nitinat River, British Columbia. *Can. J. Fish. Aquat. Sci.* **42**, 1755–1765.

Beacham, T. D., Withler, F. C., and Morley, R. B. (1985). Effect of egg size on incubation time and alevin and fry size in chum salmon (*Oncorhynchus keta*) and coho salmon (*Oncorhynchus kisutch*). *Can. J. Zool.* **63**, 847–850.

Blaxter, J. H. S. (1969). Development: Eggs and larvae. *In* "Fish Physiology" (W. S. Hoar and D. J. Randall, eds.), Vol. 3, pp. 177–252. Academic Press, New York.

Blaxter, J. H. S., and Ehrlich, K. F. (1974). Changes in behaviour during starvation of herring and plaice larvae. *In* "The Early Life History of Fish" (J. H. S. Blaxter, ed.), pp. 575–588. Springer-Verlag, Berlin and New York.

Blaxter, J. H. S., and Hempel, G. (1963). The influence of egg size on herring larvae (*Clupea harengus* L.). *J. Cons., Cons. Int. Explor. Mer* **28**, 211–240.

Blaxter, J. H. S., and Hempel, G. (1966). Utilization of yolk by herring larvae. *J. Mar. Biol. Assoc. U.K.* **46**, 219–234.

Boulekbache, H. (1981). Energy metabolism in fish development. *Am. Zool.* **21**, 377–389.

Brannon, E. L. (1965). "The Influence of Physical Factors on the Development and Weight of Sockeye Salmon Embryos and Alevins." Int. Pac. Salmon Fish. Comm., Prog. Rep. 12. Seattle, Washington.

Braum, E. (1967). The survival of fish larvae with reference to their feeding behaviour and the food supply. *In* "The Biological Basis of Freshwater Fish Production" (S. D. Gerking, ed.), pp. 113–131. Blackwell, Oxford.

Buckley, L. J. (1981). Biochemical changes during ontogenesis of cod (*Gadus morhua* L.) and winter flounder (*Pseudopleuronectes americanus*) larvae. *Rapp. P.-V. Reun., Cons. Int. Explor. Mer* **178**, 547–552.

Buddington, R. K., and Christofferson, J. P. (1985). Digestive and feeding characteristics of the chondrosteans. *Environ. Biol. Fishes* **14**, 31–41.

Buddington, R. K., and Doroshov, S. I. (1986). Development of digestive secretions in the white sturgeon (*Acipenser transmontanus*). *Comp. Biochem. Physiol. A* **83A**, 233–238.

Cetta, C. M., and Capuzzo, J. M. (1982). Physiological and biochemical aspects of embryonic and larval development of the winter flounder *Pseudopleuronectes americanus*. *Mar. Biol. (Berlin)* **71**, 327–337.

Childs, E. A., and Law, D. K. (1972). Growth characteristics of progeny of salmon with different maximum life spans. *Exp. Gerontol.* **7**, 405–407.

Crawford, R. B., and Guarino, A. M. (1985). Effects of environmental toxicants on development of a teleost embryo. *J. Environ. Pathol. Toxicol.* **6**, 185–194.

Dabrowski, K., and Luczynski, M. (1984). Utilization of body stores in embryonated ova and larvae of two coregonid species (*Coregonus lavaretus* L. and *C. albula* L.). *Comp. Biochem. Physiol.* A **79A**, 329–334.

Dabrowski, K., Kaushik, S. J., and Luquet, P. (1984). Metabolic utilization of body stores during the early life of whitefish, *Coregonus lavaretus* L. *J. Fish Biol.* **24**, 721–729.

Dean, B. (1906). "Chimaeroid Fishes and Their Development," Publ. No. 32. Carnegie Institution, Washington, D. C.

de Vlaming, V., Baltz, D., Anderson, S., Fitzgerald, R., Delahunty, G., and Barkley, M. (1983). Aspects of embryo nutrition and excretion among viviparous embiotocid teleosts: Potential endocrine involvements. *Comp. Biochem. Physiol.* A **76A**, 189–198.

Eddy, F. B. (1974). Osmotic properties of the perivitelline fluid and some properties of the chorion of Atlantic salmon eggs (*Salmo salar*). *J. Zool.* **174**, 237–243.

Ehrlich, K. F., and Muszynski, G. (1982). Effects of temperature on interactions of physiological and behavioural capacities of larval California grunion: Adaptations to the planktonic environment. *J. Exp. Mar. Biol. Ecol.* **60**, 223–244.

Eisler, R. (1957). Some effects of artifical light on salmon eggs and larvae. *Trans. Am. Fish. Soc.* **87**, 151–162.

Eldridge, M. B., Whipple, J., and Eng, D. (1981a). Endogenous energy sources as factors affecting mortality and development in striped bass (*Morone saxatilis*) eggs and larvae. *Rapp. P.-V. Reun., Cons. Int. Explor. Mer.* **178**, 568–570.

Eldridge, M. B., Whipple, J. A., Eng, D., Bowers, M. J., and Jarvis, B. M. (1981b). Effects of food and feeding factors on laboratory-reared striped bass larvae. *Trans. Am. Fish. Soc.* **110**, 111–120.

Eldridge, M. B., Whipple, J. A., and Bowers, M. J. (1982). Bioenergetics and growth of striped bass, *Morone saxatilis*, embryos and larvae. *Fish. Bull.* **80**, 461–474.

Ellertsen, B., Moksness, E., Solemdal, P., Strømme, T., Tilseth, S., Westgård, T., and Øiestad, V. (1980). Some biological aspects of cod larvae (*Gadus morhua* L.). *Fiskeridir. Skr., Ser. Havunders.* **17**, 29–47.

Ernst, V. V., Neff, J. M., and Anderson, J. W. (1977). The effects of the water-soluble fractions of No. 2 fuel oil on the early development of the estuarine fish *Fundulus grandis* Baird and Girard. *Environ. Pollut.* **14**, 25–35.

Escaffre, A. M., and Bergot, P. (1984). Utilization of the yolk in rainbow trout alevins (*Salmo gairdneri* Richardson): Effect of egg size. *Reprod. Nutr. Dev.* **24**, 449–460.

Fedorov, K. Y., and Smirnova, Z. V. (1978). Dynamics of ammonia accumulation and its effect on the development of the pink salmon, *Oncorhynchus gorbuscha*, in closed circuit incubation systems. *J. Ichthyol. (Engl. Transl.)* **18**, 288–295.

From, J., and Rasmussen, G. (1984). A growth model, gastric evacuation, and body composition in rainbow trout, *Salmo gairdneri* Richardson, 1836. *Dana* **3**, 61–139.

Fujii, T. (1960). Comparative biochemical studies on the egg-yolk proteins of various animal species. *Acta Embryol. Morphol. Exp.* **3**, 260–285.

Fujita, K. (1981). Oviphagous embryos of the pseudocarchariid shark, *Pseudocarcharias kamoharai*, from the Central Pacific. *Jpn. J. Ichthyol.* **28**, 37–44.

Gilmore, R. G., Dodrill, J. W., and Linley, P. A. (1983). Reproduction and embryonic development of the sand tiger shark, *Odontaspis taurus* (Rafinesque). *Fish. Bull.* **81**, 201–225.

Gray, J. (1926). The growth of fish. I. The relationship between embryo and yolk in *Salmo fario*. *Br. J. Exp. Biol.* **4**, 215–225.

Gray, J. (1928). The growth of fish. II. The growth-rate of the embryo of *Salmo fario*. *Br. J. Exp. Biol.* **6**, 110–124.

Grodziński, Z. (1973). Yolk of lower vertebrates. *Przegl. Zool.* 17, 159–171.

Hamai, I., Kyûshin, K., and Kinoshita, T. (1974). On the early larval growth, survival and variation of body form in the walleye pollock, *Theragra chalcogramma* (Pallas), in rearing experiment feeding the different diets. *Bull. Fac. Fish., Hokkaido Univ.* 25, 20–35.

Hamlett, W. C., and Wourms, J. P. (1984). Ultrastructure of the pre-implantation shark yolk sac placenta. *Tissue Cell* 16, 613–625.

Hamlett, W. C., Allen, D. J., Stribling, M. D., Schwartz, F. J., and DiDio, L. J. A. (1985). Permeability of external gill filaments in the embryonic shark. Electron microscopic observations using horseradish peroxidase as a macromolecular tracer. *J. Submicrosc. Cytol.* 17, 31–40.

Hamlett, W. C., Schwartz, F. J., and DiDio, L. J. (1987). Subcellular organization of the yolk syncytial–endoderm complex in the preimplantation shark yolk sac. *Cell Tissue Res.* 247, 275–285.

Hamor, T., and Garside, E. T. (1973). Peroxisome-like vesicles and oxidative activity in the zona radiata and yolk of the ovum of the Atlantic salmon (*Salmo salar* L.). *Comp. Biochem. Physiol. B* 45B, 147–151.

Hamor, T., and Garside, E. T. (1975). Regulation of oxygen consumption by incident illumination in embryonated ova of Atlantic salmon *Salmo salar* L. *Comp. Biochem. Physiol. A* 52A, 277–280.

Hamor, T., and Garside, E. T. (1977a). Quantitative composition of the fertilized ovum and constituent parts in the Atlantic salmon *Salmo salar* L. *Can. J. Zool.* 55, 1650–1655.

Hamor, T., and Garside, E. T. (1977b). Size relations and yolk utilization in embryonated ova and alevins of Atlantic salmon *Salmo salar* L. in various combinations of temperature and dissolved oxygen. *Can. J. Zool.* 55, 1892–1898.

Hansen, T. J., and Møller, D. (1985). Yolk absorption, yolk sac constrictions, mortality, and growth during first feeding of Atlantic salmon (*Salmo salar*) incubated on astroturf. *Can. J. Fish. Aquat. Sci.* 42, 1073–1078.

Harvey, H. H. (1966). Commencement of feeding in the sockeye salmon (*Oncorhynchus nerka*). *Verh.—Int. Ver. Theor. Angew. Limnol.* 16, 1044–1055.

Hayes, F. R. (1949). The growth, general chemistry, and temperature relations of salmonid eggs. *Q. Rev. Biol.* 24, 281–308.

Hayes, F. R., and Armstrong, F. H. (1942). Physical changes in the constituent parts of developing salmon eggs. *Can. J. Res.* 20, 99–114.

Hayes, L. W., Tinsley, I. J., and Lowry, R. R. (1973). Utilization of fatty acids by the developing steelhead sac-fry, *Salmo gairdneri*. *Comp. Biochem. Physiol. B* 45B, 695–707.

Heming, T. A. (1982). Effects of temperature on utilization of yolk by chinook salmon (*Oncorhynchus tshawytscha*) eggs and alevins. *Can. J. Fish. Aquat. Sci.* 39, 184–190.

Heming, T. A., and Preston, R. P. (1981). Differential effect of formalin preservation on yolk and tissue of young chinook salmon (*Oncorhynchus tshawytscha* Walbaum). *Can. J. Zool.* 59, 1608–1611.

Heming, T. A., McInerney, J. E., and Alderdice, D. F. (1982). Effects of temperature on initial feeding in alevins of chinook salmon (*Oncorhynchus tshawytscha*). *Can. J. Fish. Aquat. Sci.* 39, 1554–1562.

Henderson, V., Fisher, J. W., D'Allessandris, R., and Livingston, J. M. (1983). Effects of hydrazine on functional morphology of rainbow trout embryos and larvae. *Trans. Am. Fish. Soc.* 112, 100–104.

Henning, S. J. (1981). Postnatal development: Coordination of feeding, digestion, and metabolism. *Am. J. Physiol.* **241**, G199–G214.

Howell, W. H. (1980). Temperature effects on growth and yolk utilization in yellowtail flounder (*Limanda ferruginea*) yolk sac larvae. *Fish. Bull.* **78**, 731–739.

Hulata, G., Moav, R., and Wohlfarth, G. (1976). The effects of maternal age, relative hatching time and density of stocking on growth rate of fry in the European and Chinese races of the common carp. *J. Fish Biol.* **9**, 499–513.

Hunter, J. R. (1972). Swimming and feeding behavior of larval anchovy *Engraulis mordax*. *Fish. Bull.* **70**, 821–838.

Hunter, J. R., and Sanchez, C. (1976). Diel changes in swim bladder inflation of the larvae of the northern anchovy, *Engraulis mordax*. *Fish. Bull.* **74**, 847–855.

Hurley, D. A., and Brannon, E. L. (1969). "Effect of Feeding Before and After Yolk Absorption on the Growth of Sockeye Salmon." Int. Pac. Salmon Fish. Comm., Prog. Rep. 21.

Imada, O. (1984). Effects of light intensity, lighting period, intermittent lighting, and sugars and salts of organic acids supplemented to the culture media on the growth of cultured larvae thalli. *Bull. Jpn. Soc. Sci. Fish.* **50**, 931–936.

Jared, D. W., and Wallace, R. A. (1968). Comparative chromatography of the yolk proteins of teleosts. *Comp. Biochem. Physiol.* **24**, 437–443.

Johns, D. M., and Howell, W. H. (1980). Yolk utilization in summer flounder (*Paralichthys dentatus*) embryos and larvae reared at two temperatures. *Mar. Ecol.* **2**, 1–8.

Johns, D. M., Howell, W. H., and Klein-MacPhee, G. (1981). Yolk utilization and growth to yolk-sac absorption in summer flounder (*Paralichthys dentatus*) larvae at constant and cyclic temperatures. *Mar. Biol. (Berlin)* **63**, 301–308.

Kaitaranta, J. K., and Ackman, R. G. (1981). Total lipids and lipid classes of fish roe. *Comp. Biochem. Physiol. B* **69B**, 725–729.

Kamler, E. (1976). Variability of respiration and body composition during early developmental stages of carp. *Pol. Arch. Hydrobiol.* **23**, 431–485.

Kamler, E., and Kato, T. (1983). Efficiency of yolk utilization by *Salmo gairdneri* in relation to incubation temperature and egg size. *Pol. Arch. Hydrobiol.* **30**, 271–306.

Kaushik, S. J., Dabrowski, K., and Luquet, P. (1982). Patterns of nitrogen excretion and oxygen consumption during ontogenesis of common carp (*Cyprinus carpio*). *Can. J. Fish. Aquat. Sci.* **39**, 1095–1105.

Kazakov, R. V. (1981). The effect of the size of Atlantic salmon, *Salmo salar* L, eggs on embryos and alevins. *J. Fish Biol.* **19**, 353–360.

Kitahara, T. (1984). Behavior of carotenoids in the chum salmon *Oncorhynchus keta* during development. *Bull. Jpn. Soc. Sci. Fish.* **50**, 531–536.

Klekowski, R. Z., and Duncan, A. (1975). Physiological approach to ecological energetics. *In* "Methods for Ecological Bioenergetics" (W. Grodziński, R. Z. Klekowski, and A. Duncan, eds.), pp. 15–64. Blackwell, Oxford.

Knutsen, G. M., and Tilseth, S. (1985). Growth, development, and feeding success of Atlantic cod larvae *Gadus morhua* related to egg size. *Trans. Am. Fish. Soc.* **114**, 507–511.

Korzhuev, P. A., and Sharkova, L. B. (1967). Digestion characteristics of Caspian sturgeon. *In* "Metabolism and Biochemistry of Fishes" (G. S. Karzinkin, ed.), pp. 326–330. Nauka Press, Moscow.

Krayushkina, L. S. (1957). The histology of the digestive system of sturgeon larvae at different developmental stages. *Dokl. Akad. Nauk SSSR* **177**, 966–968.

Kühnhold, W. W. (1972). The influence of crude oils on fish fry. *In* "Marine Pollution and Sea Life" (M. Ruvio, ed.), pp. 315–318. Fishing News (Books) Ltd., London.

Kuo, C. M., Shehadeh, Z. H., and Milisen, K. K. (1973). A preliminary report on the development, growth and survival of laboratory reared larvae of the grey mullet, *Mugil cephalus* L. *J. Fish Biol.* **5**, 459–470.

Kuznetsov, V. A., and Khalitov, N. K. (1979). Alterations in fecundity and egg quality of the roach, *Rutilus rutilus*, in connection with different feeding conditions. *J. Ichthyol. (Engl. Transl.)* **18**, 63–70.

Laale, H. W. (1980). The perivitelline space and egg envelopes of bony fishes: A review. *Copeia*, pp. 210–226.

Lange, R. H. (1981). Highly conserved lipoprotein assembly in teleost and amphibian yolk-platelet crystals. *Nature (London)* **289**, 329–330.

Lange, R. H. (1982). The lipoprotein crystals of cyclostome yolk platelets (*Myxine glutinosa* L., *Lampetra planeri* (Bloch), *L. fluviatilis* (L)). *J. Ultrastruct. Res.* **79**, 1–17.

Lange, R. H., Grodziński, Z., and Kilarski, W. (1982). Yolk-platelet crystals in three ancient bony fishes: *Polypterus bichir* (Polypteri), *Amia calva* L., and *Lepisosteus osseus* (L.) (Holostei). *Cell Tissue Res.* **222**, 159–165.

Lapin, V. I., and Matsuk, V. Ye. (1979). Utilization of yolk and change in the biochemical composition of the eggs of the navaga, *Eleginus navaga*, during embryonic development. *J. Ichthyol. (Engl. Transl.)* **19**, 131–136.

Lasker, R. (1962). Efficiency and rate of yolk utilization by developing embryos and larvae of the Pacific sardine *Sardinops caerulea* (Girard). *J. Fish. Res. Board Can.* **19**, 867–875.

Laurence, G. C. (1973). Influence of temperature on energy utilization of embryonic and prolarval tautog, *Tautoga onitis. J. Fish. Res. Board Can.* **30**, 435–442.

Leduc, G. (1978). Deleterious effects of cyanide on early life stages of Atlantic salmon (*Salmo salar*). *J. Fish. Res. Board Can.* **35**, 166–174.

Li, S., and Mathias, J. A. (1982). Causes of high mortality among cultured larval walleye. *Trans. Am. Fish. Soc.* **111**, 710–721.

Lin, M., and Arnold, C. (1982). Transfer of glucose from sea water to the blood of red fish. *Proc. 66th Annu. Meet. Fed. Am. Soc. Exp. Biol.* **41**, Abstr. 489.

Luczynski, M., Dlugosz, M., Szutkiewicz, B., and Kirklewska, A. (1984). The influence of the incubation temperature on the body length and the yolk sac volume of *Coregonus albula* (L.) eleutheroembryos. *Acta Hydrochim. Hydrobiol.* **12**, 615–628.

McGurk, M. D. (1984). Effects of delayed feeding and temperature on the age of irreversible starvation and on the rates of growth and mortality of Pacific herring larvae. *Mar. Biol. (Berlin)* **84**, 13–26.

Marr, D. H. A. (1966). Influence of temperature on the efficiency of growth of salmonid embryos. *Nature (London)* **212**, (5065), 957–959.

May, R. C. (1974). Effects of temperature and salinity on yolk utilization in *Bairdiella icistia* (Jordan & Gilbert) (Pisces: Sciaenidae). *J. Exp. Mar. Biol. Ecol.* **16**, 213–225.

Monroy, A., Ishida, M., and Nakano, E. (1961). The pattern of transfer of the yolk material to the embryo during the development of the teleostean fish *Oryzias latipes. Embryol.* **6**, 151–158.

Moroz, I. Ye., and Luzhin, B. P. (1976). Dynamics of metabolism in the embryonic and early post-embryonic development of the carp *Cyprinus carpio. J. Ichthyol. (Engl. Transl.)* **16**, 964–970.

Morris, R. W. (1955). Some considerations regarding the nutrition of marine fish larvae. *J. Cons., Cons. Int. Explor. Mer* **20**, 255–265.

Mounib, M. S., and Eisan, J. S. (1969). Metabolism of pyruvate and glyoxylate by eggs of salmon (*Salmo salar*). *Comp. Biochem. Physiol.* **29**, 259–264.

Nakagawa, H. (1970). Studies on rainbow trout egg (*Salmo gairdnerii irideus*). II. Carbohydrate in the egg protein. *J. Fac. Fish. Anim. Husb., Hiroshima Univ.* **9**, 57–63.

Nakagawa, H., and Tsuchiya, Y. (1969). Studies on rainbow trout egg (*Salmo gairdnerii irideus*). I. Electrophoretic analyses of egg protein. *J. Fac. Fish. Anim. Husb., Hiroshima Univ.* **8**, 77–84.

Nakagawa, H., and Tsuchiya, Y. (1971). Studies on rainbow trout egg (*Salmo gairdnerii irideus*). III. Determination of lipid composition of oil globule and lipoprotein. *J. Fac. Fish. Anim. Husb., Hiroshima Univ.* **10**, 11–19.

Nakagawa, H., and Tsuchiya, Y. (1972). Studies on rainbow trout egg (*Salmo gairdnerii irideus*). IV. Changes of yolk content during embryogenesis. *J. Fac. Fish. Anim. Husb., Hiroshima Univ.* **11**, 111–118.

Nakagawa, H., and Tsuchiya, Y. (1974). Studies on rainbow trout egg (*Salmo gairdnerii irideus*). V. Further studies on the yolk protein during embryogenesis. *J. Fac. Fish. Anim. Husb., Hiroshima Univ.* **13**, 15–27.

Nakagawa, H., and Tsuchiya, Y. (1976). Studies on rainbow trout egg (*Salmo gairdnerii irideus*). VI. Changes of lipid composition in yolk during development. *J. Fac. Fish. Anim. Husb., Hiroshima Univ.* **15**, 35–46.

Nelson, J. A. (1982). Physiological observations on developing rainbow trout, *Salmo gairdneri* (Richardson), exposed to low pH and varied calcium ion concentrations. *J. Fish Biol.* **20**, 359–372.

Nevenzel, J. C. (1970). Occurrence, function and biosynthesis of wax esters in marine organisms. *Lipids* **5**, 308–319.

Ochiai, T., Kodera, K., Kon, T., Miyazaki, T., and Kubota, S. S. (1977). Studies on disease owing to erroneous-swallowing in ayu fry. *Fish Pathol.* **12**, 135–139.

Ohno, S., Karasaki, S., and Takata, K. (1964). Histo- and cytochemical studies on the superficial layer of yolk platelets in *Triturus* embryo. *Exp. Cell Res.* **33**, 310–318.

Peterson, R. H., and Metcalfe, J. L. (1977). Changes in specific gravity of Atlantic salmon (*Salmo salar*) alevins. *J. Fish. Res. Board Can.* **34**, 2388–2395.

Powell, A. B., and Chester, A. J. (1985). Morphometric indices of nutritional condition and sensitivity to starvation of spot larvae. *Trans. Am. Fish. Soc.* **114**, 338–347.

Quantz, G. (1985). Use of endogenous energy sources by larval turbot *Scophthalmus maximus*. *Trans. Am. Fish. Soc.* **114**, 558–563.

Rahn, C. H., Sand, D. M., and Schlenk, H. (1977). Metabolism of oleic, linolic and linolenic acids in gourami (*Trichogaster cosby*) fry and mature females. *Comp. Biochem. Physiol. B* **58B**, 17–20.

Rice, S. D., and Stokes, R. M. (1974). Metabolism of nitrogenous wastes in the eggs and alevins of rainbow trout, *Salmo gairdneri* Richardson. *In* "The Early Life History of Fish" (J. H. S. Blaxter, ed.), pp. 325–337. Springer-Verlag, Berlin and New York.

Robins, C. R., Bailey, R. M., Bond, C. E., Brooker, J. R., Lachner, E. A., Lea, R. N., and Scott, W. B. (1980). "A List of Common and Scientific Names of Fishes from the United States and Canada," 4th ed., Spec. Publ. No. 12. Am. Fish. Soc., Bethesda, Maryland.

Rogers, B. A., and Westin, D. T. (1981). Laboratory studies on effects of temperature and delayed initial feeding on development of striped bass larvae. *Trans. Am. Fish. Soc.* **110**, 100–110.

Rombough, P. J. (1985). Initial egg weight, time to maximum alevin wet weight, and optimal ponding time for chinook salmon (*Oncorhynchus tshawytscha*). *Can. J. Fish. Aquat. Sci.* **42**, 287–291.

Rosenthal, H., and Alderdice, D. F. (1976). Sublethal effects of environmental stressors, natural and pollutional, on marine fish eggs and larvae. *J. Fish. Res. Board Can.* **33**, 2047–2065.

Santerre, M. T. (1976). Effects of temperature and salinity on the eggs and early larvae of *Caranx mate* (Cuv. & Valenc.) (Pisces: Carangidae) in Hawaii. *J. Exp. Mar. Biol. Ecol.* **21**, 51–68.

Satia, B. P., Donaldson, L. R., Smith, L. S., and Nightingale, J. N. (1974). Composition of ovarian fluid and eggs of the University of Washington strain of rainbow trout (*Salmo gairdneri*). *J. Fish. Res. Board Can.* **31**, 1796–1799.

Shimizu, M., and Yamada, J. (1980). Ultrastructural aspects of yolk absorption in the vitelline syncytium of the embryonic rockfish, *Sebastes schlegeli*. *J. Ichthyol.* **27**, 56–63.

Siebers, D., and Rosenthal, H. (1977). Amino-acid absorption by developing herring eggs. *Helgol. Wiss. Meeresunters.* **29**, 464–472.

Smith, S. (1957). Early development and hatching. *In* "The Physiology of Fishes" (M. E. Brown, ed.), Vol. 1, pp. 323–359. Academic Press, New York.

Smith, S. (1958). Yolk utilization in fishes. *In* "Embryonic Nutrition" (D. Rudnick, ed.), pp. 33–53. Univ. of Chicago Press, Chicago, Illinois.

Takahashi, K., Hatta, N., Sugawara, Y., and Sato, R. (1978). Organogenesis and functional revelation of alimentary tract and kidney of chum salmon. *Tohoku J. Agric. Res.* **29**, 98–109.

Takama, K., Zoma, K., and Igarashi, H. (1969). Changes in the lipids during development of salmon eggs. *Bull. Fac. Fish., Hokkaido Univ.* **20**, 118–126.

Terner, C. (1968). Studies of metabolism in embryonic development. I. The oxidative metabolism of unfertilized and embryonated eggs of the rainbow trout. *Comp. Biochem. Physiol.* **24**, 933–940.

Terner, C. (1979). Metabolism and energy conversion during early development. *In* "Fish Physiology" (W. S. Hoar, D. J. Randall, and J. R. Brett, eds.), Vol. 8, pp. 261–278. Academic Press, New York.

Terner, C., Kumar, L. A., and Choe, T. S. (1968). Studies of metabolism in embryonic development. II. Biosynthesis of lipids in embryonated trout ova. *Comp. Biochem. Physiol.* **24**, 941–950.

Te Winkel, L. E. (1943). Observations on later phases of embryonic nutrition in *Squalus acanthias*. *J. Morphol.* **73**, 177–205.

Tilseth, S., Solberg, T. S., and Westrheim, K. (1984). Sublethal effects of the water-soluble fraction of Ekofisk crude oil on the early larval stages of cod (*Gadus morhya* L.). *Mar. Environ. Res.* **11**, 1–16.

Tsukamoto, K., and Kajihara, T. (1984). On the relation between yolk absorption and swimming activity in the ayu larvae *Plecoglossus altivelis*. *Bull. Jpn. Soc. Sci. Fish.* **50**, 59–61.

Vernier, J. M., and Sire, M. F. (1977). Plaquettes vitellines et activité hydrolasique acide au cours du développement embryonnaine de la truite arc-en-ciel. Étude ultrastructurale et biochimique. *Biol. Cell.* **29**, 99–112.

Vetter, R. D., Hodson, R. E., and Arnold, C. (1983). Energy metabolsim in a rapidly developing marine fish egg, the red drum (*Sciaenops ocellata*). *Can. J. Fish. Aquat. Sci.* **40**, 627–634.

Wallace, J. C., and Aasjord, D. (1984a). An investigation of the consequences of egg size for the culture of Arctic charr, *Salvelinus alpinus* (L.). *J. Fish Biol.* **24**, 427–435.

Wallace, J. C., and Aasjord, D. (1984b). The initial feeding of Arctic charr (*Salvelinus alpinus*) alevins at different temperatures and under different feeding regimes. *Aquaculture* **38**, 19–33.

Wallace, R. A., Jared, D. W., and Eisen, A. Z. (1966). A general method for the isolation and purification of phosvitin from vertebrate eggs. *Can. J. Biochem.* **44**, 1647–1655.

Wang, Y. L., Buddington, R. K., and Doroshov, S. I. (1987). Influence of temperature on yolk utilization by the white sturgeon, *Acipenser transmontanus. J. Fish Biol.* **30**, 263–271.

Ware, D. M. (1975). Relation between egg size, growth, and natural mortality of larval fish. *J. Fish. Res. Board Can.* **32**, 2503–2512.

Wiggins, T. A., Bender, T. R., Mudrak, V. A., and Coll, J. A. (1985). The development, feeding, growth, and survival of cultured American shad larvae through the transition from endogenous to exogenous nutrition. *Prog. Fish-Cult.* **47**, 87–93.

Williams, J. (1967). Yolk utilization. *In* "The Biochemistry of Animal Development" (R. Weber, ed.), Vol. 2, pp. 341–382. Academic Press, New York.

Wourms, J. P. (1981). Viviparity: The maternal–fetal relationship in fishes. *Am. Zool.* **21**, 473–515.

Yamada, J. (1959). On the vitelline syncytium and the absorption of the yolk in the fry of two salmonids. *Bull. Fac. Fish., Hokkaido Univ.* **10**, 205–210.

Yamamoto, K. (1982). Periblast in the egg of the eel, *Anguilla japonica. Jpn. J. Ichthyol.* **28**, 423–430.

Yarzhombek, A. A., and Maslennikova, N. V. (1971). Nitrogenous metabolites of the eggs and larvae of various fishes. *J. Ichthyol.* (*Engl. Transl.*) **11**, 276–281.

Yastrebkov, A. A. (1966). Effect of egg size upon size and growth rate of pink salmon larvae. *Tr. Murmansk Biol. Inst.* **12**, 45–53; *Fish Res. Board Can., Transl. Ser.* **1822**.

Zeitoun, I. H., Ullrey, D. E., Bergen, W. G., and Magee, W. T. (1977). DNA, RNA, protein, and free amino acids during ontogenesis of rainbow trout (*Salmo gairdneri*). *J. Fish. Res. Board Can.* **34**, 83–88.

MECHANISMS OF HATCHING IN FISH

KENJIRO YAMAGAMI

Life Science Institute
Sophia University
Chiyoda-ku, Tokyo 102, Japan

I. INTRODUCTION—EARLY STUDIES ON FISH HATCHING

Hatching is a process by which an animal changes its life from an "intracapsular" to a "free-living" type and is, therefore, of great significance in animal ontogeny. Among all animal groups, teleosts have been the most extensively studied. From a mechanismic point of view, hatching can be categorized into two types: mechanical hatching and enzymatic hatching. In the former, the egg envelope(s) is broken down, as can be seen in birds and in some insects, primarily by mechanical action such as a pressure exerted from within or mastication by the embryo (Needham, 1931; Ishida, 1948a; Davis, 1969). Similar types of egg envelope rupture have been reported in some aquatic

FISH PHYSIOLOGY, VOL. XIA

invertebrates, although the evidence for participation of enzyme(s) is increasing (Davis, 1969, 1981). In the latter, the emergence of an embryo occurs after a preceding dissolution or softening of egg envelope by an embryo-secreted hatching enzyme. Hatching mechanisms of this type were first inferred and then observed in fish about 80 years ago.

In 1900, Kerr first described in his studies on the development of the lungfish *Lepidosiren paradoxa* that the horny egg shell became quite soft so that the embryo could break it by a violent body movement. Although no experimental analyses were made at that time, he attributed this softening of the egg shell to a digestion by some ferment (enzyme) secreted by the embryo. Five years later, Bles (1905) also suggested that hatching of the amphibian *Xenopus laevis* was due to an enzyme secreted from the frontal gland of the embryo. These reasonable but somewhat speculative views were fortified when Moriwaki (1910) and Wintrebert (1912a) experimentally studied on the egg envelope-dissolving principles secreted from salmonid embryos.

Moriwaki's work was written in Japanese and published in a report of a hatchery station in Hokkaido, Japan, and was, therefore, scarcely noticed by others until Ishida (1943, 1944a,b, 1948b) brought it to scientists' attention. Moriwaki found that at the time of hatching of *Oncorhynchus keta*, the inner layer of egg envelope was dissolved by the contents of perivitelline fluid. An undigested outer layer remained like a fragile veil that was then broken by the embryo. The contents of the perivitelline fluid derived from one embryo were so powerful that they could digest more than 15 egg envelopes at a temperature as low as 8°C. He concluded that the egg envelope–dissolving substance seemed to be a kind of ferment, although a strict identification was not accomplished. Furthermore, he found a large number of unicellular glands that become differentiated on the surface of embryonic body about 10 days before hatching, and he considered that the ferment must have been secreted from the mature glands only at the time of hatching, as the perivitelline fluid obtained before the time of hatching was inactive in dissolving the egg envelope.

Likewise, Wintrebert and Bourdin made extensive studies on the hatching of fish such as rainbow trout (Wintrebert, 1912a; Bourdin, 1926a), goldfish (Wintrebert, 1912b; Bourdin, 1926b,c), perch (Wintrebert, 1926; Bourdin, 1926b,c,d), and other teleosts (Bourdin, 1926a,b,c). They found that the movement of an embryo was not necessary for hatching, as the embryo whose movement was inhibited with 0.03% chloretone was still capable of hatching. Perivitelline fluid obtained from the embryos just before hatching digested fertilized egg envelopes. Although they also noticed that a secretion from unicellular epidermal glands was responsible for the digestion of the envelope, Wintrebert (1912a) at first did not use the word ferment or enzyme for the digesting principle of rainbow trout embryos. He used this word for that of the perch embryos, *Perca fluviatilis* (Wintrebert, 1926). Thereafter, participation of "ferment" as the digesting principle became clearer (Bourdin, 1926a), and the use of the term "hatching enzyme" was settled when Needham (1931) cited their work in his *Chemical Embryology*. Studies on hatching, though not many, were also made for various animal groups other than fish by the 1940s, and hatching enzymes had been described in aquatic vertebrates (fish and amphibian) and invertebrates such as ascidians (Berrill, 1932), echinoderms (Ishida, 1936), cephalopods (Hibbard, 1937), and insects (Slifer, 1937, 1938). There have been relatively few reviews or monographs with regard to animal hatching besides those by Needham (1931, 1942). Among them are those written by Ishida (1948a,b, 1971, 1985), Hayes (1949), Smith (1957), Blaxter (1969), and Davis (1969, 1981), from which we can obtain information about hatching not only of fish but of other animals including invertebrates.

II. HATCHING-GLAND CELLS

A. Differentiation and Maturation of Hatching-Gland Cells

In the early studies of fish hatching, it was observed that many unicellular hatching glands appeared on the surface of embryos as they reached to the hatching stage. Bourdin (1926b) reported that hatching-gland cells were somewhat larger than other cells and contained many vacuoles, which were at first stainable with neutral red, but were gradually replaced by unstainable granules, while the concomitant mucous gland was an ordinary cell stained with mucicarmine. Bourdin regarded the hatching gland as being morphologically merocrine but functionally holocrine. This was confirmed later by many workers (Armstrong, 1936; Ishida, 1943, 1944b; Rosenthal and Iwai, 1979). Histochemical studies of hatching glands in *Oncorhynchus keta* were also reported by Inukai *et al.* (1939).

Differentiation of hatching-gland cells of the medaka *Oryzias latipes* was pursued histologically with light microscope by Ishida

(1943, 1944b) and later with the electron microscope by Yamamoto (1963). According to Ishida (1944b), precursory hatching-gland cells in this species become visible around the pharynx of the embryos at the stage of eye pigmentation. One day before this stage (2–3 days after fertilization), the cytoplasm of some cells in the ventral endodermal cell mass becomes stainable with eosin and contain a small number of eosinophilic granules. At the stage of eye pigmentation, many giant cells containing eosinophilic granules are seen in the posteroventral region of the eye, and then only among the mass of endodermal cells. As development proceeds, the giant cells (~14 μm in diameter) migrate forward under the brain and begin to form the foregut.

Changes in histochemical stainability of hatching enzyme granules during development of fish have been reported by several authors (Inukai *et al.*, 1939; Ouji, 1959a,b; Ouji and Iga, 1961). According to Ouji and Iga (1961), developmental changes in the carp hatching gland can be classified into several stages. At first, the precursory hatching-gland cells contain a few granules stained faintly with acid fuchsin. The number of the granules increases gradually, and some of them become stainable with iron-hematoxylin rather than acid fuchsin. As development proceeds, the number of iron-hematoxylinophilic granules increases, until almost all granules are finally stained with this dye. In the case of azan or Mallory's stain, the granules are initially stained faintly with orange G. However, they become gradually stainable with aniline blue, though faintly at first, rather than orange G. In a well-developed gland cell, all secretory granules are stained deeply with aniline blue. Just before hatching, however, the granules become stainable again with orange G rather than with aniline blue. Thus, a secretory granule changes its affinity to dyes according to its stage of differentiation or maturation, having developmental stainability of a dual nature.

Localization of well-differentiated hatching-gland cells in fish embryos differs from species to species (Yanai, 1966; Ishida, 1985). In the case of salmonid fishes, such as rainbow trout, the gland cells are distributed on the anterior surface of embryonic body and yolk sac, and on the inner surface of the pharynx and gill (Wintrebert, 1912; Ishida, 1948b; Hagenmaier, 1974c), while the distribution of gland cells in medaka is, as in some other cyprinodont fishes, generally confined to the inner surface of pharyngeal cavity. In most fish species, hatching-gland cells are distributed on the outer surface of embyronic body and/or yolk sac, and are thought to be of ectodermal origin (Yanai, 1966). In this connection, medaka is a rather exceptional fish, in the sense that the hatching glands are only in the pharyngeal wall and are of endodermal origin (Ishida, 1944a). Also, in sturgeons, which have multicellular compact hatching glands, the gland cells are formed from an anterior part of gut and originate from the endoderm (Ignat'eva, 1959). The question of the germ layer from which hatching glands originate seems to be open to further study.

The hatching-gland cells of medaka can be distinguished from other endodermal cells early in development [at st. 22 after Matui (1949), ~10–12 somites] by their relatively large size, abundance of cisternae of endoplasmic reticulum, and a large electron-dense nucleus with a large nucleolus. At stages somewhat earlier than eye pigmentation (st. 24, ~15 somites), the secretory granules (hatching enzyme granules) appear first in the cytoplasmic matrix (Yamamoto, 1963). The number of secretory granules in a gland cell increases markedly thereafter. Thus, hatching-enzyme synthesis in the gland cell seems to take place most actively around the stage of eye pigmentation, when the differentiating hatching-gland cells are increasing in size and forming a lining of pharyngeal cavity. In zebrafish embryos, *Brachydanio rerio*, the time of the first appearance of hatching enzyme granules coincides with that of eye pigmentation (Willemse and Denucé, 1973). A similar observation was also reported for rainbow trout (Hagenmaier, 1974c). According to Egami and Hama (1975), hatchability of medaka embryos was remarkably decreased when they had been irradiated either with X rays (2 kR, 250 R/min) or with γ rays (2 kR, 33.3 or 250 R/min) at the stages from optic vesicle formation to lens formation. Therefore, some irradiation-sensitive processes necessary for hatching-enzyme formation, such as mRNA synthesis, probably occur at these stages. In our preliminary studies on the isolation of hatching-enzyme granules (Iuchi and Yamagami, 1980), fraction 1 ($600g \times 10$ min pellet) and fraction 2 ($600–1000g \times 10$ min pellet) obtained from $0.3\ M$ sucrose homogenates of embryos contained the secretory granules as they exhibited an ethylenediamine tetraacetic acid-sensitive (EDTA-sensitive) proteolytic enzyme (see later). These secretory granule fractions obtained from day-3 as well as day-5 embryos exhibited a high specific activity of hatching enzyme, while those from day-2 and day-6 (posthatching) embryos showed almost no hatching enzyme activity (Fig. 1). These results also indicate that the hatching enzyme is not yet formed in day-2 embryos, which correspond to the irradiation-sensitive stage, but it is synthesized soon after these stages. More recently, Schoots *et al.* (1982b) reported that the hatching enzyme could be detected immunohistochemically in hatching-gland cells of pike embryos, *Esox lucius*, at the 10- to 20-somite stage (early stage D; Gihr, 1957). Thus, it may be inferred that hatch-

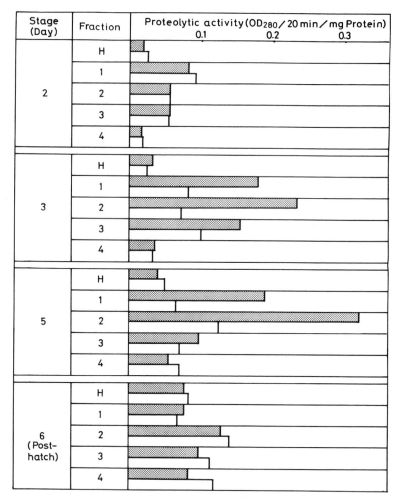

Fig. 1. Distribution of the hatching enzyme (EDTA-sensitive protease) activity among the subcellular fractions obtained from 0.3 M sucrose homogenate of medaka embryos at some developmental stages. Fraction H: whole homogenate; fraction 1: $600g \times 10$ min pellet; fraction 2: 600–$1000g \times 10$ min pellet; fraction 3: $1000g \times 10$ min to $10,000g \times 15$ min pellet; fraction 4: supertant. Proteolytic activity was assayed in principle following Kunitz (1947). Dotted column and open column refer to the activity in the absence and presence of 5 mM EDTA, respectively. Stages 3 and 4 in the original report (Iuchi and Yamagami, 1980) should read 2 and 3, respectively, as shown in this figure.

ing enzyme synthesis in fish embryos is initiated in general just after lens formation but in advance of eye pigmentation.

According to Yamamoto *et al.* (1979), the hatching gland of medaka continues to produce secretory granules until nearly the prehatching stage. A few secretory granules found in the trans face of the Golgi apparatus were less electron-dense than most other granules, probably representing an immature state. Such immature granules could be found sometimes in day-5 embryos (st. 31, 1 day before hatching). In embryos close to the hatching stage, there were two types of secretory granules in hatching gland cells; one was homogeneously electron-dense and the other consisted of an electron-dense portion and a less dense portion. In the latter, the electron dense portion often took a crescent shape in the periphery of the granule, like a shell. Such heterogeneity of electron stainability in a hatching-enzyme granule has been seen also in some cyprinid embryos, *Brachydanio rerio* and *Danio malabaricus* (Willemse and Denucé, 1973), and salmonid embryos, *Salmo gairdneri*, *S. trutta*, *Salvelinus fontinalis*, and *S. pluvius* (Yokoya and Ebina, 1976). As described above, histochemical stainability of a granule was reported to change markedly during development. Although it remains uncertain whether or not such a granule change is correlated with that of electron density, it is evident that the hatching-enzyme granules undergo some physicochemical changes during their maturation. A drastic change in the electron density of the granules in their last maturation phase seems to be closely related to the secretion process. This problem will be discussed again in the next section.

B. Ultrastructural Changes in the Hatching Gland Associated with Secretion

1. HISTOLOGICAL STUDIES

After being packaged in the secretory granules, the hatching enzyme is secreted into the perivitelline space, where it gains access to the egg envelope. In this section, the cellular and subcellular changes in the hatching gland associated with secretion will be discussed. There have so far been only a few studies on the cellular changes of the hatching gland during secretion. In their histological studies, Ishida (1944b) and Ouji (1959a,b) observed morphological changes of hatching-gland cells in *Oryzias latipes* and *Odontobutis obscura*, respectively. In the former, the nucleus of the gland cell was invisible at the time of secretion and when secretory granules were released. In the latter, the nucleus remained in the gland cell, while the granules disappeared during secretion. However, a more detailed description of gland-cell changes was possible only with the electron microscope. Yamamoto (1963) reported that there were three types of secretory granules in the hatching-gland cells of medaka embryos. Type 1 granules were homogeneously electron-dense and were predominant at earlier develomental stages. Type 2 granules were as electron-dense as type 1 but contained a crescent-shaped shell of higher electron density. Type 3 granules contained somewhat granular contents with as low an electron density as the cytoplasmic matrix; they also had an electron-dense shell around the granular contents. The granules of this type were predominant in the embryos at later developmental stages. Just before secretion, a small hole appeared at the apical end of the cell, and type 3 granules seemed to be disintegrated within the cell.

2. ELECTRICALLY INDUCED SECRETION

It is sometimes difficult to predict accurately when the hatching-gland cells of an embryo initiate secretion under natural conditions. As will be discussed in detail later, several reagents or treatments have been reported that induce hatching-enzyme secretion in fish, causing precocious hatching. Among them, an adequate dose of electric (AC) stimulation is quite effective in causing hatching-enzyme secretion in medaka as well as in rainbow trout (Iuchi and Yamagami, 1976a; Yamamoto et al., 1979). Rainbow trout embryos that would hatch normally about day 19–20 after fertilization at 15°C could be induced to hatch precociously on day 16–17 when they were stimulated with 100 V AC for 3 s 10 times with 5-min intermissions. In this case, hatching-gland cells on the surface of embryos became invisible a few minutes after the stimulation. When the dechorionated embryos were stimulated, hatching enzyme as determined by its caseinolytic activity (see later) increased in the medium (Iuchi and Yamagami, 1976a).

Medaka embryos also hatch precociously upon electric stimulation (Fig. 2). When cultured normally in a shaking incubator at 30°C, they hatch on day 6 if the day of fertilization was regarded as day 1 (Yamagami, 1960). Natural hatching of control embryos begins early on day 6 and it takes almost one more day until all the control embryos complete hatching. However, stimulation of the embryos with 100 V AC for 5 s early on day 5 (~25 h earlier than the beginning of natural

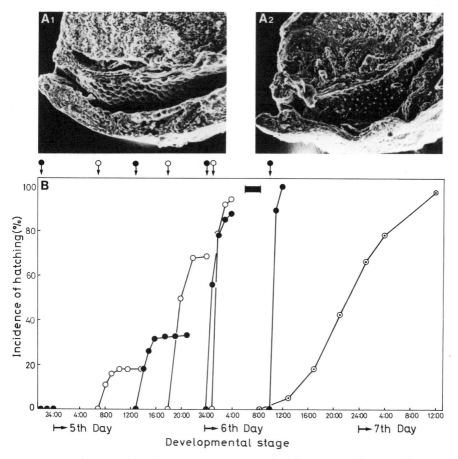

Fig. 2. Induction of hatching enzyme secretion by electric stimulation in the medaka embryos. (A) Scanning electron micrographs (SEMs) of median cuts of the head of the embryos at the prehatching stage (A_1) before and (A_2) 5 min after stimulation. Many hatching-gland cells are seen as round protrusions in the buccal wall in (A_1) but not in (A_2) (×200). (B) Incidence of hatching of the stimulated (○, ●) or unstimulated (control, ☉) embryos. The arrow indicates the time of application of the electric stimulation (AC 100V, 5 s). A bold bar in the figure indicates the prehatching stage (Yamamoto *et al.*, 1979).

hatching) gave rise to precocious hatching of some embryos. Their responsiveness increased as development proceeded, and most embryos at prehatching stages could be induced to hatch. The gland cells of the embryos at the prehatching stage are considered to have been mature, in the sense that all cells were ready to secrete the hatching enzyme upon stimulation. It was found that almost all gland cells completed their secretion 5 min after the stimulation at the latest.

Exploiting this electric stimulation, a sequential ultrastructural change of the hatching-gland cells during the course of secretion could be followed in medaka (Yamamoto *et al.*, 1979). The gland cells are arranged side by side and are covered by a sheet of squamous epithelium on the inner wall of the pharyngeal cavity. Each epithelial cell has a hexagonal contour. Three adjoining epithelial cells meet at the apical center of each underlying gland cell. Just before electrical stimulation the gland cells were full of secretory granules of homogeneous electron density, with the nucleus at the base. Near the Golgi apparatus, immature secretory granules with lower electron density were observed. Soon after the electric stimulation (usually ~30 s), a swelling of each gland cell occurred and the secretory granules within a cell became more clearly discernible as round protrusions. Every junction of three epithelial cells was separated and the apical surface of the underlying gland cell was exposed (Fig. 3). Inside the gland cell, a coalescence of electron-dense secretory granules occurred to form a large mass of secretory substance surrounded by a limiting membrane. The contents of the coalesced mass appeared to be composed of fine granules, and its electron density was reduced remarkably. The electron density of the contents of a few uncoalesced granules, except for their peripheral part, was decreased slightly. As a result, these granules appeared to have a crescent-shaped shell of high electron density. The membrane surrounding the coalesced se-

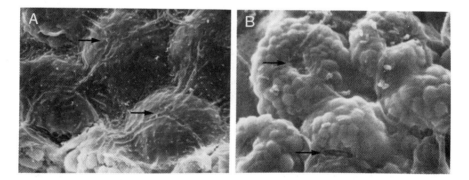

Fig. 3. SEMs of the hatching-gland cells of medaka embryos (A) before and (B) 30 s after electric stimulation. Upon stimulation, the hatching-gland cells were swollen and the secretory granules became discernible. Every junction (arrow) of adjoining epithelial cells covering the gland cells was separated (×2200) (Yamamoto *et al.*, 1979).

cretory mass became united with the cell membrane at the apex of the gland cell, forming an orifice through which the secretory substance flew out into the buccal cavity (Fig. 4). It seems that the nucleus and cytoplasm including some endoplasmic reticuli still remained in some gland cells after secretion. About 24 h after the secretion, the openings at the epithelial junctions were reclosed and the open surface of the epithelium was flat, since any swollen gland cells were now absent underneath. However, some gland cells containing an electron-dense irregular-shaped nucleus and many fragmented cisternae of rough endoplasmic reticulum but no secretory granules were found to persist under the epithelium.

3. NATURAL SECRETION

In contrast to the situation with artificially induced hatching, the hatching-gland cells in the process of natural secretion exhibited somewhat different features (Yamamoto *et al.*, 1979). As shown in Fig.

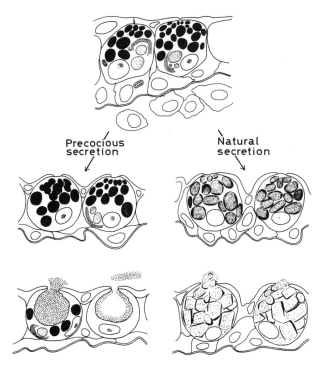

Fig. 4. Diagrammatic illustration of the ultrastructural changes in the hatching-gland cells of medaka embryos in the process of electrically induced precocious secretion and natural secretion (for explanation, see text) (Yamamoto *et al.*, 1979).

3, the gland cells were swollen and the epithelial junctions were open as in the case of electrically induced secretion. Inside the gland cell, however, a different pattern of secretory-granule change was observed. The granules did not coalesce with each other and each granule became markedly electron lucent, except at its periphery. Thus, a hatching-gland cell just before natural secretion was, as observed in earlier studies (Yamamoto, 1963), full of electron-lucent secretory granules bearing electron-dense shells. The granules became somewhat angular in shape, their membranes were dissolved partly, and their contents were mixed with cytoplasm before they were secreted from the cell. This process seemed to be different from that of exocytosis. The electron-lucent contents were composed of fine granules in this case also.

In summary, a comparison of the ultrastructural changes of hatching-gland cells during the electrically induced secretion with those during natural secretion shows two kinds of changes: those that are common to both types of secretion and those specific to each type of secretion. The common changes are swelling of gland cells, exposure of the apical center of gland cells following the separation of the epithelial junction, and reduction of electron density of secretory substance prior to secretion. By contrast, in natural secretion, no coalescence of secretory granules was observed, while in the induced secretion, many secretory granules of high electron density coalesced into a large mass of secretory substance and their electron density was decreased. A typical exocytosis was observed only in the induced precocious secretion, while the secretory granules were disintegrated and mixed with the cytoplasm of the gland cell in natural secretion.

In salmonid fishes, secretory granules become electron-lucent and fused together just before secretion. The gland cells discharge the granules together with some other cytoplasmic structures differently from ordinary exocytosis. After exhaustion of the secretory granules, the gland cells dissociate from the epithelium (Yokoya and Ebina, 1976). However, according to Schoots *et al.* (1983a), there are three types of secretion in pike embryos: (1) exocytotic discharge via a secretion vacuole, (2) exocytosis at protruded cell part, and (3) intercellular exocytosis. Among them, type 1 is predominant.

Although the reason why such different types of secretion occur in the hatching gland is obscure, the fusion of secretory granules has also been reported in the process of secretagogue-induced secretion of various cells other than hatching-gland cells (Kurosumi, 1961; Ichikawa, 1965; Amsterdam *et al.*, 1969; Kanno, 1972; Kagayama and Douglas, 1974; Lawson *et al.*, 1977). For example, in the rat peritoneal mast cell stimulated by the treatment with ferritin-conjugated sheep antibody to rat immunoglobulin (S anti-RIg-FT), an active degranulation occurs and the secretory granules coalesce into a large mass with low electron density. The membrane interaction in association with the degranulation leads to an exocytosis of the coalesced granular material (Lawson *et al.*, 1977). Thus, it seems that a fusion of granules occurs when the gland cells are *forced* to secrete somewhat rapidly by stimulants. Reduction of the electron density of secretory substances, irrespective of whether they are in granules or in vacuoles, may be related partly to hydration of the substances. According to recent work (Yamagami *et al.*, 1983), the electron density of secretory granules remained high in the fully (or overly) matured hatching glands of medaka embryos whose hatching had been retarded by an "air-incubation" (see p. 482). This fact strongly suggests that the electron-dense granules are already mature in the sense that they are ready to be secreted upon stimulation and that the reduction of electron density is not an indication of maturation but an indication of having entered into the secretion process. From the above results, it seems that some facets in the secretory changes, such as the increased fusibility of secretory granules, were manifested exaggeratedly in the electrically stimulated secretion of hatching gland compared to natural hatching. A stimulus for natural secretion may act somewhat more slowly or moderately, although its nature remains still uncertain (see later).

Even after natural secretion, some hatching-gland cells without any secretory granules but full of fragmented cisternae of endoplasmic reticulum persist under the epithelium (Yamamoto, 1963; Yamamoto *et al.*, 1979). Similar persisting gland cells in the pike reportedly degenerated sooner or later by programmed death (apoptosis) (Schoots *et al.*, 1983a).

III. HATCHING ENZYME AND CHORIOLYSIS

A. Purification and Characterization of Fish Hatching Enzymes

Dissolution of the tough egg envelope by the secreted hatching enzyme is, together with the subsequent breakage of the remnant egg envelope (outer layer of chorion) by the embryo, a major feature of hatching in fishes. Thus, the nature of the hatching enzyme and enzymatic choriolysis have been foci of interest in the study of hatching. It is known that the hatching enzyme of fish has a proteolytic activity in addition to its egg envelope-dissolving activity (choriolytic activity) (Ishida, 1944c; Kaighn, 1964). Therefore, the hatching enzyme activity can be assayed tentatively for its proteolytic activity. Assay of the proteolytic (or peptidolytic) activity of the fish hatching enzyme has been performed using different substrates such as insulin (Kaighn, 1964), casein or its derivatives (Yamagami, 1972, 1973; Hagenmaier, 1974a; Schoots and Denucé, 1981), and some synthetic peptides (Yamagami, 1973; Yasumasu et al., 1985). However, when a crude sample is used, the assay of only the proteolytic (or peptidolytic) activity is not appropriate for discriminating the real hatching enzyme from other concomitant proteases, if any. A turbidimetric method of semi-quantitative determination of choriolytic activity of medaka enzyme (Fig. 5) (Yamagami, 1970) was devised to overcome such difficulty, although the method seems not be be applicable to the rainbow trout enzyme (Ohzu et al., 1983). ^{14}C-labeled chorion was recently used as a substrate for *Fundulus* enzyme (DiMichele et al., 1981).

The purification of hatching enzyme in fish has been carried out

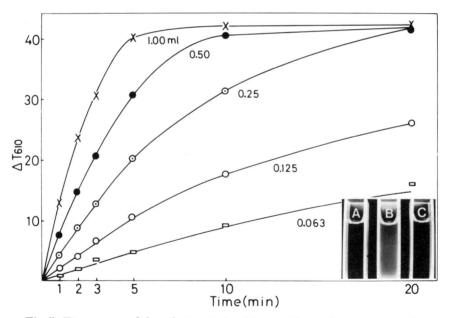

Fig. 5. Time course of choriolytic activity of the medaka hatching enzyme as determined by turbidimetry. Inset shows the cuvettes containing the reaction mixture (A and B) with or (C) without chorion paste as substrate; A, after enzymatic digestion; B, before enzymatic digestion. ΔT_{610} refers to the increase in percentage transmission of the reaction mixture, including chorion paste, at 610 nm. (Yamagami, 1970).

since the early 1960s. In most cases, the enzyme has been obtained and purified from the hatching liquid, that is, the medium in which the embryos were allowed to hatch. Kaighn (1964) tried to purify the hatching enzyme (chorionase) of *Fundulus heteroclitus* by gel filtration and sucrose density-gradient ultracentrifugation. The chorionase was sedimented between two molecular-weight markers, ribonuclease and hemoglobin, suggesting that its molecular weight was between 15,000 and 40,000. Kaighn reported that the chorionase hydrolyzed tyrosine–threonine and threonine–proline peptide bonds in the B chain of insulin. Enzyme activity was inhibited with diisopropylphosphorofluoridate (DFP), a specific modifier of a serine residue.

Ogawa and Ohi (1968) and Ohi and Ogawa (1970) fractionated an aqueous extract of manually isolated hatching glands of medaka by agar gel electrophoresis and obtained two fractions bearing chorion digesting activity and another fraction active in causing swelling of the chorion. According to a recent report (Schoots *et al.*, 1983c), the swelling of the chorion seems to be an intermediate phase in the proteolytic digestion of the chorion. Assuming that this is true, choriolytic enzyme(s) obtained from hatching-gland cells of medaka separated as three different fractions moving toward the cathode at pH 8.6 on agar gel electrophoresis. Purification of medaka hatching enzyme used Sephadex column chromatography of the ammonium sulfate precipitates of hatching liquid, followed by CM-cellulose column chromatography (Yamagami, 1972). As will be shown in Fig. 7, Sephadex G-75 column chromatography of the ammonium sulfate precipitate of hatching liquid gave two peaks of choriolytic and proteolytic activities, which were named PI enzyme (or enzyme I) and PII enzyme (or enzyme II), respectively (Yamagami, 1972, 1973). The specific activity of PII was much higher than that of PI. When the PII enzyme was fractionated by CM-cellulose column chromatography, a single peak of choriolytic and proteolytic activities coincident with a peak of protein was eluted by 0.02 M Tris HCl (pH 7.1)–0.3 M NaCl. Specific activities of this enzyme fraction (named PII-0.3) with respect to choriolytic activity and proteolytic activity were 212 and 183 times those of hatching liquid, respectively. The enzyme protein eluted as a single peak on Sephadex column chromatography and gave a single band moving toward the cathode on starch gel electrophoreses at pH 8.6 and 5.2 and on polyacrylamide gel disc electrophoresis (PAGE). However, this enzyme preparation showed some heterogeneity on sodium dodecyl sulfate (SDS) PAGE (Iuchi *et al.*, 1982). Thus, it has yet to be determined whether the additional protein(s) in PII-0.3 are merely contaminants or some fragments of chorion protein associated with the enzyme. Recently, the secretory granules of the medaka hatching

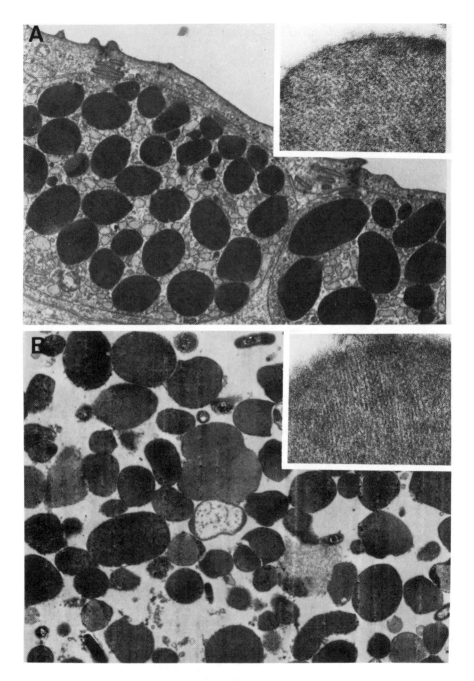

Fig. 6. Transmission electron micrographs (TEMs) of the hatching enzyme granules of medaka (A) fixed in situ and (B) isolated from the homogenate of whole prehatching embryos (×7000). Insets are higher magnifications of a part of respective granules. Note crystalline patterns ~70 Å wide in both granules (×130,000) (Iuchi *et al.*, 1982).

gland were isolated in 0.3 M sucrose (Fig. 6). The aqueous extract of the isolated granules exhibited a high choriolytic activity, representing a single band of protein on SDS-PAGE (Iuchi et al., 1982). The molecular weight of PII-0.3 enzyme as determined by Sephadex column chromatography was reported at first to be about 8000 (Yamagami, 1972). When determined on SDS-PAGE following Weber and Osborn (1969), however, it was about 21,000. The molecular weight of the enzyme in the aqueous extract of the isolated granules was also about 21,000 on SDS-PAGE (Iuchi et al., 1982). A similar discrepancy in the molecular weight was also reported for the pike hatching enzyme (Schoots and Denucé, 1981, see later). This discrepancy may be attributable partly to a high affinity of the hatching enzyme for the supporting medium of gel filtration, and this method seems to be inadequate for the estimation of molecular weight of this enzyme. It seems highly probable that the hatching enzyme of medaka is a metalloprotease but is not a serine protease nor sulfhydryl protease, as its activity is inhibited by ethylenediamine tetraacetic acid (EDTA) but neither by DFP nor by iodoacetamide (IAM) (Ohi and Ogawa, 1970; Yamagami, 1973). Low concentrations of some monovalent and divalent cations activate the enzyme slightly, while high concentrations inhibit it (Yamagami, 1973).

As Fig. 7A shows, the hatching liquid of medaka embryos contains apparently two hatching enzyme fractions, PI enzyme (enzyme I) and PII enzyme (enzyme II). It was found, however, that a part of PI enzyme could be converted to PII enzyme (which was named enzyme PI-PII) through re-salting out and rechromatography on Sephadex. Such a conversion from PI to PII was observed if rechromatography of the PI enzyme was repeated. The properties of enzyme I, enzyme II, and enzyme PI-PII in terms of the sensitivity to some inhibitors were found to be almost identical (Fig. 7C). These observations strongly suggest that PI and PII enzymes were essentially the same enzyme, but that they behaved differently because of their different states of association with some heterologous substances such as hydrolyzed chorion (Yamagami, 1975). This view has been confirmed by our recent work. PI enzyme and PII enzyme have recently been highly purified from the hatching liquid by repeating Toyopearl gel filtration chromatography at pH 10. These procedures resulted in dissociation of the bound hatching enzymes. As a result, each PI enzyme and PII enzyme was found to consist of two types of proteases; one was a protease with high choriolytic activity (HCE) and the other was a protease with low choriolytic activity (LCE) (Yasumasu et al., 1988). Their molecular weights are about 24,000 and 25,500 respectively, on SDS-PAGE after Laemmli (1970).

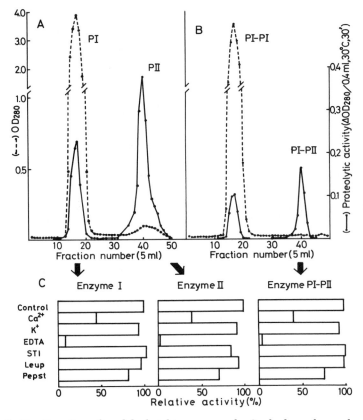

Fig. 7. Fractionation of medaka hatching enzyme by Sephadex column chromatography. (A) Elution pattern of the ammonium sulfate precipitate of the hatching liquid on Sephadex G-75 column chromatography. (B) Elution pattern of rechromatography of the ammonium sulfate precipitate of PI through the same column as in (A). (C) Comparison of the properties of enzyme I, enzyme II, and enzyme PI-PII. STI, Soybean trypsin inhibitor (33 μg/ml); Leup, leupeptin (33μg/ml); Pepst, pepstatin (0.33 μg/ml) (Yamagami, 1975).

Appearance of multiple hatching enzyme peaks on gel filtration chromatography has been reported also in some other fish species such as rainbow trout (Ohzu and Kasuya, 1979) and pike (Schoots and Denucé, 1981). It seems that such a physical heterogeneity is a characteristic of the hatching enzyme not only of fish but also of some other animal species such as sea urchin (see p. 477). An analysis of this problem will be useful for elucidation of the nature and the mechanism of action of this enzyme.

The hatching enzyme of rainbow trout was purified from the hatching liquid through the fractionation procedure similar to that of the medaka enzyme (Hagenmaier, 1974a; Ohzu and Kasuya, 1979). The enzyme protein seems to be a basic protein from its behavior on chromatography and electrophoresis. The molecular weight as determined by gel filtration chromatography was about 10,000. However, considering the probable inadequacy of the gel filtration method for determination of the molecular weight of hatching enzyme, it would be necessary to reexamine the molecular weight of the salmonid hatching enzyme with some other analytical methods. This enzyme also appears to be a metalloprotease, as it was inhibited by EDTA, ethyleneglycol bistetraacetate (EGTA), O-phenanthroline, or KCN, but not by phenylmethyl sulfonylfluoride (PMSF), tosyl-L-lysylchloromethane (TLCK), tosyl-L-phenylalanylchloromethane (TPCK), or iodoacetamide (Hagenmaier, 1974a,b). It was reported that the activity of the EGTA-inactivated enzyme was restored only by iron (Fe^{2+}) (Hagenmaier, 1974b). The optimal pH of this enzyme was found to be around 8, resembling the medaka enzyme.

Recently, the hatching enzymes of *Fundulus heteroclitus* and of the pike *Esox lucius* have been well studied from biochemical and physiological viewpoints. Using chorion labeled with [^{14}C]iodoacetamide as substrate, DiMichele *et al.* (1981) examined some characteristics of *Fundulus* chorionase. This enzyme was found to be quite stable below 30°C, like the medaka enzyme (Yamagami, 1973), and had a Q_{10} of 2.2 between 15 and 30°C. The pH optimum for the activity was between 8.0 and 8.5. This enzyme seems to be halophilic; in solutions of ionic strength below 0.05 M, approximately 50% of the activity was lost in 18 h, but addition of NaCl within 48 h restored the activity. The optimum ionic strength was between 0.1 and 0.2 M. Such a salt requirement is seen in the enzymes of medaka (Yamagami, 1973) and the marine fish *Gobius jozo* (Denucé, 1976). The *Fundulus* enzyme was found to be insensitive to IAM but sensitive to EDTA as well as to PMSF. These results show that the *Fundulus* enzyme is a serine protease and/or metalloprotease but not sulfhydrylprotease. Kaighn (1964) also reported that *Fundulus* chorionase was inhibited by DFP and was, therefore, presumably a serine protease. According to Denucé and Thijssen (1975), the hatching enzyme of zebrafish, *Brachydanio rerio*, also seems to be a serine protease.

Schoots and Denucé (1981) purified the pike hatching enzyme 1600 times from the original hatching liquid using affinity chromatography with carbobenzoxy-D-phenylalanyl-triethylenetetramine (Z-D-Phe-T) Sepharose. This enzyme is a glycoprotein containing 2% carbohydrate. The molecular weight of this enzyme was 10,000–15,000 by gel filtration but 23,500–25,400 with other methods such as PAGE, SDS-PAGE, and sedimentation analysis. The activity was inhibited by some metal chelators such as EDTA, EGTA, and O-phenanthroline but not by DFP, PMSF, iodoacetic acid, or N-ethylmaleimide (NEM). Furthermore, they concluded that this enzyme is a zinc metalloprotease based on atomic absorption spectrometry and renaturation experiments of the denatured apoenzyme.

In summary of the above results (Table I), we notice common features in some of the enzymes, although we still lack much information for drawing a precise picture of the fish hatching enzyme. The hatching enzyme is a choriolytic protease with a broad pH optimum around 8.0. It requires a metal (probably a divalent cation) for full activity, although some enzymes are reported to be inhibited by serine active site reagents. The molecular weight of the enzymes seems to be in the range of 15,000–30,000, most probably somewhat higher than 20,000.

B. Solubilization of Egg Envelope (Chorion)

Following activation or fertilization, the weak and fragile egg envelope of the unfertilized fish egg is transformed into a tough structure through a process called (water) hardening. The egg envelope (chorion) of the fertilized egg consists of a thin outer layer and a thick inner layer (Fig. 8). (Yamamoto, 1963; Lönning, 1972; Yamamoto and Yamagami, 1975), the former being divided structurally into two sublayers (Anderson, 1967; Flügel, 1967; Wourms and Sheldon, 1976; Dumont and Brummett, 1980). The salmon egg chorion is composed of a scleroprotein, which was classified as pseudokeratin (Young and Inman, 1938) and was later named ichthulokeratin (Young and Smith, 1956). There is a great similarity in amino acid composition of the chorion proteins among the eggs of salmonids, *Fundulus*, and medaka, characterized by an abundance of proline and glutamic acid (and/or glutamine). The hardening occurs mainly in the thick inner layer. This tough structure protects the embryo against mechanical, chemical, and biological harm during development but also seems to be a barrier to the embryo in terms of hatching. Usually, the hatching enzyme is secreted shortly before an actual hatching occurs. In medaka, secretion occurs less than 1 h before hatching. Thus, the thick chorion is digested by the enzyme within 1 h or so, depending on temperature. We tried to simulate the process of natural choriolysis in medaka

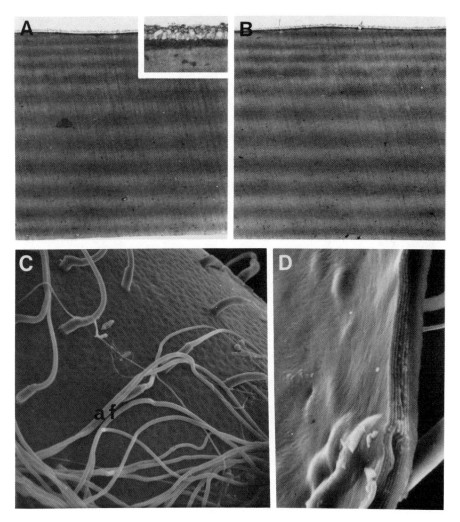

Fig. 8. Electron micrographs of intact egg envelopes of medaka embryos. (A) TEM of the egg envelope of day-1 embryo (middle blastula) (×3600). Inset shows a higher magnification of a part of the outer layer. (B) TEM of the egg envelope of day-6 embryo (~1 h before hatching) (×3600). (C) SEM of the outer surface of egg envelope of day-1 embryo (×240). (D) SEM of the inner surface of the egg envelope of day-1 embryo. af, Attaching filaments (×800) (Yamamoto and Yamagami, 1975).

eggs by incubating chorion pieces in the concentrated hatching liquid or the purified hatching enzyme solution (PII-0.3 enzyme) and to follow the sequential ultrastructural changes of the chorion pieces (Yamamoto and Yamagami, 1975).

Table I
Some Characteristics of Fish Hatching Enzymes[a]

Fish species	Molecular weight (method of determination)	Optimum pH	Effect of some inhibitors	
			Inhibited by	Not inhibited by
Medaka *Oryzias latipes*	8,000 (Gel filtration)[b] 21,000 (SDS-PAGE,W,-O)[c] 24,000 (SDS-PAGE,L)[d]	8.0–9.0 (Choriolysis)[i,j] 7.5–8.3 (Proteolysis)[k]	KCN^n H_2S^n $EDTA^{k,q}$	$DFP^{k,q}$ IAM^k $SBTI^r$ $LEUP^r$
Mummichog *Fundulus heteroclitus*	15,000–40,000 (Sedimentation analysis)[e]	8.0–8.5 (Choriolysis)[l]	$EDTA^l$ $PMSF^l$ DFP^e	IAM^l
Rainbow trout *Salmo gairdneri*	10,000 (Gel filtration)[f,g]	8.5 (Proteolysis)[f]	$EDTA^{f,o}$ $EGTA^{f,o}$ O-PHEN^o KCN^o	SBTI, LBTI^o $PMSF^{f,o}$ TPCK^o, TLCK^o IAA^o, OVOM^o
Pike *Esox lucius*	10,000–15,000 (Gel filtration)[h] 24,000 (SDS-PAGE,L)[h] 25,400 (Sedimentation analysis)[h]	7.0–9.0 (Proteolysis)[h]	$EDTA^h$, $EGTA^h$ $CDTA^h$, O-PHEN^h DTT^h, TGA^h	DFP^h, $PMSF^h$ NEM^h, $SBTI^h$ $OVOM^h$, $TPCK^h$

Species				
Goby *Gobius jozo*	—	8.1–8.4 (Choriolysis, proteolysis)[m]	EDTA[m], EGTA[m] O-PHEN[m]	SBTI[m] OVOM[m]
Zebrafish *Brachydanio rerio*	—	—	DFP[p]	

[a] Abbreviations: CDTA, cyclohexanediaminetetraacetate; DFP, diisopropylphosphorofluoridate; DTT, dithiothreitol; EDTA, ethylenediaminetetraacetate; EGTA, ethylene glycol bistetraacetate; IAA, iodoacetate; IAM, iodoacetamide; LBTI, lima bean trypsin inhibitor; LEUP, leupeptin; O-PHEN, O-phenanthroline; OVOM, ovomucoid; PMSF, phenylmethylsulfonyl fluoride; SBTI, soybean trypsin inhibitor; TLCK, tosyl-L-lysylchloromethane; TPCK, tosyl-L-phenylethyl chloromethyl ketone; SDS-PAGE,W,O, SDS–polyacrylamide gel electrophoresis following the method of Weber and Osborn (1969); SDS-PAGE,L, SDS–polyacrylamide gel electrophoresis following the method of Laemmli (1970).

[b] Yamagami (1972).
[c] Iuchi et al. (1982).
[d] Yasumasu et al. (1985).
[e] Kaighn (1964).
[f] Hagenmaier (1974a).
[g] Ohzu and Kasuya (1979).
[h] Schoots and Denucé (1981).
[i] Ishida (1944b).
[j] Yamagami (1970).
[k] Yamagami (1973).
[l] DiMichele et al. (1981).
[m] Denucé (1976).
[n] Ishida (1944c).
[o] Hagenmaier (1974b).
[p] Denucé and Thijssen (1975).
[q] Ohi and Ogawa (1970).
[r] Yamagami (1975).

As shown in Fig. 8C, there are a number of honeycomb-like patterns on the outer surface of intact chorion, as was first documented by Kamito (1928). A large number of villi are present all over the surface of chorion, and many attaching filaments, much longer than villi, are restricted to the surface of vegetal pole area of the egg chorion. The inner surface of the intact chorion appears to be smooth, showing a somewhat parallel wavy pattern. On incubation of chorion pieces in the hatching liquid or in the purified hatching enzyme solution buffered with Tris-HCl at pH 7.2, the outer surface of chorion became rougher and many irregular dents and grooves appeared as the enzymatic erosion proceeded (Fig. 9A). In contrast to the outer surface, the inner surface of the chorion showed no irregular erosions during digestion; the partially digested inner surface remained smooth and

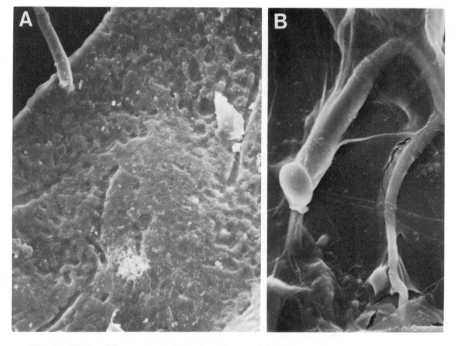

Fig. 9. SEMs of the outer and inner surfaces of the egg envelope of medaka embryos during the process of enzymatic digestion by the hatching liquid. (A) After 5-min incubation (from outside) (×240). (B) After 15-min incubation (from inside). The inner layer has been digested away and a sheet of outer layer with a villus is remaining (×800) (Yamamoto and Yamagami, 1975).

flat. After complete digestion, a thin outer layer remained, apparently only slightly digested (Fig. 9B). When examined successively by transmission electron microscopy, the thickness of the inner layer was found to decrease evenly. It has been reported, however, that in the enzymatic choriolysis *in vivo,* instead of *in vitro,* the degree of inner layer digestion varied from fish species to species depending on the thickness of the inner layer of chorion (Schoots *et al.,* 1982c). As shown in Fig. 10B, there were indications of enzymatic solubilization of inner layer at the peripheral (outer) parts, just beneath the outer layer. These areas seemed to correspond to the dents of the grooves shown in Fig. 9A and to be caused by the enzyme that had permeated through the outer layer of chorion pieces incubated in the enzyme solution. As shown in Fig. 10, the partially digested inner layer was slightly swollen, decreased in its electron density, and loosened into a fibrous network. The solubilized products of the inner layer could be fixed with glutaraldehyde and osmic acid, which suggests that the solubilized products were of high molecular weight (Fig. 10D).

This is also confirmed by analyzing the enzymatic digests of medaka chorion biochemically (Yamagami and Iuchi, 1975; Iuchi and Yamagami, 1976b). In a preliminary experiment (Yamagami, 1970), it was found unexpectedly that the hatching enzyme digests of the medaka chorion contained a small amount of free amino acids as detected by thin layer chromatography. When a large number of chorions isolated from blastulae was incubated with the purified hatching enzyme (PII-0.3), most of them were digested to a clear viscous solution, leaving the outer layers with villi and attaching filaments undigested. The solubilized material was fractionated using Sephadex G-75 column chromatography into a major fraction (PI) of high-molecular-weight glycoproteins and a minor one (PII) of lower-molecular-weight substances, that is, small peptides and/or free amino acids. The former was fractionated further into two peaks of glycoproteins on Sephadex G-200 column chromatography: one (named Fr. 1) was eluted at the void volume and the other (Fr. 2) eluted later. Both peaks are considered to be major constituents of the inner layer of the chorion. They were approximately equal in amount and were very similar to each other in amino acid composition as well as in absorption spectrum. Upon ultracentrifugal analyses, each of them exhibited symmetrical Schlieren profiles with sedimentation constants of 7.0 S for Fr. 1 and 4.5 S for Fr. 2. However, disc electrophoretic analyses revealed that Fr. 1 was highly heterogeneous, being composed of about six protein bands (C1–C6), while Fr. 2 was homogeneous (Yamagami and Iuchi,

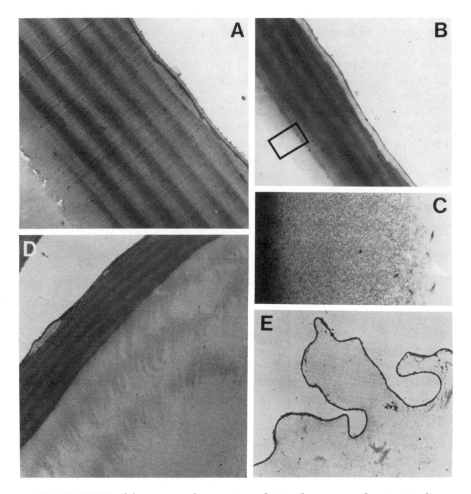

Fig. 10. TEMs of the egg envelope sections during the process of enzymatic digestion by the hatching liquid. (A) After 2-min incubation (×4500). (B) After 5-min incubation. Peripheral part of the inner layer, just beneath the outer layer, is digested by the enzyme that had permeated through the outer layer (×3500). (C) A higher magnification of a partially digested part of the inner layer as indicated by a square in (B) (×24,000). (D) After 10-min incubation. The sample was carefully fixed to avoid dispersing the solubilized material (×3500). (E) After 15-min incubation. Only a sheet of outer layer is remaining (×3500) (Yamamoto and Yamagami, 1975).

1975; Iuchi and Yamagami, 1976b). Thus the major products of enzymatic choriolysis comprise about seven high-molecular-weight proteins (Fig. 11). Denucé (1975) also found seven proteins including those of approximate molecular weight of 80,000 and 200,000 in the enzymatic hydrolysate of medaka chorion. On further examination of the pattern of Fr. 1, it was noticed that there seemed to be a regularity of chemical characteristics among the components of Fr. 1 (Iuchi and Yamagami, 1976b). After determining the molecular weights of the native forms of the Fr. 1 components, and of Fr. 2 following the method of Hedrick and Smith (1968), it was concluded that the net electric charge of each of the six components of Fr. 1 was approxi-

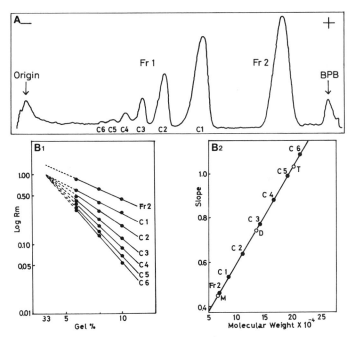

Fig. 11. Major glycoproteins solubilized from the medaka egg envelope by the action of the hatching enzyme. (A) Densitometric illustration of a polyacrylamide gel electrophoretic pattern. Major glycoproteins consist of Fr. 1 (C1–C6) and Fr. 2. (B) Molecular-weight determination of the six components of Fr. 1 and Fr. 2 according to the method of Hedrick and Smith (1968). The log R_m values of the glycoproteins were plotted against acrylamide concentrations (B_1), and their approximate molecular weights were estimated from the slope of the plots (B_2). M, D, and T refer to monomer (MW = 67,000), dimer, and trimer of bovine serum albumin, used as references (Iuchi and Yamagami, 1976b).

mately the same, but different from that of Fr. 2. The molecular weights of the six components of Fr. 1 ranged from 8.6×10^4 for C1 to 21.4×10^4 for C6 with an average molecular weight difference of about 2.6×10^4 between neighboring components, while the molecular weight of Fr. 2 was approximately 7×10^4 (Fig. 11). It may be presumed that to the smallest component of Fr. 1, C1, is added a kind of repeating unit polypeptide of about 2.6×10^4 molecular weight to form the second smallest polypeptide, C2, and to C2 is added the repeating unit to form C3, and so on. As described above, the pI values of all components of Fr. 1 seem to be identical. From these observations, it seems that these components could be named a Fr. 1 *family*. Moreover, it was found that the molar ratios of C1 to C2, C2 to C3, . . . , and of C5 to C6, as calculated from their relative molar concentrations in the chorion digests, are all about 3. This might mean that one molecule of C6 is combined with three molecules of C5 and one molecule of C5 with three molecules of C4, and so forth, and that the hatching enzyme could break the connections between the components (Yamagami, 1981). This assumption implicates some cross-linking of polypeptide chains in the inner layer of the hardened chorion. In this connection, a report of Hagenmaier *et al.* (1976) that γ-glutamyl-ε-lysine was present only in the hardened chorion proteins of rainbow trout eggs is of particular interest. An exhaustive choriolysis with a prolonged enzymatic digestion of chorion resulted in no significant change in Sephadex column chromatographic pattern and PAGE pattern of Fr. 1 and Fr. 2 (Iuchi and Yamagami, 1976b; S. Yasumasu *et al.*, unpublished).

These results suggest that the hatching enzyme digests the inner layer of chorion by hydrolyzing some restricted peptide bonds of its constituent proteins to give rise to two groups of soluble glycoprotein compounds, Fr. 1 and Fr. 2. Once these glycoproteins (C1–C6 of Fr. 1, and Fr. 2) are formed, they seem to be resistant to further enzymatic breakdown. A similar mode of choriolysis in principle may occur in the hatching of other fish species, although an accumulation of free amino acids is reported in the hatching fluid of rainbow trout (Ohzu and Kusa, 1981). Some 40 years ago, Hayes (1942, 1949) suspected that the action of hatching enzyme was not hydrolytic, as the amount of amino-N produced by the enzymatic digestion of a capsule was so small. A limited cleavage of the inner-layer proteins of chorion by the hatching enzyme would give rise to the result compatible with the Hayes's observations as well as explain the efficient and rapid solubilization of chorion by the hatching enzyme.

C. Comparative Studies of Enzymatic Hatching and Related Problems

Although the hatching enzyme or enzymatic hatching was documented first in fish, hatching has been described in many other animal species, and the number of such examples is increasing (Ishida, 1948a; Davis, 1969, 1981). Whether there is a phylogenetic correlation to the mechanisms of the enzymatic hatching in various animals is still uncertain, but it would be useful for better understanding of fish hatching to make reference to the enzymatic hatching in other animal groups. In this section a brief survey will be made of hatching in amphibians and sea urchins. Discussion will be extended to the digestion of the cocoon by cocoonase in insects and the solubilization of the vitelline envelope by sperm lysins, as these phenomena are closely related to enzymatic hatching in some respects.

1. AMPHIBIAN HATCHING

Studies on the enzymatic hatching of Amphibia, like those of fish, have a long history. After Bles (1905) described the role of frontal glands in the hatching of *Xenopus laevis* embryos, and presumed that a proteolytic enzyme was secreted from it, this gland was studied histologically in urodele as well as in anurans by other workers (Jaensch, 1921; Noble and Brady, 1930; Holtfreter, 1933; Yanai, 1950, 1953, 1959). The hatching gland of Amphibia is of ectodermal origin. Although there was a view that the anuran hatching gland originated from the neural crest (Yanai *et al.*, 1953, 1955, 1956), it was found recently that most gland cells were derived from the superficial epidermal cells situated on the neural crest (Yoshizaki, 1976; Yoshizaki and Yamamoto, 1979). Moreover, Yoshizaki (1979) succeeded in inducing the hatching-gland cells from the explanted superficial layers of the presumptive ectoderm in *Rana japonica* with LiCl. Thus, it is believed at present that the anuran hatching glands originate mostly from ectoderm other than the neural crest. In pilocarpine-induced secretion, the electron density of secretory granules of gland cells decreases and a partial coalescence of some granules occurs (Yoshizaki, 1973). As to the escaping of anuran embryos from the jelly layers, there have been some reports suggesting a nonenzymatic process (Kobayashi, 1954a,b). In the toad *Bufo vulgaris formosus,* there are four jelly layers, which are named A, B, C, and D, respectively, from the outer to the inner layers. A and B form a jelly string and the innermost

D interfaces with the vitelline envelope. When the embryos attain the late neurula stage, they escape preliminarily from the jelly string by perforating layers A and B, while each embryo remains still covered by layers C and D. Hatching from these layers occurs only when the embryos reach the tail-bud stage. Escape from layers A and B is not due to any proteolytic action but is primarily due to swelling of layer C. Kobayashi (1954b) argued that an augmented respiratory activity of the embryos was closely related to the swelling of layer C. Thus, enzymatic hatching is preceded by a nonenzymatic process. A similar observation was also made on *Xenopus laevis* embryos by Carroll and Hedrick (1974), who reported that the hatching process consisted of two temporally distinct phases, that is, phase 1 and phase 2. In phase 1, the embryo escapes from the outer jelly layers, J3 and J2, without the aid of a hatching enzyme, but probably by a physical process such as water imbibition by the inner jelly layer, J1; in phase 2, a hatching protease participates in the dissolution of the vitelline envelope. This two-step hatching process seems to be of some interest and suggests that such an analysis should be made also in fish hatching, although no thick multijelly layers are present. In salmonid embryos, the hardness of the egg envelope gradually decreases long before actual hatching (Hayes, 1942, 1949). It seems improbable that the hatching enzyme had already been secreted and participated in such envelope soften-ing. Thus, there is a possibility of participation of some factor(s) other than the hatching enzyme in a preliminary softening of the egg enve-lope in the hatching of some fish. The amphibian hatching enzyme is also a protease (Cooper, 1936; Ishida, 1947; Carroll and Hedrick, 1974; Katagiri, 1975; Yoshizaki and Katagiri, 1975; Urch and Hedrick, 1981). The *Rana chensinensis* enzyme was purified about 100-fold from its original culture medium. The molecular weight is approxi-mately 55,000–60,000 and its optimum pH is 7.4–7.8. This enzyme is not affected by Na^+, K^+ or soybean trypsin inhibitor but is strongly inhibited by Ca^{2+}, Mg^{2+}, EDTA, and DFP (Katagiri, 1975). The *Xeno-pus laevis* hatching enzyme was purified 2200-fold over the starting crude hatching media (Urch and Hedrick, 1981). This enzyme has two enzymatically active charge isomers present with molecular weights of 62,500. The activity toward its natural substrate is optimal at pH 7.7. The enzyme is inhibited by Zn^{2+} and by EDTA and seems to be a serine protease from inhibition by DFP and PMSF. From these char-acteristics, the amphibian enzyme is different from the enzymes of *Oryzias* and salmonids but somewhat similar to those of *Fundulus* and *Brachydanio*.

2. ECHINOID HATCHING

The aquatic invertebrate whose hatching enzyme has been best studied is the sea urchin. The echinoderm hatching enzyme was first documented in *Strongylocentrotus (Hemicentrotus) pulcherrimus,* by Ishida (1936), and its properties were studied by Sugawara (1943). The optimal pH of its proteolytic activity was around 8.5–9.5, and low concentrations of Ca^{2+} seemed to be necessary for its activity. Of all the animal species, purification of hatching enzyme was tried first in echinoderm by Yasumasu (1961), who obtained *Anthocidaris crassispina* enzyme in crystalline form. The optimal pH of the crystalline enzyme activity was 8.2–8.4, and nearly a half of the activity remained even after heating at 60°C for 10 min. Following these pioneering studies, there have been many studies of purification and partial characterization of hatching enzymes from various species of echinoderms. The hatching-enzyme characteristics are general in some respects but contradictory in others (as in the case of fish hatching enzyme). All the echinoderm hatching enzymes reported heretofore are found to require a suitable concentration of Ca^{2+} for their maximum activity, while the effect of Mg^{2+} on the activity is not settled (Barrett and Edwards, 1976; Takeuchi *et al.*, 1979; Nakatsuka, 1979). The enzyme seems to be retained on DEAE-cellulose at pH 8.0 or 8.2. There has been considerable variation in the reported molecular weights. On the one hand, the *S. purpuratus* enzyme was purified and the molecular weight was reported to be about 28,500 and 30,000 on SDS-PAGE and Sephadex column chromatography, respectively (Barrett and Edwards, 1976); on the other hand, the molecular weight of the *S. intermedius* enzyme, which was purified to a single band on SDS-urea-PAGE and separated from concomitant β-1,3-glucanase, esterases and most proteinases, was found to be about 44,000 on SDS-urea-PAGE and 45,000 on Sephadex column chromatography (Takeuchi *et al.*, 1979). This variation in molecular weights seems to be attributable in part to a physical heterogeneity of echinoderm hatching enzymes in the original hatching liquid; the enzyme may be combined with some heterologous molecules such as various-sized materials of fertilization envelope (Barrett *et al.*, 1971). Moreover, Nakatsuka (1985) reported recently that the sea urchin hatching enzyme was present inside the blastula in proenzyme form, which had a larger molecular weight.

Another line of hatching enzyme study in echinoderms is of the genetic control of this enzyme protein. Koshihara and Yasumasu

(1966) reported that the *Hemicentrotus pulcherrimus* enzyme could be synthesized *in vitro* using chromatin of the embryos about 3 hr before hatching as a template. Barrett and Angelo (1969) reported, however, that the echinoid hatching enzyme was entirely maternal based on their studies on reciprocal hybrid embryos, whose parent echinoid species, *S. purpuratus* and *S. franciscanus*, had the hatching enzymes of different sensitivities to the added Mn^{2+}. Showman and Whiteley (1980) have recently reported that the messenger RNA of the echinoid hatching enzyme is newly transcribed in advance of hatching, based on their well-devised experiment using the hybrid andromerogons between the two echinoid species *S. purpuratus* and *Dendraster excentricus*. The developmental stage at hatching is much earlier in echinoids than in fish; the echinoid hatching enzyme seems to be one of a few specific proteins that may be synthesized during cleavage. Therefore, it seems highly probable that the fish hatching enzyme is also synthesized under the control of not the "maternal" genome but the embryonic genome. It has not yet been observed electron microscopically that the echinoid hatching enzyme is packaged in any particular structure such as secretory granules, although Nakatsuka (1985) reported that a granular hatching enzyme could be obtained by centrifugation. Thus, the cellular site of synthesis of the echinoid hatching enzyme has not yet been identified.

3. OTHER PHENOMENA RELATED TO HATCHING

There are some enzymes similar to the hatching enzyme in a strict sense, that is, an embryonic enzyme dissolving a fertilization envelope. The best studied among them is cocoonase, which is synthesized in and secreted from the maxillary galea of the pupa of certain saturniid moths and participates in the digestion of the cocoon, making the "escape hatching" of the pupa possible (Kafatos and Williams, 1964; Kafatos, 1972). This enzyme is an organophosphate-sensitive protease, resembling trypsin in its substrate specificity, amino acid composition, and molecular weight ($\sim$24,000) (Kafatos *et al.*, 1967a,b). Cocoonase is synthesized in zymogen-producing cells of the galea and transported into zymogen-storing vacuoles (Berger and Kafatos, 1971; Selman and Kafatos, 1975). A remarkable characteristic of this enzyme is that the active enzyme is deposited on the galea as a semicrystalline encrustation after secretion. The enzyme powder is dissolved in a galeal exudate, which serves as the buffer solvent for the enzyme before being applied onto the inner surface of cocoon. Thus, it seems that a natural enzyme solution can be easily obtained from a pupa just before "escape hatching" (Kafatos *et al.*, 1967a). The second feature of this enzyme is its unique mechanism of action; this enzyme digests the cocoon by hydrolyzing not its main constituent protein, fibroin, but sericin, which glues the fibroin fibers together (Kafatos and Williams, 1964).

Another group of egg envelope-dissolving enzymes (or agents) of interest in comparison with fish hatching enzymes is the so-called egg membrane lysin or sperm lysin, although it is quite different from the hatching enzyme. Many lysins have so far been described in vertebrates and invertebrates (Hoshi, 1985). The lysin is thought to be localized in the sperm acrosome and to participate in the dissolution of the egg envelope when the sperm penetrates the envelope. Mammalian acrosin is one of the best-characterized vertebrate lysins and is similar to trypsin (Polakoski *et al.*, 1972, 1973; Zaneveld *et al.*, 1972; Polakoski and McRorie, 1973; Parrish and Polakoski, 1979). It is assumed that mammalian acrosin is functional in a form bound to the acrosomal membrane under natural condition (Brown and Hartree, 1976; Castellani-Ceresa *et al.*, 1983). There have been many studies on the egg envelope dissolution by sperm lysins in marine invertebrates (Tyler, 1939; Berg, 1950; Wada *et al.*, 1956; Haino, 1971; Haino-Fukushima, 1974; Heller and Raftery, 1973; Levine *et al.*, 1978; Levine and Walsh, 1980; Hoshi *et al.*, 1981; Sawada *et al.*, 1982, 1984; Lewis *et al.*, 1982; Ogawa and Haino-Fukushima, 1984). Among them are some reports in which the lysins of some gastropod sperm are not enzymes but rather low-molecular-weight proteins, which dissolve or loosen markedly the vitelline coat of eggs by combining with it *stoichiometrically* to form a soluble complex (Haino-Fukushima, 1974; Lewis *et al.*, 1982; Ogawa and Haino-Fukushima, 1984). It seems improbable that such a nonenzymatic action of sperm lysins is prevalent in marine invertebrates; it appears that lysins of this type are found only in some restricted animal groups such as archaeogastropods (Ogawa and Haino-Fukushima, 1984). However, the mechanism of action of this gastropod lysin gives us important information about a facet of the mechanisms of egg envelope dissolution or of the biological breakdown of a noncellular structure composed of scleroprotein. It seems that the hatching enzyme also has a high affinity for its natural substrate. Is it unreasonable to think that the mechanism of action of the archaeogastropod lysin is an extreme example of the interaction between the egg envelope-dissolving factor and its substrate?

IV. PHYSIOLOGY OF HATCHING IN FISH

A. Factors Controlling Fish Hatching

As described before, hatching of fish is a developmental stage-specific phenomenon. In fact, the embryo must have attained a particular developmental stage and have fully matured hatching-gland cells before hatching occurs. However, attainment of a specific developmental stage is not sufficient to cause actual hatching. Some triggering stimuli, either extrinsic or intrinsic, have to be received by the appropriately developed embryo in order to induce hatching enzyme secretion. Thus, as Smith (1957) pointed out, the onset of hatching in teleosts is a complex phenomena. As shown in Table II, there have been many factors or treatments that are reported to either stimulate or suppress the hatching of fish. They are believed to influence the secretion of the fish hatching enzyme. In one of the earliest studies of the factors inducing fish hatching, Armstrong (1936) argued that there were two factors involved: the lashing movement of the embryonic tail and the secreted hatching enzyme. He showed that no hatching occurred when either of these factors was inhibited. At present, it is well known that the lashing movement of embryo is effective only after the enzyme has exerted its digesting action on the inner layer of the chorion.

1. OXYGEN AVAILABILITY AND RESPIRATORY MOVEMENT

Ishida (1944b) observed that in medaka embryos, an opercular movement took place followed by disintegration of the hatching gland shortly before hatching. When the opercular movement was suppressed by treatment with 0.25 M KCl, the gland did not disintegrate. On the other hand, the beakdown of hatching glands occurred when the embryo was treated with 0.1 or 0.2% Veronal-sodium, which affected the whole body movement but not the opercular movement. It was further observed that the gland cells could be disintegrated by water flow from a capillary that had been inserted into the pharynx of the embryo. Thus, the enhancement of opercular movement of embryos seems to be one of the phenomena most closely correlated with the initiation of the hatching enzyme secretion in medaka, although it remains obscure whether or not water flow is the sole cause for the hatching-enzyme secretion. When the shaking of a large number of

Table II
Factors Influencing the Hatching-Enzyme Secretion

Stimulants	Reference	Suppressants	Reference
Hypoxia		Hyperoxia	
H_2 gas	Trifonova (1937)	O_2 gas	Milkman (1954)
Respiratory movement	Ishida (1944b)		DiMichele and Taylor (1980)
Stoppage of shaking	Yamagami (1970)		
N_2 gas	Hagenmaier (1972)	Air incubation	Taylor et al. (1977)
CN^-	Ishida (1944c)		DiMichele and Taylor (1980)
	Iuchi et al. (1985)		Yamagami et al. (1983)
	DiMichele and Taylor (1981)		
MS 222		MS 222	
($10^{-5}\ M$)		($3.8 \times 10^{-4}\ M <$)	
Epinephrine	DiMichele and Taylor (1981)		DiMichele and Taylor (1981)
			Iuchi et al. (1985)
Corticosteroid	Cloud (1981)	Tubocurarine	DiMichele and Taylor (1981)
	Schoots et al. (1982a)	Atropine	DiMichele and Taylor (1981)
Prolactin	Schoots et al. (1982a)	Tetrodotoxin	Iuchi et al. (1985)
Electric current	Iuchi and Yamagami (1976a);		
	Yamamoto et al. (1979);		
	Iuchi et al. (1985);		
	Luczynski (1984c)		
Rise in temperature	Luczynski (1984c)		
Ionophore	Schoots et al. (1981)		
	Iuchi et al. (1985)		

medaka embryos was stopped just before hatching and they were heaped up in a small beaker, their hatching was markedly accelerated (Yamagami, 1970). Such treatment is considered to cause oxygen shortage, which in turn stimulates the respiratory activity of the embryos. Besides these observations, the results of a close relationship between respiratory activity of the embryo and hatching have been accumulated. According to Trifonova (1937), hatching of salmon embryos can be accelerated if they are subjected to asphyxia by bubbling hydrogen through the hatchery water. Similarly, hatching of rainbow trout embryos is reported to have been stimulated by treatment with nitrogen gas (Hagenmaier, 1972). In the pike *Esox lucius*, it was also reported that a reduced oxygen concentration in the culture medium (lower than 3.4 ppm) accelerated hatching (Gulidov, 1969). On the contrary, the hatching of *Fundulus* embryos is markedly retarded by a fully oxygenated medium, but boiling of the medium removes the inhibitory activity (Milkman, 1954). Elevated oxygen concentration has been found to retard the hatching of pike (Gulidov, 1969) and of the bream *Abramis brama* L. (Gulidov and Popova, 1977). This line of study on *Fundulus* hatching has been carried out by DiMichele and Taylor (1980), who found that embryos incubated in water with dissolved oxygen concentrations greater than 6 ml O_2/l delayed hatching indefinitely, while they hatched normally when placed in water of 4 ml O_2/l or less. Moreover, they reported that incubation of *Fundulus* embryos in air resulted in a marked retardation of hatching (Taylor *et al.*, 1977; DiMichele and Taylor, 1980). The air-incubated (hatching-retarded) embryos hatch within a short period of time after their reimmersion in water. A similar result was observed also in medaka embryos (Yamagami *et al.*, 1983). Such a curious phenomenon was first documented by Stockard (1907; Atz, 1986). Air incubation would provide the embryos with a high pressure of oxygen, which, like a highly oxygenated medium, results in retardation of hatching enzyme secretion. Effects of the "air incubation" on hatching are of some interest also from an ecological or ethological viewpoint and will be discussed again in the following section. An apparently strange observation that potassium cyanide in low concentrations accelerates hatching in medaka (Ishida, 1944c; Iuchi *et al.*, 1985) can be reasonably understood in that this reagent causes hypoxia or anoxia in an embryo by affecting its respiratory activity. DiMichele and Powers (1982) found that the hatching time was different between two lactate dehydrogenase-B (LDH-B) genotypes of *Fundulus heteroclitus* (Place and Powers, 1978). It seemed that the difference was related to the difference in their developmental rate (DiMichele and Powers, 1984b). In the course of their studies, however, DiMichele and Powers (1984a) have proposed that hatching is stimulated when a growing respiratory demand of the embryo creates a hypoxic condition in the microenvironment surrounding the egg.

2. TEMPERATURE

Temperature is an important factor influencing the hatching-enzyme secretion. In the case of the vendace *Coregonus albula,* temperatures as low as 1–2°C not only delay hatching markedly but also reduce the survival rate to as low as 6.5–47%, while a temporary exposure of the cooled embryos to higher temperatures, such as 8–12°C, accelerates their hatching and increases the extent of hatching to 82–96% (Luczynski, 1984a). Thus, a previous cooling of the coregonid embryos, followed by an elevation of the water temperature, facilitates synchronization of hatching and controls the hatching time (Luczynski, 1984b). Such treatment seems to be of great practical value in culture of this fish (Luczynski, 1984a,b). A similar observation was also reported in the related lake whitefish *Coregonus clupeaformis* (Davis and Behmer, 1980). In this case, it seems that a rise in temperature stimulates enzyme secretion. Once the enzyme is secreted, it will solubilize the egg envelope faster at higher temperatures than at lower temperatures, bringing about earlier hatching.

3. LIGHT

Light is another environment factor that may influence fish hatching. In studies of dopaminergic regulation of fish hatching, Schoots *et al.* (1983b) reported that in medaka and zebrafish embryos cultured in a light–dark (12 h light/12 h dark) cycle, the hatching rate was significantly higher in the light period than in the dark period. In our recent experiments (K. Yamagami and T. Hamazaki, 1985, unpublished), development and hatching of the medaka embryos cultured were compared under conditions of constant light, constant darkness, or 14 h light/10 h dark (14L/10D); the results showed that the developmental rate from fertilization to hatching was not affected by any of these conditions, but hatching was significantly suppressed under the constant-dark condition. Moreover, if the embryos were allowed to develop before hatching in a 14L/10D cycle, a 14L/10D rhythm was observed in their hatching pattern. Therefore, hatching-enzyme secretion seems to be controlled by stimulation of photoreceptors such as eyes (and/or pineal gland?), probably via the central nervous system.

4. OTHER FACTORS

There have been many other agents that either stimulate or suppress hatching-enzyme secretion in fish. Epinephrin and a low concentration (10^{-5} M) of metaaminobenzoic acid ethyl ester methanesulfonate (MS-222) accelerate the secretion in *Fundulus* embryos, while tubocurarine, atropine, and a high concentration (10^{-2} M) of MS-222 act as inhibitors (DiMichele and Taylor, 1981). The effect of suitable doses of electric current (AC) as secretion stimulants has been observed in various fish species such as rainbow trout (Iuchi and Yamagami, 1976a), medaka (Yamamoto *et al.*, 1979), and coregonids (Luczynski, 1984c). Moreover, divalent ionophores such as A23187 (Schoots *et al.*, 1981) and X-537A in the presence of Ca^{2+} but not of Mg^{2+} (Iuchi *et al.*, 1985) are known to be potent stimulants of hatching-enzyme secretion (Fig. 12). There have been so many agents or treatments influencing fish hatching that little consistency or regularity can be found among them.

Recently, we classified all agents or treatments controlling secretion into two categories: one acting directly on the gland cells, and the other acting indirectly probably through the nervous system (and/or humoral system). In a recent study (Iuchi *et al.*, 1985), electric current (AC) and potassium cyanide were chosen as stimulants and tetrodotoxin and MS-222 (3.8×10^{-4} M) were employed as suppressants of hatching enzyme secretion in medaka embryos. These suppressants were considered to act by affecting the nervous system of embryos. By the use of these stimulants and suppressants in combination, it was found that electric current induced the secretion in embryos that had been treated with these nervous system–mediated suppressants, while potassium cyanide did not. As mentioned before, a low concentration of potassium cyanide would cause hypoxia, which causes the enhancement of respiratory activity of embryos on the one hand and the hatching enzyme release on the other hand, probably through the nervous system. Whether the nervous function of an embryo was impaired or not, electric stimulation acted directly on each gland cell. Therefore, all stimulants of respiratory activity seem to belong to the indirect effects, including a possibility that opercular (or body) movement may cause a disintegration of hatching-gland cells. It was reported that the hatching glands of sturgeon were innervated by a branch of the palatine nerve, but that they could secrete the hatching enzyme without nervous stimulation since the glands in tissue culture secreted spontaneously (Ignat'eva, 1959). In this case, however, a possibility still remains that some direct stimulant present in the cultured

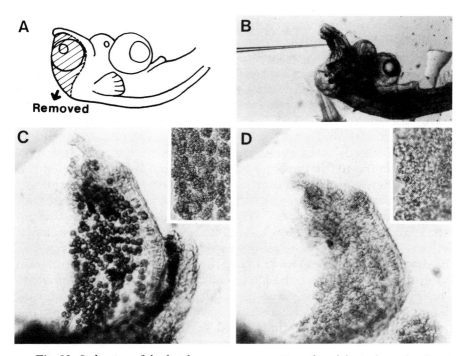

Fig. 12. Induction of the hatching enzyme secretion of medaka embryos by direct application of Ca ionophore. (A) Diagrammatic illustration showing the preparation of the lower-jaw specimen used for the experiment. (B) Injection of about 10 pl of reagent into the intercellular space among hatching gland cells ($\times$25). (C) Hatching-gland cells in the lower jaw about 3 min after injection of the control solution (200 μM $CaCl_2$–1% dimethyl sulfoxide in Mg-free saline solution) ($\times$150). (D) Hatching-gland cells in the lower jaw about 3 min after injection of the Ca ionophore solution (200 μM X-537A–200 μM $CaCl_2$–1% dimethyl sulfoxide in Mg-free saline solution) ($\times$150). Insets are higher magnifications of the gland cells ($\times$350) (Iuchi *et al.*, 1985).

tissues might have acted on the hatching-gland cells. Action of ionophores is a direct stimulation, as it is effective also on the isolated hatching-gland cells (Schoots *et al.*, 1981). Isolation of intact and unimpaired hatching-gland cells seems to be of great value for studying the secretion mechanisms (Yoshizaki *et al.*, 1980).

Light, increased temperature, and increased respiratory activity (or decreased oxygen supply) are probably the natural stimulants of hatching-enzyme secretion in fully developed embryos. What then intervenes between these stimulations and the secretion of gland cells? There have been some reports of hormonal regulation of hatching in fish. Cloud (1981) and Schoots *et al.* (1982a) found that a precocious hatching of medaka was induced by corticosteroids. The latter authors also reported that crude extracts of the prolactin lobe of the brook trout hypophysis or whole-body extracts of medaka released the hatching enzyme from isolated hatching glands of medaka. Furthermore, these authors found a dopaminergic regulation of hatching-enzyme secretion in medaka, extending their views that dopamine controls the stimulating action of prolactin (Schoots *et al.*, 1983b). However, Iuchi *et al.* (1985) were unable to induce hatching-enzyme secretion in medaka by applying highly purified chum salmon prolactin either directly to hatching glands in situ with the aid of micropipette or indirectly by injecting it into the heart of the dechorionated intact embryos at a prehatching stage. At present, it remains uncertain whether this discrepancy is due to the differences in prolactin samples. Besides humoral control, if any, it seems highly probable that the stimulation of hatching enzyme secretion is mediated by the nervous system. As mentioned before, the experiment using nervous system–mediated suppressants (Iuchi *et al.*, 1985) suggests an intervention of some nervous function between the enhancement of respiratory activity and the secretion of gland cells. Also, in the analysis of hatching enzyme release in *Fundulus* embryos whose spinal cord had been cut at different positions, DiMichele and Taylor (1981) found a close correlation between respiratory movement and hatching-enzyme secretion, suggesting an intervention of neurotransmission between these two kinds of physiological processes.

B. Ecological and Ethological Facets of Fish Hatching

In all animals, actual emergence of embryos from their envelopes takes a negligibly short time in comparison with their whole life history. However, hatching is a crucial event in their lives, and when and under what conditions hatching occurs have a great influence on their posthatching life.

As described in the preceding section, temperature, oxygen supply (or availability), and light are environmental factors influencing fish hatching in nature. Among these, oxygen supply is a factor whose influence on hatching of some fish is especially interesting from ecological and ethological viewpoints. Harrington (1959) described delayed hatching in *Fundulus confluentus* embryos that were not immersed in water but were stranded in air on the moist leaves of some littoral plants. As mentioned before, Stockard (1907) found experimentally that *Fundulus* embryos could not hatch when cultured in a moist atmosphere. Later, Taylor *et al.* (1977) reported a similar phenomenon in a related killifish species. *Fundulus heteroclitus.* According to their observations, tidal fluctuations of a marsh that is the habitat of the killifish are 1–1.5 m. Major peaks of spawning activity of this fish occur in conjunction with the night high tide for several days at the new or full moon spring tides. Many eggs were found to be laid on the inner surface of the primary leaves of a marsh plant, *Spartina arterniflora,* at a level that is exposed at low tide. The embryos stranded in air on the plant leaves develop at the same rate as those in water up to the time of hatching. However, the embryos held in air past the normal hatching time continue to develop at a reduced rate without hatching. These "latent embryos" (Harrington, 1959) begin to hatch within a very short time when they are placed back into water. Thus, it seems that the air-stranded embryos in the field remain unhatched until they are resubmerged by a subsequent high tide. These authors explain that reproduction of this type may have important benefits for survival of eggs, in addition to a reduced exposure to predators. Laying the eggs in plants high on the marsh eliminates the probability of their being dispersed to inhospitable habitats or covered by silt in the strong tidal currents, and the larvae probably enjoy an improved chance for survival by hatching and spending their first days in the protected pools at the base of plants (Taylor *et al.,* 1977). DiMichele and Taylor (1980) extended their experimental studies on the "delayed hatching" in connection with the mechanism of hatching-enzyme secretion and proposed that both water and low dissolved oxygen concentration are necessary for hatching. Similar "retarded hatching" in stranded embryos can be observed also in the medaka, *Oryzias latipes,* although no stranding of embryos occurs in the normal life history of this species (Yamagami *et al.,* 1983). When medaka embryos were kept on a moist filter paper, being partially dehydrated but receiving a fully adequate oxygen supply, they showed a marked retardation of hatching. Electron-microscopic examination revealed that the hatching gland cells of the hatching-retarded embryos seemed to be fully matured but showed no sign of initiation of secretion. The embryos in "retarded hatching" were found to hatch within a short period of time after reimmersion in water. Thus, it seems that the retardation (or arrest) of hatching would be useful for synchronous hatching. Synchronization of hatching seems to be of special significance also in hatchery culture. This has been successful, as mentioned before, through the use of delayed hatching in coregonid fish (Luczynski, 1984b).

According to Wourms (1972), various developmental arrests occur in annual fish embryos at different developmental stages. This seems to be a survival strategy, generating several subpopulations that will develop according to different schedules under varied environmental conditions. Among these arrests, there is a diapause of prehatching embryos (diapause III), which may have resulted from the intensification of "delayed hatching" or "retarded hatching." It is said that a short-term arrest phenomenon, "retarded hatching," is sometimes encountered among nonannual aphyosemions and other nonannual cyprinodonts (Wourms, 1972). Thus the hatching phenomenon seems to be closely related to an adaptive strategy in the life cycle of some fish, especially of cyprinodonts.

A famous example of "retarded hatching" of the stranded embryos associated with semilunar rhythm of spawning is the grunion *Leuresthes tenius* (Schwassmann, 1971). It is known that grunion have a spawning period from March to September, with successive spawning runs at the interval of approximately half a month. Spawning is performed at night when the fish come out of the water onto the beach to bury and fertilize their eggs in the sand. Highest tides occur with the full and new moon, and the tides are higher at night than during the daytime in the spawning season of grunion on the California coast. The spawning occurs during a receding tide series, when the high water levels are lessening each night. From 1 to 4 nights are utilized for spawning (Walker, 1952; Schwassmann, 1971). The embryos buried in the wet sand develop, but the well-developed embryos do not hatch until they are reimmersed by a following new series of high tides. Similar spawning runs on the beach sand are also reported in some other fish such as *Hubbsiella sardina, Galaxias attenuatus,* and *Enchelyopus cimbrus* (Schwassmann, 1971), and the puffer *Fugu niphobles* (Uno, 1955; Nozaki *et al.,* 1976; Kobayashi *et al.,* 1978). In the last case, however, it is not necessarily accepted that the spawned eggs are stranded alive in moist sand and hatching of the embryos is retarded until they are reimmersed in a following high tide (Nozaki *et al.,* 1976; Kobayashi *et al.,* 1978). Thus, "retarded hatching" associated with the egg stranding may be of ecological significance in some special groups of fish such as cyprinodonts and some other related fish groups.

It seems that no evidence of ecological or ethological significance has so far been recorded with regard to two other environmental factors, temperature and light, except that, as mentioned before, there seems to be a day/night rhythm in the hatching pattern of the medaka (Schoots *et al.,* 1983b; Yamagami and Hamazaki, 1985). When carefully examined, however, I believe it will be found that these factors also exert significant influence on the hatching pattern of some fish in nature.

V. EPILOGUE—PROBLEMS TO BE SOLVED IN THE FUTURE

In this chapter, some of the topics and problems of fish hatching at the molecular to ecological levels were surveyed. Actual emergence of embryos is preceded by many sequentially occurring preparatory processes, that is, fundamental processes of hatching (Fig. 13). In fish hatching, it seems that most studies have centered on purification and characterization of hatching enzyme(s) rather than on other fundamental processes such as choriolysis, hatching-enzyme secretion, hatching-enzyme synthesis associated with hatching-gland differentiation, and genetic control of these processes. Although the hatching enzyme itself should be further examined, clarification of other fundamental processes is also of great significance. Among these, the genetic control of hatching enzyme synthesis is one of the most fascinating problems from the viewpoints of cell and developmental biology. At present, however, this problem remains open to the future studies in fish, although some analyses have been done in the echinoid hatching.

As Wourms (1972) appropriately noted, hatching of fish seems to be more of a physiological than a developmental process. Neuro–humoral control mechanisms of hatching enzyme secretion still remain obscure, although they are considered to be physiologically and ecologically significant. As mentioned in Section I, hatching of an animal marks an epoch in its ontogeny as a transition from intracapsular (embryonic) to free-living (larval, or eleutheroembryonic in fish; cf. Balon, 1975) life. It is known that physiological states of an embryo

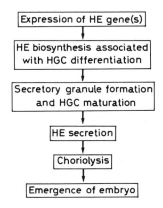

Fig. 13. Some fundamental processes of hatching in fish. HE, Hatching enzyme; HGC, hatching-gland cell.

change markedly in advance of hatching to provide for a drastic alteration of environmental conditions, such as osmotic condition or the
condition of water retention, partial pressure of oxygen, mechanical
impact exerting directly on the animal's body, etc. For example, a
transition of hemoglobins from larval to adult type has been known to
occur shortly before hatching and has been well studied in fish (Iuchi,
1985) as well as in birds at hatching (Wilt, 1967; Tobin *et al.*, 1979)
and in mammals at birth (Solomon, 1965; Wood *et al.*, 1979). Taking
account of these facts, hatching studies should not be confined to the
mechanism of embryos' emergence from their capsules, but should
also be concerned with the embryo's physiology around hatching.
Although birth and hatching are biologically different phenomena,
they must have many underlying physiological changes in common.
Thus "perihatching biology" will provide basic contributions to perinatal biology and perinatal medicine.

ACKNOWLEDGMENTS

The author wishes to thank Prof. Emer. J. Ishida, University of Tokyo, and Prof. J. L.
Hedrick, University of California, Davis, for their critically reading through the manuscript. Most of our studies herein described were performed in cooperation with Dr. M.
Yamamoto, Okayama University, and Dr. I. Iuchi, Sophia University, to whom the
author is grateful. Thanks are also extended to Dr. M. Luczynski, Academy of Agricul-
ture and Technology, Kortowo, for informing the author of some of the Russian literature. Our studies were supported in part by Grants-in-Aid for Scientific Research from
the Ministry of Education, Science and Culture, Japan.

REFERENCES

Amsterdam, A., Ohad, I., and Schramm, M. (1969). Dynamic changes in the ultrastruc-
   ture of the acinar cell of the rat parotid gland during the secretory cycle. *J. Cell Biol.*
   41, 753–773.
Anderson, E. (1967). The formation of the primary envelope during oocyte differentia-
   tion in teleosts. *J. Cell Biol.* **35**, 193–212.
Armstrong, P. B. (1936). Mechanism of hatching in *Fundulus heteroclitus. Biol. Bull.*
   (Woods Hole, Mass.) **71**, 407.
Atz, J. W. (1986). *Fundulus heteroclitus* in the laboratory: A history. *Am. Zool.* **26**, 111–
   120.
Balon, E. K. (1975). Terminology of intervals in fish development. *J. Fish. Res. Board
   Can.* **32**, 1663–1670.
Barrett, D., and Angelo, G. M. (1969). Maternal characteristics of hatching enzymes in
   hybrid sea urchin embryos. *Exp. Cell Res.* **57**, 159–166.

Barrett, D., and Edwards, B. F. (1976). Hatching enzyme of the sea urchin, *Strongylocentrotus purpuratus*. *In* "Methods in Enzymology" (L. Lorand, ed.), Vol. 45, Part B, pp. 354–372. Academic Press, New York.

Barrett, D., Edwards, B. F., Wood, D. B., and Lane, D. J. (1971). Physical heterogeneity of hatching enzyme of the sea urchin, *Strongylocentrotus purpuratus*. *Arch. Biochem. Biophys.* **143**, 261–268.

Berg, W. E. (1950). Lytic effects of sperm extracts on the eggs of *Mytilus edulis*. *Biol. Bull. (Woods Hole, Mass.)* **98**, 128–138.

Berger, E., and Kafatos, F. C. (1971). Quantitative studies of prococoonase synthesis and accumulation during development. *Dev. Biol.* **25**, 377–397.

Berrill, N. J. (1932). The mosaic development of the ascidian egg. *Biol. Bull. (Woods Hole, Mass.)* **63**, 381–386.

Blaxter, J. H. S. (1969). Development; Egg and larvae. *In* "Fish Physiology" (W. S. Hoar and D. J. Randall, eds.), Vol. 3, pp. 177–252. Academic Press, New York.

Bles, E. J. (1905). The life history of *Xenopus laevis* (Daud.). *Trans. R. Soc. Edinburgh* **41**, 789–821.

Bourdin, J. (1926a). Le mécanisme de l'éclosion chez les téléostéens. I. Etude biologique et anatomique. *C. R. Seances Soc. Biol. Ses Fil.* **95**, 1149–1151.

Bourdin, J. (1926b). Le mécanisme de l'éclosion chez les téléostéen. II. Evolution histologique des cellules séreuses cutanées provoquant l'éclosion, chez la truite. *C. R. Seances Soc. Biol. Ses Fil.* **95**, 1183–1186.

Bourdin, J. (1926c). Le mécanisme de l'éclosion chez les téléostéen. III. Morphologie et répartition des glandes séreuses du tégument. *C. R. Seances Soc. Biol. Ses Fil.* **95**, 1239–1241.

Bourdin, J. (1926d). Le mécanisme de l'éclosion chez les téléostéen. IV. Physiologie du liquid périvitellin des oeufs de téléostéens à l'éclosion. *C. R. Seances Soc. Biol. Ses Fil.* **95**, 1242–1243.

Brown, C. R., and Hartree, E. F. (1976). Effects of acrosin inhibitors on the soluble and membrane-bound forms of ram acrosin, and a reappraisal of the role of the enzyme in fertilization. *Hoppe-Seyler's Physiol. Chem.* **357**, 57–65.

Carroll, E. J., Jr., and Hedrick, J. L. (1974). Hatching in the toad *Xenopus laevis*: Morphological events and evidence for a hatching enzyme. *Dev. Biol.* **38**, 1–13.

Castellani-Ceresa, L., Berrut, G., and Colombo, R. (1983). Immunocytochemical localization of acrosin in boar spermatozoa. *J. Exp. Zool.* **227**, 297–304.

Cloud, J. G. (1981). Deoxycorticosterone-induced precocious hatching of teleost embryos. *J. Exp. Zool.* **216**, 197–199.

Cooper, K. W. (1936). Demonstration of a hatching secretion in *Rana pipiens* Schreber. *Proc. Natl. Acad. Sci. U.S.A.* **22**, 433–434.

Davis, C. C. (1969). Mechanisms of hatching in aquatic invertebrate eggs. *Oceanogr. Mar. Biol.* **6**, 325–376.

Davis, C. C. (1981). Mechanisms of hatching in aquatic invertebrate eggs. II. *Oceanogr. Mar. Biol.* **19**, 95–123.

Davis, D., and Behmer, D. J. (1980). Hatching success of lake whitefish eggs in heated water. *Prog. Fish. Cult.* **42**, 215–217.

Denucé, J. M. (1975). Chemical changes in the chorionic membrane of teleost embryos caused by hatching enzyme. *Arch. Int. Physiol. Biochim.* **83**, 179–180.

Denucé, J. M. (1976). Some characteristics of a hatching enzyme produced by the marine teleost, *Gobius jozo*. *Arch. Int. Physiol. Biochim.* **84**, 1067–1068.

Denucé, J. M., and Thijssen, F. J. W. (1975). Les protéases de la glande de l'éclosion des téléostéens: Application de la technique du substrat-film. *Arch. Biol.* **86**, 391–398.

DiMichele, L., and Powers, D. A. (1982). LDH-B genotype-specific hatching times of *Fundulus heteroclitus*. *Nature (London)* **296**, 563–564.

DiMichele, L., and Powers, D. A. (1984a). The relationship between oxygen consumption rate and hatching in *Fundulus heteroclitus*. *Physiol. Zool.* **57**, 46–51.

DiMichele, L., and Powers, D. A. (1984b). Developmental and oxygen consumption rate differences between lactate dehydrogenase-B genotypes of *Fundulus heteroclitus* and their effect on hatching time. *Physiol. Zool.* **57**, 52–56.

DiMichele, L., and Taylor, M. H. (1980). The environmental control of hatching in *Fundulus heteroclitus*. *J. Exp. Zool.* **241**, 181–187.

DiMichele, L., and Taylor, M. H. (1981). The mechanism of hatching in *Fudulus heteroclitus*: Development and physiology. *J. Exp. Zool.* **217**, 73–79.

DiMichele, L., Taylor, M. H., and Singleton, R. (1981). The hatching enzyme of *Fundulus heteroclitus*. *J. Exp. Zool.* **216**, 133–140.

Dumont, J. N., and Brummett, A. R. (1980). The vitelline envelope, chorion, and micropyle of *Fundulus heteroclitus* eggs. *Gamete Res.* **3**, 25–44.

Egami, N., and Hama, A. (1975). Dose-rate effects on the hatchability of irradiated embryos of the fish, *Oryzias latipes*. *Int. J. Radiat. Biol.* **28**, 273–278.

Flügel, H. (1967). Licht- und elektronenmikroskopische Untersuchungen an Oocyten und Eiern einiger Knochenfische. *Z. Zellforsch. Mikrosk. Anat.* **83**, 82–116.

Gihr, M. (1957). Zur Entwicklung des Hechtes. *Rev. Suisse Zool.* **64**, 355–474.

Gulidov, M. V. (1969). Embryonic development of pike (*Esox lucius* L.) at different oxygen concentrations. *Vopr. Ikhtiol.* **9**, 1046–1058.

Gulidov, M. V., and Popova, K. S. (1977). The influence of the elevated oxygen concentrations on the survival and hatching of bream (*Abramis brama* L.) embryos. *Vopr. Ikhtiol.* **17**, 188–191.

Hagenmaier, H. E. (1972). Zum Schlüpfprozess bei Fischen. II. Gewinnung und Charakterisierung des Schlüpfsekretes bei der Regenbogenforelle (*Salmo gairdneri* Rich). *Experientia* **28**, 1214–1215.

Hagenmaier, H. E. (1974a). The hatching process in fish embryos. IV. The enzymological properties of highly purified enzyme (chorionase) from the hatching fluid of the rainbow trout, *Salmo gairdneri* Rich. *Comp. Biochem. Physiol. B* **49B**, 313–324.

Hagenmaier, H. E. (1974b). The hatching process in fish embryos. V. Characterization of the hatching protease (Chorionase) from the perivitelline fluid of the rainbow trout, *Salmo gairdneri* Rich, as a metalloenzyme. *Wilhelm Roux' Arch. Entwicklungsmech. Org.* **175**, 157–162.

Hagenmaier, H. E. (1974c). Zum Schlüpfprozess bei Fischen. VI. Entwicklung, Struktur und Funktion der Schlüpfdrüsenzellen bei der Regenbogenforelle, *Salmo gairdneri* Rich. *Z. Morphol. Tiere* **79**, 233–244.

Hagenmaier, H. E., Schmitz, I., and Fohles, J. (1976). Zum Vorkommen von Isopeptidbindungen in der Eihülle der Regenbogenforelle (*Salmo gairdneri* Rich). *Hoppe-Seyler's Z. Physiol. Chem.* **357**, 1435–1438.

Haino, K. (1971). Studies on the egg-membrane lysin of *Tegula pfeifferi*: Purification and properties of the egg-membrane lysin. *Biochim. Biophys. Acta* **229**, 459–470.

Haino-Fukushima, K. (1974). Studies on the egg-membrane lysin of *Tegula pfeifferi*: The reaction mechanism of the egg-membrane lysin. *Biochim. Biophys. Acta* **352**, 179–191.

Harrington, R. W., Jr. (1959). Delayed hatching in stranded eggs of marsh killifish, *Fundulus confluentus*. *Ecology* **40**, 430–437.

Hayes, F. R. (1942). The hatching mechanism of salmon eggs. *J. Exp. Zool.* **89**, 357–373.

Hayes, F. R. (1949). The growth, general chemistry, and temperature relations of salmonid eggs. *Q. Rev. Biol.* **24**, 281–308.

Hedrick, J. L., and Smith, A. J. (1968). Size and charge isomer separation and estimation of molecular weights of proteins by disc electrophoresis. *Arch. Biochem. Biophys.* **126**, 155–164.

Heller, E., and Raftery, M. A. (1973). Isolation and purification of three egg-membrane lysins from sperm of the marine invertebrate *Megathura crenulata* (giant keyhole limpet). *Biochemistry* **12**, 4106–4113.

Hibbard, H. (1937). The hatching of the squid. *Biol. Bull. (Woods Hole, Mass.)* **73**, 385.

Holtfreter, J. (1933). Der Einfluss von Wirtsalter und verschiedenen Organbezirken auf die Differenzierung von angelagertem Gastrulaektoderm. *Wilhelm Roux' Arch. Entwicklungsmech. Org.* **127**, 619–775.

Hoshi, M. (1985). Lysins. *In* "Biology of Fertilization" (C. B. Metz, and A. Monroy, eds.), Vol. 2, pp. 431–462. Academic Press, New York.

Hoshi, M., Numakunai, T., and Sawada, H. (1981). Evidence for participation of sperm proteinases in fertilization of the solitary ascidian, *Halocynthia roretzi*: Effects of protease inhibitors. *Dev. Biol.* **86**, 117–121.

Ichikawa, A. (1965). Fine structural changes in responses to hormonal stimulation of the perfused canine pancreas. *J. Cell Biol.* **24**, 369–385.

Ignat'eva, G. M. (1959). Secretion of the hatching enzyme in the hatching gland explants of sturgeon embryos. *Dokl. Akad. Nauk SSSR* **128**, 212–215.

Inukai, T., Kunihiro, I., and Kashioka, T. (1939). Hatching mechanism of dog salmon (in Japanese). *Zool. Mag.* **51**, 108.

Ishida, J. (1936). An enzyme dissolving the fertilization membrane of sea-urchin eggs. *Annot. Zool. Jpn.* **15**, 453–457.

Ishida, J. (1943). Histology and development of the hatching gland of medaka (in Japanese). *Zool. Mag.* **55**, 172.

Ishida, J. (1944a). Mr. Ikumo Moriwaki, the first witness of the hatching enzyme (in Japanese). *Zool. Mag.* **56**, 63–66.

Ishida, J. (1944b). Hatching enzyme in the fresh-water fish, *Oryzias latipes*. *Annot. Zool. Jpn.* **22**, 137–154.

Ishida, J. (1944c). Further studies on the hatching enzyme of the fresh-water fish, *Oryzias latipes*. *Annot. Zool. Jpn.* **22**, 155–164.

Ishida, J. (1947). The hatching enzyme in amphibians (in Japanese). *Zool. Mag.* **57**, 77–78.

Ishida, J. (1948a). "Hatching of Animals" (in Japanese). Kawade Shobo Publ. Co., Tokyo.

Ishida, J. (1948b). "Hatching Enzyme" (in Japanese). Hokuryukan Publ. Co., Tokyo.

Ishida, J. (1971). Hatching enzyme. *In* "Cellular Aspects of Early Development" (in Japanese) (Jpn. Soc. Dev. Biol., ed.), pp. 226–246. Iwanami Publ. Co., Tokyo.

Ishida, J. (1985). Hatching enzyme: Past, present and future. *Zool. Sci.* **2**, 1–10.

Iuchi, I. (1985). Cellular and molecular bases of the larval-adult shift of hemoglobins in fish. *Zool. Sci.* **2**, 11–23.

Iuchi, I., and Yamagami, K. (1976a). Induction of a precocious secretion of the hatching enzyme in the rainbow trout embryo by electric stimulation. *Zool. Mag.* **85**, 273–277.

Iuchi, I., and Yamagami, K. (1976b). Major glycoproteins solubilized from the teleostean egg membrane by the action of the hatching enzyme. *Biochim. Biophys. Acta* **453**, 240–249.

Iuchi, I., and Yamagami, K. (1980). Hatching enzyme in the secretory granule of hatching gland, isolated from the homogenate of whole embryos of medaka at some developmental stages. *Annot. Zool. Jpn.* **53**, 147–155.

Iuchi, I., Yamamoto, M., and Yamagami, K. (1982). Presence of active hatching enzyme in the secretory granule of prehatching medaka embryos. *Dev. Growth Differ.* **24**, 135–143.

Iuchi, I., Hamazaki, T., and Yamagami, K. (1985). Mode of action of some stimulants of the hatching enzyme secretion in fish embryos. *Dev. Growth Differ.* **27**, 573–581.

Jaensch, A. P. (1921). Beobachtung über das Auskrieden der Larven von *Rana arvalis* und *fusca* und die Funktion des Stirndrüsenstreifen. *Anat. Anz.* **53**, 567–584.

Kafatos, F. C. (1972). The cocoonase zymogen cells of silk moth: A model of terminal cell differentiation for specific protein synthesis. *Curr. Top. Dev. Biol.* **7**, 125–191.

Kafatos, F. C., and Williams, C. M. (1964). Enzymatic mechanism for the escape of certain moth from their cocoons. *Science* **146**, 538–540.

Kafatos, F. C., Tartakoff, A. L., and Law, J. H. (1967a). Cocoonase. I. Preliminary characterization of a proteolytic enzyme from silk moths. *J. Biol. Chem.* **242**, 1477–1487.

Kafatos, F. C., Law, J. H., and Tartakoff, A. M. (1967b). Cocoonase. II. Substrate specificity, inhibitors and classification of the enzyme. *J. Biol. Chem.* **242**, 1488–1494.

Kagayama, M., and Douglas, W. W. (1974). Electron microscope evidence of calcium-induced exocytosis in mast cells treated with 48/80 or the ionophores A23187 and X-537A. *J. Cell Biol.* **62**, 519–526.

Kaighn, M. E. (1964). A biochemical study of the hatching process in *Fundulus heteroclitus. Dev. Biol.* **9**, 56–80.

Kamito, A. (1928). Early development of the Japanese killifish (*Oryzas latipes*), with notes on its habits. *J. Coll. Agric., Tokyo Imp. Univ.* **10**, 21–38.

Kanno, T. (1972). Calcium-dependent amylase release and electrophysiological measurements in cells of the pancreas. *J. Physiol. (London)* **226**, 353–371.

Katagiri, C. (1975). Properties of the hatching enzyme from frog embryo. *J. Exp. Zool.* **193**, 109–118.

Kerr, J. G. (1900). The external features in the development of *Lepidosiren paradoxa*, Fitz. *Philos. Trans. R. Soc. London, Ser. B.* **192**, 299–330.

Kobayashi, H. (1954a). Hatching mechanism in the toad, *Bufo vulgaris formosus*. 1. Observations and some experiments on perforation of gelatinous envelope and hatching. *J. Fac. Sci., Univ. Tokyo, Sect. 4* **7**, 79–87.

Kobayashi, H. (1954b). Hatching mechanism in the toad, *Bufo vulgaris formosus*. 4. Relation between the perforation of the jelly string and CO_2 production by the embryos. *J. Fac. Sci., Univ. Tokyo, Sect. 4* **7**, 107–112.

Kobayashi, Y., Kobayashi, H., Takei, Y., and Nozaki, M. (1978). Spawning habits of the puffer, *Fugu niphobles* (Jordan et Snyder) II (in Japanese). *Zool. Mag.* **87**, 44–55.

Koshihara, H., and Yasumasu, I. (1966). Synthesis of RNA and protein during early development of sea urchin (in Japanese). *Symp. Cell. Chem.* **17**, 167–175.

Kunitz, M. (1947). Crystalline soybean trypsin inhibitor. II. General properties. *J. Gen. Physiol.* **30**, 291–310.

Kurosumi, K. (1961). Electron microscopic analysis of the secretion mechanism. *Int. Rev. Cytol.* **11**, 1–124.

Laemmli, U. K. (1970). Cleavage of structural proteins during the assembly of the head of bacteriophage T$_4$. *Nature (London)* **227**, 680–685.

Lawson, D., Raff, M. C., Gomperts, B., Fewtrell, C., and Gilula, N. B. (1977). Molecular events during membrane fusion. A study of exocytosis in rat peritoneal mast cells. *J. Cell Biol.* **72**, 242–256.

Levine, A. E., and Walsh, K. A. (1980). Purification of an acrosin-like enzyme from sea urchin sperm. *J. Biol. Chem.* **255**, 4814–4820.

Levine, A. E., Walsh, K. A., and Fodor, E. J. B. (1978). Evidence of an acrosin-like enzyme in sea urchin sperm. *Dev. Biol.* **63**, 299–306.

Lewis, C. A., Talbot, C. F., and Vacquier, V. D. (1982). A protein from abalone sperm dissolves the egg vitelline layer by a nonenzymatic mechanism. *Dev. Biol.* **92**, 227–239.

Lönning, S. (1972). Comparative electron microscopic studies of teleostean eggs with special reference to the chorion. *Sarsia* **49**, 41–48.

Luczynski, M. (1984a). Improvement in the efficiency of stocking lakes with larvae of *Coregonus albula* L. by delaying hatching. *Aquaculture* **41**, 99–111.

Luczynski, M. (1984b). A technique for delaying embryogenesis of vendace (*Coregonus albula*, L.) eggs in order to synchronize mass hatching with optimal conditions for lake stocking. *Aquaculture* **41**, 113–117.

Luczynski, M. (1984c). Temperature and electric shock control the secretion of chorionase in Coregoniae embryos. *Comp. Biochem. Physiol. A* **78A**, 371–374.

Matui, K. (1949). Illustration of the normal course of development in the fish, *Oryzias latipes* (in Japanese). *Jpn. J. Exp. Morphol.* **5**, 33–38.

Milkman, R. (1954). Controlled observation of hatching in *Fundulus heteroclitus. Biol. Bull. (Woods Hole, Mass.)* **107**, 300.

Moriwaki, I. (1910). The mechanism of escape of the fry out of the egg chorion in the dog salmon (in Japanese). *Rep. Hokkaido Fish. Res. Stn., 3rd* (cited from Ishida, 1944a).

Nakatsuka, M. (1979). Salt-dependent properties of hatching enzyme from embryos of the sea urchin, *Hemicentrotus pulcherrimus. Dev., Growth Differ.* **21**, 245–253.

Nakatsuka, M. (1985). Properties of intracellular hatching enzyme in embryos of the sea urchin, *Hemicentrotus pulcherrimus. Dev., Growth Differ.* **27**, 653–661.

Needham, J. (1931). "Chemical Embryology." Cambridge Univ. Press, London and New York.

Needham, J. (1942). "Biochemistry and Morphogenesis." Cambridge Univ. Press, London and New York.

Noble, G. K., and Brady, M. K. (1930). The mechanism of hatching in marbled salamander (*Amblystoma opacum*). *Anat. Rec.* **45**, 274.

Nozaki, M., Tsutsumi, T., Kobayashi, H., Takei, Y., Ichikawa, T., Tsuneki, K., Miyagawa, K., Uemura, H., and Tatsumi, Y. (1976). Spawning habit of the puffer, *Fugu niphobles* (Jordan et Snyder) I (in Japanese). *Zool. Mag.* **85**, 156–168.

Ogawa, A., and Haino-Fukushima, K. (1984). Isolation and purification of vitelline-coat lysin from testis of *Turbo cornatus* (Mollusca). *Dev., Growth Differ.* **26**, 345–360.

Ogawa, N., and Ohi, Y. (1968). On the chorion and the hatching enzyme of the medaka, *Oryzias latipes* (in Japanese). *Zool. Mag.* **77**, 151–156.

Ohi, Y., and Ogawa, N. (1970). Electrophoretic fractionation of the hatching enzyme of the medaka, *Oryzias latipes* (in Japanese). *Zool. Mag.* **79**, 17–18.

Ohzu, E., and Kasuya, H. (1979). Studies of the hatching enzyme of the fish. I. Isolation of the hatching enzyme of the rainbow trout, *Salmo gairdnerir. Annot. Zool. Jpn.* **52**, 125–132.

Ohzu, E., and Kusa, M. (1981). Amino acid composition of the egg chorion of rainbow trout. *Annot. Zool. Jpn.* **54**, 241–244.

Ohzu, E., Yamaguchi, M., and Kusa, M. (1983). Notes on the failure in the turbidimetry of choriolytic activity on the hatching enzyme in the rainbow trout eggs. *Annot. Zool. Jpn.* **56**, 73–77.

Ouji, M. (1959a). On the nature and behaviour of secretory granules in hatching gland cells of the fresh-water teleost, *Odontobutis obscura. J. Fac. Sci., Hokkaido Univ., Ser. 6* **14**, 282–285.

Ouji, M. (1959b). Cytochemical studies on hatching gland cells in two fresh-water teleostes, *Odontobutis obscura* and *Zacco platypus. J. Fac. Sci., Hokkaido Univ., Ser. 6* **14**, 286–291.

Ouji, M., and Iga, T. (1961). On the specific affinity to some dyes of granules of the hatching gland cells in carp embryos (in Japanese, with English resume). *Zool. Mag.* **70**, 356–360.

Parrish, R. F., and Polakoski, K. L. (1979). Mammalian sperm proacrosin–acrosin system. *Int. J. Biochem.* **10**, 391–395.

Place, A. R., and Powers, D. A. (1978). Genetic basis for protein polymorphism in *Fundulus heteroclitus*. I. Lactate dehydrogenase, malate dehydrogenase, glucose phosphate isomerase, and phosphoglucomutase. *Biochem. Genet.* **16**, 577–591.

Polakoski, K. L., and McRorie, R. A. (1973). Boar acrosin. II. Classification, inhibition and specificity studies of a preteinase from sperm acrosomes. *J. Biol. Chem.* **248**, 8183–8188.

Polakoski, K. L., Zaneveld, L. J. D., and Williams, W. L. (1972). Purification of acrosin, a proteolytic enzyme from rabbit sperm acrosomes. *Biol. Reprod.* **6**, 23–29.

Polakoski, K. L., McRorie, R. A., and Williams, W. L. (1973). Boar acrosin. I. Purification and preliminary characterization of a proteinase from boar sperm acrosomes. *J. Biol. Chem.* **248**, 8178–8182.

Rosenthal, H., and Iwai, T. (1979). Hatching glands in herring embryos. *Mar. Ecol.: Prog. Ser.* **1**, 123–127.

Sawada, H., Yokosawa, H., Hoshi, M., and Ishii, S. (1982). Evidence for acrosin-like enzyme in sperm extract and its involvement in fertilization of the ascidian, *Halocynthia roretzi. Gamete Res.* **5**, 291–301.

Sawada, H., Yokosawa, H, and Ishii, S. (1984). Purification and characterization of two types of trypsin-like enzymes from sperm of the ascidian (Prochordata) *Halocynthia roretzi*. Evidence for the presence of spermosin a novel acrosin-like enzyme. *J. Biol. Chem.* **259**, 2900–2904.

Schoots, A. F. M., and Denucé, J. M. (1981). Purification and characterization of hatching enzyme of the pike (*Esox lucius*). *Int. J. Biochem.* **13**, 591–602.

Schoots, A. F. M., Sackers, R. J., De Bont, R. G., and Denucé, J. M. (1981). Ionophore A23187-induced hatching enzyme secretion in medaka embryos. *Arch. Int. Physiol. Biochim.* **89**, B77.

Schoots, A. F. M., De Bont, R. G., Van Eys, G. J. J. M., and Denucé, J. M. (1982a). Evidence for a stimulating effect of prolactin on teleostean hatching enzyme secretion. *J. Exp. Zool.* **219**, 129–132.

Schoots, A. F. M., Opstelten, R. J. G., and Denucé, J. M. (1982b). Hatching in the pike, *Esox lucius* L.: Evidence for a single hatching enzyme and its immunocytochemical localization in specialized hatching gland cells. *Dev. Biol.* **89**, 48–55.

Schoots, A. F. M., Stikkelbroeck, J. J. M., Bekhuis, J. F., and Denucé, J. M. (1982c). Hatching in teleostean fishes: Fine structural changes in the egg envelope during enzymatic breakdown *in vivo* and *in vitro. J. Ultrastruct. Res.* **80**, 185–196.

Schoots, A. F. M., Evertse, P. A. C. M., and Denucé, J. M. (1983a). Ultrastructural changes in hatching-gland cells of pike embryos (*Esox lucius* L.) and evidence for ther degeneration by apoptosis. *Cell Tissue Res.* **229**, 573–589.

Schoots, A. F. M., Meijer, R. C., and Denucé, J. M. (1983b). Dopaminergic regulation of hatching in fish embryos. *Dev. Biol.* **100**, 59–63.

Schoots, A. F. M., Sackers, R. J., Overkamp, P. S. G., and Denucé, J. M. (1983c). Hatching in the teleost, Oryzias latipes: Limited proteolysis causes egg envelope swelling. J. Exp. Zool. 226, 93–100.

Schwassmann, H. O. (1971). Biological rhythms. In "Fish Physiology" (W. S. Hoar and D. J. Randall, eds.), Vol. 6, pp. 371–428. Academic Press, New York.

Selman, K., and Kafatos, F. C. (1975). Differentiation in the cocoonase producing silk moth galea: Ultrastructural studies. Dev. Biol. 46, 132–150.

Showman, R. M., and Whiteley, A. H. (1980). The origin of echinoid hatching enzyme messenger RNA. Dev., Growth Differ. 22, 305–314.

Slifer, E. H. (1937). The origin and fate of the membranes surrounding the grasshopper egg; together with some experiments on the source of the hatching enzyme. Q. J. Microsc. Sci. [N.S.] 79, 493–506.

Slifer, E. H. (1938). A cytological study of the pleuropodia of Melanoplus differentialis (orthoptera, acrididae) which furnishes new evidence that they produce the hatching enzyme. J. Morphol. 63, 181–196.

Smith, S. (1957). Early development and hatching. In "Physiology of Fishes" (D. E. Brown, ed.), Vol. 1, pp. 323–359. Academic Press, New York.

Solomon, J. B. (1965). Development of nonenzymatic proteins in relation to functional differentiation. In "The Biochemistry of Animal Development" (R. Weber, ed.), Vol. 1, pp. 367–440. Academic Press, New York.

Stockard, C. R. (1907). The influence of external factors, chemical and physical, on the development of Fundulus heteroclitus. J. Exp. Zool. 4, 165–202.

Sugawara, H. (1943). Hatching enzyme of the sea urchin, Strongylocentrotus pulcherrimus. J. Fac. Sci., Univ. Tokyo, Sect. 4 6, 109–127.

Takeuchi, K., Yokosawa, H., and Hoshi, M. (1979). Purification and characterization of hatching enzyme of Strongylocentrotus intermedius. Eur. J. Biochem. 100, 257–265.

Taylor, M. H., DiMichele, L., and Leach, G. J. (1977). Egg stranding in the life cycle of the mummichog, Fundulus heteroclitus. Copeia, pp. 397–399.

Tobin, A. J., Chapman, B. S., Hansen, D. A., Lasky, L., and Selvig, S. E. (1979). Regulation of embryonic and adult hemoglobin synthesis in chickens. In "Cellular and Molecular Regulation of Hemoglobin Switching" (G. Stamatoyannopoulos and A. W. Nienhuis, eds.), pp. 205–212. Grune & Stratton, New York.

Trifonova, A. N. (1937). La physiologie de la différenciation et de la croissance. I. L'équilibre Pasteur–Meyerhof dans le développement des poissons. Acta Zool. (Stockholm) 18, 375–445.

Tyler, A. (1939). Extraction of an egg membrane-lysin from sperm of the giant keyhole limpet (Megathura crenulate). Proc. Natl. Acad. Sci. U.S.A. 25, 317–323.

Uno, Y. (1955). Spawning habit and early development of a puffer, Fugu (Fugu) niphobles (Jordan et Snyder). J. Tokyo Univ. Fish. 41, 169–183.

Urch, U. A., and Hedrick, J. L. (1981). Isolation and characterization of the hatching enzyme from the amphibian, Xenopus laevis. Arch. Biochem. Biophys. 206, 424–431.

Wada, S. K., Collier, J. R., and Dan, J. C. (1956). Studies on the acrosome. V. An egg-membrane lysin from the acrosomes of Mytilus edulis spermatozoa. Exp. Cell Res. 10, 168–180.

Walker, B. W. (1952). A guide to the grunion. Calif. Fish Game 38, 409–420 (cited from Schwassmann, 1971).

Weber, K., and Osborn, M. (1969). The reliability of molecular weight determinations by dodecyl sulfate–polyacrylamide gel electrophoresis. J. Biol. Chem. 244, 4406–4412.

Willemse, M. T. M., and Denucé, J. M. (1973). Hatching glands in the teleosts, *Brachydanio rerio, Danio malabaricus, Moemkhausia oligolepis* and *Barbus schuberti*. *Dev., Growth Differ.* **15**, 169–177.

Wilt, F. (1967). The control of embryonic hemoglobin synthesis. *Adv. Morphog.* **6**, 89–125.

Wintrebert, P. (1912a). Le mécanisme de l'éclosion chez la truite arc-en-ciel. *C. R. Seances Soc. Biol. Ses Fil.* **72**, 724–727.

Wintrebert, P. (1912b). Le déterminisme de l'éclosion chez le cyprin doré (*Carassius auratus* L.). *C. R. Seances Soc. Biol. Ses Fil.* **73**, 70–73.

Wintrebert, P. (1926). L'éclosion de la perche (*Perca fluviatilis* L.). *C. R. Seances Soc. Biol. Ses Fil.* **95**, 1146–1148.

Wood, W. G., Nash, J., Weatherall, D. J., Robinson, J. S., and Harrison, F. A. (1979). The sheep as an animal model for the switch from fetal to adult hemoglobins. *In* "Cellular and Molecular Regulation of Hemoglobin Switching" (G. Stamatoyannopoulos and A. W. Nienhuis, eds.), pp. 153–167. Grune & Stratton, New York.

Wourms, J. P. (1972). The developmental biology of annual fishes. III. Pre-embryonic and embryonic diapause of variable duration in the eggs of annual fishes. *J. Exp. Zool.* **182**, 389–414.

Wourms, J. P., and Sheldon, H. (1976). Annual fish oogenesis. II. Formation of the secondary egg envelope. *Dev. Biol.* **50**, 355–366.

Yamagami, K. (1960). Phosphorus metabolism in fish eggs. I. Changes in the contents of some phosphorus compounds during early development of *Oryzias latipes*. *Sci. Pap. Coll. Gen. Educ., Univ. Tokyo* **10**, 99–108.

Yamagami, K. (1970). A method for rapid and quantitative determination of the hatching enzyme (chorionase) activity of the medaka, *Oryzias latipes*. *Annot. Zool. Jpn.* **43**, 1–9.

Yamagami, K. (1972). Isolation of a choriolytic enzyme (hatching enzyme) of the teleost, *Oryzias latipes*. *Dev. Biol.* **29**, 343–348.

Yamagami, K. (1973). Some enzymological properties of a hatching enzyme (chorionase) isolated from the fresh water teleost, *Oryzias latipes*. *Comp. Biochem. Physiol. B* **46B**, 603–616.

Yamagami, K. (1975). Relationship between two kinds of hatching enzymes in the hatching liquid of the medaka, *Oryzias latipes*. *J. Exp. Zool.* **192**, 127–132.

Yamagami, K. (1981). Mechanisms of hatching in fish: Secretion of hatching enzyme and enzymatic choriolysis. *Am. Zool.* **21**, 459–471.

Yamagami, K., and Hamazaki, T. (1985). Influence of light on hatching of medaka embryos. *Zool. Sci.* **2**, 928.

Yamagami, K., and Iuchi, I. (1975). Components of glycoproteins solubilized from egg chorion by the hatching enzyme of *Oryzias latipes* (in Japanese). *Zool Mag.* **84**, 332.

Yamagami, K., Yamamoto, M., Iuchi, I., and Taguchi, S. (1983). Retardation of maturation- and secretion-associatid ultrastructural changes of hatching gland in the medaka embryos incubated in air. *Annot. Zool. Jpn.* **56**, 266–274.

Yamamoto, M. (1963). Electron microscopy of fish development. I. Fine structure of the hatching glands of embryos of the teleost, *Oryzias latipes*. *J. Fac. Sci., Univ. Tokyo, Sect. 4* **10**, 115–121.

Yamamoto, M., and Yamagami, K. (1975). Electron microscopic studies on choriolysis by the hatching enzyme of the teleost, *Oryzias latipes*. *Dev. Biol.* **43**, 313–321.

Yamamoto, M., Iuchi, I., and Yamagami, K. (1979). Ultrastructural changes of the teleostean hatching gland cell during natural and electrically induced precocious secretion. *Dev. Biol.* **68**, 162–174.

Yanai, T. (1950). Hatching glands of the toad, *Bufo vulgaris formosus* Boulenger (in Japanese). *Zool. Mag.* **59**, 230–234.

Yanai, T. (1953). Structure and development of the frontal glands of the frogs, *Rhacophorus schlegelii arborea* and *Rhacophorus schlegelii schlegelii*. *Annot. Zool. Jpn.* **26**, 85–90.

Yanai, T. (1959). The hatching glands of *Rana nigromaculata brevipoda* and of the reciprocal hybrids between *Rana nigromaculata brevipoda* and *Rana nigromaculata nigromaculata*. *Annot. Zool. Jpn.* **32**, 31–34.

Yanai, T. (1966). Hatching. *In* "Vertebrate Embryology" (in Japanese) (M. Kume, ed.), pp. 49–58. Baifukan Publ. Co., Tokyo.

Yanai, T., Ouji, M., and Kobayashi, K. (1953). On the origin of the frontal gland of amphibians. *Annot. Zool. Jpn.* **26**, 193–201.

Yanai, T., Ouji, M., and Iga, T. (1955). Experimental studies on the origin of the frontal glands of amphibians. II. Transplantation of the neural crest. *Annot. Zool. Jpn.* **28**, 227–232.

Yanai, T., Ouji, M., and Iga, T. (1956). Effects of Ringers and Holtfreters solutions of the development of the frontal gland. *Annot. Zool. Jpn.* **29**, 28–33.

Yasumasu, I. (1961). Crystallization of hatching enzyme of the sea urchin, *Anthocidaris crassispina*. *Sci. Pap. Coll. Gen. Educ., Univ. Tokyo* **11**, 275–280.

Yasumasu, S., Iuchi, I., and Yamagami, K. (1985). Some characteristics of highly purified hatching enzyme of medaka, *Oryzias latipes*. *Zool. Sci.* **2**, 958.

Yasumasu, S., Iuchi, I., and Yamagami, K. (1988). Medaka hatching enzyme consists of two kinds of proteases which act cooperatively. *Zool. Sci.* **5**, (in press).

Yokoya, S., and Ebina, Y. (1976). Hatching glands in salmonid fishes, *Salmo gairdneri*, *Salmo trutta*, *Salvelinus fontinalis* and *Salvelinus pluvius*. *Cell Tissue Res.* **172**, 529–540.

Yoshizaki, N. (1973). Ultrastructure of the hatching gland cells in the South African clawed toad, *Xenopus laevis*. *J. Fac. Sci., Hokkaido Univ., Ser. 6* **18**, 469–480.

Yoshizaki, N. (1976). Effect of actinomycin D on the differentiation of hatching gland cell and cilia cell in the frog embryo. *Dev., Growth Differ.* **18**, 133–143.

Yoshizaki, N. (1979). Induction of the frog hatching gland cell from explanted presumptive ectodermal tissue by LiCl. *Dev., Growth Differ.* **21**, 11–18.

Yoshizaki, N., and Katagiri, C. (1975). Cellular basis for the production and secretion of the hatching enzyme by frog embryos. *J. Exp. Zool.* **192**, 203–212.

Yoshizaki, N., and Yamamoto, M. (1979). A stereoscan study of the development of hatching gland cells in the embryonic epidermis of *Rana japonica*. *Acta Embryol. Exp.* **3**, 339–348.

Yoshizaki, N., Sackers, R. J., Schoots, A. F. M., and Denucé, J. M. (1980). Isolation of hatching gland cells from the teleost, *Oryzias latipes*, by centrifugation through Percoll. *J. Exp. Zool.* **213**, 427–429.

Young, E. G., and Inman, W. R. (1938). The protein of the casing of salmon eggs. *J. Biol. Chem.* **124**, 189–193.

Young, E. G., and Smith, D. G. (1956). The amino acid in the ichthulokeratin of salmon eggs. *J. Biol. Chem.* **219**, 161–164.

Zaneveld, L. J. D., Polakoski, K. L., and Williams, W. L. (1972). Properties of a proteolytic enzyme from rabbit sperm acrosomes. *Biol. Reprod.* **6**, 30–39.

AUTHOR INDEX

Numbers in italics refer to the pages on which the complete references are listed.

SYSTEMATIC INDEX

Note: Names listed are those used by the authors of the various chapters. No attempt has been made to provide the current nomenclature where taxonomic changes have occurred. Boldface letters refer to Parts A and B of Volume XI.

SUBJECT INDEX

Note: Boldface **A** refers to entries in Volume XIA; **B** refers to entries in Volume XIB.

A

Absolute aerobic scope, **A**, 99–106
Acid rain, *see* pH
Acoustic system, development, **A**, 40–41, 44
Activation
  defined, **A**, 167
  ion fluxes and electrical events, **A**, 178–183, 217–220, 238
Adelphophagy, **B**, 37, 43–48, 81–82
Adrenal gland, *see* Interrenal gland, and smolting
Aerobic scope, *see* Absolute aerobic scope
Air incubation, **A**, 482
Alarm substance, **B**, 362–363, 379
Alimentary canal, development, **B**, 376–377
Allometric growth, **A**, 12, 39
Anabolic steroids, *see* Testosterone
Androgens, in vitellogenesis, **A**, 352, 360
Annual fish, **B**, 372
Adrenocorticotropic hormone (ACTH), *see* Interrenal gland, and smolting
Apstein's stages, **A**, 8
Atricial young, **B**, 376–377
Astaxanthin, *see* Carotenoids
Atroposic model, **B**, 244–251, 256, 260, 261

B

Behavior
  development of, **B**, 346–395
  feeding, **B**, 373–377

myogenic vs. neurogenic origin, **B**, 346
  pollutants and, **A**, 273–275, 313–315
Boundary layer, **A**, 61–64
Branchiostegal rays, development, **B**, 242
Buoyancy
  lipids and, **A**, 420–421
  salinity and, **A**, 121
  of smolts, **B**, 281
  wax esters and, **A**, 382

C

Calorimetry, **A**, 83–87
Calyces nutriciae, **B**, 101
Cannibalism, **B**, 28, 43
Carbonic anhydrase, in smolting, **B**, 284
Cardiac rate, pollutant effects, **A**, 271–272, 317
Carotenoids
  in eggs, **A**, 23, 384, 392–393, 421
  in respiration, **A**, 76–77
Chemosensory system
  alarm substance and fright reaction, **B**, 362–363
  development of, **A**, 41, **B**, 356–363
  in habitat selection, **B**, 359–360
  in mate choice, **B**, 361–362
Chloride cells
  "acid rain" and, **A**, 229
  in deionized water, **A**, 229
  in embryos and larvae, **A**, 214, 224–234, **B**, 24
  in smolts, **B**, 288–289, 301, 306, 322
  structure of, **A**, 225–228

537